Biopharmaceutics Applications in Drug Development

Biopharmaceutics Applications in Drug Development

Edited by

Rajesh Krishna

Merck & Co., Inc.
Rahway, NJ, USA

and

Lawrence Yu

Center for Drug Evaluation & Research
Rockville, MD, USA

Rajesh Krishna
Merck Research Laboratories Quantitative
Clinical Pharmacology
Merck & Co., Inc.
126 East Lincoln Avenue
Rahway, NJ 07065
USA

Lawrence Yu
Food and Drug Administration
Center for Drug Evaluation & Research
5600 Fishers Lane
Rockville, MD 20855
USA

ISBN: 978-0-387-72378-5 e-ISBN: 978-0-387-72379-2

Library of Congress Control Number: 2007931199

Printed on acid-free paper.

9 8 7 6 5 4 3 2 1

springer.com

To my wife, Bhuvana, my mom, and dad for their unwavering support and lifelong inspiration (R.K.)

To my family, Jenny, Alex, and Stephanie for their understanding, support, and love (L.Y.)

Preface

Drug product performance is a vital aspect of drug development as it draws on interdisciplinary expertise from both pharmaceutics and pharmacokinetics disciplines. It is at the key interface that the discipline of biopharmaceutics has emerged. The past two decades have witnessed considerable advances in biopharmaceutics particularly with regard to bioavailability and bioequivalence, as they relate to product quality and regulatory standards of approval.

While the foundation for biopharmaceutics has been laid by pioneers in the field and has been captured in early textbooks devoted to this area, a technical gap does exist on the current and emerging applications of regulatory aspects of biopharmaceutics. The current volume presents an integrated view linking pharmaceutics and the biological consequences to drug development decision making. The book is composed of carefully crafted chapters introducing fundamental concepts, methods, and advances in the area of dissolution, absorption, and permeability and their key applications in dosage form performance, with a specific focus on the applications of biopharmaceutical strategies in the development of successful drugs using case studies.

Chapter 1 introduces the basic concepts of biopharmaceutics and discusses the role of biopharmaceutics in various stages of drug development. Chapter 2 describes the molecular and physicochemical properties influencing drug absorption. Chapter 3 examines the utilities and limitations of dissolution testing and discusses biorelevant dissolution methods. Chapter 4 introduces the principles governing drug absorption. It includes all aspects of drug absorption including transport phenomena, factors influencing drug absorption, and methods to evaluate drug absorption. The current industrial practices of evaluating permeability, absorption, and p-glycoprotein interaction are presented in Chap. 5. Chapter 6 investigates the uses of pharmaceutical excipients in drug absorption as enhancers and proposes applications in dosage form design of mucoadhesive materials. Chapter 7 comprehensively describes various transporter families and outlines the role of intestinal transporters in drug absorption. Chapter 8 critically examines bioavailability and bioequivalence from a regulatory perspective, addresses issues inherent in the assessment and demonstration of bioequivalence. The mechanisms and strategies toward BCS are further outlined in Chap. 9. Further, the chapter introduces several case studies encompassing drugs from all BCS classes and multiple formulation types as examples of where a science-based approach to dissolution method characterization has been employed to determine rational product quality testing strategies. Chapter 10 discusses the impact of food in regulatory assessment of bioequivalence and proposes recommendations on the design of appropriate biopharmaceutics studies. Chapters 11 and 12 explore the applications of *in vitro* and *in vivo* correlation (IVIVC) for parenteral (Chap. 11) and orally administered drug products (Chap. 12).

We anticipate that the book will be helpful to individuals who work in the pharmaceutical industry in areas that apply pharmaceutics, biopharmaceutics, and pharmacokinetics, individuals who interact with formulation scientists and pharmacokineticists, as well as those who are in academic and research institutions. Since the fundamentals are also reviewed, we believe that the book will appeal to advanced undergraduate students and graduate students in pharmacy, pharmacology, and allied health professions.

We welcome comments and suggestions for improvement from our readers.

Rahway, NJ *Rajesh Krishna*
Rockville, MD *Lawrence Yu*

Acknowledgments

This volume is clearly a team effort. We take this opportunity to extend our appreciation to our contributors, publishers, and reviewers. We are particularly indebted to Mr. Sankar and Mr. Venkataraman for their help during proofreading. A special thanks to Kathleen Lyons at Springer for her support and inspiring us to complete this volume in a reasonable time frame.

Contents

Contributors xxi

1 Introduction to Biopharmaceutics and its Role in Drug Development . . 1
1.1 Introduction to Biopharmaceutics 1
1.1.1 What is Biopharmaceutics? 1
1.1.2 Physical Pharmacy: Physical–Chemical Principles 2
1.1.2.1 Solubility 2
1.1.2.2 Hydrophilicity/Lipophilicity 2
1.1.2.3 Salt Forms and Polymorphs 2
1.1.2.4 Stability 3
1.1.2.5 Particle and Powder Properties 3
1.1.2.6 Ionization and p*K*a 3
1.1.3 Formulation Principles 4
1.1.4 Physiological/Biological Principles 4
1.1.4.1 Pharmacokinetics 4
1.1.5 Biopharmaceutics: Integration of Physical/Chemical and Biological/Pharmacokinetic Principles and Impact on Clinical Efficacy 6
1.1.5.1 Introduction to the Biopharmaceutics Classification System 6
1.1.5.2 Impact of Physical/Chemical Properties on Absorption and Transport 7
1.1.5.3 Strategies to Achieve Target Pharmacokinetic Profile 10
1.2 Role of Biopharmaceutics in Drug Development 14
1.2.1 Importance of Biopharmaceutics in the Overall Development Process 14
1.2.2 Discovery and Preclinical Development: Candidate Selection 15
1.2.3 Preclinical Development: Preparation for Phase I Clinical Studies 16
1.2.4 Early Clinical Development 17
1.2.5 Advanced Clinical Development 19
1.2.6 Postapproval Considerations 20

1.2.7 Regulatory Considerations 21
1.3 Summary 21

2 Molecular and Physicochemical Properties Impacting Oral Absorption of Drugs 26
2.1 Introduction 26
2.2 Molecular and Physicochemical Properties Impacting Oral Absorption 27
2.2.1 Molecular Weight, Log P, the Number of H-Bond Donors and Acceptors, Polar Surface Area, and the Number of Rotatable Bonds 27
2.2.2 Chirality 29
2.2.3 Dissolution 29
2.2.4 Solubility 30
2.2.4.1 Definition of Solubility 30
2.2.4.2 Factors Contributing to Poor Aqueous Solubility 30
2.2.4.3 pH-Solubility Profile 31
2.2.4.4 Effect of Temperature on Solubility 34
2.2.4.5 Solubility in Gastric and Intestinal Fluids 35
2.2.4.6 Solubility as a Limiting Factor to Absorption . . 36
2.2.4.7 Solubility Determination 36
2.2.4.8 Solubility Prediction 38
2.2.5 Chemical Stability 38
2.2.6 Solid State Properties 39
2.2.6.1 Polymorphism 39
2.2.6.2 Amorphous Material 40
2.2.6.3 Particle Size 41
2.3 Physicochemical Properties and Drug Delivery Systems 41
2.4 Summary 43

3 Dissolution Testing 47
3.1 Introduction 47
3.2 Significance of Dissolution in Drug Absorption 47
3.3 Theories of Dissolution 49
3.4 Factors Affecting Dissolution 51
3.4.1 Factors Related to the Physicochemical Properties of the Drug Substance 52
3.4.1.1 Solubility 52
3.4.1.2 Particle Size 52
3.4.1.3 Solid Phase Characteristics 53
3.4.1.4 Salt Effects 53
3.4.2 Factors Related to Drug Product Formulation 53
3.4.3 Factors Related to Manufacturing Processes 54
3.4.4 Factors Related to Dissolution Testing Conditions 55
3.5 Roles of Dissolution Testing 55

3.6 *In Vitro* Dissolution Testing as a Quality Control Tool 56
3.6.1 Dissolution Method for Quality Control of Immediate-Release Dosage Forms 57
3.6.1.1 Dissolution Media 57
3.6.1.2 Apparatus and Test Conditions 58
3.6.2 Dissolution Method for Quality Control of Modified-Release Dosage Forms 58
3.6.2.1 Dissolution Media 58
3.6.2.2 Apparatus and Test Conditions 59
3.6.3 Limitations of Quality Control Dissolution Tests 59
3.7 Biorelevant Dissolution Testing 60
3.7.1 *In Vivo–In Vitro* Correlations 61
3.7.2 The Importance of BCS on Biorelevant Dissolution Testing 61
3.7.3 Biorelevant Dissolution Methods 64
3.7.3.1 Biorelevant Dissolution Media for Gastric Conditions 65
3.7.3.2 Apparatus and Test Conditions for Simulating the Stomach 66
3.7.3.3 Biorelevant Dissolution Media for Intestinal Conditions 67
3.7.3.4 Apparatus and Test Conditions for Simulating Small Intestine 69
3.7.3.5 Biorelevant Methods for Extended-Release Dosage Forms 69
3.7.3.6 Remaining Challenges 69
3.8 Conclusions 70

4 Drug Absorption Principles 75
4.1 Drug Absorption and Bioavailability 75
4.2 Types of Intestinal Membrane Transport 76
4.2.1 Passive Diffusion 76
4.2.2 Carrier-Mediated Transport 78
4.2.2.1 Facilitated Diffusion 78
4.2.2.2 Active Transport 79
4.2.3 Paracellular Transport 79
4.2.4 Endocytosis 79
4.2.5 Which Absorption Path Dominates Drug Absorption? . . 80
4.3 Three Primary Factors Influence Drug Absorption 80
4.3.1 Membrane Permeability 81
4.3.1.1 Effective Permeability 81
4.3.1.2 Fraction of Drug Absorbed 81
4.3.1.3 Permeability and Absorption Rate Constant . . 82
4.3.2 Solubility 82
4.3.3 Dissolution of Solid Dosage Forms 83

4.4 Secondary Factors Influencing Drug Absorption 84
4.4.1 Biological Factors of Gastrio Intestinal Tract 84
4.4.1.1 Gastric Emptying Time 84
4.4.1.2 Surface Area 84
4.4.1.3 GI Transit Time 84
4.4.1.4 Intestinal Motility 85
4.4.1.5 Components, Volume, and Properties of Gastrointestinal Fluids 85
4.4.1.6 Food 85
4.4.1.7 Blood Flow 85
4.4.1.8 Age 86
4.4.2 Dosage Factors Influencing Absorption 86
4.5 Evaluation of Oral Drug Absorption in Humans 86
4.5.1 Drug Absorption Assessment Using *In Vivo* Data 86
4.5.1.1 Estimation of Fraction of Drug Absorbed Using Experimental Intestinal Permeability *In Vivo* 86
4.5.1.2 Estimation of Maximum Absorbable Dose Using *In Vivo* Absorption Rate Constant and Drug Solubility 88
4.5.1.3 Estimation of MAD from Drug *In Vivo* Permeability in Human and Drug Solubility . . 89
4.5.2 Drug Absorption Assessment Using *In Vitro* Data 90
4.5.2.1 *In Vitro* Testing Conditions for Determining Drug Permeability in Caco-2 Cells and *In Vitro*/*In Vivo* Permeability Correlation 90
4.5.2.2 Estimation of Fraction of Drug Absorbed In Humans Using *In Vitro* Drug Permeability in Caco-2 Cells 92
4.5.2.3 Estimation of MAD in Human Based on *In Vitro* Data 93
4.5.3 Correlation of Oral Drug Bioavailability and Intestinal Permeability Between Rat and Human 95
4.6 Summary 97

5 Evaluation of Permeability and P-glycoprotein Interactions: Industry Outlook 101
5.1 Introduction 101
5.2 Anatomy and Physiology of the Small Intestine 104
5.3 Permeability Absorption Models 105
5.3.1 Physicochemical Methods 105
5.3.1.1 Lipophilicity (Log *P*/Log *D*) 105
5.3.1.2 Absorption Potential 105
5.3.1.3 Immobilized Artificial Membrane (IAM) 106

5.3.2 *In Vitro* Methods 106
5.3.2.1 Animal Tissue-Based Methods 107
5.3.2.2 Cell-Based Methods 109
5.3.3 *In Situ* Methods 111
5.3.4 *In Vivo* Methods 112
5.3.5 *In Silico* Methods 113
5.4 Comparison of PAMPA and Caco-2 Cells 114
5.4.1 Parallel Artificial Membrane Permeability Assay 114
5.4.1.1 PAMPA Study Protocol 115
5.4.2 Caco-2 Cells 115
5.4.2.1 Caco-2 Cell Culture 116
5.4.2.2 Caco-2 cells Study Protocol 116
5.4.3 PAMPA and Caco-2 Cell: Synergies 116
5.4.4 PAMPA and Caco-2 Cell: Caveats 123
5.4.4.1 Transporter- and Paracellular-Mediated Absorption 123
5.4.4.2 Incomplete Mass-Balance Due to Nonspecific Binding 125
5.4.4.3 Inadequate Aqueous Solubility 125
5.4.4.4 Other Experimental Variability 126
5.5 P-gp Studies Using Caco-2 Cells 127
5.5.1 Experimental Factors Effecting Efflux Ratio 129
5.6 Conclusions 132

6 Excipients as Absorption Enhancers 139
6.1 Introduction 139
6.2 Basic Mechanisms in Transcellular and Paracellular Transport . . 140
6.2.1 Transcellular Transport 141
6.2.2 Paracellular Transport 142
6.2.3 Mechanisms of Action of Absorption Enhancers 142
6.2.3.1 Action on the Mucus Layer 143
6.2.3.2 Action on Membrane Components 143
6.3 Mucoadhesive Polymers as Absorption Enhancers 148
6.3.1 Theories of Mucoadhesion 148
6.3.2 Material Properties of Mucoadhesives 150
6.3.3 Classes of Mucoadhesive Polymers 152
6.3.3.1 Polyacrylates 152
6.3.3.2 Chitosan 156
6.3.3.3 N,N,N,-Trimethyl Chitosan Hydrochloride (TMC) 158
6.3.3.4 Monocarboxymethyl Chitosan 161
6.3.3.5 Thiolated Polymers 162
6.3.3.6 Solid Dosage Form Design Based on TMC and Thiolated Polymers and Their *In Vivo* Evaluation 164
6.4 Conclusions 166

7 Intestinal Transporters in Drug Absorption 175
7.1 Introduction 175
7.2 ATP Binding Cassette Transporters 179
7.2.1 P-Glycoprotein (P-gp; ABCB1) 183
7.2.1.1 The Expression of P-gp 183
7.2.1.2 The Regulation of P-gp Expression 184
7.2.1.3 P-gp Mediated Drug Transport 185
7.2.1.4 The Substrate Specificity of P-gp 185
7.2.2 Multidrug Resistance-Associated Protein Family (MRP; ABCC) 188
7.2.2.1 The Expression of MRPs 188
7.2.2.2 The Regulation of MRP Isoform Expression . . 190
7.2.2.3 The Substrate Specificity of MRP's 190
7.2.3 Breast Cancer Resistance Protein (BCRP; ABCG2) 193
7.3 Solute Carrier Transporters 195
7.3.1 Proton/Oligopeptide Transporters (POT; SLC15A) 195
7.3.1.1 Peptide Transporter Mediated Transport 197
7.3.1.2 The Substrate Specificity of Peptide Transporters 198
7.3.1.3 The Regulation of Peptide Transporters 200
7.3.2 Organic Anion Transporters (OAT, SLC22A; OATP, SLCO) 202
7.3.2.1 OAT (SLC22A) 202
7.3.2.2 OATP (SLCO) 204
7.3.3 Organic Cation Transporters (OCT, OCTN; SLC22A) . . 209
7.3.3.1 The Substrate Specificity of Organic Cation Transporters 210
7.3.3.2 Organic Cation Transporter Mediated Transport 211
7.3.3.3 The Expression of Organic Cation Transporters 211
7.3.3.4 The Regulation of Organic Cation Transporters 212
7.3.4 Nucleoside Transporters (CNT, SLC28A; ENT, SLC29A) 214
7.3.4.1 The Molecular and Structural Characteristics of Nucleoside Transporters 215
7.3.4.2 The Substrate Specificities of Nucleoside Transporters 217
7.3.4.3 The Expression of Nucleoside Transporters . . . 219
7.3.4.4 The Regulation of Nucleoside Transporters . . . 220
7.3.5 Monocarboxylate Transporters (MCT; SLC16A) 221
7.3.5.1 Molecular and Structural Characteristics of Monocarboxylate Transporters 222

7.3.5.2 The Substrate Specificity of Monocarboxylate Transporters 222
7.3.5.3 The Expression of Monocarboxylate Transporters 223
7.3.5.4 The Regulation of Monocarboxylate Transporters 224
7.4 Impact of Intestinal Transporters on Bioavailability 225

8 Bioavailability and Bioequivalence 262
8.1 Introduction 262
8.2 Bioavailability and Bioequivalence 262
8.2.1 Bioavailability and its Utility in Drug Development and Regulation 262
8.2.2 Bioequivalence and its Utility in Drug Development and Regulation 263
8.2.3 Bioavailability and Bioequivalence Studies: General Approaches 264
8.3 Pharmacokinetic Bioavailability and Bioequivalence Studies . . . 265
8.3.1 Bioavailability Studies: General Guidelines and Recommendations 265
8.3.2 Bioequivalence Studies: General Guidelines and Recommendations 267
8.3.2.1 Study Design 267
8.3.2.2 Dose 267
8.3.2.3 Subjects 268
8.3.2.4 Statistical Analysis of Bioequivalence 268
8.4 Bioequivalence: Challenging Topics 271
8.4.1 Drugs with Active Metabolites 271
8.4.2 Enantiomers vs. Racemates 273
8.4.3 Endogenous Substances 273
8.4.4 Highly Variable Drugs 274
8.4.4.1 Static Expansion of the BE Limits 274
8.4.4.2 Expansion of Bioequivalence Limits Based on Fixed Sample Size 275
8.4.4.3 Scaled Average Bioequivalence 275
8.5 Biowaivers 276
8.5.1 Solutions 276
8.5.2 Lower Strength 278
8.5.3 Biopharmaceutical Classification System 279
8.5.3.1 Biowaivers for BCS Class 2 Drugs with pH Dependent Solubility 280
8.5.3.2 Biowaivers for BCS Class 3 Drugs 280
8.6 Locally Acting Drugs 281
8.6.1 Topical Dermatological Products 281

8.6.2 Locally Acting Nasal and Oral Inhalation Drug Products . . . 283
8.6.2.1 Nasal Spray Products . . . 284
8.6.2.2 Oral Inhalation Products . . . 285
8.7 Conclusions . . . 287

9 A Biopharmaceutical Classification System Approach to Dissolution: Mechanisms and Strategies . . . 290
9.1 Introduction . . . 290
9.2 Biopharmaceutical Classification System Approach to Dissolution 290
9.3 *In Vitro–In Vivo* Dissolution Correlation . . . 292
9.4 Recent Climate: Pharmaceutical Quality Assessment . . . 294
9.5 Discussion . . . 296
9.5.1 BCS Class I and III Case Studies . . . 296
9.5.1.1 Case Study 1: Fast Release (>85% Release in 15 min) with Disintegration Controlled Dissolution . . . 298
9.5.1.2 Case Study 2: <85% Release in 15 min with Disintegration/Erosion Controlled Dissolution . . . 299
9.5.1.3 Case Study 3: Dissolution Mechanism not Dependent on Disintegration/Erosion . . . 301
9.5.2 BCS Class II and IV Case Studies . . . 302
9.5.2.1 Case Study 4: Liquid Filled (True Solution) Capsules . . . 303
9.5.2.2 Case Studies 5, 6, and 7: Intrinsic Rate of Drug Solubilization Controlled Dissolution . . . 303
9.5.2.3 Case Studies 8 and 9: Mixed Contribution of Formulation Colligative Properties and Intrinsic Rate of Drug Solubilization . . . 306
9.5.2.4 Case Study 10: API with High Solubility at Gastric pHs . . . 309
9.5.3 Controlled Release Dosage Form Case Study . . . 310
9.5.4 Pharmaceutical Quality Assessment Implications of Dissolution . . . 313
9.6 Conclusion . . . 314

10 Food Effects on Drug Bioavailability: Implications for New and Generic Drug Development . . . 317
10.1 Introduction . . . 317
10.1.1 Objectives . . . 317
10.1.2 Oral Bioavailability Defined . . . 317
10.1.3 How Food Can Affect Drug Bioavailability . . . 317
10.2 Food Interactions with Drug Substance . . . 318

10.2.1 Pharmacokinetic Parameters Used to Characterize Food Effects on Drug Bioavailability 318
10.2.2 Prolonged Rate of Drug Absorption in the Presence of Food 318
10.2.3 Decreased Drug Absorption in the Presence of Food . . . 319
10.2.3.1 Overview 319
10.2.3.2 Instability in Gastric Acids 319
10.2.3.3 Physical or Chemical Binding with Food Components 319
10.2.3.4 Increased First-Pass Metabolism and Clearance 320
10.2.4 Increased Drug Absorption in the Presence of Food 320
10.2.4.1 Inhibition of First-Pass Effect 320
10.2.4.2 Physicochemical and Physiological Effects . . . 321
10.2.4.3 Effects of Bile Release 322
10.2.4.4 Effects of Longer Gastric Residence Time . . . 322
10.2.5 Drug Absorption Unaffected by Food 322
10.2.6 FDA Guidance for Industry on Characterizing Food Effects in Drug Development 323
10.2.6.1 Objectives 323
10.2.6.2 Recommended Designs for Food-Effect Bioavailability Studies 323
10.2.6.3 Recommendations for Drug Product Labeling 323
10.3 Food Interactions with Drug Product 324
10.3.1 Introduction 324
10.3.2 Issues with Modified-Release Drug Products: Potential for Dose-Dumping 325
10.3.3 Issues with Modified-Release Drug Products: Formulation-Dependant Food Effects 326
10.3.3.1 *In Vitro* Drug Release Predictive of Food Effects 326
10.3.3.2 *In Vitro* Drug Release Profiles Not Predictive of Food Effects 326
10.3.4 Implications for Development of Generic Modified-Release Drug Products 327
10.3.4.1 Introduction 327
10.3.4.2 Role of *In Vivo* Fed Bioequivalence Studies . . 327
10.3.5 Implications for Development of Generic Immediate-Release Drug Products 328
10.3.5.1 BCS Class I Drugs 328
10.3.5.2 Label-Driven Criteria for Requesting Fed Bioequivalence Studies 329
10.3.6 Recommendations for Designing Fed Bioequivalence Studies 330

10.3.7 Food Effects and Generic Drug Product Labeling 331
10.3.8 Sprinkle Studies in New and Generic Drug Product Development 331
10.3.8.1 Sprinkle Studies in Development of New Modified-Release Capsules 331
10.3.8.2 Sprinkle Studies in Development of Generic Modified-Release Capsules 331
10.3.8.3 Example 332
10.4 Summary and Conclusions 332

11 *In Vitro–In Vivo* Correlation on Parenteral Dosage Forms 336
11.1 IVIVC Definition 336
11.2 Modified Release Parenteral Products 336
11.3 Factors to Consider for Meaningful IVIVC 337
11.3.1 Product Related Factors 337
11.3.2 Factors Affecting *In Vitro* Release 338
11.3.2.1 Accelerated *In Vitro* Release Testing 341
11.3.3 Mathematical Models of *In Vitro* Drug Release 341
11.3.4 Factors Affecting *In Vivo* Release 343
11.4 *In Vitro–In Vivo* Correlation 344
11.5 Microspheres 345
11.6 Liposomes 347
11.7 Emulsions 349
11.8 Hydrogels, Implants 350
11.9 Dendrimers 351

12 *In Vitro–In Vivo* Correlation in Dosage Form Development: Case Studies 359
12.1 Introduction 359
12.2 IVIVC in Drug Product Development: A Four-Tier Approach . . . 360
12.3 Case Studies 363
12.3.1 Tier 1 – Discovery and Early Preclinical Development: Assessing Developability and Formulation Principles 363
12.3.2 Tier 2 – Preclinical Product Development: Selection of a Meaningful Dissolution Method 367
12.3.3 Tier 3 – Full Development: Deconvolution of Human Pharmacokinetic Data and Comparison with *In Vitro* Dissolution Data 372
12.4 Deconvolution and Convolution 372
12.4.1 Tier 4: Application of IVIVC in LCM 376
12.5 Conclusions 380

Index 383

Contributors

Praveen V. Balimane, Bristol-Myers Squibb Company, Princeton, NJ
Nancy P. Barbour, Bristol-Myers Squibb Company, New Brunswick, NJ
Rajinder K. Bhardwaj, Ernest Mario School of Pharmacy, Rutgers, The State University of New Jersey, Piscataway, NJ
Current Address: Bristol-Myers Squibb Company, Wallingford, CT
William E. Bowen, Merck Research Laboratories, West Point, PA
Diane J. Burgess, School of Pharmacy, University of Connecticut, Storrs, CT
Xianhua Cao, College of Pharmacy, The Ohio State University, Columbus, OH
Stephen M. Carl, Ernest Mario School of Pharmacy, Rutgers, The State University of New Jersey, Piscataway, NJ
Current Address: School of Pharmacy, Purdue University, West Lafayette, IN
Saeho Chong, Bristol-Myers Squibb Company, Princeton, NJ
Dale P. Conner, Office of Generic Drugs, U.S. Food and Drug Administration, Rockville, MD
Thomas J. Cook, Ernest Mario School of Pharmacy, Rutgers, The State University of New Jersey, Piscataway, NJ
Barbara Myers Davit, Office of Generic Drugs, U.S. Food and Drug Administration, Rockville, MD
Kimberly Gallagher, Merck Research Laboratories, West Point, PA
Sam H. Haidar, Office of Generic Drugs, U.S. Food and Drug Administration, Rockville, MD
Dea R. Herrera-Ruiz, Facultad de Farmacia, Universidad Autónoma del Estado de Morelos, Cuernavaca, México
Brian Hill, Merck Research Laboratories, West Point, PA
Hans E. Junginger, Faculty of Pharmaceutical Sciences, Naresuan University, Phitsanulok, Thailand
Gregory T. Knipp, Ernest Mario School of Pharmacy, Rutgers, The State University of New Jersey, Piscataway, NJ
Current Address: School of Pharmacy, Purdue University, West Lafayette, IN
Hyojong (Hue) Kwon, Office of Generic Drugs, U.S. Food and Drug Administration, Rockville, MD
Sau Lawrence Lee, Office of Generic Drugs, U.S. Food and Drug Administration, Rockville, MD
Shoufeng Li, Novartis Pharmaceuticals Corporation, East Hanover, NJ

Robert Lionberger, Office of Generic Drugs, U.S. Food and Drug Administration, Rockville, MD
Robert A. Lipper, Bristol-Myers Squibb Company, New Brunswick, NJ
Yun Mao, Merck Research Laboratories, West Point, PA
Eric D. Nelson, Merck Research Laboratories, West Point, PA
Andre S. Raw, Office of Generic Drugs, U.S. Food and Drug Administration, Rockville, MD
Robert A. Reed, Merck Research Laboratories, West Point, PA
Alan E. Royce, Novartis Pharmaceuticals Corporation, East Hanover, NJ
Abu T. M. Serajuddin, Novartis Pharmaceuticals Corporation, East Hanover, NJ
Duxin Sun, College of Pharmacy, The Ohio State University, Columbus, OH
Denise L. Thomas, Merck Research Laboratories, West Point, PA
Mark Thompson, Merck Research Laboratories, West Point, PA
Wei-Qin (Tony) Tong, Novartis Pharmaceuticals Corporation, East Hanover, NJ
Nicholi Vorsa, Philip E. Marucci Center, Rutgers, The State University of New Jersey, Chatsworth, NJ
Qingxi Wang, Merck Research Laboratories, West Point, PA
W. Peter Wuelfing, Merck Research Laboratories, West Point, PA
Yan Xu, Ernest Mario School of Pharmacy, Rutgers, The State University of New Jersey, Piscataway, NJ
Lawrence Yu, Office of Generic Drugs, U.S. Food and Drug Administration, Rockville, MD
Banu S. Zolnik, National Cancer Institute, Frederick, MD

1
Introduction to Biopharmaceutics and its Role in Drug Development

Nancy P. Barbour and Robert A. Lipper

1.1 Introduction to Biopharmaceutics

1.1.1 What is Biopharmaceutics?

In the world of drug development, the meaning of the term "biopharmaceutics" often evokes confusion, even among scientists and professionals who work in the field. "Pharmaceutics" narrowly defined is a field of science that involves the preparation, use, or dispensing of medicines (Woolf, 1981). Addition of the prefix "bio," coming from the Greek "bios," relating to living organisms or tissues (Woolf, 1981), expands this field into the science of preparing, using, and administering drugs to living organisms or tissues. Inherent in the concept of biopharmaceutics as discussed here is the interdependence of biological aspects of the living organism (the patient) and the physical–chemical principles that govern the preparation and behavior of the medicinal agent or drug product. This philosophy was pioneered in the mid-twentieth century by the first generation of what we refer to now as biopharmaceutical scientists: those who recognized the importance of absorption, distribution, metabolism, and elimination (ADME) on the clinical performance of medicinal agents as well as the impact of the physical–chemical properties of the materials on their *in vivo* performance. As a result, biopharmaceutics has evolved into a broad-based discipline that encompasses fundamental principles from basic scientific and related disciplines, including chemistry, physiology, physics, statistics, engineering, mathematics, microbiology, enzymology, and cell biology. The biopharmaceutical scientist, therefore, must have sufficient understanding of all of these scientific fields in order to be most effective in a drug development role. A scientist educated in the field of biopharmaceutics or biopharmaceutical sciences could have expertise in a number of interrelated specialty disciplines including formulation, pharmacokinetics (PK), cell-based transport, drug delivery, or physical pharmacy. For the subsequent discussion we will look broadly at the areas of physical pharmacy (pharmaceutics) and PK and their roles and interdependencies in the drug development process.

1.1.2 Physical Pharmacy: Physical–Chemical Principles

Physical pharmacy is a term that came into common use in the pharmacy community in the mid-twentieth century, and the field has grown and evolved over the years. Essentially, physical pharmacy is a collection of basic chemistry concepts that are firmly rooted in thermodynamics and chemical kinetics. Scientists in the mid-twentieth century pioneered research in the areas of physical–chemical properties of drugs and their influence on biological performance (Reinstein and Higuchi, 1958; Higuchi, 1958, 1976; Higuchi *et al.*, 1956, 1958, 1963; Kostenbauder and Higuchi, 1957; Shefter and Higuchi, 1963; Agharkar *et al.*, 1976; Shek *et al.*, 1976). Key aspects of physical–chemical properties discussed in greater detail in Chap. 2, briefly include the following.

1.1.2.1 Solubility

Solubility is a thermodynamic parameter that defines the amount of material (in this case a drug) that can dissolve in a given solvent at equilibrium. Solubility is one of the most critical and commonly studied physical–chemical attributes of drug candidates. The amount of drug in solution as a function of time prior to reaching equilibrium is often referred to as the "kinetic solubility," which can be exploited in pharmaceutical applications to manipulate drug delivery. A compound's solubility impacts its usefulness as a medicinal agent and also influences how a compound is formulated, administered, and absorbed. A thorough review of the scientific fundamentals of solubility theory has been presented previously (Flynn, 1984).

1.1.2.2 Hydrophilicity/Lipophilicity

The partition or distribution coefficient of a drug candidate (log P or log D) is a relative measure of a compound's tendency to partition between hydrophilic and lipophilic solvents and thus indicates the hydrophilic/lipophilic nature of the material. The relative lipophilicity is important with respect to biopharmaceutics since it affects partitioning into biological membranes and therefore influences permeability through membranes as well as binding and distribution into tissues *in vivo* (Ishii *et al.*, 1995; Lipka *et al.*, 1996; Merino *et al.*, 1995).

1.1.2.3 Salt Forms and Polymorphs

Drug substances can often exist in multiple solid-state forms, including salts (for ionizable compounds only), solvates, hydrates, polymorphs, co-crystals or amorphous materials. The solid form of the compound affects the solid-state properties including solubility, dissolution rate, stability, and hygroscopicity, and can also impact drug product manufacturability and clinical performance (Singhal and Curatolo, 2004). There are numerous examples in the literature of the impact of pH and salt form on solubility, and how this phenomenon can be utilized to manipulate the solubility behavior of a drug compound (Li *et al.*, 2005; Agharkar *et al.*,

1976; Morris, 1994). For example, salts can be chosen to impart greater solubility to improve dissolution rate of an active pharmaceutical ingredient (API). Polymorphs and solvated forms of drug candidates can also affect not only the stability and manufacturability of a drug substance but also potentially impact biopharmaceutical performance due to their differing solubilities (Raw and Yu, 2004).

1.1.2.4 Stability

The chemical stability of a drug is important in order to avoid generation of undesirable impurities, which could have pharmacologic activity and/or toxicologic implications, in the drug substance or drug product. Chemical stability of the API in a dosage form influences shelf-life and storage conditions of drug products to minimize generation of undesirable impurities. The pH-stability profile is also important from a physiological perspective considering the range of pH values that a pharmaceutical material may encounter *in vivo*, particularly in the GI tract. Sufficient stability is required for the compound as well during the course of administration. Physical stability refers to changes in the drug substance solid-state form including polymorphic transitions, solvation/desolvation, or salt disproportionation. As mentioned previously, changes in drug substance form can lead to changes in physical properties such as solubility and dissolution rate. At the product (dosage form) level, physical stability refers broadly to mechanical property integrity (hardness, friability, swelling) and potential impact of changes on product performance.

1.1.2.5 Particle and Powder Properties

Bulk properties of a pharmaceutical powder include particle size, density, flow, wettability, and surface area. Some are important from the perspective of a manufacturing process (e.g., density and flow) while others could potentially impact drug product dissolution rate (particle size, wettability, and surface area) without changing equilibrium solubility.

1.1.2.6 Ionization and p*K*a

The ionization constant is a fundamental property of the chemical compound that influences all of the physical–chemical properties discussed above. The presence of an ionizable group (within the physiologically relevant pH range) leads to pH-solubility effects, which can be used to manipulate the physical properties and biological behavior of a drug. For an ionizable compound, the aqueous solubility of the ionized species is typically higher than the unionized due to the greater polarity afforded by the presence of the ionized functional group. The ionizable functional group and the magnitude of the p*K*a determine whether a compound is ionized across the physiological pH range, or if conversion between ionized/nonionized species occurs in the GI tract, and if so, which region. The p*K*a also affects the available choices of counterions for potential salt forms that are suitable from a physical perspective.

1.1.3 Formulation Principles

The goal of a formulation scientist is to manipulate the properties and environment of the API to optimize its delivery to the target tissue by a specific route of administration and to do so in a manner compatible with large-scale product manufacture. Excipients are added to solubilize, stabilize, modify dissolution rate, improve ease of administration (e.g., swallowing or taste-masking), enable manufacturing (e.g., ensure sufficient compactibility to make tablets, improve powder flow in a manufacturing line), control release rate (immediate vs. prolonged vs. enteric), or inhibit precipitation (Gennaro, 1995). The formulation is key to a compound's biopharmaceutical profile since the composition, dosage form type, manufacturing process, and delivery route are intimately linked to pharmacokinetic results. A PK assessment cannot be complete without inclusion of the relevant formulation parameters to establish the appropriate context.

1.1.4 Physiological/Biological Principles

1.1.4.1 Pharmacokinetics

The other broad discipline in biopharmaceutics is PK, which is the study of the time course of ADME (Gibaldi and Perrier, 1982; Rowland and Tozer, 1989). Just as the physical–chemical and formulation principles are intimately linked with the pharmacokinetic profile, the PK profile is directly related to the pharmacologic activity of a drug. For the purpose of this discussion, we will use PK and ADME interchangeably.

Absorption

In most cases, a drug must be absorbed across a biological membrane in order to reach the general circulation and/or elicit a pharmacologic response. Even drugs that are dosed intravenously may need to cross the vascular endothelium to reach the target tissue or distribute into blood cells. Often multiple membranes are encountered as a drug traverses the absorptive layer and diffuses into the blood stream. Transport across these membranes is a complex process, impacted by ionization equilibria, partitioning into and diffusion across a lipophilic membrane and potential interaction with transporter systems (influx and/or efflux).

Membrane transport can occur either passively or actively (Rowland and Tozer, 1989). Passive transport (diffusion) is the movement of molecules from a region of high concentration to one of low concentration. The membrane permeability, which is directly related to the relative lipophilicity of the drug, is a major factor affecting the rate and extent of absorption for a given compound, and for GI absorption the concentration gradient is related to the solubility of the compound in the intestinal brush border microenvironment (Rowland and Tozer, 1989).

Active transport is an energy-consuming process whereby membrane-bound transporters bind and transport materials across membranes, even against a concentration gradient. Physiologically, these active transporters exist to promote absorption of nutrients and hence are typically related to food substances such as peptides, amino acids, carbohydrates, and vitamins. They can lead to absorption efficiency that is significantly greater than what would be predicted based on a passive diffusion mechanism. In recent years many of these transporters have been characterized with respect to structure, cellular location, and substrate specificity (Katsura and Inui, 2003; Sai, 2005). Conversely, active transport mechanisms also exist to transport materials out of cells (efflux pumps). The most well-studied efflux pumps are in the class of ATP-binding cassette (ABC) transporter proteins, including p-glycoprotein (P-gp) and the multidrug resistance protein (MRP) family (Kivisto *et al.*, 2004; Leslie *et al.*, 2005). These natural transporters are cellular defenses that exist to prevent entry of unwanted potentially toxic materials into the systemic circulation, and they can also work against the movement of drug molecules. The reader is referred to Chap. 7, which discusses role of such transporters in absorption processes in detail.

The concepts of permeability, absorption, and bioavailability (BA) are sometimes used interchangeably, while in fact each represents a different aspect related to membrane transport. Permeability refers to the ability of a compound to cross a membrane. A permeable compound may diffuse across the intestinal epithelium only to be actively transported out of the cell. This compound is permeable, yet not absorbed. Likewise, a drug may pass through the intestinal epithelium, indicating absorption, yet be metabolized in the gut wall or the liver prior to reaching the peripheral circulation. This drug is absorbed, yet it is not bioavailable. The relevance of this will be discussed in subsequent Chaps. 4 and 5.

Distribution

Distribution is a measure of the relative concentrations of a drug in different body tissues as a function of time (Rowland and Tozer, 1989) and is related to its ability to diffuse from the blood stream, tissue perfusion, relative lipophilicity, and tissue/plasma protein binding. The apparent volume of distribution (V_d) is reflective of the extent of tissue distribution. Drug distribution *in vivo* is often related to the drug's chemical structure. It can be measured and manipulated during the course of compound optimization by addition or deletion of certain functional groups or structural features. However, formulations typically cannot have significant impact on a drug's distribution properties without chemical alterations such as conjugation or use of specific drug targeting technology.

Metabolism and Elimination

Metabolism is one of the most important mechanisms that the body has for detoxifying and eliminating drugs and other foreign substances. Drugs delivered by the oral route must pass through the liver before reaching the general circulation.

Metabolism at this point is called “first-pass metabolism,” which can limit systemic exposure for drugs despite good absorption. Oxidation, reduction, hydrolysis, and conjugation are the most common metabolic pathways, generally leading to more hydrophilic compounds that can be readily excreted renally. Cytochrome P450 (CYP) enzymes are a family of drug metabolizing enzymes that are responsible for the majority of drugs' metabolism as well as many drug–drug interactions (Shou *et al.*, 2001; Meyer, 1996). Although the primary role of metabolism is to facilitate elimination of drugs from the body, secondary effects include transformation of drugs into other active or toxic species, which could be desirable in the case of prodrugs (Stella *et al.*, 1985) or undesirable with respect to toxic metabolites (Kalgutkar *et al.*, 2005). The reader is encouraged to refer to authoritative texts in this field.

Elimination of drugs from the body can occur *via* metabolism, excretion (renal, biliary, respiratory), or a combination of both mechanisms. As with distribution, these phases of the drug's PK profile are inherent to the chemical structure of the drug and are optimized (along with pharmacologic potency and fundamental safety) during the drug discovery process.

1.1.5 Biopharmaceutics: Integration of Physical/Chemical and Biological/Pharmacokinetic Principles and Impact on Clinical Efficacy

In the previous overview discussion, we highlighted some principles governing physical pharmacy, formulation, and PK; the assessment of any one of these is dependent on the context of the others. This interplay is often complex. The integration of these various principles is necessary to define fully the biopharmaceutical profile for a new drug candidate and to evaluate the utility of a particular compound to treat the intended disease. The suitability of any given parameter is always dependent on one or more other, related parameters. For example, the target solubility for a new compound depends on the dose (Curatolo, 1998; Hilgers *et al.*, 2003), which depends on the receptor affinity and BA, which are related to the lipophilicity, which is in turn is related to the solubility. Another common goal is to define compounds with good receptor binding, which is often increased by higher lipophilicity, which can negatively impact absorption and effective dose. Failure to consider all of these factors and their interrelationships can likely lead to the selection of chemical compounds that may not be useful as drugs, or to misleading conclusions regarding interpretation of a clinical issue. Hence, the answer to a question regarding acceptable biopharmaceutical properties is often “it depends.” This point is illustrated in the following general examples of integration of biopharmaceutical principles.

1.1.5.1 Introduction to the Biopharmaceutics Classification System

The biopharmaceutics classification system (BCS) was originally proposed based on the understanding that absorption of drugs in the GI tract *via* passive diffusion is governed primarily by the amount of drug in solution at the luminal–epithelial border and the ability of that drug to diffuse across the intestinal endothelium (Amidon *et al.*, 1995). Flux of a compound is dependent on the diffusivity (permeability) and concentration gradient (solubility). The BCS categorizes solubility and permeability of drugs as either high or low and considers the dose and ionization of the drug in the GI tract. A strict definition of permeability is difficult considering the factors in the GI tract that influence apparent permeability (efflux pumps, metabolism, region), and therefore permeability can be estimated from either *in vitro* transport in cell culture models of intestinal transport or from *in vivo* data on drug absorption. The BCS also recognizes the importance of the dose of a drug, as a high dose drug with low solubility is more likely to exhibit absorption difficulties than a drug with the same solubility and low dose. Conversely, high permeability of a compound may be able to overcome perceived issues with low solubility. Hence, some drugs with extremely low solubility can nevertheless show high systemic BA due to high permeability. The relative balance of these properties influences whether the absorption rate of the drug is controlled primarily by solubility, dissolution rate, or membrane transport.

The BCS can be constructively used to assess the potential for impact of various factors, including formulation variables and physiological changes, on pharmacologic performance. For example, BA of a drug that is highly soluble in the full pH range of the GI tract (BCS Class I or III) would not be expected to be sensitive to formulation factors in an immediate-release dosage form that shows rapid dissolution. Conversely, drugs with low solubility (BCS Class II or IV) have greater potential for effects of particle size, dissolution rate, or excipients on PK behavior. Drugs with low permeability are more likely to show variable absorption, whereas absorption of high permeability drugs could show a dependence on solubility since the rate-limiting step in this scenario is dissolution. The BCS classification of a drug has regulatory implications as well, as current guidances define whether the compound requires additional bioequivalence studies or whether biowaivers may be possible for new strengths or modified formulations (FDA, 2005; Ahr *et al.*, 2000).

The BCS system can also be used by a formulator to provide guidance on the formulation strategy for a new compound. Class I drugs are less likely to require novel drug delivery approaches and have greater potential for equivalence among formulations, whereas Class IV drugs often pose significant challenges to overcome limitations in both solubility and permeability. For the latter, exploration of formulations that include solubilizing agents to enhance microenvironmental solubility or utilization of high energy solid-state forms to affect kinetic solubility could be warranted. Key to all of this is the dose.

1.1.5.2 Impact of Physical/Chemical Properties on Absorption and Transport

The oral absorption process is complex, but for many molecules it can be simplified into a general process that, for a passive diffusion mechanism, requires dissolution followed by partitioning into and transport across the intestinal epithelium. This particular aspect of ADME is most amenable to manipulation by the pharmaceutical scientist to influence PK profile and alter *in vivo* performance for an orally administered drug. Once absorbed, the drug's distribution, metabolism, and elimination are dependent on the chemical structure and physiology.

GI Transit and Ionization

Throughout the GI tract, an ionizable drug can undergo multiple transitions depending on its functional groups and p*K*a values. The state of ionization of an ionizable compound strongly influences passage across membranes as well as solubility. For a compound to be transported efficiently across a biological membrane by a passive transcellular route, the drug must be in solution and non-ionized. These two factors normally work in opposition to each other since non-ionized molecules tend to have greater lipophilicity, which favors membrane partitioning, yet lower solubility relative to ionized species. A weak monoprotic acid with a p*K*a in the range of 4–5 would be non-ionized in the stomach and as such would be at the lower range of its solubility. Once it transits to the small intestine, the drug would be predominantly ionized and have greater solubility. For a weakly basic amine, the ionization state would be reversed, with the drug predominantly ionized and most soluble in the stomach milieu, and non-ionized and less soluble in the small intestine. This might seem to suggest inherent differences in exposure for weak acids vs. bases, but this is not necessarily the case since, as noted previously, solubility is only part of the absorption equation.

Permeability is the other key determinant of exposure following oral dosing. For an ionizable compound, the ionized and non-ionized species both exist in solution, with the relative ratio determined by the pH and p*K*a. As the non-ionized species is absorbed, it is continually "regenerated" as the molecule drives toward a state of equilibrium that is never reached in the dynamic environment of the GI tract.

The dynamic pH environment of the GI tract impacts the utility of salts of ionizable drugs to improve oral absorption. Although a salt form typically has greater aqueous solubility than the corresponding free form, it may not always be the best choice for clinical development. Depending on the p*K*a, pH-solubility factors can lead to variability *in vivo* due to conversion to insoluble salts (e.g., with coadministration of calcium-containing foods), precipitation of insoluble free acids or free bases, or potential drug interactions with concomitantly administered drugs that affect gastric pH (Zhou *et al.*, 2005).

Dissolution and Relationship to BA

Systemic exposure to a drug after oral administration is the culmination of a multistep process that starts with disintegration and dissolution of the dosage form in the stomach contents. Dissolution of a drug *in vivo* is required for intestinal absorption and is impacted by multiple factors, including the solubility of the drug, release rate from the dosage form, and subsequent phase conversions, precipitation, *in situ* salt formation, micellar solubilization in the small intestine by bile salts, and pH gradients.

An integral part of the formulation development cycle is development of analytical test methods to assure quality and integrity of the product intended for human use. Dissolution or drug release *in vitro* in aqueous media under controlled pH conditions, often with added surfactants to solubilize poorly soluble drugs, is a commonly used technique to evaluate oral drug product performance. This *in vitro* dissolution test is relevant as a tool to evaluate the relative performance of different prototype formulations during the formulation development and selection process, and once a product is in clinical testing to assure consistency of the manufacturing process. The development of an appropriate dissolution method should be an iterative process that is done in parallel with the formulation development since choice of dissolution apparatus, media, and other parameters will be dependent on the solubility of the API, the nature of the excipients and dosage form, and the BCS class of the drug. For a method to be useful during formulation development, it should be discriminating, i.e., be able to distinguish differences among formulation and/or process parameters that could impact the choice or *in vivo* performance of the formulation. On the other hand, care should be taken to avoid developing an overly discriminating method that detects differences that are artifactual and/or have no relevance to the use of the product by the patient.

An *in vitro* dissolution test may also be used to assess *in vivo* biopharmaceutical performance if it is physiologically relevant, i.e., is shown to be predictive of *in vivo* behavior. Determining the physiological relevance, however, is difficult with many drugs because of the interplay of multiple factors in a human body that affect drug absorption. The relationship of solubility to absorption in the gut is complex because of the varying composition of the GI fluid and the dynamic environment governing dissolution and absorption. The solubility determined experimentally in a compositionally defined system such as a simple buffer or solvent is a thermodynamic value that reflects the amount of drug in solution at equilibrium (which may take minutes, hours, or days to achieve). In contrast, the GI tract often contains water, fats, pH-modifiers, salts, surfactants, emulsifiers, enzymes, and food components that together determine the effective GI solubility, which may be significantly different from the solubility in an aqueous buffer. This composition also changes with time as the material moves through regions of varying pH (e.g., stomach to small intestine), in a fed or fasted state, and with secretion of pancreatic enzymes and bile salts. Consideration of these additional variables has led to the development of alternative methods to assess solubility and dissolution in biorelevant media such as simulated GI fluids (Nicolaides *et al.*, 1999; Dressman *et al.*, 1998) and to compartmentalized dissolution simulation systems (Parrott and Lave, 2002; Gu *et al.*, 2005).

The only definitive way to establish physiological relevance of *in vitro* dissolution data is to perform a human PK study to correlate dissolution rate using a given method with the resultant PK profile. Ideally, a clear *in vitro–in vivo* correlation (IVIVC) can be made, but in many cases this may be elusive. The BCS class of the drug can be used to predict which compounds could potentially achieve a meaningful IVIVC. Class I and III drugs, because of their high solubility and expected rapid dissolution, would not be expected to show meaningful IVIVCs;

Class IV compounds typically exhibit variable predictability in IVIVCs due to the fact that dissolution and/or permeability could be rate-limiting factors for absorption depending on the particular compound. Class II drugs, however, are most likely to exhibit these relationships since absorption tends to be rapid, leaving dissolution as the rate-controlling step in the process (Amidon *et al.*, 1995; Lennernas and Abrahamsson, 2005; Blume and Schug, 1999). These concepts are elaborated in later chapters.

Maximum Absorbable Dose Concept

A question often raised by scientists designing new drug candidates is "how much is enough?" with respect to solubility and permeability of a compound. In the current climate of drug discovery, key criteria for identification of clinical candidates include potent and selective binding to the target of interest, adequate safety, lack of CYP interactions, and appropriate pharmacokinetic profile to achieve the desired clinical effect. The concept of maximum absorbable dose (MAD), utilizes absorption rate constant, small intestine residence time, intestinal volume, and solubility (Johnson and Swindell, 1996). This concept mathematically illustrates once again the basic tenet of the BCS that passive GI absorption results from the interplay of permeability and solubility. The maximum amount of drug that could be expected to be absorbed based on these two parameters provides guidance as to whether the solubility and permeability are adequate. For example, for a drug with a solubility of 10 μg/mL, the estimated MAD, assuming no limitations due to site-specific absorption, could range from 0.9 mg (low permeability drug) to 90 mg (high permeability drug). Therefore, potent low dose drugs or highly permeable drugs can tolerate what may appear at first glance to be unacceptable solubility.

Impact of Active Transport Mechanisms

Active transport mechanisms are less predictable than passive transport due to the requirement for binding to a cellular membrane ligand. The involvement of active transport systems can lead to erroneous conclusions concerning the permeability of a drug if a passive diffusion mechanism has been assumed. Drug interactions are also possible among drugs that are actively transported, possibly leading to significant changes in pharmacokinetic behavior upon coadministration. Great advancements in the area of active transport mechanisms have been made over the past several years. *In vitro* assays are now frequently used during the early stages of drug development to screen for desirable and undesirable interactions with active transporters, yet more work is required to understand the nature of transporters fundamentally and their ultimate utility in predicting and manipulating PK behavior (Kunta and Sinko, 2004).

1.1.5.3 Strategies to Achieve Target Pharmacokinetic Profile

Although many biopharmaceutical properties are determined by the chemical structure of the compound, there are multiple strategies available for exploiting the properties of any given molecule to try to achieve the desired clinical behavior. The choice of paths to explore is dependent on the nature and extent of the delivery issue to be solved. For example, if poor BA is caused by high first-pass metabolism, delivery *via* a non-oral route may yield sufficient blood levels for activity. Likewise, non-linear PK caused by interactions with active transport systems would not be resolved by improving the dissolution rate of the dosage form.

Route of Delivery

The target PK profile and resultant therapeutic effect (including onset and duration of activity) of any drug are influenced by the route of administration. Oral dosing is normally preferred for a chronically administered medication due to ease of dosing and general patient acceptability. However, compounds limited by solubility, permeability, or first-pass metabolism may not be amenable to the oral route. Delivery alternatives include other transmucosal routes or parenteral administration. Each has its own advantages and limitations. Intravenous administration leads to immediate blood levels and is often used to treat serious acute symptoms such as seizures or strokes. Rapid absorption can also be achieved by non-oral transmucosal routes, including nasal, sublingual, buccal, or inhalation (Chen *et al.*, 2005; Shyu *et al.*, 1993; Song *et al.*, 2004; Berridge *et al.*, 2000). Local treatment *via* ocular, inhalation, nasal, or vaginal routes may be advantageous compared to systemic delivery due to increased potency at the target and decreased systemic toxicity (Rohatagi *et al.*, 1999). In addition, other properties of the molecule may dictate the routes that are possible. For example, protein/peptide drugs are highly susceptible to degradation upon oral administration and are not likely to diffuse across the intestinal barrier unless by a specific active transporter. As a result, these compounds are often dosed parenterally, and more extensive research is being conducted with oral and alternate non-oral routes, including inhalation (Adessi and Soto, 2002). Metabolism can also influence the choice of route of administration. Drugs that undergo high first-pass metabolism may be much more bioavailable by non-oral routes such as rectal, buccal, or nasal (Hao and Heng, 2003; Song, 2004), leading to pharmacologically relevant blood levels that cannot be achieved with oral dosing.

Chemical Modification

An alternative to formulation approaches to modify PK is chemical modification. A prodrug, for example, is a compound that has been designed with a metabolically labile functional group that imparts desired biopharmaceutical characteristics. Prodrugs by themselves are not pharmacologically active but revert *in vivo* to the active moiety through either targeted chemical or enzymatic mechanisms in the general circulation or specific tissue. This type of strategy has been used in many different ways, including modification of physical–chemical properties to improve delivery (Varia and Stella, 1984; Pochopin *et al.*, 1994; Prokai-Tatrai and Prokai, 2003), targeting to a specific enzyme or transporter (Yang *et al.*, 2001; Han and Amidon, 2000; Majumdar *et al.*, 2004), antibody-directed targeting (Jung, 2001), e-directed targeting (Chen and Waxman, 2002; Lee *et al.*, 2002). Although proaches and applications are varied, they all are rationally chosen to modify a particular biopharmaceutical property while relying on *in vivo* generation of the parent molecule to elicit the intended pharmacologic response.

Although prodrugs have been successful in achieving intended drug delivery objectives, they have certain limitations. They may inadvertently lead to unintended consequences if not designed with a full understanding of the fundamental mechanisms of the drug's biopharmaceutical and pharmacologic behavior. For example, a strategy for increasing the oral BA of a poorly soluble drug might be to add an enzymatically labile hydrophilic functional group such as an amino acid or phosphate to modify solubility and/or dissolution rate in the intestinal lumen. This is logical for a compound with good absorption but for which BA is limited by a solubility/dissolution mechanism. However, an unintended consequence may arise if the slow absorption process is rate-limiting for systemic clearance (i.e., flip-flop kinetics) (Rowland and Tozer, 1989). In this case, the terminal elimination rate constant is actually controlled by the absorption rate, and alteration in absorption rate through prodrug modification could unmask a previously unrecognized rapid systemic clearance. This case also highlights the risks in the interpretation of data from extravascular administration and the importance of intravenous data to determine fundamental PK properties such as clearance and volume of distribution. Other unintended consequences of prodrugs could include pharmacologic activity of the prodrug itself, alterations in metabolic or elimination pathways, or drug–drug interactions. This being said, prodrugs have their place in the toolkit of the pharmaceutical scientist and can be used under the right circumstances to enable the clinical utility of a drug candidate.

Strategies to Improve Oral Absorption

Considering the frequency of use of the oral administration route and the multiple factors, both chemical and physiological, affecting oral absorption, a tremendous amount of research on strategies to improve oral absorption has been and continues to be conducted. One area of active research is modification of *effective* solubility and dissolution. Many of the assumptions with respect to dissolution and impact on oral absorption are based on a thermodynamic parameter such as equilibrium solubility. In reality, the GI tract is a dynamic system that is also highly influenced by kinetic as well as thermodynamic factors. Passive drug transport requires a compound to be in solution, and in some cases absorption rates and extents can be higher or lower than predicted by equilibrium solubility values. In the dynamic environment of the GI tract, the kinetic solubility, i.e., concentration of drug in solution as a function of time, may be a more relevant indicator of absorption behavior than the equilibrium solubility, considering the time frame of *in vivo* dissolution and absorption. Importantly, kinetic solubility can be manipulated by the formulator to improve drug product performance.

Commonly used approaches to increase effective solubility include high-energy amorphous solid systems, lipid dispersions, precipitation-resistant solutions, or micellar systems (Verreck *et al.*, 2004; Singhal and Curatolo, 2004; Dannenfelser *et al.*, 2004; Leuner and Dressman, 2000). Amorphous drugs are high energy solid systems that are capable of reaching higher kinetic solubility values (supersaturation) than would be expected from the equilibrium solubility of a crystalline material. This higher initial solubility may be sufficient to assure increased and more rapid absorption for a drug with good permeability. A caution with this approach is the risk that a more thermodynamically stable form may crystallize at any time during processing or storage, and this would have a major impact on the product performance *in vivo*.

Solutions or dispersions in lipid-based matrices have also been extensively evaluated as means to improve oral BA. Presenting the drug to the GI tract in solution removes the dissolution step, and lipid-based or amphiphilic excipients can be used to enhance solubility and dissolution rate for a hydrophobic drug. As with amorphous high energy systems, a risk with solutions and dispersions is the potential for conversion to a less soluble polymorphic form in the dosage form over time leading to potential quality issues. Addition of nucleation inhibitors such as polymers can minimize the potential for form conversion, but the preferred approach is to formulate in a system that is thermodynamically stable. This requires an exhaustive screening for polymorphs and solvates, but even with an extensive body of knowledge on known crystalline forms, the potential may exist for new forms to appear. The potential for precipitation upon dilution in the GI tract must also be considered for these types of systems, and there are ways to formulate thermodynamically stable systems such as microemulsions that are infinitely dilutable in an aqueous environment (Yang *et al.*, 2004; Ritschel, 1996). The physiological factors affecting *in vivo* stability of dispersions and other lipid-based formulations must also be considered, since enzymes such as lipase may compromise the utility of lipid systems (Porter *et al.*, 2004).

Immediate vs. Modified Release

Immediate release solid oral dosage forms are typically designed to disintegrate rapidly and have the API dissolve rapidly leading to rapid absorption. This type of strategy is most useful in those cases when rapid drug levels are desirable (e.g., pain relief), when therapeutic action is dependent on achievement of high C_{max} values, or when safety is not adversely affected by high peak blood levels (i.e., the drug has a high therapeutic index). Drug formulations can be modified in many ways to modulate (up or down) the release rate of drug to achieve the desired PK profile. Within the realm of immediate release products, strategies that could be employed include decreasing disintegration and dissolution rates in order to blunt a high C_{max} or use of micronized or nanomilled drug substance to increase surface area and dissolution rate. Prolonged release dosage forms (oral, subcutaneous, or intramuscular) (Anderson and Sorenson, 1994) can be utilized to modify the rate of release and duration of action for compounds with shorter-than-desired half-lives, or to decrease the frequency of dosing to improve patient compliance. Technologies for controlled or modified-release are numerous and must be tailored to the drug and desired PK profile. These include but are not limited to slowly eroding matrices that gradually release the drug during the entire course of GI transit; diffusion-controlled or osmotically driven systems to approximate zero-order release; and enteric-coated dosage forms, which have an outer coating barrier that is stable under acidic conditions but dissolves in the higher pH of the small intestine, effectively protecting an acid-labile drug from the low pH environment of the stomach. The choice of a particular delivery technology is tied to the properties of the material and the rationale for exploring modified-release. As with everything else that has been discussed with respect to biopharmaceutics properties, there is not a single drug delivery platform that will serve as a standard template for modified release dosage forms.

While the options for formulation are numerous, the practical options for any specific drug candidate are dictated by the physical–chemical properties of the drug and the dose. As a rule of thumb, the formulator's toolkit of delivery technologies is inversely related to the dose of the compound. Transdermal, inhalation, and nasal transport are limited to doses in the microgram to low milligram range because of transport capacity, while subcutaneous and intramuscular delivery are limited by injection volume and therefore dose and solubility. While the formulator can work to manipulate the behavior of the API in the drug product to control the delivery, the PK parameters, particularly, clearance, distribution, and metabolism, are intrinsic properties of the compound and cannot be readily manipulated directly, i.e., without some type of chemical modification of the drug candidate or coadministration of compounds that interfere with biological mechanisms (e.g., enzyme inhibitors).

Several chapters will discuss many of these concepts in further detail including dissolution (Chap. 3), BA (Chaps. 8 and 10), and excipients (Chap. 6). Chapter 9 will cover these concepts in the application of BCS to dissolution.

1.2 Role of Biopharmaceutics in Drug Development

1.2.1 Importance of Biopharmaceutics in the Overall Development Process

Biopharmaceutics is an integral component of the overall development cycle of a drug. Evaluation begins during the drug discovery process, proceeds through compound selection, preclinical efficacy and safety testing, formulation development, clinical efficacy studies, and postapproval stages. At each stage, biopharmaceutical scientists interface with colleagues in multiple disciplines including discovery chemistry and biology, drug safety assessment, clinical development, pharmaceutical development, regulatory affairs, marketing, and manufacturing. The ensuing section will discuss the general activities and impact at each stage of development and provide an overall view of the role of biopharmaceutics at various stages of drug development.

1.2.2 Discovery and Preclinical Development: Candidate Selection

The preclinical development stage encompasses aspects of both drug discovery and drug development. The process to identify a potential drug candidate is an iterative one, as discovery scientists strive to synthesize candidate compounds with appropriate activity and maximal potency at the intended target, maximal safety profile, and desirable ADME properties. The definition of "desirable" properties will be variable considering the therapeutic target and class of compounds, but typical goals are to minimize frequency of dosing, maximize BA, avoid interactions with efflux transport systems (e.g., P-gp) and metabolic enzymes (CYPs), reach target organ or tissue (particularly important for CNS activity), and avoid adverse effects (e.g., for oncology compounds to maximize delivery to the tumor and minimize to healthy tissues). Depending on the desired therapeutic action, the target blood concentration–time profile must be considered with respect to C_{max}, t_{max}, AUC, clearance, accumulation, and dose proportionality. Species effects are also an important consideration since ADME can often be species-specific and therefore the performance in humans may not be readily predictable from animal data.

In vitro/ex vivo techniques to assess ADME properties include *in vitro* CYP screens to assess potential for metabolic liabilities and drug interactions, transporter screens against known targets and efflux pumps, *in vitro* metabolism in the presence of isolated hepatocytes or microsomes (various species to assess interspecies differences), and transport across cell culture model systems as surrogates for passive membrane transport. These screens are used to eliminate candidates with a high potential for ADME liabilities that could negatively impact utility in a clinical setting. Preclinical ADME studies *in vivo* using various animal models are also necessary to assess blood concentration–time profiles, AUC, BA, C_{max}, t_{max}, dose proportionality, accumulation upon multiple dosing or enzyme induction. Intravenous delivery is necessary for determining absolute BA, clearance, and volume of distribution. Specialized studies can be designed to better understand the fundamental mechanisms of intestinal absorption, including bile-duct cannulation to look for biliary excretion and portal vein studies to evaluate extent of absorption and first-pass metabolism.

The physical–chemical properties of the drug candidate, such as solubility, stability, and lipophilicity, influence the *in vivo* performance and must be considered for any drug candidate (Venkatesh and Lipper, 2000). The solubility affects the choice of dosing vehicle used in preclinical testing and is often a major challenge, with many drug candidates having solubility at best in the low μg/mL range and requiring nonaqueous solvents for administration. In some cases, pharmacologic effects resulting from the dosing vehicle can become dose-limiting or confound the *in vivo* results. Stability of compounds is another factor that must be evaluated as it affects the integrity of the material being dosed, could potentially lead to generation of degradants with distinct pharmacologic action or toxicity, and also impacts the handling and shelf-life of a pharmaceutical product. As with solubility, standard criteria for acceptable stability are difficult to define absolutely. Specific requirements are defined depending on the route of administration, safety concerns with degradants, and potential for stabilization of the drug compound in a formulation using appropriate excipients.

Often referred to as "developability" or "drugability," these biopharmaceutical criteria have become increasingly important in the choice of drug candidates (Sun *et al.*, 2004). While achieving high *in vitro* target potency is critical, highly potent compounds with poor biopharmaceutical performance may not be able to achieve the desired therapeutic effect under practical dosing conditions. As discussed previously, there is no set of standard criteria for developable candidates, but rather the complete package of data must be assessed by the entire project team so that they consider all of the interrelated factors and can ultimately decide whether a particular compound with specific and selective receptor binding activity also has potential to be a safe and efficacious therapeutic agent for treatment of a disease in a patient. In addition, the therapeutic area and medical need influence recommendations on developable candidates (e.g., dosing frequency). For example, dosing four times daily may be acceptable for a life-threatening illness for which no other treatment is available, while it may not be acceptable for a chronic-use medication for which patient compliance is critical.

1.2.3 Preclinical Development: Preparation for Phase I Clinical Studies

Once a drug candidate is chosen for clinical development, additional biopharmaceutical assessment is conducted to build on existing knowledge and experience. A clinical candidate must be tested in formal animal safety studies in multiple species in order to establish a safety profile and provide guidance on the choice of clinical doses. For these studies, the dose range is typically much higher than that expected to be used in humans, and this aspect offers some specific challenges with respect to dosing and dose proportionality. Solutions are highly desirable for dosing because they are homogeneous systems that are easy to administer to animals (particularly rodents), offer dose flexibility, and have the potential for maximizing *in vivo* exposure by avoiding issues with dissolution. However, poorly soluble compounds may lack sufficient solubility to prepare highly concentrated solutions, and pharmaceutically acceptable non-aqueous vehicles or suspensions must be used if a liquid vehicle is necessary. Regardless of the formulation used, maximizing exposure in these studies is important. Because of the high doses that may be used to establish the safety multiples relative to clinical doses, there is potential for saturating transport mechanisms leading to decreased relative exposure with increasing dose. Saturation of metabolic processes could also lead to the opposite problem with sudden increases in relative exposure with increasing dose.

Preclinical *in vivo* biopharmaceutics studies can also be conducted, if necessary, to evaluate the relative *in vivo* performance of different API forms (including free acids/bases, salts, polymorphs) and clinical formulations. As discussed above, exposure can be significantly affected by solubility and rate of dissolution, which in turn are influenced by the form of the drug substance and formulation used. *In vitro* dissolution is a first step in screening API forms and potential clinical formulations. From the formulation perspective, the scientist may be interested in the relative differences in exposure between two different types of formulations (e.g., solution vs. tablet, or tablet vs. liquid-filled capsule) in order to provide insight into the critical factors affecting performance for that particular compound. Absolute predictability of drug product performance in humans based on animal data is not possible considering differences in metabolism and absorption from one species to another. However, preclinical screens are useful for assessing rank orders and gaining insight into the significance of factors such as particle size or dosage form type. An additional consideration for Phase I clinical studies is the relationship between the formulations used in safety and clinical studies and their respective PK behavior since the data from the preclinical safety studies are critical for defining the initial clinical development plan and starting clinical dose (which is determined based on the relative safety multiples established in preclinical safety studies).

1.2.4 Early Clinical Development

The primary goals in early clinical development are to establish safety, PK, and pharmacodynamics, and also to provide guidance on a dose range expected to be efficacious, in both single-dose and multiple-dose studies. The dose range for Phase I studies is usually fairly wide because of the uncertainties with respect to interspecies scaling and lack of predictability based on preclinical data. The plasma drug concentration time profiles are used to determine AUC, half-life, C_{max}, t_{max}, dose-proportionality, and extent of accumulation upon multiple dosing. In the absence of PK data from intravenous dosing, interpretation of PK data from a non-intravenous dosing route must be done carefully to avoid erroneous conclusions.

Two general types of biopharmaceutic studies are often conducted in order to assess the comparability and suitability of products for their intended clinical use. A relative BA study is a relative comparison of two or more formulations with respect to PK properties, normally AUC, C_{max}, t_{max} and half-life. Such BA studies are usually done early in a drug's development cycle before significant experience has been gained in human subjects, normally to assess the relative performance of a new formulation as compared with a reference. For example, a solution dosage form may be used for Phase I studies because of the need for dosing flexibility, but eventually a switch to a solid dosage form is desired. In order to compare the relative exposure from each formulation at a given dose (or range of doses), a two-way crossover BA study could be performed in a small number of subjects and the BA of the test formulation determined relative to the reference formulation. Because of the limited number of subjects used in this type of study, the study tends not to be sufficiently powered to establish statistical equivalence among various formulations but the data can be used to guide development decisions or to support a formulation change in a non-pivotal clinical study. Relative BA studies can also be used to evaluate the effect of an alternate route of administration on the drug's PK profile, evaluate drug product variables (e.g., particle size of the API) on clinical performance or to screen for effects of physiological factors (fed vs. fasted state, gastric pH effects) affecting drug absorption.

Another type of study that may be conducted in the course of drug development is an absolute BA study, which is a comparison of AUC of a test formulation to the intravenous route, which is considered to have a BA of 100%. These types of studies are extremely valuable yet not always done because of limitations related to feasibility of developing an intravenous formulation for a highly insoluble drug substance.

A bioequivalence study, on the other hand, is a distinct type of BA study with the objective of assessing statistical equivalence among different treatment groups. These studies are typically done at or before a stage of development in which the clinical data will be generated to establish the efficacy of the drug. The criteria for establishing bioequivalence are much more strict than with a relative BA study and include a statistical assessment of PK parameters including AUC, C_{max}, t_{max}, and half-life. The number of subjects required for this type of study is higher than that required for a relative BA study. The actual number for any individual drug candidate is dependent on the desired statistical power as well as the variance of the measures (e.g., AUC). For example, a drug with a large degree of variability in AUC would require more subjects to establish bioequivalence than a drug with less variability, and a desire for a higher degree of statistical confidence in the results (power of the study) would also require inclusion of a greater number of subjects. Bioequivalence studies may be conducted to switch a formulation during a Phase III clinical study, to establish equivalence of a generic product to the respective branded product, or to support manufacturing changes postapproval (FDA, 1995; FDA, 2000). The reader is referred to Chapter 8.

Pharmacokinetic studies are also done at various stages of the drug development process to assess factors other than formulation that could impact a drug's physiological behavior. As mentioned previously, the GI tract is a complex system that includes not only biological membranes but also fluid, pH modifiers, food, enzymes, and bile salts. The interplay of these variables can alter the way a drug is absorbed from a dosage form. PK studies to assess effects of food (fasted vs. high or low-fat meal) are used to establish whether any dosing restrictions relative to meal time need to be included on a drug product's label. Dosing with a meal can impact absorption either positively or negatively depending on the nature of the drug and the mechanism of interaction. For example, a high fat meal or secretion of bile salts in the small intestine may serve to solubilize a lipidic drug. The GI pH can also be altered in the presence of food and could potentially impact the disintegration/dissolution of a pH-sensitive API. Food-effect studies are described in Chapter 10.

The pH in the GI tract can not only be affected by food but also by physiological differences among patients (normal stomach pH varies normally between pH 1 and 5) or concomitant administration of pH-modifying agents (Lui *et al.*, 1986). The previous discussion of compound ionization and absorption highlights the need to understand and control any pH effects that influence the dissolution and absorption rates. For compounds with potential for showing pH-dependent absorption, a human PK study to evaluate the effect of pH modification (e.g., using preadministration of an H2-receptor antagonist) on AUC and C_{max} may be appropriate. These PK studies to screen for pH effects can be performed preclinically in animals, and/or during clinical development. The data from such a study can be used to guide additional formulation optimization to minimize the pH-liability.

A drug–drug interaction study is another type of clinical PK study that is typically performed on a clinical drug candidate to assess the impact of concomitant administration of other drugs on the PK behavior of the drug of interest. Interactions can arise due to metabolic factors (CYP450 interactions, enzyme induction) or competition for an active transporter. Results from *in vitro* screens can be used to assess the risk of drug interactions due to a CYP-related mechanism and to design meaningful clinical drug–drug interaction studies.

A variety of additional specialized PK studies can be performed to evaluate differences in physiology in special populations on drug product performance. Examples of special populations include children, renally impaired patients, and the elderly, in which the PK may be significantly altered relative to typical adult human subjects based on differences in metabolism and clearance. In these populations, dose and/or dosing regimens may need to be adjusted to account for any differences. The description of these types of studies are beyond the scope of this chapter.

During the early phases of drug development, numerous studies are done to build a fundamental understanding of the qualitative and quantitative nature of what the body does to a drug (PK) in addition to what the drug does to the body (safety and efficacy). The biopharmaceutics knowledge gained in early development can be used as a basis for designing clinical efficacy trials. A fundamental understanding of the biopharmaceutics properties early in the drug development process allows the development scientist to evaluate a comprehensive and integrated set of data and design development strategies that are meaningful and appropriate for any individual compound.

1.2.5 Advanced Clinical Development

As a compound moves from Phase I into Phase II and eventually into Phase III, the objectives of the clinical development program evolve from primarily safety and PK to safety and efficacy. The data collected during earlier studies are used to define a potentially efficacious clinical dose range and dosing regimen, identify any special patient populations, and guide selection of a drug product to be used in pivotal registrational clinical studies. As a result, the biopharmaceutics focus shifts from an exploratory to a registrational paradigm in which the objective is to establish consistency, robustness, and predictability of the formulations.

Considering the previous discussion of the dependency of efficacy on PK and dependency of PK on formulation, changes in critical properties of a drug substance or formulation may have consequences for a clinical study (Ahr *et al.*, 2000). PK studies conducted in Phase I and II are used to establish a body of knowledge surrounding the intrinsic properties of the medicinal agent (e.g., clearance) as well as the dependency of the performance on the actual product used. Since the outcome of PK and clinical studies depends on the product used, any changes to that product must be adequately qualified to establish its acceptability for use in the clinic. The definition of "qualification" in this sense is consistent with much of the discussion here: it depends on the drug and the available body of knowledge about that drug and its formulations. If an IVIVC has been established, dissolution equivalence may be sufficient; for a Class I drug, demonstration of rapid and complete dissolution across the physiological pH range may serve to qualify a new dosage form. For Class II and IV drugs, qualification in a bioequivalence study prior to use in a large-scale clinical study may be necessary. The design of the study would be driven by the extent of clinical experience already gained with a given drug product and the extent of change to the product. Changing from a capsule to tablet dosage form is a significant change that would likely require a bioequivalence study, while addition of a non-functional coating for ease of swallowing may not be considered to be sufficiently impactful to affect the product performance.

1.2.6 Postapproval Considerations

As a product proceeds through the registrational process and into commercial manufacturing, additional considerations with respect to biopharmaceutics arise. A product approval is based on the evidence that a drug is safe and effective when administered according to the product labeling. Upon review of a product insert or other reference literature, the reader would find an extensive discussion of the properties of the drug product, including details on the ingredients, dosage form, available strengths, and pharmacokinetic properties, in addition to indications and dosing information. Once a product is approved by a regulatory agency, any changes to the formulation, manufacturing process or site, or dosing regimen must be assessed for impact on the biopharmaceutical behavior. Regulatory guidances are available that discuss the requirements to support a postapproval manufacturing change, the extent of which depend on the scope of the change intended. For example, a minor change may require the sponsor to inform the regulatory agency prior to implementation, and a more significant change could require human PK data, submission of other supporting data, and agency review to assure that the changes do not impact the drug's performance in humans. Inherent in this assessment is the assumption that pharmacokinetic equivalence will be predictive of clinical equivalence.

Another major source of change in a postapproval environment is product enhancements or extensions, including different dosage forms (e.g., capsule to tablet or oral liquid), new strengths, modified-release (e.g., for less frequent dosing to improve patient compliance), or alternate routes of administration (e.g., addition of an injectable dosage form for use as a loading dose or for emergency use, or long acting depot injection). The data requirements for these new products vary. A change to a new oral solid dosage form may require a demonstration of bioequivalence while a new route of administration may necessitate additional human safety, PK, and efficacy data. The biopharmaceutical profile of a particular drug is one of the important determinants in the design of the studies used to support approval of product enhancements. The reader is referred to the available regulatory documents to provide guidance for requirements for specific products and markets. Specific applications of *in vitro–in vivo* correlations in dosage form development are further discussed in Chaps. 11 (parenterals) and 12 (oral products).

1.2.7 Regulatory Considerations

Across the globe, numerous regulatory bodies are responsible for assuring safety, quality, and efficacy of medicinal products. Significant progress has been made over the years toward harmonization of requirements for regulatory filings through the work of the International Council on Harmonization (ICH). This work is continuing, and there is also an ongoing paradigm shift in the US FDA concerning CMC regulatory packages and agency reviews. The CMC regulatory paradigm is evolving into a system emphasizing the establishment of fundamental understanding of product critical quality attributes, which are those critical aspects of the drug product that impact the performance in the patient and may be influenced by robustness of the manufacturing process. The new process acknowledges that the concept of product quality must be based upon clinical relevance, and the previous discussion has highlighted the relevance of biopharmaceutics to clinical performance. Importantly, the biopharmaceutics knowledge base contributes to the establishment of a product's "design space," reflecting the ranges of multiple, interrelated material properties and manufacturing parameters within which acceptable product performance is assured with a high degree of confidence.

1.3 Summary

The previous discussion highlighted the fundamental principles of biopharmaceutics and illustrated examples of their applicability in the drug development process. The current level of scientific understanding of the field is substantial but continues to expand. The nature of the drug molecules and types of issues encountered during development are diverse, so there is no standard approach that can be applied to every compound. However, as the state of knowledge increases, the biopharmaceutical scientist becomes better able to apply the right tools to any compound. A good scientific understanding of physical–chemical principles, PK, and physiology as well as the integration of these areas is a key to the efficient development of quality products for the benefit of patients.

References

Adessi, C. and Soto, C. (2002). Converting a peptide into a drug: strategies to improve stability and bioavailability. *Curr. Med. Chem.* 9:963–978.

Agharkar, S., Lindenbaum, S., and Higuchi, T. (1976). Enhancement of solubility of drug salts by hydrophilic counterions: properties of organic salts of an antimalarial drug. *J. Pharm. Sci.* 65:747–749.

Ahr, G., Voith, B., and Kuhlmann, J. (2000). Guidances related to bioavailability and bioequivalence: European industry perspective. *Eur. J. Drug Metab. Pharmacokinet.* 25:25–27.

Amidon, G. L., Lennernas, H., Shah, V. P., and Crison, J. R. (1995). A theoretical basis for a biopharmaceutic drug classification: the correlation of in vitro drug product dissolution and in vivo bioavailability. *Pharm. Res.* 12:413–420.

Anderson, P. M. and Sorenson, M. A. (1994). Effects of route and formulation on clinical pharmacokinetics of interleukin-2. *Clin. Pharmacokinet.* 27:19–31.

Berridge, M. S., Lee, Z., and Heald, D. L. (2000). Regional distribution and kinetics of inhaled pharmaceuticals. *Curr. Pharm. Des.* 6:1631–1651.

Blume, H. H. and Schug, B. S. (1999). The biopharmaceutics classification system (BCS): class III drugs – better candidates for BA/BE waiver? *Eur. J. Pharm. Sci.* 9:117–121.

Chen, L. and Waxman, D. J. (2002). Cytochrome P450 gene-directed enzyme prodrug therapy (GDEPT) for cancer. *Curr. Pharm. Des.* 8:1405–1416.

Chen, J., Jiang, X. G., Jiang, W. M., Gao, X. L., and Mei, N. (2005). Intranasal absorption of rizatriptan – in vivo pharmacokinetics and bioavailability study in humans. *Pharmazie* 60:39–41.

Curatolo, W. (1998). Physical chemical properties of oral drug candidates in the discovery and exploratory development settings. *Pharm. Sci. Technol. Today* 1:387–393.

Dannenfelser, R. M., He, H., Joshi, Y., Bateman, S., and Serajuddin, A. T. (2004). Development of clinical dosage forms for a poorly water soluble drug I: application of polyethylene glycol–polysorbate 80 solid dispersion carrier system. *J. Pharm. Sci.* 93:1165–1175.

Dressman, J. B., Amidon, G. L., Reppas, C., and Shah, V. P. (1998). Dissolution testing as a prognostic tool for oral drug absorption: immediate release dosage forms. *Pharm. Res.* 15:11–22.

FDA (U.S. Food and Drug Administration) Guidance Documents for Industry, 1995, Rockville. Guidance for Industry. Immediate Release Solid Oral Dosage Forms. Scale-Up and Postapproval Changes: Chemistry, Manufacturing, and Controls, In Vitro Dissolution Testing, and In Vivo Bioequivalence Documentation (November 1995); http://www.fda.gov/cder/guidance/cmc5.pdf.

FDA (U.S. Food and Drug Administration) Guidance for Industry, 2000, Rockville. Bioavailability and Bioequivalence Studies for Orally Administered Drug Products – General Considerations (October 2000); http://www.fda.gov/cder/guidance/3615fnl.pdf.

FDA (U.S. Food and Drug Administration) Guidance for Industry, 2005, Rockville. The Biopharmaceutics Classification System (BCS) Guidance (June 2005); http://www.fda.gov/cder/OPS/BCS_guidance.htm.

Flynn, G. L. (1984). Solubility concepts and their applications to the formulation of pharmaceutical systems. Part I. Theoretical foundations. *J. Parenter. Sci. Technol.* 38:202–209.

Gennaro, A. R. (ed.). (1995). *Remington: The Science and Practice of Pharmacy*, 19th ed. Mack Publishing Co., Easton, PA.

Gibaldi, M. and Perrier, D. (1982). *Pharmacokinetics*, 2nd ed. Marcel Dekker Inc., New York.

Gu, C. H., Rao, D., Gandhi, R. B., Hilden, J., and Raghavan, K. (2005). Using a novel multi-compartment dissolution system to predict the effect of gastric pH on the oral absorption of weak bases with poor intrinsic solubility. *J. Pharm. Sci.* 94:199–208.

Han, H. K. and Amidon, G. L. (2000). Targeted prodrug design to optimize drug delivery. *AAPS PharmSci. (online)*. 2(1): article 6.

Hao, J. and Heng, P. W. (2003). Buccal delivery systems. *Drug Dev. Ind. Pharm.* 29:821–832.

Higuchi, T. (1958). Some physical chemical aspects of suspension formulation. *J. Am. Pharm. Assoc.* 47:657–660.

Higuchi, T. (1976). Pharmacy, pharmaceutics and modern drug delivery. *Am. J. Hosp. Pharm.* 33:795–800.

Higuchi, T., Szulczewski, D. H., and Yunker, M. (1956). Physical chemical properties and ultraviolet spectral characteristics of amyl nitrite. *J. Am. Pharm. Assoc.* 45:776–779.

Higuchi, W. I., Parrott, E. L., Wurster, D. E., and Higuchi, T. (1958). Investigation of drug release from solids. II. Theoretical and experimental study of influences of bases and buffers on rates of dissolution of acidic solids. *J. Am. Pharm. Assoc.* 47:376–383.

Higuchi, W. I., Lau, P. K., Higuchi, T., and Shell, J. W. (1963). Polymorphism and drug availability. Solubility relationships in the methylprednisolone system. *J. Pharm. Sci.* 52:150–153.

Hilgers, A. R., Smith, D. P., Biermacher, J. J., Day, J. S., Jensen, J. L., Sims, S. M., Adams, W. J., Friis, J. M., Palandra, J., Hosley, J. D., Shobe, E. M., and Burton, P. S. (2003). Predicting oral absorption of drugs: a case study with a novel class of antimicrobial agents. *Pharm. Res.* 20:1149–1155.

Ishii, K., Katayama, Y., Itai, S., Ito, Y., and Hayashi, H. (1995). A new pharmacokinetic model including in vivo dissolution and gastrointestinal transit parameters. *Biol. Pharm. Bull.* 18:882–886.

Johnson, K. C. and Swindell, A. C. (1996). Guidance in the setting of drug particle size specifications to minimize variability in absorption. *Pharm. Res.* 13:1795–1798.

Jung, M. (2001). Antibody directed enzyme prodrug therapy (ADEPT) and related approaches for anticancer therapy. *Mini. Rev. Med. Chem.* 1:399–407.

Kalgutkar, A. S., Vaz, A. D., Lame, M. E., Henne, K. R., Soglia, J., Zhao, S. X., Abramov, Y. A., Lombardo, F., Collin, C., Hendsch, Z. S., and Hop, C. E. (2005). Bioactivation of the nontricyclic antidepressant nefazodone to a reactive quinone-imine species in human liver microsomes and recombinant cytochrome P450 3A4. *Drug Metab. Dispos.* 33:243–253.

Katsura, T. and Inui, K. (2003). Intestinal absorption of drugs mediated by drug transporters: mechanisms and regulation. *Drug Metab. Pharmacokinet.* 18:1–15.

Kivisto, K. T., Niemi, M., and Fromm, M. F. (2004). Functional interaction of intestinal CYP3A4 and P-glycoprotein. *Fundam. Clin. Pharmacol.* 18:621–626.

Kostenbauder, H. B. and Higuchi, T. (1957). A note on the water solubility of some *N*, *N*,-dialkylamides. *J. Am. Pharm. Assoc.* 46:205–206.

Kunta, J. R. and Sinko, P. J. (2004). Intestinal drug transporters: in vivo function and clinical importance. *Curr. Drug Metab.* 5:109–124.

Lee, H. J., Cooperwood, J. S., You, Z., and Ko, D. H. (2002). Prodrug and antedrug: two diametrical approaches in designing safer drugs. *Arch. Pharm. Res.* 25:111–136.

Lennernas, H. and Abrahamsson, B. (2005). The use of biopharmaceutic classification of drugs in drug discovery and development: current status and future extension. *J. Pharm. Pharmacol.* 57:273–285.

Leslie, E. M., Deeley, R. G., and Cole, S. P. (2005). Multidrug resistance proteins: role of P-glycoprotein, MRP1, MRP2, and BCRP (ABCG2) in tissue defense. *Toxicol. Appl. Pharmacol.* 204:216–237.

Leuner, C. and Dressman, J. (2000). Improving drug solubility for oral delivery using solid dispersions. *Eur. J. Pharm. Biopharm.* 50:47–60.

Li, S., Wong, S., Sethia, S., Almoazen, H., Joshi, Y. M., and Serajuddin, A. T. (2005). Investigation of solubility and dissolution of a free base and two different salt forms as a function of pH. *Pharm. Res.* 22:628–635.

Lipka, E., Crison, J., and Amidon, G. L. (1996). Transmembrane transport of peptide type compounds: prospects for oral delivery. *J. Control Release* 39:121–129.

Lui, C. Y., Amidon, G. L., Berardi, R. R., Fleisher, D., Youngberg, C., and Dressman, J. B. (1986). Comparison of gastrointestinal pH in dogs and humans: implications on the use of the beagle dog as a model for oral absorption in humans. *J. Pharm. Sci.* 75:271–274.

Majumdar, S., Duvvuri, S., and Mitra, A. K. (2004). Membrane transporter/receptor-targeted prodrug design: strategies for human and veterinary drug development. *Adv. Drug Deliv. Rev.* 56:1437–1452.

Merino,V., Freixas, J., del Val, B. M., Garrigues, T. M., Moreno, J., and Pla-Delfina, J. M. (1995). Biophysical models as an approach to study passive absorption in drug development: 6-fluoroquinolones. *J. Pharm. Sci.* 84:777–782.

Meyer, U. A. (1996). Overview of enzymes of drug metabolism. *J. Pharmacokinet. Biopharm.* 24:449–459.

Morris, K. R. (1994). An integrated approach to the selection of optimal salt form for a new drug candidate. *Int. J. Pharm.* 105:209–217.

Nicolaides, E., Galia, E., Efthymiopoulos, C., Dressman, J. B., and Reppas, C. (1999). Forecasting the in vivo performance of four low solubility drugs from their in vitro dissolution data. *Pharm. Res.* 16:1876–1882.

Parrott, N. and Lave, T. (2002). Prediction of intestinal absorption: comparative assessment of GASTROPLUS and IDEA. *Eur. J. Pharm. Sci.* 17:51–61.

Pochopin, N. L., Charman, W. N., and Stella, V. J. (1994). Pharmacokinetics of dapsone and amino acid prodrugs of dapsone. *Drug Metab. Dispos.* 22:770–775.

Porter, C. J., Kaukonen, A. M., Boyd, B. J., Edwards, G. A., and Charman, W. N. (2004). Susceptibility to lipase-mediated digestion reduces the oral bioavailability of danazol after administration as a medium-chain lipid-based microemulsion formulation. *Pharm. Res.* 21:1405–1412.

Prokai-Tatrai, K. and Prokai, L. (2003). Modifying peptide properties by prodrug design for enhanced transport into the CNS. *Prog. Drug Res.* 61:155–188.

Raw, A. S. and Yu, L. X. (ed.). (2004). Pharmaceutical solid polymorphism in drug development and regulation. *Adv. Drug Deliv. Rev.* 56:235–418.

Reinstein, J. A. and Higuchi, T. (1958). Examination of the physical chemical basis for the isoniazid-*p*-aminosalicylic acid combination. *J. Am. Pharm. Assoc.* 47:749–752.

Ritschel, W. A. (1996). Microemulsion technology in the reformulation of cyclosporine: the reason behind the pharmacokinetic properties of Neoral. *Clin. Transplant.* 10:364–373.

Rohatagi, S., Rhodes, G. R., and Chaikin, P. (1999). Absolute oral versus inhaled bioavailability: significance for inhaled drugs with special reference to inhaled glucocorticoids. *J. Clin. Pharmacol.* 39:661–663.

Rowland, M. and Tozer, T. N. (1989). *Clinical Pharmacokinetics: Concepts and Applications*, 2nd ed. Lea & Febiger, Philadelphia.

Sai, Y. (2005). Biochemical and molecular pharmacological aspects of transporters as determinants of drug disposition. *Drug Metab. Pharmacokinet.* 20:91–99.

Shefter, E. and Higuchi, T. (1963). Dissolution behavior of crystalline solvated and nonsolvated forms of some pharmaceuticals. *J. Pharm. Sci.* 52:781–791.

Shek, E., Higuchi, T., and Bodor, N. (1976). Improved delivery through biological membranes. 3. Delivery of *N*-methylpyridinium-2-carbaldoxime chloride through the blood–brain barrier in its dihydropyridine pro-drug form. *J. Med. Chem.* 19:113–117.

Shou, M., Lin, Y., Lu, P., Tang, C., Mei, Q., Cui, D., Tang, W., Ngui, J. S., Lin, C. C., Singh, R., Wong, B. K., Yergey, J. A., Lin, J. H., Pearson, P. G., Baillie, T. A., Rodrigues, A. D., and Rushmore, T. H. (2001). Enzyme kinetics of cytochrome P450-mediated reactions. *Curr. Drug Metab.* 2:17–36.

Shyu, W. C., Mayol, R. F., Pfeffer, M., Pittman, K. A., Gammans, R. E., and Barbhaiya, R. H. (1993). Biopharmaceutical evaluation of transnasal, sublingual, and buccal disk dosage forms of butorphanol. *Biopharm. Drug Dispos.* 14:371–379.

Singhal, D. and Curatolo, W. (2004). Drug polymorphism and dosage form design: a practical perspective. *Adv. Drug Deliv. Rev.* 56:335–347.

Song, Y., Wang, Y., Thakur, R., Meidan, V. M., and Michniak, B. (2004). Mucosal drug delivery: membranes, methodologies, and applications. *Crit. Rev. Ther. Drug Carrier Syst.* 21:195–256.

Stella, V. J., Charman, W. N., and Naringrekar, V. H. (1985). Prodrugs. Do they have advantages in clinical practice? *Drugs* 29:455–473.

Sun, D., Yu, L. X., Hussain, M. A., Wall, D. A., Smith, R. L., and Amidon, G. L. (2004). In vitro testing of drug absorption for drug 'developability' assessment: forming an interface between in vitro preclinical data and clinical outcome. *Curr. Opin. Drug Discov. Devel.* 7:75–85.

Varia, S. A. and Stella, V. J. (1984). Phenytoin prodrugs V: in vivo evaluation of some water-soluble phenytoin prodrugs in dogs. *J. Pharm. Sci.* 73:1080–1087.

Venkatesh, S. and Lipper, R. A. (2000). Role of the development scientist in compound lead selection and optimization. *J. Pharm. Sci.* 89:145–154.

Verreck, G., Vandecruys, R., De Conde, V., Baert, L., Peeters, J., and Brewster, M. E. (2004). The use of three different solid dispersion formulations – melt extrusion, film-coated beads, and a glass thermoplastic system – to improve the bioavailability of a novel microsomal triglyceride transfer protein inhibitor. *J. Pharm. Sci.* 93:1217–1228.

Woolf, H. B. (ed.). (1981). *Webster's New Collegiate Dictionary*, G. & C. Merriam Co., Springfield.

Yang, C., Tirucherai, G. S., and Mitra, A. K. (2001). Prodrug based optimal drug delivery via membrane transporter/receptor. *Expert Opin. Biol. Ther.* 1:159–175.

Zhou, R., Moench, P., Heran, C., Lu, X., Mathias, N., Faria, T. N., Wall, D. A., Hussain, M. A., Smith, R. L., and Sun, D. (2005). pH-dependent dissolution in vitro and absorption in vivo of weakly basic drugs: development of a canine model. *Pharm. Res.* 22:188–192.

2
Molecular and Physicochemical Properties Impacting Oral Absorption of Drugs

Wei-Qin (Tony) Tong

2.1 Introduction

Oral administration is still regarded as the most commonly accepted route of drug administration offering numerous advantages including convenience, ease of compliance, and cost-effectiveness. Not surprisingly, desirable oral bioavailability is one of the most important considerations for the successful development of bioactive molecules. Poor oral bioavailability affects drug performance and leads to high intra- and inter-patient variability.

The advent of combinatorial chemistry and high throughput screening in recent years have resulted in unfavorable changes in the molecular and physicochemical properties of drug candidates. Gaining sufficient understanding of the properties that impact oral bioavailability has become vitally important in designing new drug candidates and formulations that can successfully deliver them (Curatolo, 1998).

In the last few years, there have been numerous attempts to predict physicochemical properties that are most desirable for a good drug candidate. Analysis of the structures of orally administered drugs, and of drug candidates, as pioneered by Lipinski and his colleagues (1997), has been the primary guide to correlating physicochemical properties with successful drug development candidates. Time-related differences in the physical properties of oral drugs have also been analyzed (Wenlock *et al.*, 2003). These analyses have led to several sets of rules relating to lipiphilicity, molecular weight (MW), the number of hydrogen-bond donors and acceptors, polar surface area, and molecular rigidity as indicated by the number of rotatable bonds (Vieth *et al.*, 2004).

While these rules provide good guidance in molecular design and lead optimization, the advances in the understanding of the causes of low oral bioavailability have led to significant improvements in drug delivery technologies.

This chapter summarizes the molecular and physicochemical properties that influence oral absorption and reviews the impact of these properties on the formulation strategies that can be applied to improve oral absorption.

2.2 Molecular and Physicochemical Properties Impacting Oral Absorption

2.2.1 Molecular Weight, Log P, the Number of H-Bond Donors and Acceptors, Polar Surface Area, and the Number of Rotatable Bonds

In the last few years, many detailed statistical analyses have been reported aiming to have a detailed understanding of the molecular-level properties that are important for optimal oral bioavailability. Lipinski and colleagues (1997) introduced the so-called "rule of 5," which states that poor absorption or permeations more likely when the MW is over 500; the log *P* is over 5; the hydrogen-bond donors are more than 5; and the hydrogen-bond acceptors are more than 10.

More recently, Veber *et al.* (2002) found polar surface area and molecular flexibility to be important predictors of good oral bioavailability, interestingly independent of MW. After analyzing the oral bioavailability results measured in rats for over 1,100 drug candidates studied at SmithKline Beecham, they found an unexpected positive influence of increasing molecular rigidity as measured by the number of rotatable bonds and the expected negative impact of increasing polar surface area. Figure 2.1 shows the fraction of compounds with an oral bioavailability of 20% or greater as a function of MW and rotatable bond count. It is clear that the effect of molecular rigidity on oral bioavailability is rather independent of MW.

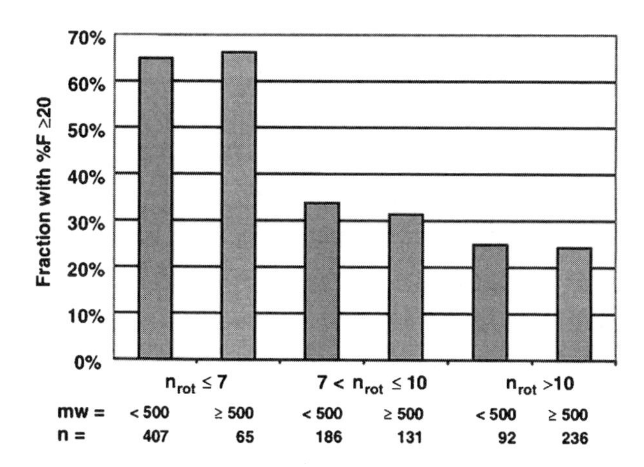

FIGURE 2.1. Fraction of compounds with a rat oral bioavailability of 20% or greater as a function of MW and number of rotatable bond (n_{rot})

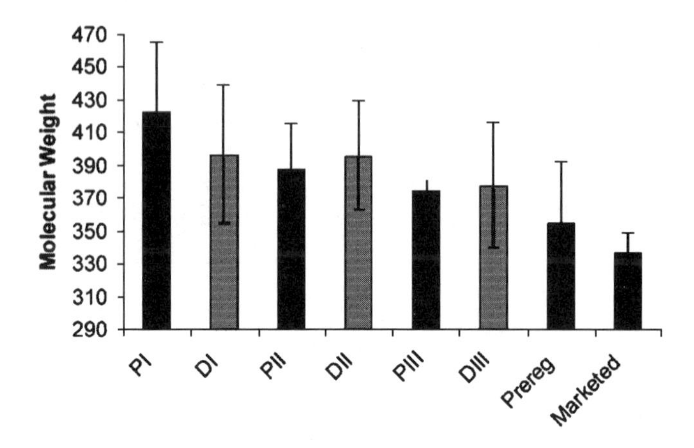

FIGURE 2.2. Mean MW for drugs in different phases

Given that the major contributors to the polar surface area are hydrogen-bond donors or acceptors and also the rotatable bond count in general increases with MW, these findings are still consistent with the rule of 5. However, these results do suggest that candidate design directed at reduced flexibility and polar surface area without having to reduce MW could still be successful in achieving a high oral bioavailability.

Since expectations for drug candidates tailored for the pharmacological targets have changed significantly over the years, do the necessary characteristics for oral delivery also change? Several statistical analyses of over 1,000 marketed drugs indicate that lower MW, balanced log P, and greater rigidity remain important features of oral drug molecules (Wenlock *et al.*, 2003). When the physicochemical properties of oral drugs in different phases of clinical development are compared to those already marketed (Fig. 2.2), it is interesting that the mean MW of orally administered drugs in development decreases in later phases and gradually converges toward the mean MW of marketed oral drugs. It is clear that the most lipophilic compounds get discontinued from development, suggesting that physicochemical properties are intimately linked to physiological control.

To ensure that these drug-like properties are part of the drug design, approaches to address drug "developability" have been emphasized over the last decade (Kerns and Di, 2003; Huang and Tong, 2004; Borchardt *et al.*, 2006). The empirical rules such as the rule of 5 are now widely applied by many pharmaceutical companies as an alert system for compounds with potential solubility and permeability problems. Since lead optimization often leads to further increases in molecular size and structural complexity, controlling molecular properties within the drug-like domain while optimizing binding efficiency with good selectivity for the desired target remain a major challenge to drug design.

2.2.2 *Chirality*

The impact of stereochemistry on the bioavailability of drug molecules has been emphasized by Jamali (1992). Typically, stereoisomers have very similar physicochemical properties, and passive processes such as diffusion across the gastrointestinal membrane are not governed by any specific enantioselective mechanisms. However, the drug absorption process is likely to be stereospecific when mediated by carrier molecules. In addition, when chiral excipients are used in the formulation of enantiomers, interaction between them may result in stereoselective release from the dosage form (Duddu *et al.*, 1993). Many enantioselective drugs are usually marketed as racemates, although their therapeutic benefits are attributed only to specific enantiomers.

2.2.3 *Dissolution*

The availability of a drug in the body depends on its ability to dissolve in the gastrointestinal (GI) fluids. If the rate of dissolution is the rate-limiting step in drug absorption, any factor affecting the dissolution rate will have an impact on bioavailability.

The dissolution rate of suspended, poorly soluble drugs according to the well known diffusion-layer model, modified Noyes–Whitney equation (Horter and Dressman, 1997; Nystrom, 1998) is as follows:

$$\frac{\mathrm{d}c}{\mathrm{d}t} = \frac{DA}{hv}(C_\mathrm{s} - C) \tag{2.1}$$

where D is the diffusion coefficient, h is the thickness of the diffusion layer at the solid liquid interface, A is the surface area of drug exposed to the dissolution media, v is the volume of the dissolution media, C_s is the concentration of a saturated solution of the solute in the dissolution medium at the experimental temperature, and C is the concentration of drug in solution at time t. The dissolution rate is given by $\mathrm{d}c/\mathrm{d}t$.

According to this model, dissolution rate is a function of:

- Drug solubility
- The diffusional transport of dissolved molecules away from the solid surface through a thin region of more or less stagnant solvent which surrounds the drug particles
- The solid surface area that is effectively in contact with the solvent

The physicochemical properties that influence the kinetics of drug dissolution can all be attributed to their effects on one or more of these factors. The detailed discussion on the theory of dissolution is covered in Chap. 3, thus will not be discussed here.

2.2.4 *Solubility*

Solubility is one of the most important properties impacting bioavailability because of its role in dissolution. It is one of two factors defining the biopharmaceutics classification system (BCS) (Amidon *et al.*, 1995).

Before the advent of combinatorial chemistry and high throughput screening in the late 1980s and early 1990s, most compounds that were considered poorly soluble had solubility in the range of 10–100 μg/mL (Lipinski *et al.*, 1997). Practically, no marketed drug had solubility below 10 μg/mL. Today, compounds with solubility in the range of 1–10 μg/mL and even <1 μg/mL are very common. Having a good understanding of factors affecting solubility is crucial to our ability to address deficiencies in formulation caused by poor solubility.

2.2.4.1 Definition of Solubility

Based on the official IUPAC definition, Lorimer and Cohen-Adad (2003) more recently defined solubility in more broad terms:

Solubility was defined as the analytical composition of a mixture or solution that is saturated with one of the components of the mixture or solution, expressed in terms of the proportion of the designated component in the designated mixture or solution.

The term "saturated" implies equilibrium with respect to the processes of dissolution and crystallization for solubility of a solid in a liquid, and of phase transfer for solubility of a liquid in another liquid. The equilibrium may be stable or metastable, that is, the composition of a system may maintain a particular value for a long time, yet may shift suddenly or gradually to a more stable state if subjected to a specific disturbance. For example, the solid may be amorphous and may convert to a more stable crystalline form, bringing the system from a metastable equilibrium to a stable one.

In recent pharmaceutical literature, the terms "equilibrium solubility" and "kinetic solubility" are often used for the systems with stable and metastable equilibria, respectively (Lipinski *et al.*, 1997; Huang and Tong, 2004; Borchardt *et al.*, 2006).

2.2.4.2 Factors Contributing to Poor Aqueous Solubility

Two main factors governing aqueous solubility are heat of solvation and heat of fusion (Grant and Highuchi, 1990; Jain and Yalkowsky, 2000). The octanol/water partition coefficient (log P) is a good measure of the solvation energy, which is the energy associated with dissolving solute in water. Lipophilic compounds do not like to interact with water, thus the heat of solvation is small and not enough to overcome the strong hydrogen bonds between water molecules, leading to poor solubility. If the compound is crystalline, additional energy, characterized as heat of fusion, is also required to liberate the molecule from its crystal lattice before it can dissolve. Melting point is the property that is most useful in term of characterizing crystal packing interactions. Compounds with high melting point and large heat of fusion will have poor aqueous solubility unless this large heat of fusion is surpassed by the heat of solvation.

Many high throughput solubility screening methods do not use crystalline drug substance as the starting material. Some start the solubility study by adding dimethyl sulfoxide or DMSO solution in various solubility media. If the material remains as amorphous during the time solubility sample is equilibrated, the impact of heat of fusion on solubility can not be assessed. This is why many of these screening methods can not obtain results that correlate well with the results measured by the traditional shake flask method using crystalline material.

2.2.4.3 pH-Solubility Profile

Majority of drugs are ionizable; therefore, their solubilities are affected by the solution pH and the counter ions that can form salts with them. The solubility theory has been well reviewed in several recent publications (Pudipeddi *et al.*, 2002; Tong, 2000), thus only a brief discussion of pH-solubility principles related to their importance to oral absorption will be presented here.

The equilibrium for the dissociation of the monoprotonated conjugate acid of a basic compound may be expressed by:

$$BH^+ + H_2O \overset{K'_a}{\rightleftharpoons} B + H_3O^+ \tag{2.2}$$

where BH^+ is the protonated species, B is the free base, and K'_a is the apparent dissociation constant of BH^+, which is defined as follows:

$$K'_a = \frac{\left[H_3O^+\right][B]}{\left[BH^+\right]} \tag{2.3}$$

Generally, the relationships drawn in (2.2) and (2.3) must be satisfied for all weak electrolytes in equilibrium irrespective of pH and the degree of saturation. At any pH, the total concentration of a compound, S_T, is the sum of the individual concentrations of its respective species:

$$S_T = \left[BH^+\right] + [B] \tag{2.4}$$

In a saturated solution of arbitrary pH, this total concentration, S_T, is the sum of the solubility of one of the species and the concentration of the other necessary to satisfy the mass balance.

At low pH where the solubility of BH^+ is limiting, the following relationship holds:

$$S_{T,pH<pH_{max}} = [BH^+]_s + B = [BH^+]_s\left(1 + \frac{K'_a}{\left[H_3O^+\right]}\right) \tag{2.5}$$

where pH_{max} refers to the pH of maximum solubility and the subscript $pH < pH_{max}$ indicates that this equation is valid only for pH values less than pH_{max}. The subscript s indicates a saturated species. A similar equation can be written for solutions at pH values greater than pH_{max} where the free base solubility is limiting:

$$S_{T,\,pH>pH_{max}} = [BH^+] + [B]_s = [B]_s\left(1 + \frac{[H_3O^+]}{K'_a}\right) \tag{2.6}$$

Each of these equations describes an independent curve that is limited by the solubility of one of the two species.

The pH-solubility profile is nonuniformly continuous at the juncture of the respective solubility curves. This occurs at the precise pH where the species are simultaneously saturated, previously designated as the pH_{max}.

Figure 2.3 is a typical pH-solubility profile for a poorly soluble basic drug (Tong, 2000). It is constructed by assuming that the solubilities of the hydrochloride salt and the free base (B) are 1 and 0.001 mg/mL, respectively, and the pKa' of the compound is equal to 6.5. The solubility under curve I is limited by the solubility of the salt, whereas the solubility under curve II is limited by the solubility of the free base.

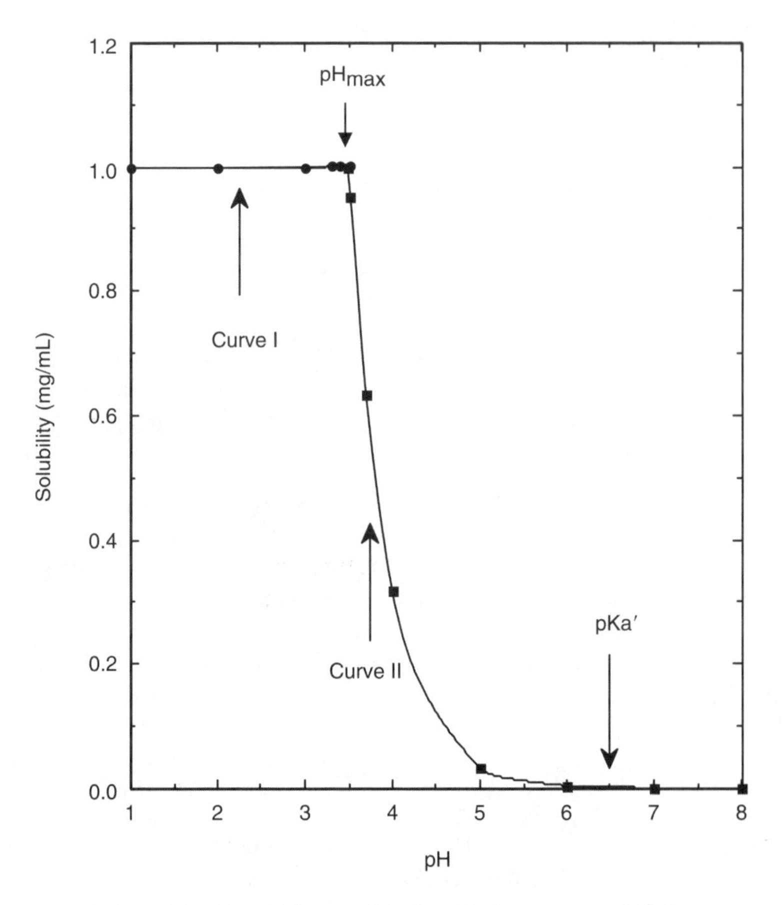

FIGURE 2.3. pH-solubility profile of an ideal compound BH^+Cl^-

The dissociation constant (pKa′), the intrinsic solubility of the unionized form, and the solubility of the salt are three determining factors defining the pH-solubility profile. All other factors being equal, each upward or downward shift in the pKa′ is matched exactly by an upward or downward shift in pH_{max}. If the solubility of the free base is very small relative to that of the salt, the free base limiting curve (curve II) of the overall pH-solubility profile cuts deeply into the acidic pH range, resulting pH_{max} of several pH units smaller than pKa′. Every one-order-of-magnitude increase in the intrinsic solubility of the free base increases the pH_{max} by one unit, whereas every one-order-of-magnitude increase in the solubility of the salt results in a decrease in the pH_{max} by one unit. These effects are illustrated in Fig. 2.4 (Li *et al.*, 2005).

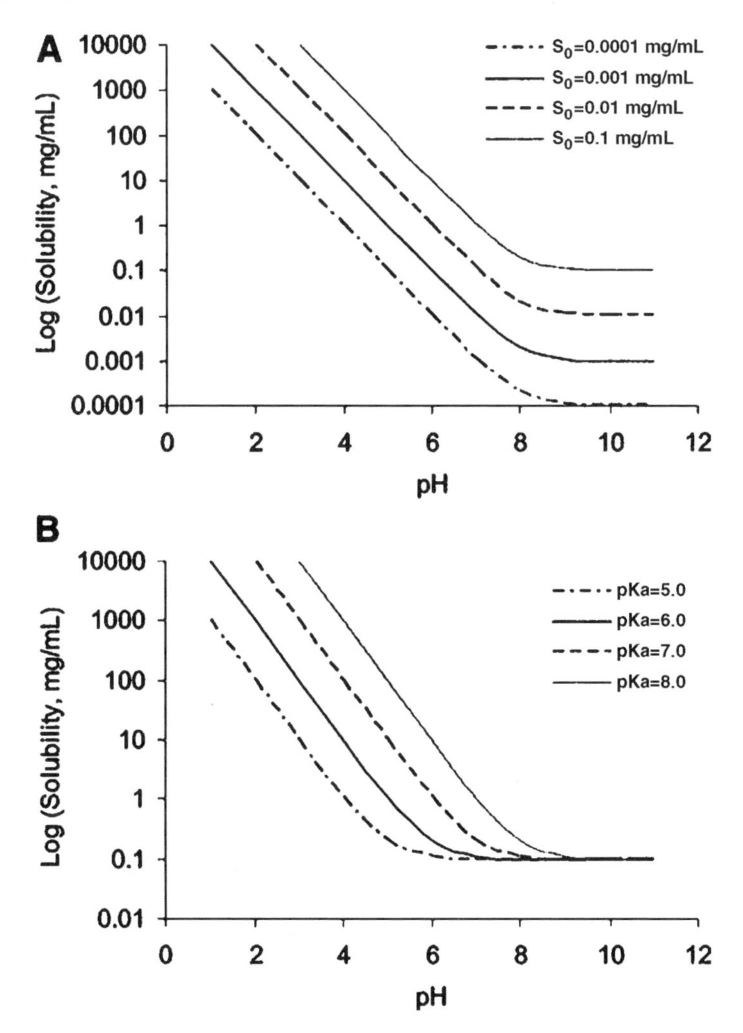

FIGURE 2.4. Effect of relevant parameters on pH-solubility curves: (A) effect of intrinsic solubility S_O, (B) effect of pKa

For acidic compounds, the pH-solubility curve is the mirror image of the curve for basic compounds. Curve I, where solubility is limited by the salt, is on the right side and curve II, where solubility is limited by the free acid, is on the left side with lower pH values (Yalkowsky, 1999; Pudipeddi *et al.*, 2002).

In the stomach, the acidic pH and high concentration of the chloride ion can be problematic for many basic compounds if their hydrochloride salts are poorly soluble. A more soluble salt may be advantageous from the absorption point of view since the conversion to the less soluble hydrochloride salt may not happen right away, maintaining a certain degree of supersaturation on the surface (Li *et al.*, 2005). The conversion to the less soluble hydrochloride salt may be avoided by formulation means such as enteric coating (Tong, 2000). In the intestine, the presence of bile salts and other components such as fat and lipid typically can improve the intrinsic solubility of the free base, shifting pH_{max} to higher value.

The pH-solubility profile is also an important consideration in designing robust formulations. When the microenvironmental pH of a salt of a weakly acidic drug is less than the pH_{max}, conversion of the salt to the free acid may occur upon storage or formulation, leading to potential undesirable changes in product performance both *in vitro* and *in vivo*.

2.2.4.4 Effect of Temperature on Solubility

The van't Hoff equation defines the relationship between solubility and temperature:

$$\ln s = \Delta H/R(1/T) + \text{constant}$$

where s is the molar solubility at temperature T (K) and R is the ideal gas law constant. ΔH is the heat of solution, representing the heat released or absorbed when a mole of solute is dissolved in a large quantity of solvent. For most organic compounds, the heat of solution is about 10 kcal/mol, suggesting that solubility differences between 25 and 37 °C are typically about two times. Practically, most of the solubility studies are done at room temperature for convenience. The two times solubility difference may not be significant when using solubility as criteria to rank order compounds for developability assessment. However, the temperature effect needs to be carefully studied to support formulation development, especially for liquid dosage form. After all, the solubility differences caused by most polymorphic changes are typically only less than two times (Pudipeddi and Serajuddin, 2005).

Additionally, the dependence of solubility on temperature will most likely change for different solubilizing systems (Tong, 2000). For example, temperature changes may affect the micellar size and the degree of drug uptake, leading to a dependence of solubilization on temperature. For solubilizing systems containing complexing agents, because the standard enthalpy change accompanying the complexing process is generally negative, increasing temperature will reduce the degree of complexation. For cosolvent systems, because the heat of solution in different solvent systems is generally different, the temperature effect on solubility in these systems is also different. Detailed solubility mapping in the solvents of interest, including the effect of pH (for ionizable compounds), temperature, and cosolvent compositions is typically required in order to develop a robust formulation, such as a soft gel formulation (Winnike, 2005).

2.2.4.5 Solubility in Gastric and Intestinal Fluids

In the stomach and intestine, drug solubility can be enhanced by the food and bile components such as bile salts, lecithin, and monooleins. Depending on the properties of the drugs, the degree of solubilization differs. Increases in solubility of up to 100-fold upon addition of physiological concentrations of bile salts to aqueous media have been reported for some hydrophobic compounds. Based on the solubility studies of 11 steroidal and nonsteroidal compounds, Michani *et al.* (1996) found that the solubility enhancement by bile salts is a function of the $\log P$. Other factors that may affect the extent of solubilization include MW of the drug and specific interactions between drug and bile salts (Horter and Dressman, 1997).

Supersaturation in the intestinal fluid is an important property that can play a significant role in drug absorption. This is because for compounds with poor intrinsic solubility in the intestinal fluid, solubility is often a limiting factor for absorption. Creating or maintaining supersaturation in the intestinal fluid is necessary to enhance absorption of these compounds. For example, hydroxypropylmethylcellulose (HPMC-AC) has been shown to significantly enhance the absorption of several poorly soluble compounds (Shanker, 2005). Several dissolution systems that are shown to be able to better estimate or predict the supersaturation phenomenon have recently been reported (Kostewicz *et al.*, 2004; Gu *et al.*, 2005).

To realistically estimate the impact of solubility on absorption, solubilities and the degree of supersaturation in more physiologically relevant media should be determined. Two media that were developed based on literature and experimental data in dogs and humans have been used extensively in the industry and academic research (Greenwood, 1994; Dressman, 2000). The compositions of these media, FaSSIF, simulating fasted state of intestinal fluid and FeSSIF, simulating fed state of intestinal fluid, are given below (Table 2.1).

TABLE 2.1. Compositions of FaSSIF and FeSSIF

	FaSSIF	FeSSIF
KH_2PO_4	0.39% (w/w)	na
Acetic acid	na	0.865% (w/w)
Na taurocholate	3 mM	15 mM
Lecithin	0.75 mM	3.75 mM
KCl	0.77%	1.52% (w/w)
pH	6.5	5.0

na: not applicable

2.2.4.6 Solubility as a Limiting Factor to Absorption

Since oral drug absorption is a consecutive and continuous process of dissolution and permeation, poor absorption is traditionally considered to be caused by poor dissolution and/or poor permeation. However, with the more recent drug candidates becoming less soluble, there are cases where the dissolution of a compound is relatively fast and membrane permeability is also already relatively high, but the oral absorption is still poor. In these cases, solubility becomes the limiting factor to absorption since the gut is already saturated and further increase of dose does not increase the absolute amount of drug absorbed.

Lipinski (1997) noted that solubility is not likely to be the limiting factor for absorption for an orally administered drug with a dose of 1 mg/kg, if the solubility is greater than 65 μg/mL, but is likely to limit absorption if the solubility is less than 10 μg/mL (Lipinski *et al.*, 1997). These estimates are supported by the concept of maximum absorbable dose (MAD) (Johnson and Swindell, 1996; Curatolo, 1998). MAD is a conceptual tool that relates the solubility requirement for oral absorption to the dose, permeability and GI volume and transit time. It is defined as:

$$\mathrm{MAD(mg)} = S(\mathrm{mg/mL}) \times K_{\mathrm{a}}(1/\mathrm{min}) \times \mathrm{SIWV(mL)} \times \mathrm{SITT(min)} \quad (2.7)$$

where S is solubility at pH 6.5 reflecting typical small intestine condition; K_a is trans-intestinal absorption rate constant determined by a rat intestinal perfusion experiment; SIWV is small intestinal water volume, generally considered to be 250 mL; and SITT is small intestinal transit time, typically around 270 min. A more simplified and conservative approach is adopted by the FDA to define a BCS (Amidon *et al.*, 1995). This system defines low solubility compounds as those whose aqueous solubility in 250 mL of pH 1–7.5 aqueous solution is less than the total dose.

It is important to remember that the BCS was created more as a guideline to determine whether an *in vitro* and *in vivo* correlation (IVIVC) can be expected and whether a biowaiver could be made on the basis of dissolution tests. Many compounds that are classified as low solubility (classes II and IV) have been shown to be well absorbed. Based on historical data, even the MAD can be considered as a very conservative and simple approach. Certain compounds such as basic compounds with low pKa may have poor solubility in the intestinal fluid, but they may be soluble in the stomach and may be able to maintain supersaturation in the intestine. Other compounds may have much improved solubility in the intestinal fluids such as FaSSIF and FeSSIF compared to pure buffer solutions. In these cases, it may be more realistic to use the kinetic solubility in either FaSSIF or FeSSIF to estimate the MAD.

2.2.4.7 Solubility Determination

The most traditional method for measuring equilibrium solubility is the shake flask method. An excess amount of material is equilibrated in a vial or flask with the solubility medium. The vial or flask is shaken or stirred under a controlled temperature, and the amount of drug is determined at various time intervals by analysis of the supernatant fluid. The equilibrium is considered reached when solubility is not changing any more in two consecutive samples. The residual solid from the solubility studies should be examined for any form changes. Care should be taken when studying the residual solid to make sure a hydrate is not missed. For this reason, a powder X-ray diffraction (PXRD) run with both wet and dry samples is very useful.

For poorly soluble compounds, the time required to reach equilibrium may be rather long due to the poor dissolution rate. There are several practical ways to improve the saturation rate, mainly by manipulating the dissolution rate. Using excess amount of material can increase the effective surface area for dissolution. The surface area of the solid can also be increased by preprocessing the solubility samples. Both vortexing with a small teflon ball in the suspension and sonication are very effective techniques for this purpose. Adding amorphous samples to the solubility sample may create temporary supersaturation, making the dissolution rate a nonlimiting step in reaching equilibrium.

Extra care must be given to determine solubility of salts to avoid the impact of potential conversion of salts to the free form. This conversion is common for salts of compounds with very low intrinsic solubility and weak basicity or acidity. One way to avoid this problem is to determine solubility in a diluted acidic solution using the same acid that formed the salt with the base (Tong, 2000). The solubility can then be calculated by correcting for the common ion effect from the acid. It is only in a suspension with a pH value that is below pH_{max} for basic compounds (or above pH_{max} for acidic compounds) that the solubility is limited by the solubility of the salt. In case the solid salts are not available, solubility of salts may be estimated by the *in situ* salt screening method (Tong and Whitesell, 1998).

For discovery, sometimes a nonequilibrium solubility, often called "kinetic solubility," which is determined by adding a compound's DMSO solution to aqueous buffers, may be useful. This is because many experiments in drug discovery are conducted using compound's DMSO solution. Additionally, the kinetic solubility may help identify poorly soluble compounds early since it is rather unlikely that compounds with poor kinetic solubility will show much improved equilibrium solubility later on when the solid crystalline material is used to measure the solubility.

Several high throughput assays for the kinetic solubility have been described using different analytical methods (Lipinski *et al.*, 1997; Bevan and Lloyd, 2000; Avdeef, 2001). Compared to equilibrium method, some differences in solubility results are expected since for compounds with poor solubility, because of high crystallinity, kinetic methods will obviously overestimate the solubility. Thus, understanding the usefulness and limitation of the kinetic solubility data is important when interpreting results and making important decision in drug candidate selection.

2.2.4.8 Solubility Prediction

Solubility prediction may be useful in early drug discovery phases when solubility measurement is not yet possible. The predicted solubility data may provide guidance in screening of computer designed combinatorial libraries and in lead optimization.

Unfortunately, despite all the effort in the last few decades, there is not a simple reliable method yet up to today for predicting solubility (Taskinen and Yliruusi, 2003). Yalkowsky and Valvani (1980) have introduced a model for solubility of nonelectrolytes which contains only two parameters, log P for liquid phase effects and the melting point (MP) for solid phase effects. Although the model has shown to give reasonable predictions for diverse organic compounds (Jain and Yalkowsky, 2000; Ran *et al.*, 2001), it requires an experimental parameter, the melting point, which is as difficult a problem for prediction as solubility. According to this model, solubility of the solid nonelectrolyte (S) can be calculated from MP and log P by the following equation:

$$\text{Log}\, S = 0.5 - 0.01(\text{MP} - 25) - \text{Log}\, P \tag{2.8}$$

Several neural network methods have been developed to predict solubility and other physicochemical properties. While some methods provide as accurate data as experimental results, others do not (Glomme *et al.*, 2004; Huuskonen *et al.*, 1998). The lack of adequate diversity in the training set is believed to be the main reason for the inaccuracy. As more experimental data with more structure diversity and good quality become publicly available, it is reasonable to assume that a better solubility prediction model will emerge in the future.

2.2.5 Chemical Stability

Chemical stability is an important physicochemical property impacting bioavailability because in order for any compound to be bioavailable, it needs to be stable before it is absorbed in the gastric and intestinal fluid. The compound also needs to be stable for the shelf-life of the products and so the performance of the products including bioavailability will not be compromised.

For ionizable drugs, stability as a function of pH, pH-stability profile, is an important property for understanding the impact of stability on absorption and for developing stable formulations. For poorly soluble compounds, studying the stability as a function of pH is complicated by the solubility limitation in certain pH regions. Solubilizing agents may have to be used so that enough compounds can be in solution for stability studies. Acidic instability of many poorly soluble compounds may not present too big of an issue since many of them are not soluble in acidic media such as gastric fluids. However, it is still important to know if the compound can undergo acidic degradation or not since certain solubilization techniques may have to be used for improving bioavailability of these compounds, for example, for higher doses required for toxicological studies. Many of these solubilization techniques such as lipid-based systems can improve the compound's solubility in gastric fluid.

The main chemical reactions that lead to degradation of the administered drug in the GI include hydrolysis, oxidations, and reductions. They are often catalyzed by the pH conditions and/or enzymes in the small intestine, or the bacterial flora of the lower intestinal tract.

Some common classes of drugs that are subject to hydrolysis include ester, thiol ester, amide, sulfonamide, imide, lactam, lactone, and halogenated aliphatic (Stewart and Tucker, 1985). For compounds that desire modified or controlled release profiles, colonic stability is an important factor to consider. Some examples of drugs that are biotransformed by the large intestinal flora include atropine, digoxin, indomethacin, phenacetin, and sulfinpyrazone (Macheras *et al.*, 1995).

Studies to investigate the effect of stability on absorption should be rather straightforward. After all, the compound only needs to be stable for at most 24 h in the gastric and intestinal fluid at 37 °C. Methods using high throughput with up to 96-well format equipped with robots have been reported (Kerns, 2001).

2.2.6 *Solid State Properties*

Different lattice energies associated with physical forms such as polymorphs, solvates, and amorphous give rise to measurable differences in physicochemical properties, including solubility and stability (Grant and Highuchi, 1990). Thus understanding solid state properties is important in designing robust formulations with optimal biopharmaceutical properties.

2.2.6.1 Polymorphism

Polymorphism is defined as a solid crystalline phase of a given compound resulting from the possibility of at least two different arrangements of that compound in the solid state (Haleblain and McCrone, 1969). In addition to polymorphs, compound may also exist as solvates and hydrates (sometimes also called pseudo-polymorphs), where solvent or water is included in the crystal lattice, and amorphous forms, where no long-range order exists.

In general, polymorphs and pseudo-polymorphs of a given drug have different solubility, crystal shape, dissolution rate, and thus possibly absorption rate. Conversion of a drug substance to a more thermodynamically stable form in the formulation can sometimes significantly increase the development cost or even result in product failure. Therefore, it is generally accepted that the thermodynamically most stable form is identified and chosen for development.

An early study by Aguiar *et al.* (1967) on chloramphenicol palmitate suspension is a classic example showing the effect of polymorphs on bioavailability. Since then, there are only limited number of studies reported. This may partially be due to the fact that most of metastable forms are not chosen for development, and therefore are not tested in human. Recently, Pudipeddi and Serajuddin (2005) reviewed a large number of literature reports on solubility or dissolution of polymorphs and found that the ratio of polymorph solubility is typically less than 2, although occasionally higher ratios can be observed. Higher and more spread out ratios were found when anhydrates were compared to hydrates. For a compound that has low bioavailability due to solubility or dissolution rate-limited absorption, a drop of two times in solubility may directly result in a two time reduction in bioavailability. On the other hand, different crystal forms of highly soluble drugs should not affect bioavailability since solubility or dissolution rate is not likely to be the limiting factor for absorption for these compounds.

Identifying the thermodynamically most stable form and the potential for the hydrate formation is probably the most important part of studying polymorphism. There are many ways to screen for polymorphs, solvates, and hydrate. Crystallization from solution and recrystallization from neat drug substance are two most commonly applied methods (Guillory, 1999). Several high throughput methods with automated or semi-automated sample handling and characterization for crystal screening have been reported. While these new methodologies provide extra capacities for solid form discovery, it still requires detailed characterization and analysis to understand the interrelationship of various forms. One of the methods that are very useful in identifying the lowest energy form is the slurry experiment. Typically, an aqueous based solvent and a water-free solvent are chosen for these studies. After all the forms are suspended in these two solvents for a certain period of time, a conversion of various forms to the lowest energy form or the hydrate will typically occur.

2.2.6.2 Amorphous Material

Theoretically, the amorphous form of a material has the highest free energy, thus should have the biggest impact on solubility and bioavailability. A review of the data in the literature indicates that improvements in solubility resulting from the use of amorphous material can range from less than two-fold to greater than 100-fold (Hancock and Zografi, 1997). Elamin *et al.* (1994) even showed that low levels of amorphous character induced in griseofulvin by milling, which were undetectable by DSC, can readily result in solubility differences of two-fold or more. Thus, it is important to understand the impact of pharmaceutical processing such as wet granulation and roller compaction on the crystallinity of drug substance in order to have a robust process with adequate control for product quality.

Since amorphous material typically is not stable in any solvents, measuring the true equilibrium solubility of amorphous material is very difficult. Thus, the solubility advantage determined experimentally is typically less than that predicted from simple thermodynamic considerations. Hancock and Parks (2000) suggested that the true solubility advantage for amorphous material could be as high as over 1,000-fold.

To take advantage of the amorphous solubility, the amorphous material needs to be physically and chemically stable in a given dosage form. To overcome the challenge, the amorphous material is typically formulated in a solid dispersion formulation. Polymers are typically employed to increase the glass transition temperature (T_g). It is typically believed that a 50°C difference between the T_g of a solid dispersion and storage temperature is required in order to minimize the mobility and reduce the risk for crystallization.

Recently, Vasanthavada *et al.* (2004, 2005) showed that when a drug in its amorphous stage is miscible with a polymer in a solid dispersion, it can be physically stable even under accelerated stability condition (high temperature and high humidity). They developed a method to determine the "extent of molecular miscibility," referred to as "solid solubility," and demonstrated that hydrogen bonding is the main contributor to the solid solubility for the indoprofen and poly vinyl pyrrolidone (PVP) system. The method developed should be an excellent tool useful for identifying carriers that can form physically most stable solid dispersions.

2.2.6.3 Particle Size

Particle size is an important physical property impacting oral absorption because it is directly related to surface area available for dissolution. For compounds whose bioavailability is limited by the dissolution rate, it is obvious that particle size reduction should enhance absorption.

Further increase in absorption can be realized by reducing particles to submicron size (Rabinow, 2004). However, when the particles are reduced to submicron size, they tend to agglomerate to reduce the free energy of the system. This tendency is resisted by the addition of surface-active agents, which reduce the interfacial tension and therefore also the free energy of the system.

A variety of techniques with different operating principles and features are available for measuring particle size distribution. Sieving or screening and microscopy are commonly used for large particles typically used for solid dosage form development. Laser light diffraction can be used for particle size ranging from 0.02 up to 2,000 μm. However, the limitation of this method is that it is not measuring a single particle rather it is measuring an ensemble of particles. For particles at nanomicron ranges, some special techniques such as field emission low-voltage scanning electron microscopy and photon-correlation spectroscopy have been used.

2.3 Physicochemical Properties and Drug Delivery Systems

Oral bioavailability depends on several factors, mainly solubility and dissolution rate in the gastric and intestinal fluids, permeability, and metabolic stability. As discussed earlier, the effect of physicochemical properties of drugs on bioavailability is mainly on the availability of the drug at the absorption sites. To date, formulation strategies have been far more successful in improving the bioavailability of compounds with poor solubility, poor dissolution rate, and poor chemical stability in acidic environments. The effect on permeability is mainly by the molecular properties, and is more effectively addressed by molecular design than by drug delivery systems. Although many studies have been reported to enhance permeability through the use of absorption enhancers such as medium chain fatty acids, bile salts, surfactants, liposacharides, and chitosans, because of the safety concerns associated with the effect of these absorption enhancers on the cell membranes, their applications in drug products are still very limited (Gomez-Orellana, 2005). With many currently ongoing research aiming to better understand the limiting factors to permeability such as active transporters and P-gp-pump, it is expected that more successful and safer permeability enhancers are to be discovered in the future.

Figure 2.5 is a formulation decision tree modified from Rabinow's review (2004). This decision illustrates the impact of physicochemical properties on the formulation strategies. Having a good understanding of the key physicochemical properties of the drug substance should not only help us understand the causes of low oral absorption but also guide us in defining the appropriate formulation strategies to improve bioavailability.

For ionizable compounds, forming a more soluble salt is obviously the first consideration for solubility enhancement. For compounds with high log P,

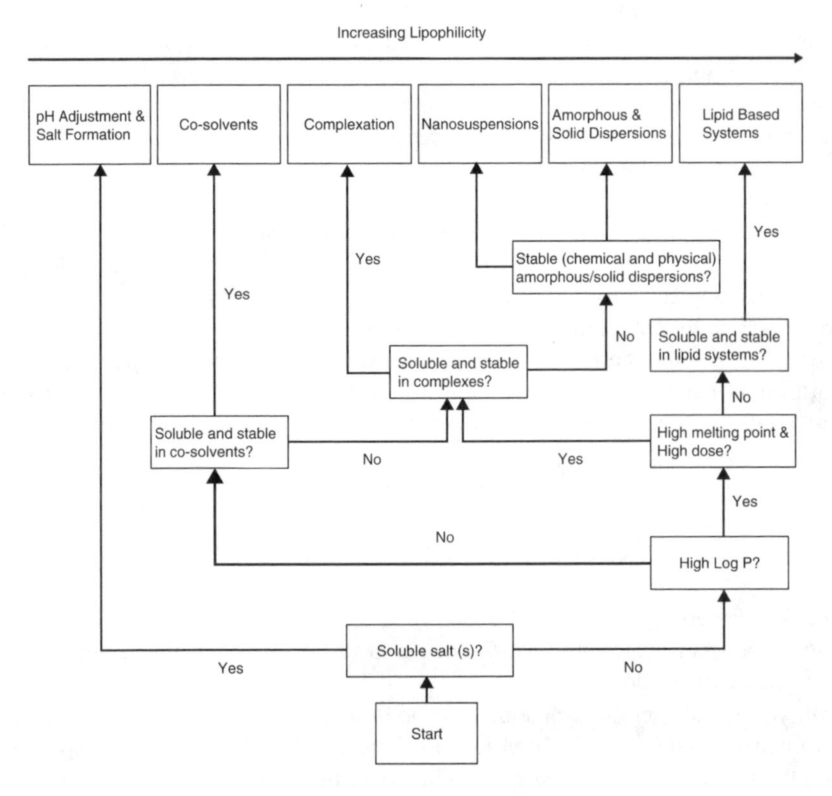

FIGURE 2.5. A decision tree for selection of formulation approach based on physicochemical properties of drug substance

lipid-based systems typically work well. However, for compounds whose solubility is low due to high crystallinity indicated by high melting point (MP), destroying crystalline structure by making them as amorphous or metastable forms should be the focus of solubilization. Additional factors to consider include dose level, chemical and physical stability.

For compounds that are not stable in gastric fluids, their hydrolysis may be prevented or reduced by using enteric coating, inclusion of pH modifier in the formulation to modify the environmental pH, a less soluble salt, or pro-drugs. Compared to gastric degradation, compounds with intestinal stability problems are more difficult to formulate. Although there may be formulation options such as pH modifiers or enzyme inhibitors these problems are better avoided by discovering more stable derivatives or pro-drugs.

For compounds with poor aqueous solubility in the intestinal fluids, enhancing solubility by lipid-based systems or maintaining supersaturation by the inclusion of crystallization inhibitors such as polymers have been proven to be successful.

2.4 Summary

Oral absorption and bioavailability are significantly affected by many molecular and physicochemical properties. High MW, high log P, high flexibility and polar surface, and high number of hydrogen donor and acceptors are found to contribute to poor absorption. Poor solubility/dissolution rate and poor stability in the gastric and intestinal fluids are the key physiochemical properties limiting oral absorption. Having a good understanding of these molecular and physicochemical properties and their impact on absorption is essential in identifying the appropriate formulation and drug delivery strategies to enhance oral absorption.

References

Aguiar A.J., Krc J., Kindel A.W., and Samyn J.C., 1967. Effect of polymorphism on the absorption of chloramphenicol from chloramphenicol palmitate. J. Pharm. Sci., 56, 847–853.

Amidon G.L., Lennernas H., Shah V.P., and Crison J.R., 1995. A theoretical basis for a biopharmaceutic drug classification: the correlation of in vitro drug product dissolution and in vivo bioavailability. Pharm. Res., 12(3), 413–420.

Avdeef A., 2001. High throughput measurements of solubility profiles. In Pharmacokinetic Optimization in Drug Research; Biological, Physiochemical, and Computational Strategies. Edited by Testa B., et al., Verlag Helvitica Chimica Acta., Zurich, 305–326.

Bevan C. and Lloyd R.S., 2000. A high-throughput screening method for the determination of aqueous drug solubility using laser nepholometry in microtiter plates. Anal. Chem., 72, 1781–1787.

Borchardt R.T., Kerns E.H., Hageman M.J., Thakker D.R., and Stevens J.L., 2006. Optimizing the "Drug-Like" Properties of Leads in Drug Discovery. AAPS Press, Arlington, VA.

Connors K.A., Amidon G.L., and Stella V.J., 1986. Chemical Stability of Pharmaceuticals. A Handbook for Pharmacists. 2nd Edition. Wiley, New York.

Curatolo W., 1998. Physicochemical properties of oral drug candidates in the discovery and exploratory development settings. Pharm. Sci. Technol. Today, 1, 387–393.

Dressman J.B., 2000. Dissolution testing of immediate-release products and its application to forcasting in vivo performance. In Oral Drug Absorption, Prediction and Assessment. Edited by Dressman J.B. and Lennernas H., Marcel Dekker, New York, 155–181.

Duddu S., Vakilynejad M., Jamili F., and Grant D., 1993. Stereoselective dissolution of propranolol hydrochloride from hydroxypropyl methylcellulose matrices. Pharm. Res., 10, 1648–1653.

Glomme A., März J., and Dressman J.B., 2004. Comparison of a miniaturized shake-flask solubility method with automated potentiometric acid/base titrations and calculated solubilities. J. Pharm. Sci., 94, 1–16.

Gomez-Orellana I., 2005. Strategies to improve oral drug bioavailability. Expert Opin. Drug Deliv., 2(3), 419–433.

Grant D.J.W. and Highuchi T., 1990. Solubility Behavior of Organic Compounds. Wiley, New York.

Gu C.H., Rao D., Gandhi R.B., Hilden J., and Raghavan K., 2005. Using a novel multicompartment dissolution system to predict the effect of gastric pH on the oral absorption of weak bases with poor intrinsic solubility. J. Pharm. Sci., 94, 199–208.

Haleblian J. and McCrone W., 1969. Pharmaceutical applications of polymorphism. J Pharm. Sci., 58, 911–929.

Hancock B.C. and Parks M., 2000. What is the true solubility advantage for amorphous pharmaceuticals? Pharm. Res., 17(4), 397–404.

Hancock B.C. and Zografi G., 1997. Characteristics and significance of the amorphous stage in pharmaceutical systems. J. Pharm. Sci., 86, 1–12.

Horter D. and Dressman J.B., 1997. Influence of physicochemical properties on dissolution of drugs in the gastrointestinal tract. Adv. Drug Deliv. Rev., 25, 3–14.

Huang L.F. and Tong W.Q., 2004. Impact of solid state properties on developability assessment of drug candidates. Adv. Drug Deliv. Rev., 56, 321–334.

Huuskonen J., Salo M., and Taskinen J., 1998. Aqueous solubility prediction of drugs based on molecular topology and neural network modeling. J. Chem. Inf. Comput. Sci., 38, 450–456.

Jain N. and Yalkowsky S.H., 2000. Estimation of the aqueous solubility I: application to organic non-electrolytes. J. Pharm. Sci., 90, 234–252.

Jamali F., 1992. Stereochemistry and bioequivalance. J. Clin. Pharmacol., 32, 930–934.

Johnson K.C. and Swindell A.C., 1996. Guidance in the setting of drug particle size specifications to minimize variability in absorption. Pharm. Res., 13, 1795–1798.

Kerns E.H., 2001. High throughput physicochemical profiling for drug discovery. J. Pharm. Sci., 90(11), 1838–1858.

Kerns E.H. and Di L., 2003. Pharmaceutical profiling in drug discovery. Drug Discov. Today, 8(7), 316–323.

Kostewicz E.S., Wunderlich M., Brauns U., Becker R., Bock T., and Dressman J.B., 2004. Predicting the precipitation of poorly soluble weak bases upon entry in the small intestine. J. Pharm. Pharmacol., 56, 43–51.

Leeson P.D. and Davis A.M., 2004. Time-related differences in the physical properties of oral drugs. J. Med. Chem., 47, 6338–6348.

Li S.F., Wong S.M., Sethia S., Almoazen H., Joshi Y., and Serajuddin A.T.M., 2005. Investigation of solubility and dissolution of a free base and two different salt forms as a function of pH. Pharm. Res., 22(4), 628–635.

Lipinski C.A., Lombardo F., Dominy B.W., and Feeney P.J., 1997. Experimental and computational approaches to estimate solubility and permeability in drug discovery and development settings. Adv. Drug Deliv. Rev., 23, 3–25.

Lobenberg R., Amidon G.L., and Vieira M., 2000. Solubility as a limiting factor to drug absorption. In Oral Drug Absorption, Prediciton and Assessment. Edited by Dressman J.B and Lennernas H., Marcel Dekker, New York, 137–153.

Macheras P., Reppas C., and Dressman J.B., 1995. Biopharmaceutics of Orally Administered Drugs. Ellis Horwood Limited.

Nystrom C., 1998. Dissolution properties of poorly soluble drugs: theoretical background and possibilities to improve the dissolution behavior. In Emulsions and Nalosuspensions for the Formulation of Poorly Soluble Drugs. Edited by Muller R.H., Benita S., and Bohm B., Medpharm Scientific Publishers, Stuttgart.

Pudipeddi M., Serajuddin A.T.M., Grant D.J.W., and Stahl P.H., 2002. Solubility and dissolution of weak acids, bases and salts. In Handbook of Pharmaceutical Salts, Properties, Selection, and Use. Edited by Stahl P.H. and Wermuth C.G., Verlag Helvetica Chimica Acta, Zurich, 19–39.

Rabinow B.E., 2004. Nanosuspensions in drug delivery. Nature Rev., 3(9), 785–796.

Ran Y., Jain N., and Yalkowsky S.H., 2001. Prediction of aqueous solubility of organic compounds by the general solubility equition (GSE). J. Chem. Inf. Comput. Sci., 41, 1280–1217.

Shanker R., 2005. Current concepts in the science of solid dispersions. Second Annual Simonelli Conference in Pharmaceutical Sciences. Long Island University.

Taskinen J. and Yliruusi J., 2003. Prediction of physicochemical properties based on neutal network modelling. Adv. Drug Deliv. Rev., 55, 1163–1183.

Tong W.Q., 2000. Preformulation aspects of insoluble compounds. In Water Insoluble Drug Formulation. Edited by Liu R., Interpharm Press, Denver, Colorado, 65–95.

Tong W.Q. and Whitesell G., 1998. In situ salt screening – a useful technique for discovery support and preformulation studies. Pharm. Dev. Tech., 3(2), 215–213.

Vasanthavada M., Tong W.Q., Joshi Y., and Kislalioglu M.S., 2004. Phase behavior of amorphous molecular dispersions I: determination of the degree and mechanism of solid solubility. Pharm. Res., 21(9), 1589–1597.

Vasanthavada M., Tong W.Q., Joshi Y., and Kislalioglu M.S., 2005. Phase behavior of amorphous molecular dispersions II: role of hydrogen bonding in solid solubility and phase separation kinetics. Pharm. Res., 22(3), 440–448.

Veber D.F., Johnson S.R., Cheng H.Y., Smith B.R., Ward K.W., and Kopple K.D., 2002. Molecular properties that influence the oral bioavailability of drug candidates. J. Med. Chem., 45, 2615–2623.

Vieth M., Siegel M.G., Higgs R.E., Watson I.A., Robertson D.H., Savin K.A., Durst G.L., and Hipskind P.A., 2004. Characteristic physical properties and structural fragments of marketed oral drugs. J. Med. Chem., 47, 224–232.

Waterbeemd H., Smith D.A., and Jones B.C., 2001. Lipophilicity in PK design: methyl, ethyl, futile. J. Comput. Aided Mol. Des., 15(3), 273–286.

Wenlock M.C., Austin R.P., Barton P., Davis A.M., and Lesson P., 2003. A comparison of physiochemical property profiles of development and marketed oral drugs. J. Med. Chem., 46, 1250–1256.

Winnike R., 2005. Solubility assessment in pharmaceutical development, practical considerations for solubility profiling and solubilization techniques. AAPS Short Course on Fundamentals of Preformulation in Pharmaceutical Product Development. Nashville, TN, Nov. 5, 2005.

Yalkowsky S.H., 1999. Solubility and Solubilization in Aqueous Media. American Chemical Society, Washington D.C.

Yalkowsky S.H. and Valvani S.C., 1980. Solubility and partitioning I: solubility of nonelectrolytes in water. J. Pharm. Sci., 69, 912–922.

3 Dissolution Testing[1]

Sau Lawrence Lee, Andre S. Raw, and Lawrence Yu

3.1 Introduction

Ever since dissolution was known to have a significant effect on bioavailability and clinical performance, dissolution analysis of pharmaceutical solids has become one of the most important tests in drug product development and manufacturing, as well as in regulatory assessment of drug product quality. Not only can dissolution testing provide information regarding the rate and extent of drug absorption in the body, it can also assess the effects of drug substance biopharmaceutical properties and formulation principles on the release properties of a drug product. Nevertheless, despite the wide use of dissolution testing by the pharmaceutical industry and regulatory agencies, the fundamentals and utilities of dissolution testing are still not fully understood. The objective of this chapter is to provide a concise review of dissolution methods that are used for quality control (QC) and bioavailability assessment, highlight issues regarding their utilities and limitations, and review challenges of improving some of these current dissolution methods, particularly those used for assessing *in vivo* drug product performance. In this chapter, we first provide some background information on dissolution, including the significance of dissolution in drug absorption, theories of dissolution, and factors affecting dissolution testing. Second, we examine the current roles of dissolution testing. Third, we evaluate the utilities and limitations of dissolution as a QC tool under the current industry setting. Finally, we conclude this chapter by discussing the biopharmaceutics classification system (BCS) and biorelevant dissolution methods.

3.2 Significance of Dissolution in Drug Absorption

Oral administration of solid formulations has been the most common route of administration for almost a century. However, the importance of dissolution processes in the oral drug absorption was only recognized about 50 years ago when Nelson published his finding that showed a relationship between the blood

---

[1] The opinions expressed in this chapter by the authors do not necessarily reflect the views or policies of the Food and Drug Administration (FDA).

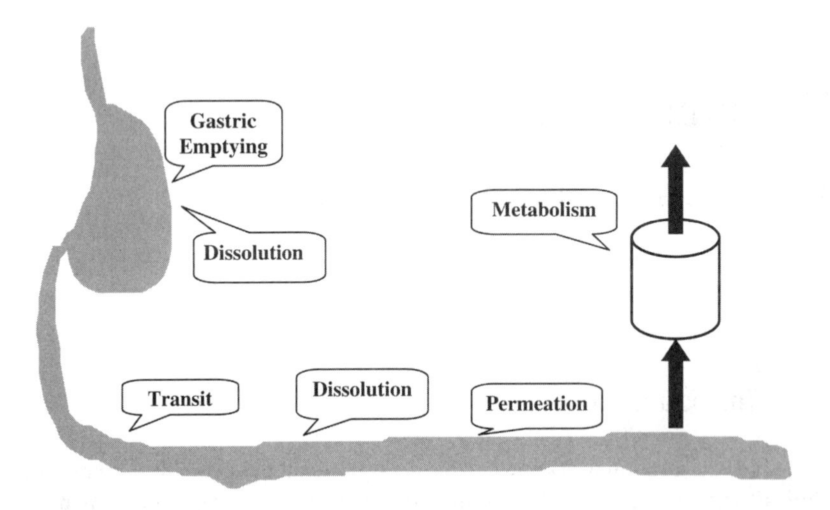

FIGURE 3.1. Schematic representation of the simplified oral drug absorption process that consists of transit (gastric emptying), dissolution, permeation, and first pass metabolism

levels of orally administered theophylline salts and their *in vitro* dissolution rate (Nelson, 1957). The need for dissolution testing can be understood easily by considering the importance of dissolution on oral drug absorption, which is described below.

When a systemically acting drug is administered in solid dosage forms, such as a tablet or capsule, its absorption into the systemic circulation can be generally described by four consecutive steps (Fig. 3.1). The first step involves delivery of the drug into its absorption site through gastric emptying and intestinal transit flow. It is followed by the second step in which dissolution takes place in the stomach and/or in the small intestine. It should be noted that the first two steps need not to be sequential and that lymphatic absorption is not considered. The third step is characterized by the permeation of the dissolved drug across the gastrointestinal (GI) membrane. Finally, the absorbed drug passes through the liver (first pass metabolism) and reaches the systemic circulation. Although this is a simplified description of the drug absorption process, it shows that transit (gastric emptying), dissolution, absorption across intestinal membrane, and metabolism constitute the fundamental processes of oral drug absorption. If the dissolution process is slow relative to the other three processes, which is usually the case for most poorly soluble drugs formulated in a conventional dosage form, dissolution will be the rate limiting step. As a result, the dissolution rate will determine the overall rate and extent of drug absorption into a systemic circulation, and hence bioavailability.

3.3 Theories of Dissolution

Dissolution is generally defined as a process by which a solid substance is solubilized into the solvent to yield a solution. This process is fundamentally controlled by the affinity between the solid substance and the solvent and consists of two consecutive steps. The first step involves the liberation of molecules from the solid phase to the liquid layer near the solid surface (an interfacial reaction between the solid surface and the solvent). It is followed by the transport of solutes from the solid–liquid interface into the bulk solution. The dissolution of solid substance is generally modeled based upon the relative significance of these two transport steps.

The diffusion layer model proposed originally by Nernst and Brunner (Brunner, 1904; Nernst, 1904) is widely used to describe the dissolution of pure solid substances. In this model, it is assumed that a diffusion layer (or a stagnant liquid film layer) of the thickness h is surrounding the surface of a dissolving particle. The reaction at the solid–liquid interface is assumed to be instantaneous. Thus, equilibrium exists at the interface, and hence the concentration of the surface is the saturated solubility of the substance (C_S). Once the solute molecules diffuse through the film layer and reach the liquid film–solvent interface, rapid mixing takes place, resulting in a uniform bulk concentration (C). Based upon this description (see Fig. 3.2a), the dissolution rate is determined entirely by Brownian motion diffusion of the molecules in the diffusion layer.

To model the diffusion process through the liquid film, Fick's first law, which relates flux of a solute to its concentration gradient, can be applied:

$$J = -D\frac{\mathrm{d}C}{\mathrm{d}x}, \tag{3.1}$$

where J is the amount of solute passing through a unit area perpendicular to the surface per unit time. D is the diffusion coefficient, and $\mathrm{d}C/\mathrm{d}x$ is the concentration gradient, which represents a driving force for diffusion. At steady state, (3.1) becomes

$$J = -D\frac{C - C_S}{h}, \tag{3.2}$$

where C is the bulk concentration, C_S is the saturation concentration, and h is the thickness of the stagnant diffusion layer. Based on (3.2), the dissolution rate, which is proportional to the flux of solutes across the diffusion layer, can be described by

$$V\frac{\mathrm{d}C}{\mathrm{d}t} = SJ \tag{3.3}$$

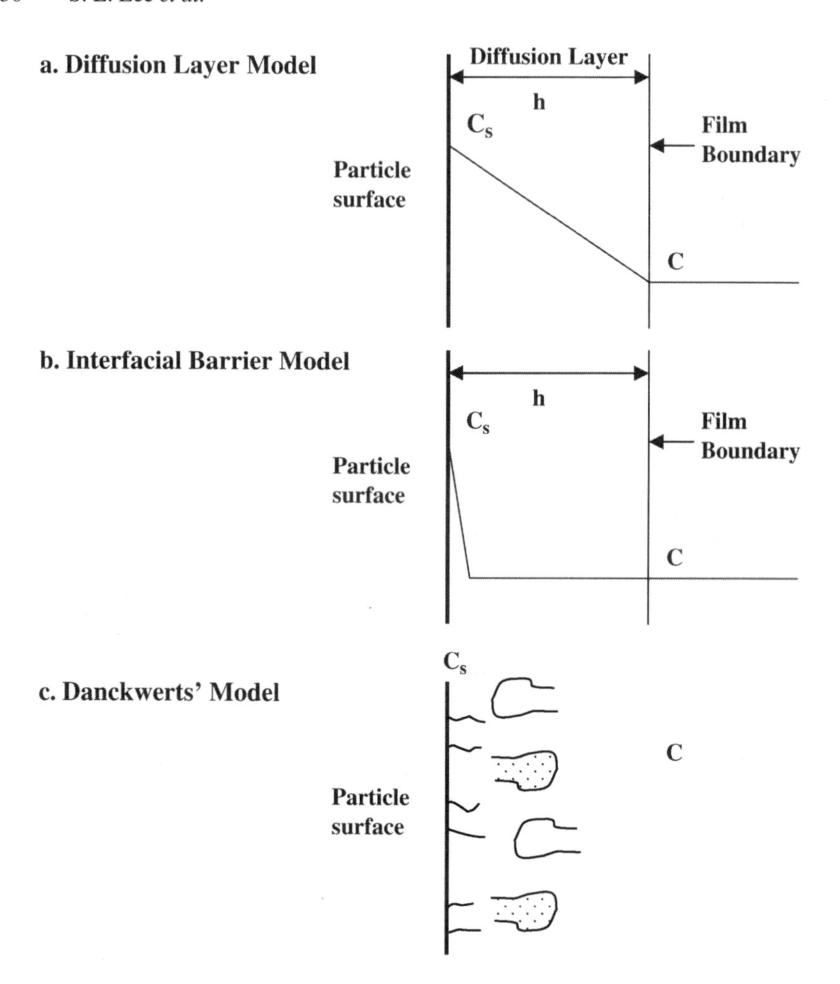

FIGURE 3.2. Schematic illustration of (**a**) the diffusion layer model, (**b**) the interfacial barrier model, and (**c**) the Danckwerts model

or

$$\frac{\mathrm{d}C}{\mathrm{d}t} = \frac{SD}{Vh}(C_\mathrm{S} - C), \tag{3.4}$$

where S is the total surface area of particles, and V is the volume of dissolution medium. The term $C_\mathrm{S} - C$ represents the concentration gradient within the stagnant diffusion layer with thickness h. This equation is known as the Nernst–Brunner equation (Brunner and Tolloczko, 1900; Nernst, 1904).

In addition to film theory, two other theories were also used to describe the dissolution process. These theories include the interfacial barrier model (Higuchi, 1961) and the Danckwerts model (Danckwerts, 1951). In contrast to the film model, the interfacial barrier model assumes that the reaction at the solid surface is significantly slower than the diffusion across the interface. Therefore, no equilibrium exists at the surface, and the liberation of solutes at the solid–liquid interface controls the overall rate of the transport process. This model is illustrated in Fig. 3.2b. Based on this model, the dissolution rate is given by

$$G = k_{\mathrm{i}}(C_{\mathrm{s}} - C), \tag{3.5}$$

where G is the dissolution rate per unit area, and k_i is the interfacial transport coefficient.

Under the assumption that the solid surface reaction is instantaneous, the Danckwerts model suggests that the transport of solute is achieved by the macroscopic packets that reach the solid surface, absorb solutes at the surface, and deliver them to the bulk solution. This transport phenomenon is depicted in Fig. 3.2c. The dissolution rate is expressed as

$$\frac{\mathrm{d}m}{\mathrm{d}t} = S(\gamma D)^{1/2}(C_{\mathrm{s}} - C), \tag{3.6}$$

where m is the mass of dissolved substances and γ is the interfacial tension.

These three models have been employed alone or in combination to describe the mechanism of dissolution. Nevertheless, the diffusion layer model is the simplest and most commonly used to describe the dissolution process of a pure substance among these three models.

3.4 Factors Affecting Dissolution

Several physicochemical processes need to be considered along with the drug substance dissolution process to determine the overall dissolution rate of drugs from solid dosage forms under standardized conditions. The dissolution process for a solid dosage form (or a drug product) in solution starts with the wetting and the penetration of the dissolution medium into the solid formulation. It is generally followed by disintegration and/or deaggregation into granules or fine particles. However, this step is not a prerequisite for dissolution. The final step involves solubilization (or dissolution) of the drug substance into the dissolution medium. A schematic diagram illustrating the processes involved in the dissolution of solid dosage forms is shown in Fig. 3.3. It should be noted that these steps can also occur simultaneously during the dissolution process. For most poorly soluble drugs, dissolution is considered to be *dissolution controlled*, since solubilization of drug particles is slow relative to disintegration or deaggregation of the dosage form. If the step of disintegration or deaggregation is rate-limiting, dissolution is considered to be *disintegration controlled*. The factors that affect the dissolution rate of solid dosage forms can be classified under four main categories: (1) factors related to the physicochemical properties of the drug substance, (2) factors related to drug product formulations, (3) factors related to manufacturing processes, and (4) factors related to dissolution testing conditions.

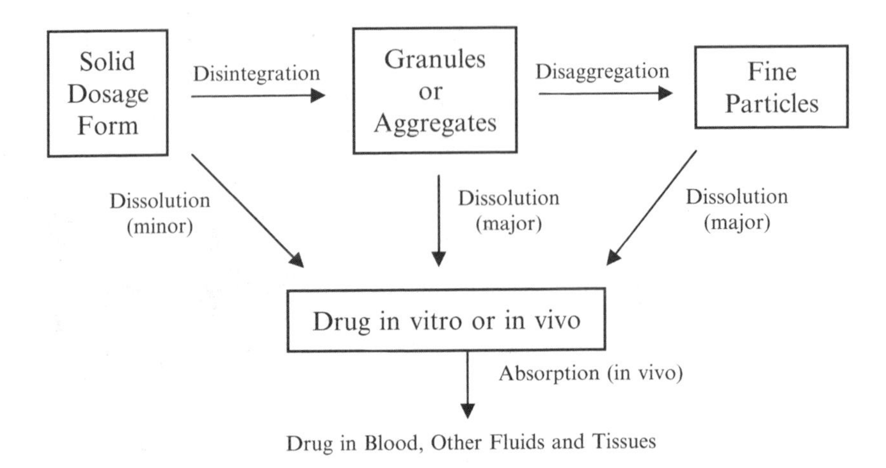

FIGURE 3.3. Schematic illustration of a dissolution process of a solid dosage form (modified from Wagner 1970)

3.4.1 Factors Related to the Physicochemical Properties of the Drug Substance

The importance of the physicochemical properties on the dissolution of a drug substance into the dissolution medium is best illustrated by (3.4)–(3.6). Despite the fact that these three equations are derived from different diffusion mechanisms, they clearly show that the dissolution rate depends on the solubility and surface area of a drug substance.

3.4.1.1 Solubility

From (3.4)–(3.6), it is evident that compounds with high solubility generally exhibit higher dissolution rates. The solubility of ionizable drugs, such as weak acids and bases, depends upon both the pH of the medium and the p*K*a of the compound. Therefore, it is important to ascertain the aqueous solubility of the drug substance over the physiologically relevant pH range of 1–7.5 in order to predict the effect of solubility on dissolution. The study of Yu *et al.* shows that there is a good relationship between solubility and disk intrinsic dissolution rate unless an extremely high or low dose is used. Solubility data may also be used as a rough predictor for indicating any potential problems with oral absorption. For example, when the dose/solubility of the drug, which provides an estimate of the fluid volume required to dissolve an individual dose, exceeds about 1 L, *in vivo* dissolution is often considered problematic (Yu and Amidon, 1999).

3.4.1.2 Particle Size

According to (3.4) and (3.6), the dissolution rate is directly proportional to the surface area of the drug. Reducing particle size leads to an increase in the surface area exposed to the dissolution medium, resulting in a greater dissolution rate. Thus, the dissolution rate of poorly soluble drugs can often be enhanced markedly by undergoing size reduction (e.g., through micronization). This is evidenced in the case of glyburide tablets (Stavchansky and McGinity, 1989). However, particle size reduction does not always improve the dissolution rate. This is in part attributed to adsorption of air on the surface of hydrophobic drugs, which inhibits the wetting and hence reduces the effective surface area. In addition, fine particles tend to agglomerate in order to minimize the surface energy, which also leads to a decrease in the effective surface area for dissolution.

3.4.1.3 Solid Phase Characteristics

A drug substance may exist in different solid-state forms (polymorphism). These different forms can be generally classified into three distinct classes including (1) crystalline phases that have different arrangements and/or conformations of the molecules in the crystal lattice, (2) solvates that contain either stoichiometric or nonstoichiometric amount of a solvent, and (3) amorphous phases that do not possess a distinguishable crystal lattice. Differences in the lattice energies among various polymorphic forms can result in differences in the solubilities. Sometimes, the solubility of different drug substance polymorphs can vary significantly. For example, the solubility of amorphous forms can be several hundred times greater than that of the corresponding crystalline state. These solubility differences may alter drug product *in vivo* dissolution, hence affecting oral drug absorption.

3.4.1.4 Salt Effects

Salt formation is frequently used to increase the solubility of a weak acid and base. The solubility enhancement of a drug substance by salt formation is related to several factors including the thermodynamically favored aqueous solvation of cations or anions used to create the salt of the active moiety, the differing energies of the salt crystal lattice, and the ability of the salt to alter the resultant pH. In addition, even if the salt formation has no impact on the solubility of the drug, the dissolution rate of the salt will often be enhanced due to the difference in the pH of the thin diffusion layer surrounding the drug particles (Stavchansky and McGinity, 1989).

3.4.2 Factors Related to Drug Product Formulation

In addition to the physicochemical properties of a drug substance, inactive ingredients (or excipients) may influence the dissolution of a drug product. The effect of these excipients on the drug product dissolution rate depends on the dosage form. For immediate-release dosage forms, excipients are often used to improve the drug release from the formulation or the solubilization of a drug substance. For instance, disintegrants such as starch are often used to facilitate the break up of a tablet and promote deaggregation into granules or particles after administration (Peck *et al.*, 1989). For poorly soluble drugs, incorporation of surfactants (e.g., polysorbate) into the formulation may increase the dissolution rate of these products. The mechanism by which surfactants enhance the dissolution rate is to improve the solubility of the drug substance by promoting drug wetting, by forming micelles, and by decreasing the surface tension of hydrophobic drug particles with the dissolution medium (Banaker, 1991). Furthermore, coprecipitation with polyvinylpyrrolidine (PVP) has been shown to significantly influence the dissolution (Corrigan, 1985). This enhancing effect on the dissolution rate can be attributed to the formation of an energetic amorphous phase or molecular dispersion. However, some excipients may have an adverse effect on the dissolution rate. For example, lubricants such as stearates, which are used to reduce friction between the granulation and die wall during compression and ejection, are often hydrophobic in nature. Thus, these hydrophobic lubricants may affect the wettability of a drug product (Pinnamaneni *et al.*, 2002).

For modified-release drug products, specific excipients are selected to control the rate and extent of drug release from the formulation matrix, and/or to target the delivery to selective sites in the GI tract. For instance, in matrix-based formulations, the active ingredient is embedded in a polymer matrix, which controls drug release through using mechanisms such as swelling, diffusion, erosion, or combinations (Gandhi *et al.*, 1999). In designing these complex formulations, in addition to the characteristics of modifying release excipients, the physicochemical properties of a drug substance, the interactions between the drug substance and excipients, the type of the release mechanism and the target release profile must be taken into consideration.

3.4.3 Factors Related to Manufacturing Processes

Many manufacturing process factors can have an impact on the dissolution characteristics of solid dosage forms. Very often, an appropriate unit operation is selected to enhance the dissolution rates of a drug product. Wet granulation, in general, has been shown to improve the wettability of poorly soluble drugs by incorporating hydrophilic properties into the surface of granules, hence resulting in a greater dissolution rate (Bandelin, 1990). Based upon the propensity for directly compressed tablets to deaggregate into finer drug particles, direct compression may be chosen over granulation for improving dissolution (Shangraw, 1990). Manufacturing variables may also have both positive and negative effects upon drug product dissolution. In tablet compression, there are always two competing factors: the positive effect due to the increase in the surface area by breaking into smaller particles, and the negative effect due to the enhancement in particle bonding that inhibits solvent penetration. For instance, high compression may reduce the wettability of the tablet, since the formation of firmer and effective sealing layer by the lubricant is likely to occur under the high pressure that is usually accompanied by high temperature. The possible influences of the force used to compress a mixture of the drug and excipients into a tablet on the dissolution rate are summarized in Fig. 3.4.

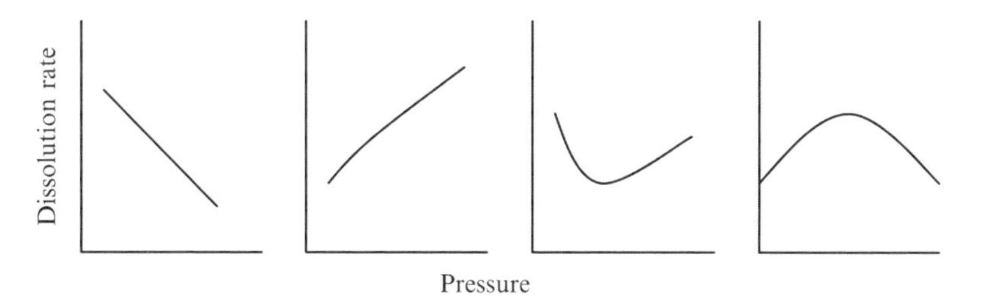

FIGURE 3.4. Possible effects of compression force on the dissolution rate (modified from Finholt, 1974)

3.4.4 Factors Related to Dissolution Testing Conditions

External factors, such as temperature and viscosity of the dissolution medium can influence the dissolution rate of a drug substance or a drug product. This is in part due to their effect on the diffusivity of a drug molecule. According to the Stokes–Einstein equation, the diffusion coefficient of a spherical molecule in solution is given by

$$D = \frac{kT}{6\pi \eta r}, \tag{3.7}$$

where T is the temperature, r is the radius of a molecule in solution, η is the viscosity of the solution, and k is the Boltzmann constant. This equation indicates that diffusion is enhanced with increasing temperature but is reduced with increasing viscosity.

Solution hydrodynamics also play an important role in determining the dissolution rate. One possible mechanism by which solution hydrodynamics influences the dissolution rate is through their effect on the stationary diffusion layer around the drug molecule, as shown in (3.7). Since the thickness of this layer at the surface of the drug is determined by the shear force exerted by the fluid, an increase in the agitation (or stirring) rate may cause h to decrease, resulting in the improvement of drug dissolution. This hydrodynamic effect is demonstrated in the dissolution study of aspirin tablets, which shows that the dissolution half life of an aspirin tablet decreases with increasing agitation intensity (Levy *et al.*, 1965). In addition, Armenante and Muzzio studied the velocity and shear stress/strain distribution in the USP Apparatus II (paddle). Their result shows that the flow rate and shear rate vary significantly at different locations near the vessel bottom of the Apparatus II, thus resulting in different dissolution rates (Armenante and Muzzio, 2005).

3.5 Roles of Dissolution Testing

Dissolution testing plays many key roles in the development and production of solid dosage forms. At the early stage of the drug research and development (Phases 0 and 1), dissolution testing is used for active pharmaceutical ingredient

(API) characterization and formulation screening. It is also employed to develop and evaluate the performance of new formulations by examining drug release from dosage forms, evaluating the stability of these formulations, monitoring and assessing the formulation consistency and changes.

In addition to the use of dissolution testing in formulation optimization, process development and scale up during Phases II and III, appropriate dissolution methods are developed to obtain an *in vitro–in vivo* correlation (IVIVC) and other biorelevant information that will guide bioavailability and/or bioequivalence assessment of drug products.

For the release of drug products, dissolution testing serves as an important QC tool which is used to verify manufacturing and product consistency. It is also employed to evaluate the quality of the product during its shelf life, as well as to assess postapproval changes and examine the need for bioequivalence studies (FDA, 1997c).

Because of the diverse roles of dissolution testing in drug development and manufacturing, it is often preferable to develop a single dissolution test that can evaluate product quality and consistency, as well as predict *in vivo* performance. However, developing such a dissolution method remains a significant challenge. Under most circumstances, this goal is not achievable since dissolution tests used for QC and *in vivo* drug product performance assessment have very contrasting characteristics, which is discussed below. Under the current industry setting, the design of dissolution testing used for QC is primarily based upon the selection of discriminatory media, apparatus, and conditions that can be used routinely for QC purposes. Nevertheless, there is an increasing demand for the development of biorelevant dissolution methods that can provide some predictive estimates of the drug release with respect to the *in vivo* drug product performance. The remaining chapter will be devoted to the review of dissolution methods that are currently employed for QC and bioavailability assessment, as well as the discussion of scientific and regulatory issues associated with these two kinds of dissolution methods. Meanwhile, the significance of the BCS is also emphasized in relation to its use in the design of biorelevant testing.

3.6 *In Vitro* Dissolution Testing as a Quality Control Tool

The purpose of the dissolution test often dictates the choice of dissolution media. In principle, dissolution testing should be carried out under physiological conditions if possible, allowing interpretation of dissolution data with respect to the *in vivo* performance of a drug product. However, strict adherence to the GI environment is not necessary for routine dissolution testing. In fact, as mentioned previously, under the current setting of the pharmaceutical industry, the development of dissolution methods for QC focuses more on discriminatory capability, ruggedness, and stability. Particularly, as a QC and testing tool, it is critical to develop a dissolution method, which can consistently deliver a reliable test result and also assess drug product quality attributes (e.g., particle size, polymorphic form, or excipients) that are sensitive to formulation and manufacturing changes.

For QC, dissolution tests are developed and optimized to target and assess product attributes by monitoring their effect on the rate and extent to which the drug is released from the formulation. The design of a dissolution test used for QC is therefore often dictated by the physicochemical properties (particularly solubility) of a drug substance and its formulation. The details regarding QC dissolution testing of two solid dosage forms, immediate-release dosage forms and modified-release dosage forms are discussed below. In general, for QC purposes, the use of the simplest dissolution medium is preferred whenever possible, regardless of the dosage form.

3.6.1 Dissolution Method for Quality Control of Immediate-Release Dosage Forms

3.6.1.1 Dissolution Media

Aqueous test media are generally preferred (USP, 2004; FDA, 1997b). Although the design of a dissolution test used for QC is mainly based upon the physicochemical properties of the drug substance and the characteristics of the dosage form, it is important to select dissolution media to at least reflect the pH effect in the GI environment. For this reason, the pH of these media should be within the physiologic pH range of 1.2–6.8, where pH 1.2 and pH 6.8 represent the pH values under the gastric and intestinal conditions, respectively (FDA, 1997a). Hydrochloric acid, acetate, or phosphate buffer in the physiological pH range are commonly used and accepted as a dissolution medium for QC. The use of pure water in dissolution testing is usually not recommended primarily due to its limited buffering capacity. The volume of these dissolution media should be based upon the drug solubility, but is generally 500, 900, or 1,000 mL (FDA, 1997a). Sink conditions[2] are often recommended, since the dissolution tests used for QC are intended to provide conditions under which the majority of the drug (≥90%) can be released.

For some poorly soluble drugs that cannot dissolve adequately in aqueous solutions within the physiologic pH range, surfactants may be required to provide sink conditions and achieve a complete drug dissolution within reasonable time. The surfactants, such as sodium lauryl sulfate (SLS) and Tween, can be used to improve the dissolution rate by acting as a wetting agent and/or increasing the solubility of poorly soluble compounds through reduction of the interfacial tension and induction of micellar formation (Shah *et al.*, 1989; Sievert and Siewert, 1998). They may also be used to improve the correlation between the *in vitro* dissolution data and *in vivo* drug product performance (Brown *et al.*, 2004), as will be explained below. The level and solubilizing capacity of a surfactant are critical

[2] The term sink conditions is generally referred to as the condition where the medium volume is at least greater than three times that needed to form a saturated solution of a drug substance.

to QC. When the level and/or the solubilizing capacity of the surfactant is too high, the dissolution media may not be able to adequately discriminate differences among formulations, such as changes in the polymorphic form or particle size, as suggested in ICH Q6A. For hard and soft gelatin capsules as well as gelatin-coated tablets, a specific amount of enzyme(s) may be added to the dissolution medium to prevent pellicle formation.

3.6.1.2 Apparatus and Test Conditions

The most commonly used dissolution apparatus for solid oral dosage forms are the basket method (USP Apparatus I), the paddle method (USP Apparatus II), the reciprocating cylinder (USP Apparatus III) and the flow-through cell system (USP Apparatus IV). The first two apparatus are commonly used for dissolution testing of immediate-release dosage forms. The major advantage of these two devices is that they are simple, robust, and well standardized. The reciprocating cylinder apparatus has also been used for the dissolution testing of immediate-release products of highly soluble drugs, such as metoprolol and ranitidine, and some immediate-release products of poorly soluble drugs, such as acyclovir (Yu *et al.*, 2002). However, this apparatus should be considered only when the basket and paddle method are shown to be unsatisfactory. Due to the potential need for the large volume of medium, the flow-through cell system is not suitable for a dissolution test that is used routinely for the QC purpose. Nevertheless, the reciprocating cylinder device and the flow-through cell system may offer some advantages for their use in a biorelevant method, as will be discussed below.

For QC or drug product release testing, mild agitation conditions should be maintained during dissolution testing using the basket and paddle methods to allow maximum discriminatory power. If the rotational speed is too low, coning may occur, which leads to a low dissolution rate. However, if the rate of rotation is too fast, the test will not be able to discriminate the differences between the acceptable and not acceptable formulations or batches. The common stirring speed used for Apparatus I is 50–100 rpm, while with Apparatus II the common stirring speed is 50–75 rpm (FDA, 1997a). All dissolution tests should be performed at physiological temperature ($37 \pm 0.5\,^{\circ}C$). The test duration ranges from 15 min to 1 h.

3.6.2 Dissolution Method for Quality Control of Modified-Release Dosage Forms

3.6.2.1 Dissolution Media

The media used for modified-release dosage forms are generally the same as those used for immediate-release dosage forms. However, as opposed to the dissolution test used for immediate-release products that always uses one pH, more than one dissolution media with different pH values may be employed for testing of extended-release dosage forms to simulate the change in pH along the GI tract.

3.6.2.2 Apparatus and Test Conditions

The most common types of apparatus used for routine quality testing of extended-release products are the basket and paddle methods. The reciprocating cylinder may be used particularly for enteric-coated or extended-release dosage forms, when the pH of the medium needs to be changed in order to mimic the pH changes in the GI tract. The operating conditions for the basket and paddle methods are very similar to those used for immediate-release dosage forms, with an exception of the test duration, which can be as long as 12 h for extended-release products.

3.6.3 Limitations of Quality Control Dissolution Tests

There are some issues regarding the current use of dissolution tests that were developed for QC. Because these dissolution tests are developed to provide a maximum discriminatory power to assess any formulation changes and manufacturing process deviations, they are often overly discriminating, meaning that the differences detected by these dissolution tests may not have any clinical relevance. For instance, in the FDA-sponsored studies of metoprolol (Rekhi *et al.*, 1997), although the slow-dissolving tablets of metoprolol failed the USP dissolution test, the *in vivo* pharmacokinetic studies showed that all metoprolol tablets were bioequivalent with their corresponding formulations regardless of their *in vitro* dissolution rates. Thus, these clinically insignificant differences detected by the overly discriminating dissolution test often lead to the rejection of batches that may have an acceptable clinical performance. In addition, dissolution specifications, which are established based upon acceptable clinical, pivotal bioavailability, and/or bioequivalence batches using such overly discriminating dissolution tests, may not truly reflect the *in vivo* performance of a drug product. As a consequence, without a detailed knowledge on how dissolution affects the bioavailability of the drug product, these specifications are usually set to be very tight to assure the product quality and consistency by identifying any possible subtle changes in the product attributes before *in vivo* performance is affected. These shortcomings further facilitate the need for the development of biorelevant dissolution tests.

Dissolution tests used for QC can also be subjected to the limitation of being nondiscriminating. This limitation becomes evident if testing conditions are not selected appropriately (e.g., the agitation rate or surfactant level). This situation is best illustrated by the case of mebendazole (Swanepoel *et al.*, 2003). Mebendazole, which is a broad-spectrum anthelmintic drug, exists in three polymorphic forms (A, B, C) that display solubility and therapeutic differences. Among these three forms, polymorph C is therapeutically favored. Despite these differences, the three polymorphs produced similar dissolution profiles using a dissolution method that employed 0.1N HCl with 1% SLS. Specifically, all these dissolution profiles met the specification in which 75% of the drug dissolved within 120 min. It has been understood that the use of a large amount of SLS in the dissolution medium eliminates the differences in the dissolution rates of mebendazole polymorphs.

The precision and accuracy of dissolution testing are often very sensitive to several subtle operational controls. These include, but are not limited to the eccentricity of the agitating element, vibration, stirring element alignment, stirring rate, dosage form position, sampling probes, position, and filters. These factors may have a significant effect on the dissolution measurement if they are not controlled properly. For instance, the study of nondisintegrating double layered tablets containing salicylic acid indicates that the stirring rate and basket placement influence the drug dissolution in the basket apparatus (Howard *et al.*, 1979; Mauger *et al.*, 1979). In addition, the hydrodynamics in the paddle apparatus have been shown to be very complex and vary with site in the vessel (McCarthy *et al.*, 2004). Therefore, the exact location where the tablet lands after it is dropped into the vessel may have a considerable influence on the velocity profile around the tablet and hence its dissolution behavior.

3.7 Biorelevant Dissolution Testing

In order to achieve an adequate estimate of *in vivo* release behavior for solid dosage forms, the relevant physiological conditions, in addition to the physicochemical properties of a drug substance and its formulation, should be taken into serious consideration during the development of a biorelevant dissolution testing system. Specifically, the biorelevant dissolution method should be able to simulate the *in vivo* environment where the majority of the drug is released from the formulation. In principle, the design of such a system should, at minimum, account for the following factors to reflect the physiological conditions in the GI tract:

1. pH Conditions
2. Key aspects of the composition of the GI contents (e.g., osmolarity, ionic strength, surface tension, bile salts, and phospholipids)
3. Volume of the GI contents
4. Transit times
5. Motility pattern
6. Dosing conditions (e.g., administered with food)

To reflect the effect of these factors on the drug release, it is important to utilize the dissolution media that mimic the conditions in the GI tract and the apparatus that can simulate the dynamic environment that the dosage form experiences in the GI tract. Thus, in comparison to dissolution methods used for QC, which at best simulate pH effects and/or osmolality on the drug release under *in vivo* conditions, biorelevant dissolution media are generally more complex and are often not suitable for the purpose of QC.

Prior to the discussion of biorelevant dissolution methods, it is important to first review the concept and application of the *in vitro–in vivo* correlation and the importance of the BCS on the biorelevant dissolution testing development. The biorelevant dissolution media, apparatus and test conditions will be discussed with emphasis on their relevance to the physiological factors, including the pH, composition of the GI fluids, volume, GI hydrodynamics/motility, and food effect. The remaining challenges regarding the future development of biorelevant dissolution testing will also be highlighted.

3.7.1 *In Vivo–In Vitro Correlations*

The major objective of using biorelevant dissolution methods is to establish *in vivo–in vitro* correlation (IVIVC) so that *in vitro* dissolution data can be used to predict bioavailability. The term *in vivo–in vitro* correlation is defined as a predictive mathematical model describing the relationship between an *in vitro* property of a dosage form and a relevant *in vivo* response. In general, the physicochemical property or *in vitro* property of a dosage form is the *in vitro* dissolution profile. The biological property or *in vivo* response is the plasma concentration profile. Four correlation levels are defined in the FDA guidance (FDA, 1997b), as described below:

1. Level A: a point-to-point relationship between *in vitro* dissolution rate and *in vivo* input rate of the drug from the dosage form.
2. Level B: a comparison of the mean *in vitro* dissolution time to *in vivo* residence time or the mean *in vivo* dissolution time.
3. Level C: a single point relationship between a dissolution parameter ($t_{50\%}$, $t_{90\%}$, etc.).
4. Multiple-level C: a correlation that relates one or several pharmacokinetic parameters of interest to the amount of drug dissolved at several time points of the dissolution profile.

Among all levels of correlation, Level A is the most meaningful for predicting purposes, since it provides a relationship that directly links *in vivo* drug absorption to *in vitro* dissolution. This level of correlation should be valid for a reasonably wide range of values of formulation and manufacturing parameters that are essential for the drug release characteristics. This level can be used as a surrogate for *in vivo* performance of a drug product. Therefore, *in vitro* dissolution data, without any additional *in vivo* data, can be employed to justify a change made in manufacturing sites, raw material supplies, minor formulation modifications, strength of a dosage form, etc. However, the lower levels of correlation (B and C) are usually not very useful for regulatory purposes, and are used primarily for the development of formulation or processing procedures.

3.7.2 *The Importance of BCS on Biorelevant Dissolution Testing*

The BCS, which was proposed by Amidon *et al.* (1995), emphasizes the contribution of three fundamental factors, dissolution, solubility, and intestinal permeability, to the rate and extent of drug absorption for solid oral dosage forms. The BCS identifies three dimensionless numbers as key parameters including absorption number (An), dissolution number (Dn), and dose number (Do) to represent the effects of dissolution, solubility and intestinal permeability on the absorption process. These three dimensionless numbers are defined as:

$$An = \frac{P_{\text{eff}}}{R} t_{\text{res}}, \tag{3.8}$$

$$Dn = \frac{DC_{\text{s}}}{r_0} \frac{4\pi r_0^2}{\frac{4}{3}\pi r_0^3 \rho} t_{\text{res}}, \tag{3.9}$$

$$Do = \frac{M_0/V_0}{C_{\text{s}}}, \tag{3.10}$$

where P_{eff} is the effective permeability, r_0 is the initial particle radius, t_{res} is the mean residence time for the drug in the intestinal segment ($\pi R^2 L/Q_{\text{flow}}$, where R is the radius, L is the length of the segment, and Q_{flow} is the flow rate of fluid in the small intestine), D is the diffusion coefficient, ρ is the density, V_0 is the initial gastric volume, and M_0 is the amount of drug that is administered, and C_{s} is the saturated solubility.

According to the BCS, drug compounds are classified based upon their solubility and permeability described as follows:

Class I: High Permeability, High Solubility
Class II: High Permeability, Low Solubility
Class III: Low Permeability, High Solubility
Class IV: Low Permeability, Low Solubility

In this system, a compound is considered highly soluble when the highest dose strength is soluble in ≥250 mL[3] water over a range of pH from 1.0 to 7.5. For a highly permeable drug substance, the extent of absorption in humans is ≥90% of an administrated dose, based on mass-balance or in comparison to an intravenous reference dose. When ≥85% of the label amount of drug substance dissolves within 30 min using USP apparatus I (100 rpm) or II (50 rpm) in a volume of ≤900 mL in each of the following media: (1) 0.1N HCl or USP simulated gastric fluid (SGF) without enzymes, (2) a pH 4.5 buffer, and (3) a pH 6.8 buffer or USP simulated intestinal fluid (SIF) without enzymes, a corresponding drug product is considered to be rapidly dissolving.

Although the BCS has been developed primarily for regulatory applicants and particularly for oral immediate-release drug products, it has important implications in governing the dissolution test design during the drug development. Most importantly, it provides a general guideline for determining the conditions under which IVIVC is expected, as summarized in Table 3.1 (Amidon *et al.*, 1995). In other words, the BCS can provide an early insight into whether it is possible

[3] This volume is derived based on typical bioequivalence study protocols that prescribe administration of a drug product to fasting human volunteers with a glass of water (about 8 oz).

TABLE 3.1. *In Vitro–in vivo* correlation expectations for immediate-release products (Amidon *et al.* 1995)

Class	Solubility	Permeability	IVIVC expectation
I	High	High	IVIVC is expected if dissolution rate is slower than gastric emptying rate. Otherwise limited or no correlation is expected
II	Low	High	IVIVC is expected if *in vitro* dissolution rate is similar to *in vivo* dissolution rate, unless dose is very high
III	High	Low	Limited or no IVIVC is expected since absorption (permeability) is rate determining
IV	Low	Low	Limited or no IVIVC is expected

to develop a dissolution method capable of predicting *in vivo* drug absorption for immediate-release products, based primarily upon the solubility, permeability, and dissolution data.

BCS Class I compounds (e.g., metoprolol) have a high absorption number (*An*) and a high dissolution number (*Dn*), indicating that the rate determining step for drug absorption is likely to be dissolution or gastric emptying. This class of drugs is generally well absorbed if the drug is stable or does not undergo first pass metabolism. For immediate-release products of Class I compounds, the absorption rate is likely dominated by the gastric emptying time, and no direct correlation between *in vivo* data and *in vitro* dissolution data is expected. Thus, dissolution tests for such IR drug products should be designed mainly to confirm that the drug is released rapidly from the dosage form under the test conditions described above. A dissolution specification for which 85% of drug contained in the IR dosage form is dissolved in less than 15 min may be sufficient to ensure bioavailability, since the mean gastric half emptying time is 15–20 min (Amidon *et al.*, 1995; CDER/FDA, 1997). For BCS Class I drugs, which are formulated in extended-release dosage forms and have permeability that is site independent, dissolution becomes more important and IVIVC (e.g., level A) may be expected.

Class II drugs (e.g., phenytoin) have a high absorption number (*An*) and a low dissolution number (*Dn*). Dissolution is the rate limiting step for drug absorption. The influence of dissolution on absorption of BCS Class II drugs can be classified into two scenarios: solubility-limited absorption or dissolution-limited absorption (Yu, 1999). These two scenarios are best illustrated by grisefulvin and digoxin. In the case of solubility-limited absorption, grisefulvin exhibits a high dose number (*Do*) and a low dissolution number (*Dn*). Although in theory, absorption of grisefulvin can be improved by taking more water with the administered dose (decreasing *Do*), this approach is impractical due to the limitation in the physiological and anatomical capacity of the stomach for water. Thus, the only practical way to improve the absorption of grisefulvin is to decrease *Do* and increase *Dn* by enhancing its solubility through appropriate formulation approaches such as solid dispersion. On the other hand, in the case of dissolution-limited absorption, digoxin has a low dose number (*Do*) and a low dissolution number (*Dn*). Despite the small volume (21 mL) of fluids required to dissolve a typical dose of digoxin (0.5 mg), this drug dissolves too slowly for the absorption to take place at the site(s) of uptake. However, its dissolution rate can be improved simply by increasing *Dn* through the reduction in particle size. Thus, for BCS Class II drugs, a strong correlation between *in vitro* dissolution data and *in vivo* performance (e.g., Level A) is likely to be established. When a BCS Class II drug is formulated as an extended-release product, an IVIVC may also be expected.

For BCS Class III drugs (e.g., cimetidine), permeability is likely to be a dominant factor in determining the rate and extent of drug absorption. Hence, developing a dissolution test that can predict the *in vivo* performance of products containing these compounds is generally not possible. Since BCS Class IV drugs, which are low in both solubility and permeability, present significant problems for effective oral delivery, this class of drugs is generally more difficult to develop in comparison to BCS Class I, II, and III drugs.

In spite of its usefulness in the drug product development and regulatory recommendations regarding biowaivers for *in vivo* bioequivalence studies, the BCS also has its limitations. Drug instability in the GI tract, first pass metabolism, and complexation phenomena of drugs with the GI contents may have significant influence upon bioavailability, but are not addressed by the BCS. Furthermore, the BCS is often considered to be a conservative measure with regard to highly soluble drugs, since they are required to show high solubility across the range of pH from 1.2 to 7.5. It is important to note that the solubility of a weak acid and weak base depends on pH. The solubility of weak bases is generally higher in the stomach than in the small intestine. Therefore, a low solubility at high pH may not inhibit absorption of weak bases as the absorption may already be complete prior to entering the low solubility, high pH GI region. In contrast, low solubility at low pH may not present a problem for the absorption of weak acids since high solubility and high permeability in the small intestine are sufficient for their complete absorption.

3.7.3 Biorelevant Dissolution Methods

Unlike dissolution methods used for QC in which their design is primarily based upon drug substance physicochemical properties and formulation principles, biorelevant dissolution methods are designed to closely simulate physiological conditions in the GI tract. However, it should be noted that the physicochemical properties of the drug substance (e.g., solubility) and its formulation (e.g., immediate- or extended-release dosage forms) play a key role in selecting an appropriate type of biorelevant dissolution medium (e.g., gastric or intestinal medium), apparatus (e.g., a single vessel or multiple vessels), and test conditions (e.g., agitation speed and duration of a dissolution test), since these drug substance and formulation characteristics impact the location where the drug dissolution takes place in the GI tract. For instance, weak acids that are not soluble in the stomach (low pH) are usually very soluble in the small intestine (high pH).

In addition, for drugs that are unstable in an acidic method, a delayed-release formulation can be employed to ensure that the drug release occurs only in the high pH GI region (the small intestine).

3.7.3.1 Biorelevant Dissolution Media for Gastric Conditions

For BCS Class I drugs that are formulated in immediate-release dosage forms or any products that dissolve rapidly and completely in an acidic medium, it is logical to use a dissolution medium that reflects the gastric conditions. The minimum physiological parameters that need to be considered here include pH, surfactants, and enzymes. Food effects may also be considered if significant food effects are observed *in vivo*. It should be noted that biorelevant dissolution media or methods are designed primarily to mimic GI conditions in healthy subjects under the fasted and fed state, since *in vivo* bioequivalence studies are generally performed using these healthy subjects.

The pH in the stomach has a significant influence on the dissolution rate due to its effect on the solubility of a drug substance. In the fasted state of young healthy subjects, values of gastric pH are generally between 1.4 and 2.1 (Dressman *et al.*, 1990). However, the fasted state gastric pH values are found to be higher in subjects who are either over 65 years old or receiving gastric acid blocker therapy (Russell *et al.*, 1993; Christiansen, 1968). The gastric pH values also increase immediately following meal ingestion (pH 3–7) (Dressman *et al.*, 1998). The gastric pH resumes the fasted state values in approximately 2–3 h depending on the size and content of the meal (Dressman *et al.*, 1998).

The surface tension of gastric fluid is lower than that of water, and it was measured in the 35–50 mN m^{-1} range (Finholt and Solvang, 1968; Finholt *et al.*, 1978; Efentakis and Dressman, 1998). Although the decrease in surface tension suggests the presence of surfactants in the stomach, substances that lower the surface tension *in vivo* have not been identified unequivocally. The enzyme, pepsin, is also found to be in gastric fluid. The presence of this enzyme in the stomach causes a major problem for protein and polypeptide stability in addition to the acidity of the gastric environment.

Based upon the physiological factors described above, to simulate gastric conditions in the fasted state, the pH values of a gastric dissolution medium should be in the pH range of 1.5–2.5. In addition, surfactants, such as SLS, should be added into the medium to lower its surface tension close to the *in vivo* values. As mentioned earlier, for some capsules, an enzyme (pepsin) can be added to the medium to ensure timely dissolution of the shell by preventing pellicle formation. A sample composition for SGF in the fasted state is shown in Table 3.2 (Dressman, 2000). Due to its simplicity, this medium can also be used for QC dissolution testing.

To simulate the fed state in the stomach, the use of milk (Macheras *et al.*, 1987) and Ensure® (Ashby *et al.*, 1989) may be appropriate, since these media offer appropriate ratios of fat to protein and fat to carbohydrate. However, these two media are not suitable for routine quality assurance testing due to the difficulties in filtering and separating the drug substance from the medium for analysis.

TABLE 3.2. Sample composition for simulating gastric conditions in the fasted state (Klein 2005)

SGF composition	
Sodium chloride	0.6 g
Hydrochloric acid	2.1 g
Triton × 100	0.3 g
Deionized water	*qs ad* 300 mL

3.7.3.2 Apparatus and Test Conditions for Simulating the Stomach

The basket and paddle methods are frequently used, in conjunction with biorelevant media for gastric conditions, to simulate the drug release in the stomach under fasted and fed conditions. Since these two devices consist of a single vessel for each dosage form and are operated with a fixed volume of a single medium, they are best suited for drug products in which the majority of drug release occurs in the same section of the GI tract. Although the relationship between *in vivo* hydrodynamics or motility and rotational speed is still not well understood, the range of 50–100 rpm, which is established empirically, appears to give data that can be used to establish IVIVC.

In comparison to the basket and paddle methods, the reciprocating cylinder (USP Apparatus 3) and the flow-through cell (USP Apparatus 4) may offer some advantages regarding simulating gastric conditions. Apparatus 3, which originates from the official disintegration tester (Borst *et al.*, 1997), can be used to improve the study of food effects in the stomach by simulating changes in the composition and motility with time due to gastric secretion and digestion using a series of different media and agitation rates in the vessels. Similarly, Apparatus 4 also provides the possibility of changing the composition of a medium and flow rate during the test.

Regarding the volume of these gastric media, it depends on the volume of administered fluids and endogenous secretion. For instance, in the fasted state, gastric juice secretion is usually low. Therefore, by considering the quantity of fluid that is ingested with the dosage form, the medium volume should be in the order of 200–300 mL. The duration of a dissolution test should reflect the time available for dissolution in the stomach that is a function of the emptying pattern, which can vary considerably depending on the size of the solid particles as well as the size and the content of the meal (Meyer *et al.*, 1988; Moore *et al.*, 1981). If a drug formulated in an immediate-release dosage form is administered in the fasted state and is well absorbed from the upper small intestine, it is appropriate to run the dissolution test with SGF for 15–30 min. Nevertheless, there are still discrepancies in various pharmacopoeia regarding the duration of a dissolution test for immediate-release drug products (e.g., 30–120 min).

3.7.3.3 Biorelevant Dissolution Media for Intestinal Conditions

For poorly soluble drugs (e.g., BCS Class II drugs that are neutral or weak acids), it may be more appropriate to use a dissolution media that mimics the intestinal conditions. The dissolution of drug products in the small intestine is influenced by physiological factors including but not limited to pH, endogenous secretions from the pancreas and gall bladder (e.g., bile salts, lecithin, and digestion enzymes), and food effects. The physiological aspects related to drug absorption in the colon will not be addressed here and can be found elsewhere (Dressman *et al.*, 1997).

The pH values of intestinal conditions are considerably higher than those of gastric conditions, and were measured to be in the range of 5.5–6.0 for the duodenum, 6.5 in the jejunum, 7 in the proximal, and 7.5 in the distal ileum (Dressman *et al.*, 1998). The high pH values in the small intestine are attributed to the neutralization effect of bicarbonate ion secreted by the pancreas. It should be noted that pH values gradually increase from the duodenum to the ileum, resulting in a pH gradient in the small intestine. In the fed state, the pH values in the duodenum (4.2–6.1), jejunum (5.2–6.2), and ileum (6.8–7.8) are generally lower than those in the fasted state (Dressman *et al.*, 1990; Ovesen *et al.*, 1986; Fordtran and Locklear, 1966).

In the small intestine, secretion of bile from the gallbladder in the duodenum leads to a high concentration of bile salts and phospholipids (lecithin), resulting in the formation of mixed micelles even in the fasted state. These bile salts and lecithin may have a significant enhancing effect upon the dissolution rate of poorly soluble drugs by improving the wettability of solids and by increasing the solubility of a drug substance into mixed micelles (Mithani *et al.*, 1996). As for gastric secretion, the rate of bile secretion also depends strongly on the prandial state, in which the concentration of bile salts and lecithin further increases in the presence of food. Since the majority of bile salts ($>90\%$) is reabsorbed by the active transport mechanism (Davenport, 1982), a decreasing gradient of bile salts is observed along the small intestine. The digestion enzymes, such as lipases, peptidases, amylases, and proteases, are also secreted by the pancreas in the small intestine in response to food ingestion. Lipases and peptidases present a stability problem for some drugs and hence may influence the dissolution process.

A commonly used medium for simulating fasting conditions in the proximal small intestine is fasted state simulated intestinal fluid (FaSSIF). As evidenced in the above discussion, one apparent difference between the SGF and FaSSIF is that this simulated intestinal medium contains bile salts and lecithin. Thus, the dissolution rate of poorly soluble, lipophilic drugs may be improved greatly in this medium in comparison to the dissolution rate observed in simple aqueous solutions. The composition of this medium is given in Table 3.3 (Dressman, 2000) and it was based on experimental data in dogs and humans for the concentration of bile components, pH value, buffer capacity, and osmolality (Greenwood, 1994). The pH value was chosen to be 6.5, which closely resembles the values measured from the midduodenum to the proximal ileum. Sodium taurocholate was often used as a representative bile salt since cholic acid is one of the more common bile salts in human bile (Carey and Small, 1972). In addition, because the p*K*a of

TABLE 3.3. Sample composition for simulating the fasted state conditions in the small intestine (note that the recommended volume for dissolution studies is 1 L) (Klein 2005)

FaSSIF composition	
Sodium taurocholate	3 mM
Lecithin	0.75 mM
NaH_2PO_4	3.9 g
KCl	7.7 g
NaOH	*qs ad* pH 6.5
Deionized water	*qs ad* 1 L

TABLE 3.4. Sample composition for simulating the fed state conditions in the small intestine (note that the recommended volume for dissolution studies is 1 L) (Klein 2005)

FeSSIF composition	
Sodium taurocholate	15 mM
Lecithin	3.75 mM
Acetic acid	8.65 g
KCl	15.2 g
NaOH	*qs ad* pH 5.0
Deionized water	*qs ad* 1 L

taurine conjugate is very low, precipitation and an alteration in the micellar size with small variations in pH values are unlikely to occur within the pH range in the proximal small intestine (pH 4.2–7). The ratio of phospholipids to bile salts employed in these media is approximately 1:3, which reflects the *in vivo* ratio that is generally found to be between 1:2 and 1:5 (Dressman *et al.*, 1998).

In comparison to the fasted state, a dissolution medium simulating intestinal conditions in the fed state should assume a lower pH value, higher buffer capacity, and osmolarity (Greenwood, 1994). In addition, as described earlier, lipids in food further simulate the release of bile salts and phospholipids, which certainly have major effects on the dissolution rate of the drug. Most of these factors should be taken into consideration during the development of a dissolution medium for simulating the proximal small intestinal conditions in the fed state. The sample composition of the fed state simulating intestinal fluid (FeSSIF) is given in Table 3.4 (Dressman, 2000). It should be noted in Table 3.4 that an acetic buffer is used here instead of the phosphate buffer to achieve the higher capacity and osmolarity while maintaining the lower pH value, and that taurocholate and lecithin are present in considerably higher concentrations than those in the fasted state medium.

3.7.3.4 Apparatus and Test Conditions for Simulating Small Intestine

Using biorelevant media that mimic intestinal conditions (e.g., FaSSIF and FeSSIF), the basket and paddle methods can also be employed to study the drug release in the small intestine. The advantages and disadvantages of these two apparatus used for simulating intestinal conditions are similar to those used for simulating gastric conditions. Since the relationship between in vivo hydrodynamics (or motility) and rotational speed is not known, the agitation rate (50–100 rpm) is once again determined empirically to give data that provide the best IVIVC.

With the possibility of varying the composition of media and the agitation rate (or the flow rate), both the reciprocating cylinder and flow-through cell systems can be used to simulate the pH and composition changes from the duodenum to the ileum. Furthermore, the flow-through cell system can be operated as an open system, allowing removal of dissolved drugs and hence providing sink conditions for poorly soluble drugs to mimic conditions in the small intestine. However, the open system mode requires a large volume of media. Therefore, its practical use is severely limited to product development, especially when biorelevant media are used.

With regard to the volume of the fasted state simulated intestinal medium, pharmacokinetic studies in the fasted state show that by ingesting 200–250 mL of water with the dosage form, a total volume of 300–500 mL will become available in the proximal small intestine. Based upon this evidence, a volume of 500 mL is recommended for the FaSSIF. The total volume of the fed state simulated intestinal medium should take into consideration the volume of coadministered fluid, the volume of fluid ingested meal, and the secretions of the stomach, pancreas, and bile (Fordtran and Locklear, 1966). As a result, in comparison to FeSSIF, a larger volume (up to 1 L) is generally required for dissolution testing using FeSSIF.

3.7.3.5 Biorelevant Methods for Extended-Release Dosage Forms

For extended-release drug products, the dissolution method must capture, at minimum, the changes in composition, pH, and residence times along the GI tract, since absorption of these dosage forms takes place throughout the entire intestine. Thus, the reciprocating cylinder and flow-through cell systems can be used, in conjunction with different biorelevant dissolution media, to assess the *in vivo* release behavior of extended-release dosage forms.

3.7.3.6 Remaining Challenges

Currently, similar to the dissolution test used for QC, the biorelevant dissolution method is generally drug product specific. In other words, no universal biorelevant methods have been devised. If the same drug is formulated differently, even a subtle difference in the formulation may require the development of different *in vitro* dissolution methodology in order to obtain an IVIVC. Although some progress has been made in understanding the GI tract environment (e.g., pH, composition, and volume), the establishment of IVIVC is still primarily based upon a trial and error approach (Zhang and Yu, 2004). Furthermore, none of the dissolution media, apparatus, and test conditions described previously for gastric and intestinal conditions reflects all physiological parameters that are important for determining the effects of composition, food, motility patterns, and transit times on drug release in the stomach and small intestine. In addition, transient changes in composition, motility, and volume in both the fasted and fed states are not fully captured by the current biorelevant dissolution methods.

Therefore, developing biorelevant methods that truly capture the drug release behavior under *in vivo* conditions remains extremely challenging, since the physiological environment of the GI tract is still not fully understood. For instance, despite the fact that the hydrodynamics in the GI tract are known to play an important role in dissolution, they have not been studied in detail. Thus, to devise such dissolution methods, we must seek a complete understanding of how all the key factors such as composition, hydrodynamics, volume, and transit times affect the dissolution of drugs in the GI tract. We can then utilize this knowledge in the design of biorelevant dissolution testing.

3.8 Conclusions

Dissolution testing is critical to the drug product development and production. It is routinely used in QC as well as research and development. The objective of dissolution testing in QC is to assure batch to batch consistency and detect manufacturing deviations. In research and development, dissolution testing is used to evaluate the performance of new formulations by measuring the rate of drug release from dosage forms, examining the stability of these formulations, and assessing formulation changes. More importantly, dissolution testing is employed to provide some predictive estimates of the drug release under physiological conditions by establishing IVIVCs. For QC, dissolution tests are developed and optimized to target and assess specific product properties (e.g., particle size and excipient composition) by monitoring their effects on the rate and extent to which the drug is released from the formulation. The design of a dissolution test used for QC is, therefore, dependent on the drug substance physicochemical properties (e.g., solubility) and formulation principles (e.g., extended-release dosage forms). On the other hand, in addition to the physicochemical properties of the drug substance and formulation characteristics, physiological factors also play an important role in the design of biorelevant dissolution methods, since these methods are developed mainly to simulate relevant conditions where the drug is being released from the formulation in the GI tract. Although progress has been made in developing dissolution media that reflect gastric (e.g., SGF) and intestinal conditions (e.g., FaSSIF), developing dissolution methods, which reflect all physiological parameters (e.g., motility pattern and transit times) that may influence the drug release behavior in the GI tract, remains as a significant challenge.

References

Amidon, G.L., Lennernäs, H., Shah, V.P., and Crison, J.R. (1995). A theoretical basis for a biopharmaceutics drug classification: the correlation of in vivo drug product dissolution and in vivo bioavailability. *Pharm. Res.* 12:413–420.

Armenante, P., and Muzzio, F. (2005). Inherent method variability in dissolution testing: the effect of hydrodynamics in the USP II Apparatus. *A Technical Report Submitted to the Food and Drug Administration.*

Ashby, L.J., Beezer, A.E., and Buckton, G. (1989). In vitro dissolution testing of oral controlled release preparations in the presence of artificial foodstuffs. I. Exploration of alternative methodology: microcalorimetry. *Int. J. Pharm.* 51:245–251.

Bandelin, F.J. (1990). Compressed Tablets by Wet Granulation. In: Lieberman, H.A., Lachman, L., and Schwartz, J.B. (eds.), *Pharmaceutical Dosage Forms: Tablets*, Volume 1. Marcel Dekker, Inc., New York, pp. 199–302.

Brown, C.K., Chokshi, H.P., Nickerson, B., Reed, R.A., Rohrs, B.R., and Shah, P.A. (2004). Acceptable analytical practices for dissolution testing of poorly soluble compounds. *Pharm. Tech.* 56–65.

Brunner, E. (1904). Reaktionsgeschwindigkeit in heterogenen Systemen. *Z. Phys. Chem.* 43:56–102.

Brunner, L., and Tolloczko, S. (1900). Über die Auflösungsgeschwindingkeit fester Körper. *Z. Physiol. Chem.* 35:283–290.

Borst, I., Ugwu, S., and Beckett, A.H. (1997). New and extended application for USP drug release Apparatus 3. *Dissolut. Technol.* 1–6.

Carey, M.C., and Small, D.M. (1972). Micelle formation by bile salts. Physical–chemical and thermodynamic considerations. *Arch. Intern. Med.* 130:506–527.

Christiansen, P. (1968). The incidence of achlorhydria in healthy subjects and patients with gastrointestinal diseases. *Scan. J. Gastroenterol.* 3:497–508.

Corrigan, O.I. (1985). Mechanisms of dissolution of fast release solid dispersions. *Drug Dev. Ind. Pharm.* 11:697–724.

Danckwerts, P.V. (1951). Significance of liquid-film coefficients in gas absorption. *Ind. Eng. Chem.* 43:1460–1467.

Davenport, H.W. (1982). *Physiology of the Digestive Tract*, 5th ed. Year Book Medical Publishers, Inc., Chicago.

Dressman, J.B. (2000). Dissolution testing of immediate-release products and its application to forecasting in vivo performance. In: Dressman, J.B., and Lennernäs, H. (eds.), *Oral Drug Absorption: Prediction and Assessment*, Volume 106. Marcel Dekker, Inc., New York, pp. 155–181.

Dressman, J.B., Berardi, R.R., Dermentzoglou, L.C., Russell, T.L., Schmaltz, S.P., Barnett, J.L., and Jarvenpaa, K.M. (1990). Upper gastrointestinal (GI) pH in young healthy men and women. *Pharm. Res.* 7:756–761.

Dressman, J.B., Amidon, G.L., Reppas, C., and Shah, V.P. (1998). Dissolution testing as a prognostic tool for oral drug absorption: immediate release dosage forms. *Pharm. Res.* 15:11–22.

Efentakis, M., and Dressman, J.B. (1998). Gastric juice as a dissolution medium: surface tension and pH. *Eur. J. Drug Metab. Pharmacokinet.* 23:97–102.

FDA (1997a). Center for Drug Evaluation and Research, *Guidance for industry*. Dissolution testing of immediate release solid oral dosage forms.

FDA (1997b). Center for Drug Evaluation and Research, *Guidance for industry*. Extended release oral dosage forms: development, evaluation, and application of in vitro/in vivo correlations.

FDA (2000). Center for Drug Evaluation and Research, *Guidance for industry*. Waiver of in vivo bioavailability and bioequivalence studies for immediate-release solid oral dosage forms based on a biopharmaceutics classification system.

Finholt, P. (1974). Influence of formulation on dissolution rate. In Leeson, L.J., and Carstensen, J.T. (eds.), *Dissolution Technology*, American Pharmaceutical Association, Washington, DC, pp. 106–146.

Finholt, P., and Solvang, S. (1968). Dissolution kinetics of drugs in human gastric juice – the role of surface tension. *J. Pharm. Sci.* 57:1322–1326.

Finholt, P., Gundersen, H., Smit, A., and Petersen, H. (1978). Surfactant tension of human gastric juice. *Medd. Norsk. Farm. Selsk.* 41:1–14.

Fordtran, J.S., and Locklear, T.W. (1966). Ionic constituents and osmolality of gastric and small-intestinal fluids after eating. *Am. J. Dig. Dis.* 11:503–521.

Fraser, E.J., Leach, R.H., and Poston, J.W. (1972). Bioavailability of digoxin. *Lancet* 2:541.

Gandhi, R., Lal Kaul, C., and Panchagnula, R. (1999). Extrusion and spheronization in the development of oral controlled-release dosage forms. *Pharm. Sci. Technol. Today* 4:160–170.

Gordon, M.S., and Rudraraju, V.S. (1993). Effect of the mode super disintegrant incorporation on dissolution in wet granulated tablets. *J. Pharm. Sci.* 82:220–226.

Greenwood, D.E. (1994). Small intestinal pH and buffer capacity: implication for dissolution of ionizable compounds. *Doctoral Dissertation*. The University of Michigan, Ann Arbor.

Hanson, W.A. (1982). *Handbook of Dissolution Testing*, Volume 49. Pharmaceutical Technology Publications, Oregon.

Higuchi, T. (1961). Rate of release of medicaments from ointment bases containing drugs in suspension. *J. Pharm. Sci.* 50:874–875.

Howard, S.A., Mauger, J.W., Khwangsopha, A., and Pasquerelli, D.A. (1979). Tablet position and basket type effects in spin-filter dissolution device. *J. Pharm. Sci.* 68: 1542–1545.

Klein, S. (2005). Biorelevant dissolution test methods for modified release dosage forms. *Doctoral Thesis*. Johann Wolfgang Goethe University Frankfurt, Shaker-Verlag.

Levy, G., Leonards, J.R., and Procknal, J.A. (1965). Development of in vitro dissolution tests which correlate quantitatively with dissolution rate-limited drug absorption in man. *J. Pharm. Sci.* 54:1719–1722.

Loftsson, T., Hreinsdóttir, D., and Másson, M. (2005). Evaluation of cyclodextrin solubilization of drugs. *Inter. J. Pharm.* 302:18–28.

Macheras, P., Koupparis, M., and Apostelelli, E. (1987). Dissolution of four controlled-release theophylline formulations in milk. *Int. J. Pharm.* 36:73–79.

Mauger, J.W., Howard, S.A., and Khwangsopha, A. (1979). Hydrodynamic characterization of a spin-filter dissolution device. *J. Pharm. Sci.* 68: 1084–1087.

McCarthy, L.G., Bradley, G., Sexton, J.C., Corrigan, O.I., and Healy, A.M. (2004). Computational fluid dynamics modeling of the paddle dissolution apparatus: agitation rate, mixing patterns, and fluid velocities. *AAPS PharmSciTech.* 5:1–10.

Meyer, J.H., Elashoff, J., Porter-Fink, V., Dressman, J., and Amidon, G.L. (1988). Human postprandial gastric emptying of 1–3 millimeter spheres. *Gastroenterology* 94: 1315–1325.

Miller, D.A., McConville, J.T., Yang, W., Williams III, R.O., and McGinity J.W. (2006). Hot-melt extrusion for enhanced delivery of drug particles. *J. Pharm. Sci.* 96:361–376.
Mithani, S.D., Bakatselou, V., TenHoor, C.N., and Dressman, J.B. (1996). Estimation of the increase in solubility of drugs as a function of bile salt concentration. *Pharm. Res.* 13:163–167.
Moore, J.G., Christian, P.E., and Coleman, R.E. (1981). Gastric emptying of varying meal weight and composition in man. Evaluation by dual liquid- and solid-phase isotopic method. *Dig. Dis. Sci.* 26:16–22.
Nelson, E. (1957). Solution rate of theophylline salts and effects from oral administration. *J. Am. Pharm. Assoc.* 46:607–614.
Nernst, W. (1904). Theorie der reaktionsgeschwindigkeit in heterogenen systemen. *Z. Physiol. Chem.* 47:52–55.
Ovesen, L., Bendtsen, F., Tage-Jensen, U., Pedersen, N.T., Gram, B.R., and Rune, S.L. (1986). Intraluminal pH in the stomach, duodenum, and proximal jejunum in normal subjects and patients with exocrine pancreatic insufficiency. *Gastroenterology* 90: 958–962.
Peck, G.E., Baley, G.J., McCurdy, V.E., and Banker, G.S. (1989). Tablet Formulation and Design. In: Lieberman, H.A., Lachman, L., and Schwartz, J.B. (eds.), *Pharmaceutical Dosage Forms: Tablets*, Volume 1. Marcel Dekker, Inc., New York, pp. 73–130.
Pinnamaneni, S., Das, N.G., and Das, S.K. (2002). Formulation approaches for orally administered poorly soluble drugs. *Pharmazie* 57:291–300.
Rekhi, G.S., Eddington, N.D., Fossler, M.J., Schwartz, P., Lesko, L.J., and Augsburger, L.L. (1997). Evaluation of in vitro release rate and in vivo absorption characteristics of four metoprolol tartrate immediate-release tablet formulations. *Pharm. Dev. Technol.* 2:11–24.
Russell, T.L., Berardi, R.R., Barnett, J.L., Dermentzoglou, L.C., Jarvenpaa, K.M., Schmaltz, S.P., and Dressman, J.B. (1993). Upper gastrointestinal pH in seventy-nine healthy, elderly North American men and women. *Pharm. Res.* 10:187–196.
Schott, H., Kwan, L.C., and Feldman, S. (1982). The role of surfactants in the release of very slightly soluble drugs from tablets. *J. Pharm. Sci.* 71:1038–1045.
Shah, V.P., Konecny, J.J., Everett, R.L., McCullough, B., Noorizadeh, A.C., and Skelly, J.P. (1989). In vitro dissolution profile of water-insoluble drug dosage forms in the presence of surfactants. *Pharm. Res.* 6:12–18.
Shangraw, R. F. (1990). Compressed Tablets by Direct Compression. In: Lieberman, H.A., Lachman, L., and Schwartz, J.B. (eds.), *Pharmaceutical Dosage Forms: Tablets*, Volume 1, Marcel Dekker, Inc., New York, pp. 195–246.
Sievert, B., and Siewert, M. (1998). Dissolution tests for ER products. *Dissolut. Technol.* 5:1–7.
Stavchansky, R., and McGinity, J. (1989). Bioavailability and Tablet Technology. In: Lieberman, H.A., Lachman, L., and Schwartz, J.B. (eds.), *Pharmaceutical Dosage Forms*: *Tablets*, Volume 2, Marcel Dekker, Inc., New York, pp. 349–553.
Swanepoel, E., Liebenberg, W., and de Villiers, M.M. (2003). Quality evaluation of generic drugs by dissolution test: changing the USP dissolution medium to distinguish between active and non-active mebendazole polymorphs. *Eur. J. Pharm. Biopharm.* 55:345–349.
Wagner, J.G. (1970). Rate of dissolution in vivo and in vitro, part II. *Drug Intell. Clin. Pharm.* 4:32.
Yu, L.X. (1999). An integrated absorption model for determining causes of poor oral drug absorption, *Pharm. Res.* 16:1883–1887.

Yu, L.X., and Amidon, G.L. (1999). Analytical solutions to mass transfer. In Amidon, G.L., Lee, P.I., and Topp, E.M. (eds.), *Transport Processes in Pharmaceutical Systems*, Volume 102, Marcel Dekker, Inc., New York, pp. 23–54.

Yu, L.X., Wang, J.T., and Hussain, A.S. (2002). Evaluation of USP Apparatus 3 for Dissolution Testing of Immediate-Release Products. *AAPS PharmSci*. 4:1–5.

Zhang, H., and Yu, L.X. (2004). Dissolution testing for solid oral drug products: theoretical consideration. *Am. Pharm. Rev*. 26–31.

4 Drug Absorption Principles

Xianhua Cao, Lawrence X. Yu, and Duxin Sun

4.1 Drug Absorption and Bioavailability

Pharmacokinetics describes drug absorption, distribution, metabolism, and excretion processes. Absorption is the rate and extent at which drugs reach the systemic circulation from the site of administration. Distribution of a drug includes all the processes that are involved from the time when the drug reaches the circulation to the time when it (or a metabolite of the drug) leaves the body. Metabolism involves all the biochemical processes that result in a chemical change to the drug compound including both the metabolism in the gut wall, the liver, and blood circulation. Excretion is the process in which the drug is eliminated from the systemic circulation into bile, urine, feces, sweat, and air (Allen, 1982). The reader is referred to authoritative texts in this area for a detailed review.

Bioavailability means the rate and extent to which the API or active moiety is absorbed from a drug product and becomes available at the site of action (Atkinson, 2001; Chiou, 2001; Toutain and Bousquet-Melou, 2004). Drug absorption plays an important role in bioavailability (F) determination since the drug absorption contributes importantly to the time and extent that drug targets exposure to therapeutic drugs *in vivo*. For drug products that are not intended to be absorbed into the bloodstream, bioavailability may be assessed by the measurements intended to reflect the rate and extent to which active ingredient or active moiety becomes available at the site of action. Bioavailability can be mathematically represented by the equation: $F = F_a \times F_g \times F_h$, in which F_a is the fraction of drug absorbed, F_g is the fraction that escapes metabolism in the gastrointestinal tract, and F_h is the fraction that escapes first pass hepatic metabolism (Kwan, 1997; Sun *et al.*, 2004). Based on the above equation, one of the main factors governing the bioavailability of a compound is the fraction of the drug absorbed.

Oral drug absorption process occurs mainly in small intestinal regions, which includes passive transcellular diffusion, carrier-mediated transport processes, paracellular transport, and endocytosis. In general, lipophilic compounds are usually absorbed by passive diffusion through the intestinal epithelium. Many hydrophilic compounds are absorbed through a carrier-mediated process, while some small hydrophilic compounds may be transported through the paracellular junction. Under physiological conditions, the fastest absorption process may dominate the absorption for a particular compound (Sun *et al.*, 2004; Cao *et al.*, 2005).

Absorption of a compound is governed by many processes. Two fundamental parameters govern drug absorption: drug solubility and gastrointestinal permeability (Amidon *et al.*, 1995). If both drug solubility and permeability are enhanced, there will be a great increase in the rate and extent of oral absorption. Therefore, the oral bioavailability of a drug is largely a function of its solubility characteristics in gastrointestinal fluids, absorption into the systemic circulation, and metabolic stability.

4.2 Types of Intestinal Membrane Transport

Intestinal membrane transport include paracellular and transcellular transport (Fig. 4.1). Transcellular transport can be further divided into passive diffusion, endocytosis, and carrier-mediated transport. Paracellular transport refers to the passage of solute without passage through the epithelium cells (Higuchi and Ho, 1988; Narawane and Lee, 1994; Ungell *et al.*, 1998; Oh *et al.*, 1999).

4.2.1 Passive Diffusion

Hydrophobic molecules can pass through the lipid bilayers by random molecular motions. The direction of mass transfer of molecules or substances by passive diffusion depends on the concentration gradient on the two sides of the membrane. Lipophilic compounds are generally absorbed by passive diffusion through the

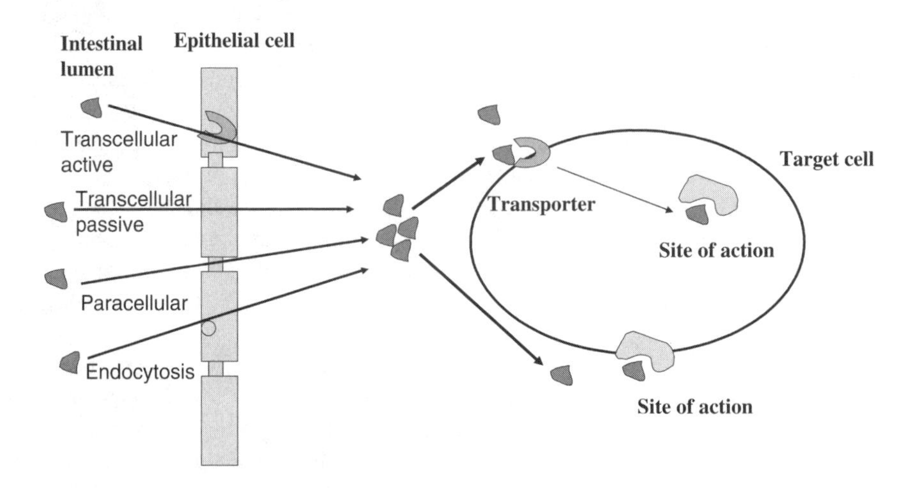

FIGURE 4.1. Drug transport and site of action (*See* Color Plate I)

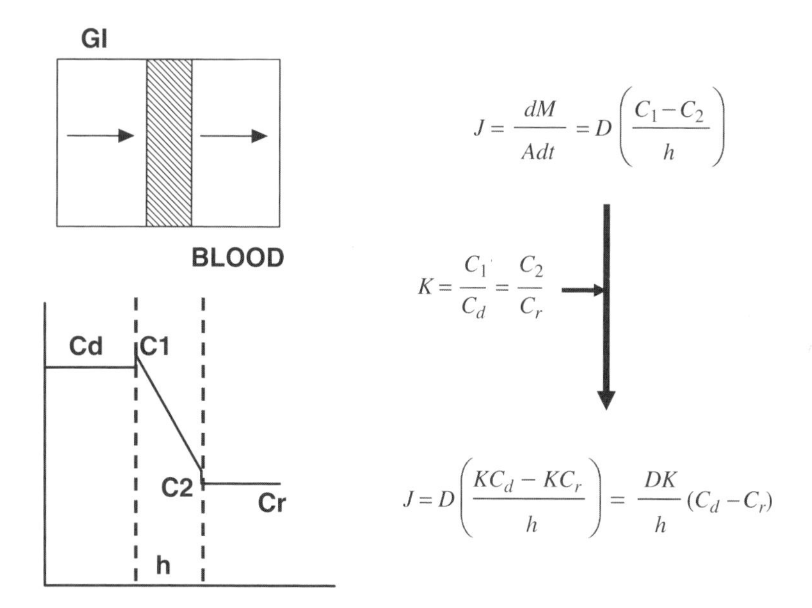

FIGURE 4.2. Fick's first law of diffusion

intestinal epithelium. The passive diffusion of the molecules is governed by Fick's first law (Lennernas, 1998; Yoon and Burgess, 1998; Chidambaram and Burgess, 2000).

Fick's first law of diffusion (Fig. 4.2)

$$J = \frac{\mathrm{d}M}{A\,\mathrm{d}t} = \frac{D(C_1 - C_2)}{h}, \tag{4.1}$$

where J is the flux (amount of material flowing through a unit cross section); M, the drug mass (g, mol); A, the surface area (cm^2); t, the time (s); D, the diffusion coefficient (diffusivity, cm^2 s^{-1}) ; C_1, the drug concentration at membrane wall in intestinal lumen (mol l^{-1}); C_2, the drug concentration at membrane wall in blood side (mol l^{-1}); and h is the membrane thickness (cm).

The assumptions made by this model are the following: (1) steady state flux. The transfer of drugs reaches to steady state very fast and (2) the steady state follows sink conditions: both sides of the membrane are well stirred and homogenous.

Define partition coefficient K as $K = C_1/C_\mathrm{d} = C_2/C_\mathrm{r}$ (C_d as drug concentration in the gastric intestinal (GI) lumen, and C_r is the drug concentration in the blood), we can get (4.2)

$$J = D\left(\frac{KC_\mathrm{d} - KC_\mathrm{r}}{h}\right) = \frac{DK}{h}(C_\mathrm{d} - C_\mathrm{r}). \tag{4.2}$$

If $C_\mathrm{d} \gg C_\mathrm{r}$, then

$$J = \frac{DK}{h}C_\mathrm{d}. \tag{4.3}$$

Define permeability coefficient P as $P = DK/h$ (unit cm s^{-1}), then

$$J = PC.$$

Consider the absorptive surface area, we can get the final (4.4)

$$\mathrm{d}M/\mathrm{d}t = PA(C_\mathrm{d} - C_\mathrm{r}). \tag{4.4}$$

4.2.2 Carrier-Mediated Transport

Intestinal epithelial cell membranes are highly polarized. Apical membrane faces the external lumen with many microvilli to increase the membrane surface area. Many membrane transporters are located in this side facilitating absorption for most nutrients and many drugs, while basolateral membrane is toward blood (Rouge *et al.*, 1996; Shin *et al.*, 2003; Anderle *et al.*, 2004) (Fig. 4.3).

Depending on the direction and category of transported solutes, drug carrier to mediate transport can also be classified into uniporter, symporter, and antiporter. Uniporter is the carrier-mediated transport with single solute; symporter facilitates the transport of two solutes with same direction, while antiporter facilitates the transport of two solutes with opposite directions. Based on the concentration gradient of the solutes and energy involved in the process, drug carrier can be classified into facilitated diffusion and active transport.

4.2.2.1 Facilitated Diffusion

Carrier proteins are involved in facilitated diffusion. This process does not need energy. Similar to passive diffusion, transport direction of facilitated diffusion depends on the solutes concentration gradient (from higher concentration to lower

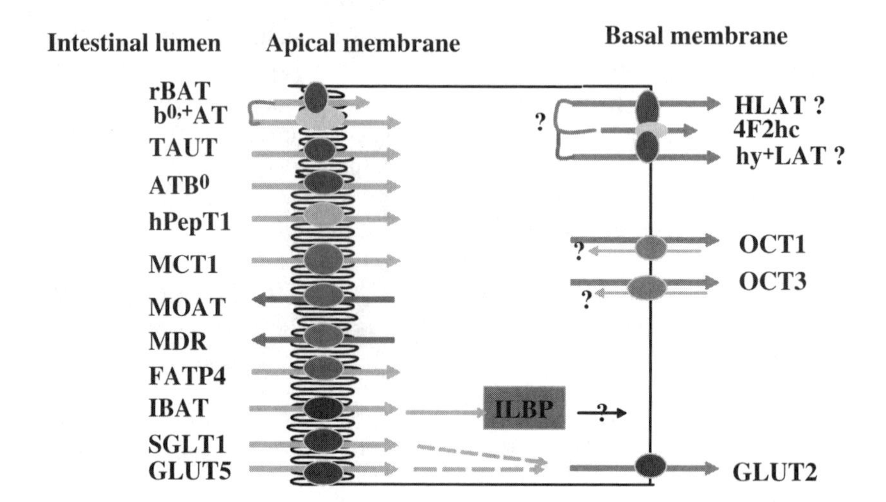

FIGURE 4.3. Apical and basolateral transporters coupling for absorption (*See* Color Plate II)

concentration) (Cainelli *et al.*, 1974; Feher, 1983). However, facilitated diffusion has higher transport rate than what would be expected from passive diffusion alone.

4.2.2.2 Active Transport

Active transport is the primary mode by which molecules are transported against electrical and/or chemical concentration gradients. The process involves a membrane bound protein molecule that binds reversibly to the solute molecule at a specific site. The complex then undergoes a change in conformation that translocates the solute to the other side of the membrane. Factors that can affect this transport include energy, temperature, and stereospecificity of the molecule. Similar to enzyme kinetics, active transport also exhibits saturable kinetics and can be inhibited by similar structural analogs.

$$J = \frac{J_{\mathrm{max}} C}{K_{\mathrm{m}} + C}, \tag{4.5}$$

where J is the drug flux (mg s^{-1}); J_{max}, the maximum drug flux; C, the drug concentration (mg ml^{-1}); and K_{m}, the drug affinity to carrier (mg ml^{-1}).

At low concentration, $C \ll K_{\mathrm{m}}$, first order absorption prevails

$$J = \frac{J_{\mathrm{max}}}{K_{\mathrm{m}}} C. \tag{4.6}$$

At high concentration, $C \gg K_{\mathrm{m}}$, zero-order absorption prevails

$$J = J_{\mathrm{max}}. \tag{4.7}$$

In contrast to passive diffusion, drugs with active transport absorption mechanism may have a concentration dependent and/or dose-dependent absorption (Fig. 4.4). Drug flux can be competitively inhibited by other substrates.

4.2.3 *Paracellular Transport*

Paracellular transport refers to transport solutes in between cells, without passage through the epithelial cells themselves. It is now well recognized that the intercellular junctions between epithelial cells of capillaries are "leaky," allowing paracellular transport of small molecules (Daugherty and Mrsny, 1999; Trischitta *et al.*, 2001). Paracellular transport is passive transport, follows drug concentration gradients, and does not require energy.

4.2.4 *Endocytosis*

Endocytosis is a process in which a substance or compound gains entry into a cell without passing through the lipid cell membrane. Based on the mechanisms and molecules involved, this process can be subdivided into different types: pinocytosis, phagocytosis, and receptor-mediated endocytosis. In each case, endocytosis

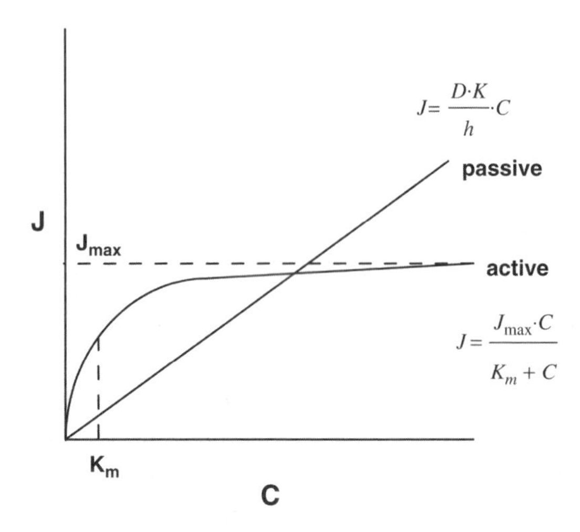

FIGURE 4.4. Active transport shows nonlinear pharmacokinetics

results in the formation of an intracellular vesicle by the invagination of the plasma membrane and membrane fusion. Drug molecules can be transported into the cells by this process (Hansen *et al.*, 2005; Liang *et al.*, 2006).

4.2.5 Which Absorption Path Dominates Drug Absorption?

Although different mechanisms of oral drug absorption have been shown in small intestinal regions, under physiological conditions, several routes may contribute to drug absorption at the same time. Usually, the fastest route dominates the absorption of a particular compound (Burton *et al.*, 2002; Cao *et al.*, 2005). In general, passive diffusion is the main mechanism for absorption of many lipophilic compounds, while the carrier-mediated process governs the absorption of transporter substrates. In some cases, paracellular junction is the route for the absorption of some small hydrophilic compounds with molecular weight less than 300.

4.3 Three Primary Factors Influence Drug Absorption

Permeability, solubility, and dissolution are the three primary factors that influence drug absorption (Narawane and Lee, 1994; Lennernas, 1998; Zhou *et al.*, 2005). Permeability reflects the physiological properties of membrane to the solutes. Fraction of drug absorbed is determined by the drug permeability through intestinal wall. Solubility is one of the physicochemical properties of drug molecules to affect drug absorption. The drug molecules have to be dissolved in solution in order for absorption to occur in the intestinal tract. Dissolution is the dosage form variable to determine the rate and extent of drug dissolved in solution.

4.3.1 *Membrane Permeability*

4.3.1.1 Effective Permeability

Passive permeability (P) of molecules across a membrane can be expressed as

$$P_{\text{passive}} = \frac{J}{C} = \frac{DK}{h}, \tag{4.8}$$

where K is the partition coefficient, D is the diffusion coefficient, and h is the thickness of the cell membrane. The diffusion coefficient (D) depends on the molecular weight or size of a molecule. K is a measure of the solubility of the substance in lipid. Therefore, the passive permeability is related to membrane and drug properties. For a specific drug, the passive membrane permeability should be a constant P_{m} and independent to drug concentration.

The permeability for active absorption can be presented by

$$P_{\text{active}} = \frac{J}{C} = \frac{J_{\max} C}{K_{\text{m}} + C}\frac{1}{C} = \frac{J_{\max}}{K_{\text{m}} + C}, \tag{4.9}$$

where J is the drug flux, $J_{\max}$ is the maximum drug flux, C is the drug concentration, and K_{m} is the drug affinity to the carrier. Obviously, active permeability is dependent on drug concentration.

Therefore, the total effective permeability is dependent on drug concentration for drugs that absorbed through both passive diffusion and active transport, and it can be expressed as follows

$$P_{\text{eff}} = P_{\text{passive}} + P_{\text{active}} = P_{\text{m}} + \frac{J_{\max}}{K_{\text{m}} + C}. \tag{4.10}$$

However, at very low concentration, $C \ll K_{\text{m}}$, drug permeability is independent to drug concentration

$$P_{\text{eff}} = P_{\text{m}} + \frac{J_{\max}}{K_{\text{m}}}. \tag{4.11}$$

At high concentration, $C \gg K_{\text{m}}$, drug permeability is dependent on drug concentration.

$$P_{\text{active}} = \frac{J}{C} = \frac{J_{\max} C}{K_{\text{m}} + C}\frac{1}{C} = \frac{J_{\max}}{C} \approx 0, \tag{4.12}$$

$$P_{\text{eff}} = P_{\text{m}}. \tag{4.13}$$

Therefore, the permeability vs. concentration plot can be generated as in Fig. 4.5.

4.3.1.2 Fraction of Drug Absorbed

Drug permeability through intestinal wall will determine the fraction of drug absorbed (F_{a}). F_{a} can be estimated by drug permeability through intestinal wall

$$F_{\text{a}} = 1 - \text{e}^{-2\text{An}}; \quad \text{An} = \frac{T_{\text{res}}}{T_{\text{abs}}}; \quad T_{\text{abs}} = \frac{R}{P_{\text{eff}}}, \tag{4.14}$$

where, T_{res} is the small intestine transit time (~3 h), T_{abs} is the absorptive time (h), R is the radius of small intestine (2 cm), and P_{eff} is the drug permeability through intestinal wall.

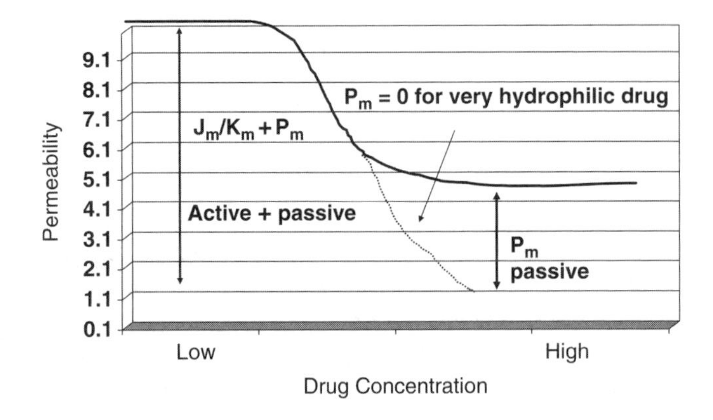

FIGURE 4.5. Active and passive permeability at low and high drug concentration

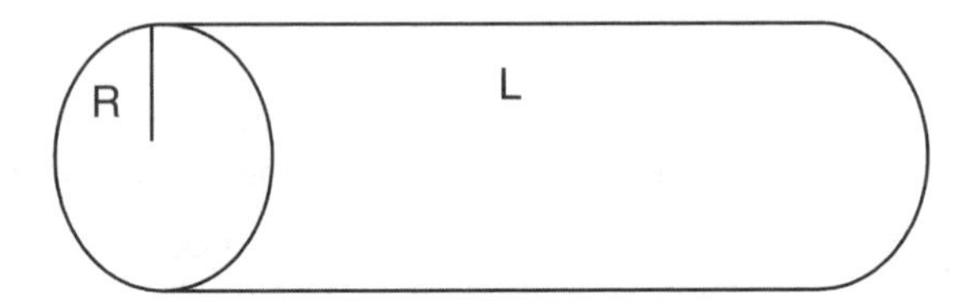

FIGURE 4.6. Model for intestinal absorption compartment

4.3.1.3 Permeability and Absorption Rate Constant

Absorption rate constant can be expressed as

$$K_a = P\frac{A}{V} = P\frac{2\pi RL}{\pi R^2 L} = \frac{2P}{R}, \tag{4.15}$$

where K_a is the absorption rate constant with unit $1\,s^{-1}$, P is the permeability ($cm\,s^{-1}$), A is the membrane surface area (cm^2), and V is the volume of absorption compartment (cm^3) (Fig. 4.6). However this equation tends to overestimate absorption by 12.5-fold, so $K_a = P/(2\pi R)$ may be more realistic.

4.3.2 *Solubility*

Solubility is the most important physicochemical property of drug molecules, which can affect the drug absorption. The drug molecules have to be dissolved in the solution for the absorption to occur in the intestinal tract. The solubility of a solute is the maximum quantity of solute that can dissolve in a certain quantity of solvent or quantity of solution at a specified temperature. The extent of ionization and oil/water partition coefficient K of the drug contribute to both drug solubility and membrane permeability. In general, low K indicates high solubility in water and high K indicates high solubility in lipid. However, the drug molecules with high lipid solubility usually possess high membrane permeability.

Color Plate I

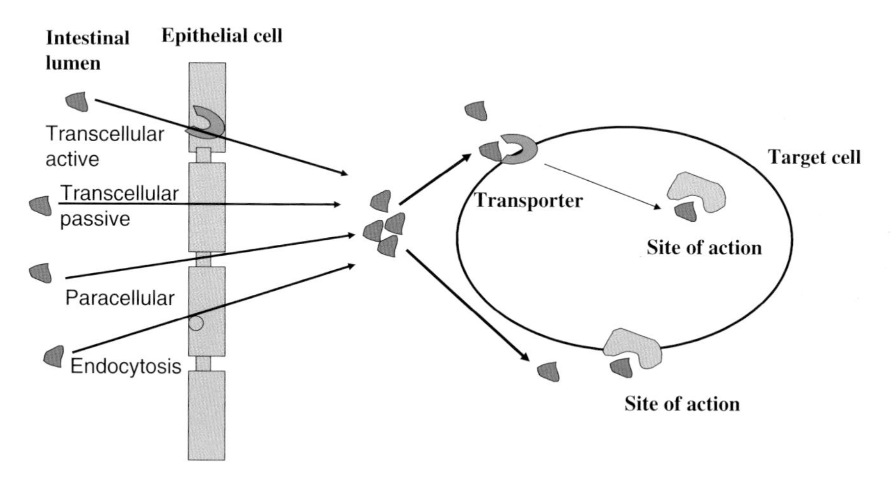

FIGURE 4.1. Drug transport and site of action (*See* page 76)

Color Plate II

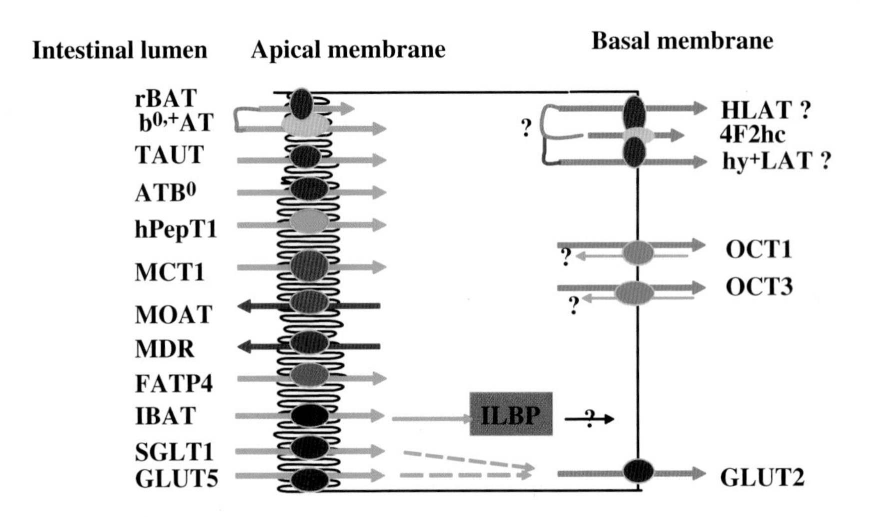

FIGURE 4.3. Apical and basolateral transporters coupling for absorption (*See* page 78)

Ionization and pH play an important role in drug water solubility (Zhou *et al.*, 2005). Ionized form is usually more water soluble than unionized form, but unionized form is easier for absorption in the GI tract by passive diffusion than ionized form. For weakly basic drugs, more unionized form would be predominant in intestine at high pH (5–8), which favors absorption. For weakly acid drug, more ionized form would be predominant in intestine. Although in theory that ionized weak acid is not favorable for absorption in intestine, the larger surface area of intestine will compensate this weakness to produce complete absorption for many weakly acidic drug.

4.3.3 *Dissolution of Solid Dosage Forms*

If drugs are administered in solid dosage forms, they must be dissolved in the GI tract before absorption can take place. For drugs with low solubility and high dose, the dissolution will be slow, and the dissolution rate will be the rate-limiting step for absorption. Factors that affect dissolution will control the whole absorption process.

Noyes–Whitney equation can be used to describe the dissolution rate as following

$$\frac{\mathrm{d}m}{\mathrm{d}t} = A\frac{D}{h}(C_\mathrm{s} - C) \quad C = 0 \text{ at sink condition,} \tag{4.16}$$

where $\mathrm{d}m/\mathrm{d}t$ is the rate of solid dissolution, A is the solid surface area, D is the diffusion coefficient, h is the thickness of unstirred boundary layer, C_S is the drug aqueous solubility, and C is the concentration at h (Fig. 4.7).

For drugs with low solubility, formulation strategies such as micronization (increases A), ionization (increases C_S), solubilization (surfactants), and

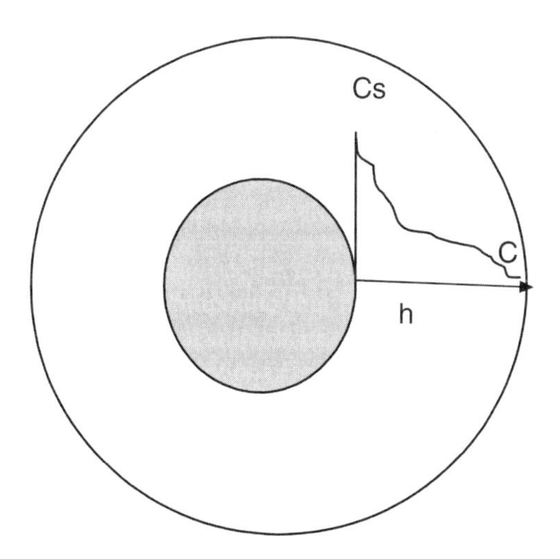

FIGURE 4.7. Model for dissolution of solid drugs

disintegrants can be used to enhance dissolution and fraction of drug absorbed (Anderson and Pitman, 1980; Frenning and Stromme, 2003; Schreiner *et al.*, 2005; Jinno *et al.*, 2006).

4.4 Secondary Factors Influencing Drug Absorption

4.4.1 Biological Factors of Gastrio Intestinal Tract

GI tract plays important roles in secretion, digestion, and absorption. Many biological factors, such as gastric emptying, gastric and intestinal pH, GI content, GI motility, GI surface area, and blood flow (Fleisher *et al.*, 1990) can affect drug absorption.

4.4.1.1 Gastric Emptying Time

Gastric emptying time refers to the time needed for the stomach to empty the total initial stomach contents. During digestion, gastric emptying depends on the tone of proximal stomach and pylorus, which is under reflex and hormonal control. Generally, anything that slows down gastric emptying is likely to slow down the rate (not extent) of drug absorption, and thus affect onset of the therapeutic response. A lot of factors promote gastric emptying, such as hunger, lying on right side, noncaloric liquid intake, drugs (metoclopramide, prokinetic drugs), and some excipients. On the other hand, factors, such as meals (especially with fatty, bulky, and viscous food), lying on left side, and other drugs (tricyclic antidepressants, anticholinergics, and alcohol) retard gastric emptying. Gastric emptying of solution-type dosage forms and suspensions of fine drug particles is generally much faster and less variable than that of solid, nondisintegrating dosage forms and aggregated particles. For drugs with high solubility and high membrane permeability, gastric emptying rate will control the absorption rate and onset of the drugs. There will be a direct relation between gastric-emptying rate and maximal plasma concentration, and an inverse relation between gastric-emptying rate and the time required to achieve maximal plasma concentrations.

4.4.1.2 Surface Area

Surface area of different regions of GI influences drug absorption. Small intestine has largest effective surface area for drug absorption due to the presence of folds of mucosa, villi, and microvilli. For carrier-mediated drug absorption, small intestine is also the most important region for most drug transporters that are also expressed in this area. In contrast, stomach and large intestine have no villi, microvilli, or less transporter expression.

4.4.1.3 GI Transit Time

GI transit time or mean resident time (MRT) can also influence oral drug absorption. Increase in the GI residence time (or decrease of motility) leads to enhanced drug absorption potential. Stomach MRT is about 1.3 h while the small intestine MRT is around 3 h. The longer MRT in small intestine will contribute to a higher drug absorption potential.

4.4.1.4 Intestinal Motility

Intestinal motility is another factor that influences oral drug absorption. Intestinal movement includes propulsion and mixing. Propulsive movement determines the intestinal transit time and is important for slow release dosage forms, enteric-coated drug that is only released in intestine, slowly dissolving drugs, and carrier-mediated absorption. Mixing movement increases dissolution rate where the drug molecule contacts with endothelial surface area for absorption.

4.4.1.5 Components, Volume, and Properties of Gastrointestinal Fluids

Components, volume, and properties of gastrointestinal fluids especially GI pH will change the drug's ionization, solubility, dissolution rate, and therefore affect drug absorption. The rate of dissolution from a dosage form, particularly tablets and capsules, is dependent on pH. Acidic drugs dissolve most readily in alkaline media and will have a greater dissolution in the intestinal fluids than in gastric fluids. Basic drugs will dissolve most readily in acidic solution, and thus the dissolution will be greater in gastric fluids than in intestinal fluids. GI pH depends on general health of the individual, disease conditions, age, type of food, and drug therapy. Antichlolinergic drugs and H_2-blockers increase gastric pH and significantly decrease bioavailability of some weakly basic drugs with pH-dependent solubility.

4.4.1.6 Food

Food influences drug absorption in different ways. High fat food may stimulate bile salt secretion, increases drug solubility and dissolution, and increases bioavailability for certain drugs with low solubility. High protein may increase gastric pH, thus decrease dissolution of weak basic drugs and bioavailability. High calorie food decreases gastric emptying rate, delays the rate of absorption, and delays the onset of therapeutic drugs. At the same time, food components may compete for drug absorption that is mediated by transporters. For instance, grapefruit juice inhibits efflux pump (P-gp) and increases bioavailability of P-gp substrates. In addition, food components may form complex with drugs (complexation) and decrease drug absorption and bioavailability as seen in the example that tetracycline forms a complex with calcium in milk to hinder its absorption.

4.4.1.7 Blood Flow

Blood flow in the GI tract also plays an important role in drug absorption. GI tract is highly vascularized and receives 28% of the cardiac output. The higher blood flow promotes the higher drug absorption, especially for those active-absorption mediated and highly permeable drugs.

4.4.1.8 Age

Age can also influence the drug absorption. Newborns, for example, have less acidic gastric fluids, smaller gut fluid volume, slower gastric emptying rate, less intestinal surface area and blood flow, and thus have relatively lower drug absorption.

4.4.2 Dosage Factors Influencing Absorption

Dosage form factors include excipients and dosage forms, which may affect drug absorption (Rouge *et al.*, 1996; Badawy Sherif *et al.*, 2006). The disintegrants can enhance the dissolution rate of the drugs and increase absorption. Surfactants such as Tween-80 may increase drug solubility of poorly soluble drugs, and increase drug absorption through enhancement of the drug permeability. The coating of enteric-coated tablets such as cellulose acetate can only be dissolved in the intestine at high pH (> 5), which protects the drug from degradation in gastric condition and against drug stimulation of gastric mucosa. In such cases, the controlled release dosage form will have a completely different absorption profile as compared with immediate release dosage forms.

4.5 Evaluation of Oral Drug Absorption in Humans

4.5.1 Drug Absorption Assessment Using In Vivo Data

4.5.1.1 Estimation of Fraction of Drug Absorbed Using Experimental Intestinal Permeability *In Vivo*

An *in vivo* method has been successfully established to measure human intestinal permeability by *in situ* intestinal perfusion (Lennernas *et al.*, 1997; Sun *et al.*, 2002, 2002; Cao *et al.*, 2006). A perfusion tube, as illustrated in Fig. 4.8, is placed in the human jejunum to allow drug passage through a 10-cm intestinal segment. The drug concentration is measured at the inlet and outlet of the perfusion tube. The drug permeability is then calculated with the following equation

$$P_{\text{eff, human}} = Q(1 - C_{\text{out}}/C_{\text{in}})/2\pi RL, \tag{4.17}$$

where $P_{\text{eff, human}}$ is drug permeability in the human intestine, Q is the perfusion flow rate (2 min ml^{-1}), C_{in} is inlet drug concentration of the perfusion tube, C_{out} is outlet drug concentration of the perfusion tube, R is human small intestine radius (2 cm), and L is the 10-cm perfusion segment. When the permeability is plotted against the fraction of drug absorbed, the relationship can be established (4.18) (Fig. 4.9) (Amidon *et al.*, 1988, 1995; Oh *et al.*, 1993)

$$F_{\text{a}} = 1 - \exp(-2\text{An}) = 1 - \exp(-2P_{\text{eff, human}}T_{\text{res}}/R), \tag{4.18}$$

where F_{a} is the fraction of drug absorbed, $P_{\text{eff, human}}$ is drug permeability in human intestine, T_{res} is transit time in human small intestine (3 h), R is the radius of human small intestine (2 cm).

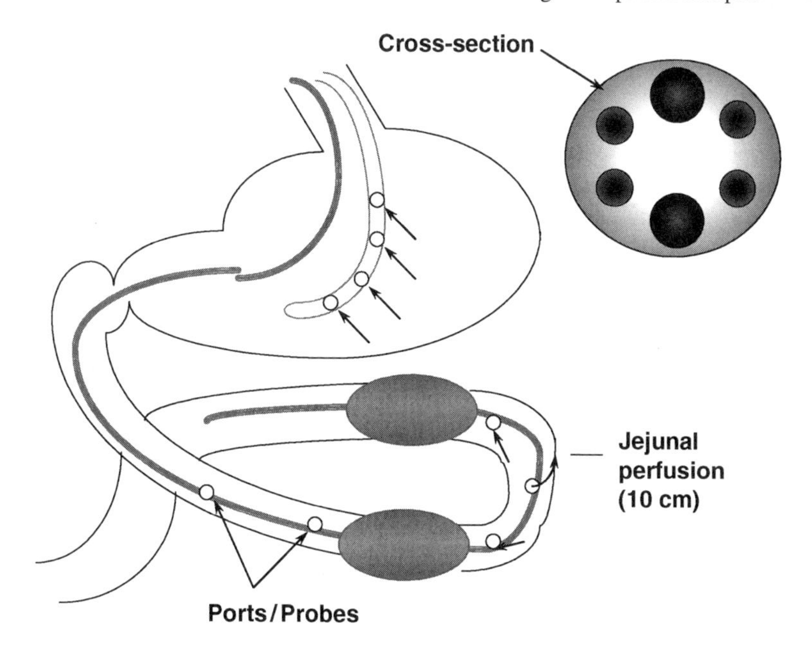

FIGURE 4.8. Perfusion tube for *in situ* human intestinal permeability measurement

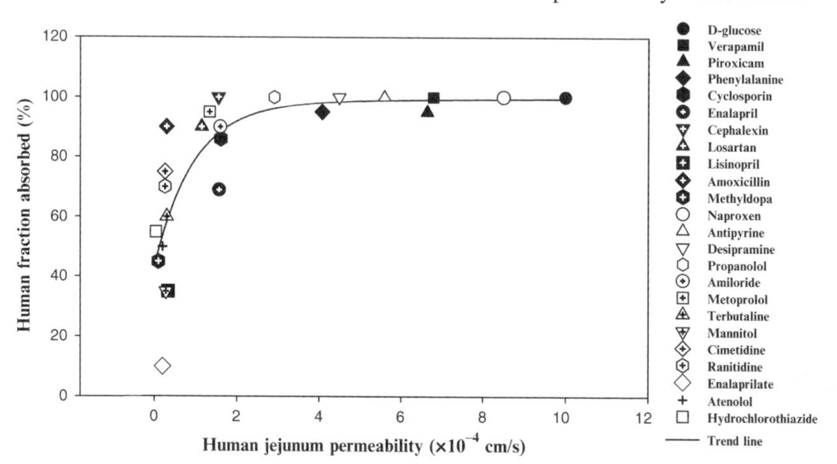

FIGURE 4.9. Prediction of the fraction of drug absorbed using human jejunum permeability. Drugs are labeled with different symbols. *Closed symbols* are drugs absorbed through carrier-mediated process, while *open symbols* are drugs absorbed through passive diffusion (Sun *et al.*, 2002)

$$\text{An} = P_{\text{eff, human}} \times T_{\text{res}}/R.$$

However, when *in situ* intestinal perfusion is performed, low drug concentrations are used for permeability measurements. In this case, the drug concentration is always below its solubility limit. Since the fraction of drug absorbed is a function of its solubility and permeability, (4.18) is not suitable for predicting the fraction of drug absorbed when high drug concentration above *in vivo* solubility limit is used in the experiment. This model has been further modified to overcome this problem by utilizing different calculation methods according to the drug's solubility (Yu *et al.*, 1996; Yu and Amidon, 1999)

$$F_a = 1 - \exp(-2\text{An}), \quad \text{when} \quad C_{in} < S, \quad C_{out} < S, \tag{4.19}$$

$$F_a = 2\text{An}/D_0, \quad \text{when} \quad C_{in} > S, \quad C_{out} > S, \tag{4.20}$$

$$F_a = 1 - 1/[D_0 \exp(-2\text{An} + D_0 - 1)], \quad \text{when} \quad C_{in} > S, \quad C_{out} < S, \tag{4.21}$$

where F_a is the fraction of drug absorbed, $\text{An} = P_{\text{eff, human}} \times T_{res}/R$, C_{in} is inlet drug concentration of the perfusion tube, C_{out} is outlet drug concentration of the perfusion tube, $P_{\text{eff, human}}$ is drug intestinal permeability in human, T_{res} is transit time in human small intestine (3 h), D_0 is dose number [D_0 = (dose/volume)/S], and S is drug solubility. The challenge for this method is that drug intestinal permeability has to be obtained *in vivo* in human, which is very difficult and not available during early stages of drug discovery and development. Meanwhile the relationship between C_{out} (or C_{in}) and solubility is also difficult to determine *in vivo*.

4.5.1.2 Estimation of Maximum Absorbable Dose Using *In Vivo* Absorption Rate Constant and Drug Solubility

Another method has been proposed to estimate maximum absorbable dose (MAD) based on the *in vivo* absorption rate constant (Curatolo, 1987) with the following equation

$$\text{MAD} = SK_aVT, \tag{4.22}$$

where S is drug solubility, K_a is absorption rate constant, V is intake water volume (250 ml), and T is transit time in small intestine (3 h). For instance, MAD could be estimated using different K_a values (Table 4.1). However, K_a has to be obtained from *in vivo* pharmacokinetic studies in animals or humans, which are usually not available during early stages of drug discovery and development. Alternatively K_a can be estimated by *in vivo* drug permeability if it is available by (4.23)

$$K_a = P_{\text{eff, human}}(A/V) = P_{\text{eff, human}}(2\pi RL/\pi R^2 L) = P_{\text{eff, human}}(2/R), \tag{4.23}$$

where A is the surface area, V is the volume, R is the radius, and L is the length of small intestine.

However, it is also difficult to estimate the appropriate volume for the calculation in this method. Although standard water intake is 250 ml, the daily gastric secretion volume is 2,000 ml; intestine secretion volume is in the range of 1,500–2,000 ml; and bile and pancreatic secretion is 500–1,500 ml (Dressman *et al.*, 1998).

TABLE 4.1. Estimation of maximum absorbable dose (MAD) using absorption rate constant (K_a) in human with following equation: $MAD = SK_aVT$

K_a (min^{-1})	Solubility ($mg\ ml^{-1}$)	MAD (mg)
0.001	0.001	0.045
0.001	0.01	0.45
0.001	0.1	4.5
0.001	1	45
0.01	0.001	0.45
0.01	0.01	4.5
0.01	0.1	45
0.01	1	450
0.1	0.001	4.5
0.1	0.01	45
0.1	0.1	450
0.1	1	4,500

4.5.1.3 Estimation of MAD from Drug *In Vivo* Permeability in Humans and Drug Solubility

At the steady state of *in situ* human intestinal perfusion, drug flux J is a function of permeability, drug concentration, and absorption surface area (Amidon *et al.*, 1988, 1995; Oh *et al.*, 1993),

$$J = dm/dt = P_{eff,\,human} S\, dA. \tag{4.24}$$

Then,

$$MAD = P_{eff,\,human} SAT = P_{eff,\,human} S 2\pi RLT, \tag{4.25}$$

where J is drug flux, $P_{eff,\,human}$ is drug permeability in human intestine, S is drug solubility, A is absorption surface area, T is transit time in small intestine (3 h), R is the radius of small intestine (2 cm), and L is the length of small intestine (6 m). It is worth noting that the small intestine surface area for drug absorption should include surface area of villi and microvilli, but the surface area calculated in (4.24) is only the intestinal tube surface area without such consideration. However, since the permeability obtained in the *in situ* perfusion is calculated by (4.17), where the surface area also does not include villi and microvilli, the error is cancelled in the MAD calculation in (4.25), and it does not affect the MAD estimation if human intestinal permeability is used. The examples for estimation of MAD using permeability with (4.25), or using calculated K_a from human permeability with (4.22) and (4.23) are summarized in Table 4.2.

In comparison of the examples in Tables 4.1 and 4.2, it seems that MAD might be underestimated using the absorption rate constant in (4.22) due to the assumption of 250 ml of volume in the calculation. MAD might be overestimated using permeability in (4.9) due to the assumption that the drug is absorbed at the maximum concentration (at its solubility) in the whole small intestinal region (6 m)

TABLE 4.2. Estimation of MAD using drug intestinal permeability in human with following equation: MAD $= P_{\text{eff, human}} S 2\pi RLT$, MAD $= P_{\text{eff, human}} S A_{\text{eff}} T$, or with calculated absorption rate constant (K_{a}) with following equations: $K_{\text{a}} = P_{\text{eff, human}}(2/R)$ and MAD $= S K_{\text{a}} V T$

$P_{\text{eff, human}}$ ($\times 10^{-4}$ cm s^{-1})	Solubility (mg ml^{-1})	MAD (mg) calculated from $P_{\text{eff, human}}$	MAD (mg) calculated from effective absorption surface area	Calculated K_{a} from $P_{\text{eff, human}}$ (min^{-1})	MAD (mg) from calculated K_{a}
0.1	0.001	0.813	0.086	0.0006	0.027
0.1	0.01	8.13	0.864	0.0006	0.27
0.1	0.1	81.3	8.64	0.0006	2.7
0.1	1	813	86.4	0.0006	27
1	0.001	8.13	0.864	0.006	0.27
1	0.01	81.3	8.64	0.006	2.7
1	0.1	813	86.4	0.006	27
1	1	8,138	864	0.006	270
10	0.001	81.3	8.64	0.06	2.7
10	0.01	813	86.4	0.06	27
10	0.1	8,138	864	0.06	270
10	1	81,388	8,640	0.06	2,700

with maximum surface area over the entire 3 h absorption period, while in reality only partial small intestine is used at a given time. Therefore, the effective absorption surface area of 800 cm^2 is proposed to calculate MAD (Curatolo, 1987). The examples for estimation of MAD using this effective surface area are also summarized in Table 4.2. MAD using the effective absorption surface area seems more appropriate. If the MAD based on permeability and solubility is below the required clinical dose, formulation development, and delivery would be very challenging.

4.5.2 Drug Absorption Assessment Using In Vitro Data

When MAD is estimated with *in vivo* data, either the *in vivo* absorption rate constant, or the drug *in vivo* intestinal permeability is required for the calculation. However, during early stages of drug discovery and development, *in vivo* data are usually unavailable. The challenge is to optimize the process for selecting compounds to evaluate *in vivo* human studies based on *in vitro* data. Fortunately, drug permeability in Caco-2 cells and drug solubility are routinely screened in the pharmaceutical industry. These data can be utilized to predict fraction of drug absorbed and MAD in humans to identify the best candidates for further clinical development.

4.5.2.1 *In Vitro* Testing Conditions for Determining Drug Permeability in Caco-2 Cells and *In Vitro*/*In Vivo* Permeability Correlation

Many laboratories have established methods for measuring drug permeability in Caco-2 cells with different testing conditions (Chong *et al.*, 1996; Yee, 1997;

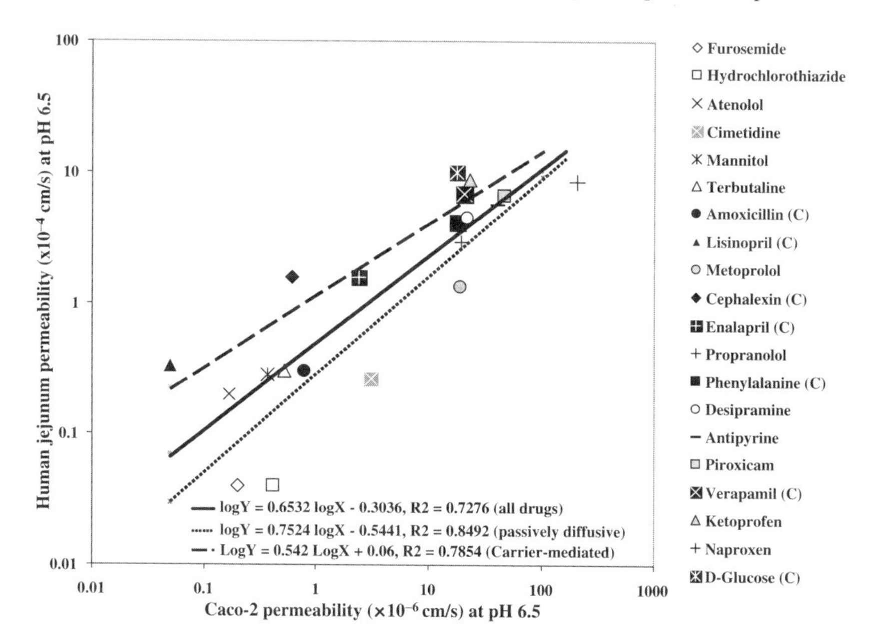

FIGURE 4.10. *In vitro/in vivo* permeability correlation of 20 drugs at pH 6.5. Correlation coefficient ($R^2 = 0.7276$) was calculated from the permeability of all 20 drugs. Correlation coefficient ($R^2 = 0.8492$) was calculated from the permeability of the following drugs: furosemide, hydrochlorothiazide, atenolol, cimetidine, mannitol, terbutaline, metoprolol, propranolol, desipramine, antipyrine, piroxicam, ketoprofen, and naproxen. Correlation coefficient ($R^2 = 0.7854$) was calculated from the permeability of the following drugs: cephalexin, enalapril, lisinopril, losartan, amoxicillin, phenylalanine, L-leucine, L-dopa, D-glucose, cyclosporin, and verapamil. Drugs are labeled with different symbols. *Black symbols* are drugs absorbed through carrier-mediated process, while *gray* and *open symbols* are drugs absorbed through passive diffusion (Sun *et al.*, 2002)

Pade and Stavchansky, 1998; Yamashita *et al.*, 2000). Some laboratories use buffer with pH 7.4 in both apical and basolateral sides of the Caco-2 cells, while others use pH 6.5 buffer at the apical side and pH 7.4 buffer at the basolateral side. When correlation analysis was performed between *in vitro* drug permeability in Caco-2 cells and *in vivo* drug permeability in humans, a better correlation was observed between human *in vivo* permeability and Caco-2 permeability measured at pH 6.5 than at pH 7.4 (Figs. 4.10 and 4.11). The correlation coefficient (R^2) of *in vitro* and *in vivo* permeability of 24 drugs assayed at pH 7.4 was 0.5126 in (4.26), while the *in vitro* and *in vivo* permeability correlation coefficient (R^2) of the 20 drugs determined at pH 6.5 was 0.7276 in (4.27) (Sun *et al.*, 2002).

$$\text{Log}\, P_{\text{eff, human}} = 0.4926\, \text{Log}\, P_{\text{eff, Caco-2}} - 0.1454, \qquad (4.26)$$

$$\text{Log}\, P_{\text{eff, human}} = 0.6532\, \text{Log}\, P_{\text{eff, Caco-2}} - 0.3036. \qquad (4.27)$$

However, if the drugs were absorbed through a carrier-mediated processes, such as cephalexin, enalapril, cyclosporin, amoxicillin, lisinopril, losartan, phenylalanine,

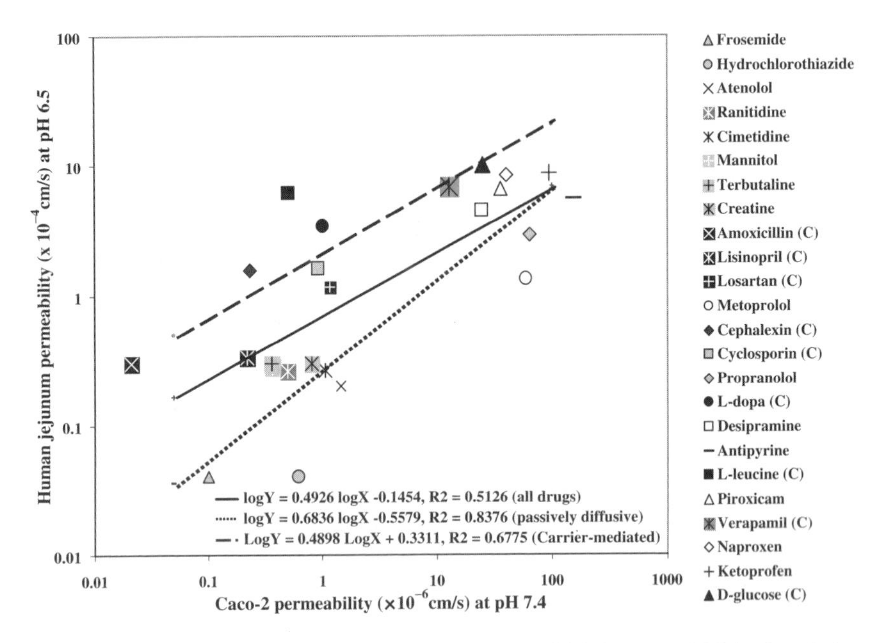

FIGURE 4.11. *In vitro*/*in vivo* permeability correlation of 24 drugs at pH 7.4. Correlation coefficient ($R^2 = 0.5126$) was calculated from the permeability of all 24 drugs. Correlation coefficient ($R^2 = 0.8376$) was calculated from the permeability of the following drugs: furosemide, hydrochlorothiazide, atenolol, ranitidine, cimetidine, mannitol, terbutaline, creatine, metoprolol, propranolol, desipramine, antipyrine, piroxicam, ketoprofen, and naproxen. Correlation coefficient ($R^2 = 0.6775$) was calculated from the permeability of the following drugs: cephalexin, enalapril, lisinopril, losartan, amoxicillin, phenylalanine, L-leucine, L-dopa, D-glucose, cyclosporin, and verapamil. Drugs are labeled with different symbols. *Black symbols* are drugs absorbed through carrier-mediated process, while *gray* and *open symbols* are drugs absorbed through passive diffusion (Sun *et al.*, 2002)

verapamil, L-dopa, D-glucose, and L-leucine were excluded, the *in vitro*/in vivo permeability correlation improves at both pHs, such that the permeability correlation coefficient (R^2) of 15 passively diffused drugs at pH 7.4 and 13 passively diffused drugs at pH 6.5 were 0.8376 in (4.28) and 0.8492 in (4.29), respectively.

$$\text{Log } P_{\text{eff, human}} = 0.6836 \text{ Log } P_{\text{eff, Caco-2}} - 0.5579, \tag{4.28}$$

$$\text{Log } P_{\text{eff, human}} = 0.7524 \text{ Log } P_{\text{eff, Caco-2}} - 0.5441. \tag{4.29}$$

4.5.2.2 Estimation of Fraction of Drug Absorbed In Humans Using *In Vitro* Drug Permeability in Caco-2 Cells

When *in vitro* drug permeability in Caco-2 cells is plotted against drug fraction absorbed in humans, a relationship could also be established as shown in Figs. 4.12 and 4.13 (Sun *et al.*, 2002). As these data clearly indicate, it might be difficult to predict the fraction of drug absorbed for the drugs with low Caco-2 permeability.

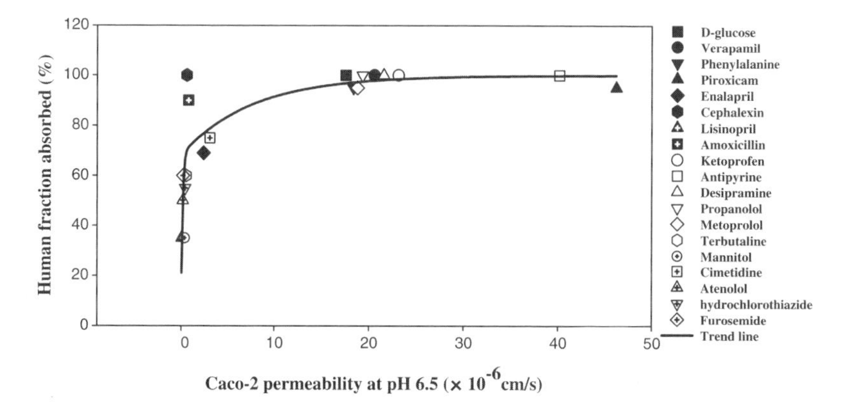

FIGURE 4.12. Prediction of the fraction of drug absorbed using Caco-2 permeability at pH 6.5. Drugs are labeled with different symbols. *Closed symbols* are drugs absorbed through carrier-mediated process, while *open symbols* are drugs absorbed through passive diffusion (Sun *et al.*, 2002)

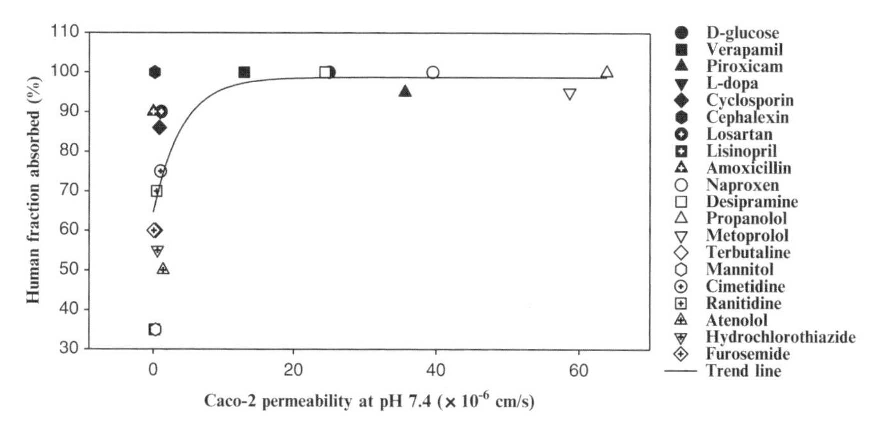

FIGURE 4.13. Prediction of the fraction of drug absorbed using Caco-2 permeability at pH 7.4. Drugs are labeled with different symbols. *Closed symbols* are drugs absorbed through carrier-mediated process, while *open symbols* are drugs absorbed through passive diffusion (Sun *et al.*, 2002)

More discrepancy was also observed for the drugs with carrier-mediated absorption routes especially when drug permeability in Caco-2 cells was obtained at pH 7.4 in the apical side.

4.5.2.3 Estimation of MAD in Human Based on *In Vitro* Data

Since an *in vitro* and *in vivo* drug permeability correlation has been established in (4.26), *in vivo* drug permeability in human could be easily estimated by *in vitro* drug permeability in Caco-2 cells. Although some of the transporter substrates showed high discrepancy from the *in vitro*/*in vivo* permeability correlation when Caco-2 permeability was obtained at pH 7.4, the overall correlation has shown reasonable prediction when Caco-2 permeability was obtained at pH 6.5 (Sun *et al.*, 2002). Since MAD could be estimated using *in vivo* drug permeability in human with (4.25), the MAD could be estimated with *in vitro* drug permeability in Caco-2 cells in the following (4.30)

$$\mathrm{MAD} = P_{\mathrm{eff,\,human}} S A_{\mathrm{eff}} T = 10^{(0.6532\,\mathrm{Log}\,P_{\mathrm{eff,\,Caco}} - 0.3036)} S A_{\mathrm{eff}} T, \qquad (4.30)$$

where $P_{\mathrm{eff,\,human}}$ is the *in vivo* drug permeability in human, $P_{\mathrm{eff,\,Caco}}$ is the *in vitro* drug permeability in Caco-2 cells, S is the drug solubility, A_{eff} is the effective absorption surface area without considering villi and microvilli, and T is transit time in small intestine (3 h). As discussed earlier, the error associated when not considering surface area of villi and microvilli is cancelled in the MAD calculation using permeability in (4.25) and (4.30). In addition, the surface area of microvilli in Caco-2 cells is also irrelevant in calculation of MAD, since the human permeability is calculated with Caco-2 permeability by the correlation analysis.

Alternatively, the MAD can be estimated using (4.22) and (4.23), where K_{a} can be estimated with human permeability *in vivo* or Caco-2 permeability *in vitro* with the following (4.31)

$$K_{\mathrm{a}} = P_{\mathrm{eff,\,human}}(2/R) = (2/R)10^{(0.6532\,\mathrm{Log}\,P_{\mathrm{eff,\,Caco}} - 0.3036)}, \qquad (4.31)$$

where $P_{\mathrm{eff,\,human}}$ is drug *in vivo* permeability in human, $P_{\mathrm{eff,\,Caco}}$ is drug *in vitro* permeability in Caco-2 cells, and R is the radius of small intestine (2 cm). The examples of MAD estimation using *in vitro* drug permeability in Caco-2 cells are summarized in Table 4.3.

TABLE 4.3. Estimation of MAD using drug permeability in Caco-2 cells with following equation: $\mathrm{MAD} = 10^{(0.6532\,\mathrm{Log}\,P_{\mathrm{eff,\,caco}} - 0.3036)} S A_{\mathrm{eff}} T$, or with calculated absorption rate constant (K_{a}) with following equations: $K_{\mathrm{a}} = (2/R)10^{(0.6532\,\mathrm{Log}\,P_{\mathrm{eff,\,caco}} - 0.3036)}$ and $\mathrm{MAD} = S K_{\mathrm{a}} V T$

$P_{\mathrm{eff,\,Caco-2}}$ ($\times 10^{-6}$ cm s^{-1})	Solubility (mg ml^{-1})	MAD (mg) calculated from $P_{\mathrm{eff,\,Caco-2}}$	Calculated K_{a} from $P_{\mathrm{eff,\,Caco-2}}$ (min^{-1})	MAD (mg) from calculated K_{a}
0.1	0.001	0.0955	0.000663	0.0298
0.1	0.01	0.955	0.000663	0.298
0.1	0.1	9.545	0.000663	2.98
0.1	1	95.45	0.000663	29.8
1	0.001	0.429	0.002982	0.134
1	0.01	4.294	0.002982	1.34
1	0.1	42.94	0.002982	13.4
1	1	429.4	0.002982	134
10	0.001	1.932	0.01342	0.603
10	0.01	19.32	0.01342	6.03
10	0.1	193.2	0.01342	60.3
10	1	1,932	0.01342	603
100	0.001	8.696	0.060388	2.17
100	0.01	86.96	0.060388	27.17
100	0.1	869.6	0.060388	271.7
100	1	8,696	0.060388	2,717

4.5.3 Correlation of Oral Drug Bioavailability and Intestinal Permeability Between Rat and Human

Animal models are widely used to evaluate drug pharmacokinetics and drug absorption. However, the correlation of oral bioavailability (F) values of 48 drugs in rat and human has been studied and no correlation ($r^2 = 0.29$) was found due to low correlation of drug metabolism in rat and human (Cao *et al.*, 2006). Results of the F values comparison are shown in Fig. 4.14. In contrast, Chiou and Buehler observed low correlation in the bioavailability of 35 drugs between monkey and human with $r^2 = 0.502$ (Chiou and Buehler, 2002), which may be due to the closer physiological similarity between monkey and human. These data indicate that oral bioavailability in rat could not be used to predict oral drug bioavailability in human.

Due to the structural similarities of intestinal membrane, drug absorption in animal models may be used to predict drug absorption in human. In order to depict the oral drug absorption process, *in situ* intestinal permeabilities of 17 drugs with different absorption mechanisms were evaluated in rat and human jejunum (Cao *et al.*, 2006). Since permeability is one of the primary factors governing absorption (Amidon *et al.*, 1995), studying the permeability correlation is useful when predicting human absorption from rat permeability. The tested drugs are absorbed by carrier-mediated processes as well as passive diffusion. For instance, valacyclovir, enalapril, and cephalexin are all absorbed through a peptide transporter (hPepT1). Leucine, phenylalanine, L-Dopa, and methyldopa are absorbed through amino acid transporters. Verapamil is a P-gp substrate. Cimetidine is an organic cation transporter substrate. Propranolol, atenolol, and furosemide are all absorbed through passive diffusion. The drug permeabilities in the rat jejunum were then correlated

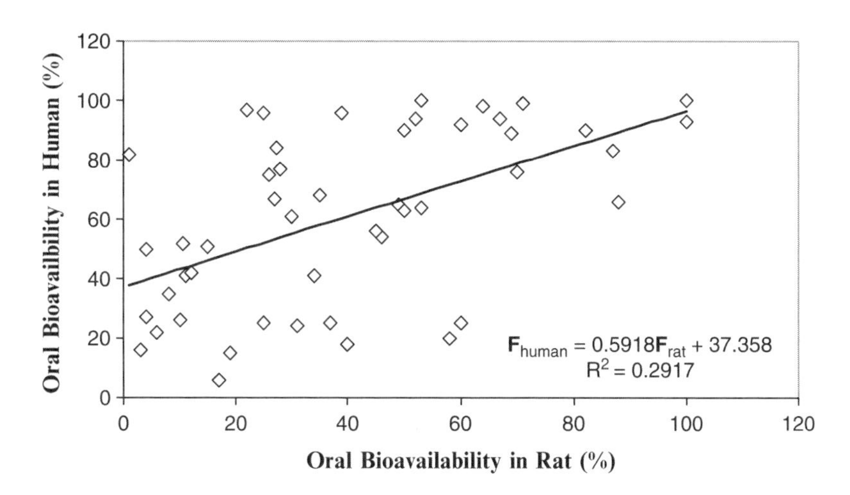

FIGURE 4.14. Correlation of oral bioavailability between rat and human. Total of 48 drugs were plotted. The equation describes the correlations for rat oral bioavailability (F_{rat}) and human oral bioavailability (F_{human}) (Cao *et al.*, 2006)

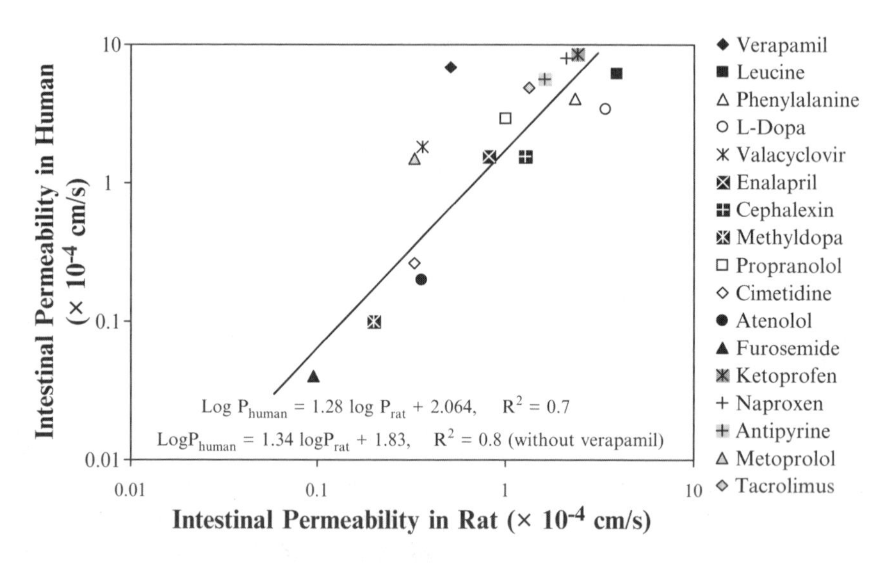

FIGURE 4.15. Correlation of drug permeability in rat jejunum and in human jejunum. Permeability coefficients (P_{eff}) were determined by *in situ* intestinal perfusion. The equations describe the correlations for rat permeability (P_{rat}) and human permeability (P_{human}) (Cao *et al.*, 2006)

with the drug permeabilities in the human jejunum (Fig. 4.15). It showed that drug permeability in the rat is generally five to ten-fold lower than the permeability in the human. However, both carrier-mediated and passively diffusing drugs showed a reasonable correlation ($r^2 = 0.7$). Interestingly, verapamil (a P-gp substrate) permeability in human deviates from the correlation curve. The permeability correlation between human and rat is highly increased ($r^2 = 0.8$) when verapamil is excluded in the analysis.

This study is in agreement with the other report, that the percentage of absorption of 98 drugs was correlated between rat and human with a correlation of $r^2 = 0.88$ (Zhao *et al.*, 2003). *In vivo* absorption in rats could be a useful method to predict the extent of absorption in humans. The permeability in rat for water soluble and poor water soluble compounds was used to predict the fraction of drug absorbed in humans (Watanabe *et al.*, 2004). In another study, a high correlation was found for a variety of compounds displaying various physicochemical and pharmacologic activities between the two species in the dose-independent absorption range (Chiou and Barve, 1998). However, a previous study reported that effective permeability estimates of passively absorbed solutes correlate highly in rat and human jejunum while carrier-mediated transport requires scaling between the models because the substrate specificity and/or transport maximum may differ (Fagerholm *et al.*, 1996). These discrepancies might be due to the different numbers of transporter substrates that are used in the correlation analysis. However, all of these studies indicate that reasonable permeability correlation between human and rat can be used to predict drug absorption in humans.

To understand the underlying mechanisms in the similarity in drug intestinal absorption between humans and rats, correlation analysis of the expression levels of transporters and metabolizing enzymes between rat and human intestine were further conducted (Cao *et al.*, 2006). Moderate correlations (with $r^2 > 0.56$) were found for the expression levels of transporters in the duodenum of human and rat. Although there is discrepancy observed in the expression of MDR1, MRP3, GLUT1, and GLUT3, other transporters (such as PepT1, SGLT-1, GLUT5, MRP2, NT2, and high affinity glutamate transporter) and the overall drug transporters expression share similar expression levels in both human and rat intestine with regional dependent expression patterns, which has high expression in the small intestine and low expression in the colon. These data provide the molecular mechanisms for the similarity and correlation of drug absorption (F_a) in the small intestine between rat and human. In contrast, the expression of metabolizing enzymes (CYP3A4/CYP3A9 and UDPG) showed 12- to 193-fold difference between human and rat intestine with distinct regional dependent expression patterns. No correlation was found for the expressions of metabolizing enzymes between rat and human intestine, which indicate the difference in drug metabolism in two different species and the challenges in predicting F_g and F from rat to human.

4.6 Summary

Drug absorption is a complicated process in which many physiological and physiochemical factors are involved. Understanding the principles of drug absorption benefits the designing of formulation strategies to enhance the bioavailability and *in vivo* drug activity. In summary, drug absorption mechanisms include passive diffusion and active transport. Permeability, solubility, and dissolution, GI physiological conditions, and dosage forms can influence the drug absorption rate. In general, if a drug has high water solubility and low membrane permeability (hydrophilic drugs), permeability usually limits absorption, unless it is carrier mediated or paracellular absorption dependent. Strategies which can enhance the drug permeability in dosage design could be used to increase this permeability controlled drug absorption. In contrast, if a drug has low solubility and high permeability (lipophilic drugs), solubility (and dissolution) usually limits absorption. Formulation strategies should be optimized in the dosage form to enhance the solubility (and dissolution) controlled drug absorption. If neither of the above two properties limits the absorption such as for drugs with high solubility and high permeability, then gastric emptying rate limits the drug absorption. Both *in vivo* and *in vitro* methods have been explored to assess drug absorption in human. Rat and human show similar drug absorption profiles and similar transporter expression patterns in the small intestine, while the two species exhibit distinct expression levels and patterns for metabolizing enzymes in the intestine. These data provide the molecular mechanisms for the similarity and correlation of drug absorption (F_a) in the small intestine between rat and human. Therefore, rat can be used to predict oral drug absorption (F_a) in the small intestine of human, but not to predict drug metabolism (F_g and F_h) and oral bioavailability (F) in human.

References

Allen L, MacKichan J and Ritschel WA (1982) Manual of symbols, equations & definitions in pharmacokinetics. *J Clin Pharmacol* **22**:1S–23S.

Amidon GL, Sinko PJ and Fleisher D (1988) Estimating human oral fraction dose absorbed: a correlation using rat intestinal membrane permeability for passive and carrier-mediated compounds. *Pharm Res* **5**:651–654.

Amidon GL, Lennernas H, Shah VP and Crison JR (1995) A theoretical basis for a biopharmaceutic drug classification: the correlation of in vitro drug product dissolution and in vivo bioavailability. *Pharm Res* **12**:413–420.

Anderle P, Huang Y and Sadee W (2004) Intestinal membrane transport of drugs and nutrients: genomics of membrane transporters using expression microarrays. *Eur J Pharm Sci* **21**:17–24.

Anderson JR and Pitman IH (1980) Solubility and dissolution rate studies of ergotamine tartrate. *J Pharm Sci* **69**:832–835.

Atkinson AJ Jr (2001) Drug absorption and bioavailability. *Princ Clin Pharmacol* 31–41.

Badawy SIF, Gray DB, Zhao F, Sun D, Schuster AE and Hussain MA (2006) Formulation of solid dosage forms to overcome gastric pH interaction of the factor Xa inhibitor, BMS-561389. *Pharm Res* **23**:989–996.

Burton PS, Goodwin JT, Vidmar TJ and Amore BM (2002) Predicting drug absorption: how nature made it a difficult problem. *J Pharmacol Exp Ther* **303**:889–895.

Cainelli SR, Chui A, McClure JD Jr and Hunter FR (1974) Facilitated diffusion in erythrocytes of mammals. *Comp Biochem Physiol A* **48**:815–825.

Cao X, Yu LX, Barbaciru C, Landowski CP, Shin HC, Gibbs S, Miller HA, Amidon GL and Sun D (2005) Permeability dominates in vivo intestinal absorption of P-gp substrate with high solubility and high permeability. *Mol Pharm* **2**:329–340.

Cao X, Gibbs ST, Fang L, Miller HA, Landowski CP, Shin HC, Lennernas H, Zhong Y, Amidon GL, Yu LX and Sun D (2006) Why is it challenging to predict intestinal drug absorption and oral bioavailability in human using rat model. *Pharm Res* **23**:1675–1686.

Chidambaram N and Burgess DJ (2000) Mathematical modeling of surface-active and non-surface-active drug transport in emulsion systems. *AAPS Pharm Sci* **2**:E31.

Chiou WL (2001) The rate and extent of oral bioavailability versus the rate and extent of oral absorption: clarification and recommendation of terminology. *J Pharmacokinet Pharmacodyn* **28**:3–6.

Chiou WL and Barve A (1998) Linear correlation of the fraction of oral dose absorbed of 64 drugs between humans and rats. *Pharm Res* **15**:1792–1795.

Chiou WL and Buehler PW (2002) Comparison of oral absorption and bioavailability of drugs between monkey and human. *Pharm Res* **19**:868–874.

Chong S, Dando SA, Soucek KM and Morrison RA (1996) In vitro permeability through caco-2 cells is not quantitatively predictive of in vivo absorption for peptide-like drugs absorbed via the dipeptide transporter system. *Pharm Res* **13**:120–123.

Curatolo W (1987) The lipoidal permeability barriers of the skin and alimentary tract. *Pharm Res* **4**:271–277.

Daugherty AL and Mrsny RJ (1999) Transcellular uptake mechanisms of the intestinal epithelial barrier Part one. *Pharm Sci Technol Today* **4**:144–151.

Dressman JB, Amidon GL, Reppas C and Shah VP (1998) Dissolution testing as a prognostic tool for oral drug absorption: immediate release dosage forms. *Pharm Res* **15**:11–22.

Fagerholm U, Johansson M and Lennernas H (1996) Comparison between permeability coefficients in rat and human jejunum. *Pharm Res* **13**:1336–1342.

Feher JJ (1983) Facilitated calcium diffusion by intestinal calcium-binding protein. *Am J Physiol* **244**:C303–307.

Fleisher D, Lippert CL, Sheth N, Reppas C and Wlodyga J (1990) Nutrient effects on intestinal drug absorption. *J Control Release* **11**:41–49.

Frenning G and Stromme M (2003) Drug release modeled by dissolution, diffusion, and immobilization. *Int J Pharm* **250**:137–145.

Hansen JE, Weisbart RH and Nishimura RN (2005) Antibody mediated transduction of therapeutic proteins into living cells. *Scientific World Journal* **5**:782–788.

Higuchi WI and Ho NFH (1988) Membrane transfer of drugs. *Int J Pharm* **2**:10–15.

Jinno J, Kamada N, Miyake M, Yamada K, Mukai T, Odomi M, Toguchi H, Liversidge GG, Higaki K and Kimura T (2006) Effect of particle size reduction on dissolution and oral absorption of a poorly water-soluble drug, cilostazol, in beagle dogs. *J Control Release* **111**:56–64.

Kwan KC (1997) Oral bioavailability and first-pass effects. *Drug Metab Dispos* **25**:1329–1336.

Lennernas H (1998) Human intestinal permeability. *J Pharm Sci* **87**:403–410.

Lennernas H, Nylander S and Ungell AL (1997) Jejunal permeability: a comparison between the Using chamber technique and the single-pass perfusion in humans. *Pharm Res* **14**:667–671.

Liang XJ, Mukherjee S, Shen DW, Maxfield FR and Gottesman MM (2006) Endocytic recycling compartments altered in cisplatin-resistant cancer cells. *Cancer Res* **66**: 2346–2353.

Narawane L and Lee VHL (1994) Absorption barriers. *Drug Target Deliv* **3**:1–66.

Oh DM, Curl RL and Amidon GL (1993) Estimating the fraction dose absorbed from suspensions of poorly soluble compounds in humans: a mathematical model. *Pharm Res* **10**:264–270.

Oh DM, Han HK and Amidon GL (1999) Drug transport and targeting. Intestinal transport. *Pharm Biotechnol* **12**:59–88.

Pade V and Stavchansky S (1998) Link between drug absorption solubility and permeability measurements in Caco-2 cells. *J Pharm Sci* **87**:1604–1607.

Rouge N, Buri P and Doelker E (1996) Drug absorption sites in the gastrointestinal tract and dosage forms for site-specific delivery. *Int J Pharm* **136**:117–139.

Schreiner T, Schaefer UF and Loth H (2005) Immediate drug release from solid oral dosage forms. *J Pharm Sci* **94**:120–133.

Shin H-C, Landowski CP, Sun D, Vig BS, Kim I, Mittal S, Lane M, Rosania G, Drach JC and Amidon GL (2003) Functional expression and characterization of a sodium-dependent nucleoside transporter hCNT2 cloned from human duodenum. *Biochem Biophys Res Commun* **307**:696–703.

Sun D, Lennernas H, Welage LS, Barnett JL, Landowski CP, Foster D, Fleisher D, Lee KD and Amidon GL (2002) Comparison of human duodenum and Caco-2 gene expression profiles for 12,000 gene sequences tags and correlation with permeability of 26 drugs. *Pharm Res* **19**:1400–1416.

Sun D, Yu LX, Hussain MA, Wall DA, Smith RL and Amidon GL (2004) In vitro testing of drug absorption for drug 'developability' assessment: forming an interface between in vitro preclinical data and clinical outcome. *Curr Opin Drug Discov Devel* **7**:75–85.

Toutain PL and Bousquet-Melou A (2004) Bioavailability and its assessment. *J Vet Pharmacol Ther* **27**:455–466.

Trischitta F, Denaro MG, Faggio C and Lionetto MG (2001) Ca++ regulation of paracellular permeability in the middle intestine of the eel, *Anguilla anguilla. J Comp Physiol [B]* **171**:85–90.

Ungell AL, Nylander S, Bergstrand S, Sjoeberg A and Lennernaes H (1998) Membrane transport of drugs in different regions of the intestinal tract of the Rat. *J Pharm Sci* **87**:360–366.

Watanabe E, Takahashi M and Hayashi M (2004) A possibility to predict the absorbability of poorly water-soluble drugs in humans based on rat intestinal permeability assessed by an in vitro chamber method. *Eur J Pharm Biopharm* **58**:659–665.

Yamashita S, Furubayashi T, Kataoka M, Sakane T, Sezaki H and Tokuda H (2000) Optimized conditions for prediction of intestinal drug permeability using Caco-2 cells. *Eur J Pharm Sci* **10**:195–204.

Yee S (1997) In vitro permeability across Caco-2 cells (colonic) can predict in vivo (small intestinal) absorption in man–fact or myth. *Pharm Res* **14**:763–766.

Yoon KA and Burgess DJ (1998) Mathematical modelling of drug transport in emulsion systems. *J Pharm Pharmacol* **50**:601–610.

Yu LX and Amidon GL (1999) A compartmental absorption and transit model for estimating oral drug absorption. *Int J Pharm* **186**:119–125.

Yu LX, Lipka E, Crison JR and Amidon GL (1996) Transport approaches to the biopharmaceutical design of oral drug delivery systems: prediction of intestinal absorption. *Adv Drug Deliv Rev* **19**:359–376.

Zhao YH, Abraham MH, Le J, Hersey A, Luscombe CN, Beck G, Sherborne B and Cooper I (2003) Evaluation of rat intestinal absorption data and correlation with human intestinal absorption. *Eur J Med Chem* **38**:233–243.

Zhou R, Moench P, Heran C, Lu X, Mathias N, Faria Teresa N, Wall Doris A, Hussain Munir A, Smith RL and Sun D (2005) pH-dependent dissolution in vitro and absorption in vivo of weakly basic drugs: development of a canine model. *Pharm Res* **22**:188–192.

5 Evaluation of Permeability and P-glycoprotein Interactions: Industry Outlook

Praveen V. Balimane and Saeho Chong

5.1 Introduction

New drug discovery and development is becoming an increasingly risky and costly endeavor. A recent report has tagged the final price of bringing a drug to the market at greater than a billion US dollars with an estimated research time running into multiple years (2004). Despite the considerable investment in terms of finance and resources, the number of drug approvals per year have held steady for the last few years. The advent of combinatorial chemistry, automation, and high-throughput screening (HTS) has afforded the opportunity to test thousands of compounds, but the success rate of progressing from initial clinical testing to final approval has remained disappointingly low. Greater than 90% of the compounds entering Phase-I clinical testing fail to reach the patients and as high as 50% entering Phase-III do not make the cut (Kola and Landis, 2004).

Historically, drug discovery adopted a linear design; the new chemical entities were first selected on the basis of their pharmacological activity/potency followed by their sequential profiling to assess the absorption, distribution, metabolism, elimination, and toxicity (ADMET) characteristics. Such a strategy was generally more time and resource intensive and left little room for errors during the discovery process. Today, the new drug design effort integrates a parallel matrix approach where the pharmacological efficacy is screened in parallel with ADMET profiling, providing information in a timely manner to maximize the chance for selecting superior drug candidates with better quality for further development. Therefore, the availability of highly accurate, low cost and HTS techniques that can provide fast and reliable data on the developability characteristics of drug candidates is crucial. Screening a large number of drug candidates for biopharmaceutical properties (e.g., solubility, intestinal permeability, CYP inhibition, metabolic stability, and more recently drug–drug interaction potential involving drug transporters) has become a major challenge. Determination of permeability property

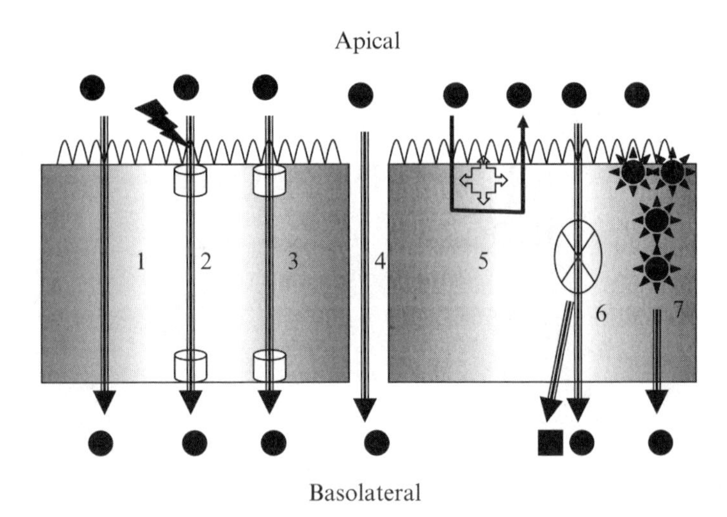

FIGURE 5.1. Multiple pathways for intestinal absorption of a compound. (1) passive, transcellular (2) active or secondary active, (3) facilitated diffusion, (4) passive, paracellular, (5) absorption limited by P-gp and/or other efflux transporters, (6) intestinal first-pass metabolism followed by absorption of parent and metabolite, (7) receptor mediated transport

and the drug–transporter interaction of drug candidates is fast becoming key characterization studies performed during the lead selection and lead optimization.

Drug absorption across the intestinal membrane is a complex multipathway process as shown in Fig. 5.1. Passive absorption occurs most commonly through the cell membrane of enterocytes (transcellular route) or *via* the tight junctions between the enterocytes (paracellular route). Carrier-mediated absorption occurs *via* an active (or secondary active process) or by facilitated diffusion. Various efflux transporters such as P-gp, BCRP, MRP2 are also functional that could limit the absorption. Intestinal enzymes could be involved in metabolizing drugs to alternate moieties that might be absorbed. Finally, receptor mediated endocytosis could also play a role. Because of the multivariate processes involved in intestinal absorption of drugs, it is often difficult to use a single model to accurately predict the *in vivo* permeability characteristics.

Currently, a variety of experimental models are available when evaluating intestinal permeability of drug candidates (Hillgren *et al.*, 1995; Balimane *et al.*, 2000, 2006; Hidalgo, 2001). A few commonly used models include: *in vitro* methods; artificial lipid membrane such as parallel artificial membrane permeability assay (PAMPA); cell-based systems such as Caco-2 cells, Mardin-Darby canine kidney (MDCK) cells, etc.; tissue-based Ussing chamber; *in situ* methods, intestinal single pass perfusion, and *in vivo* methods, whole animal absorption studies. Typically, a combination of these models is used routinely in assessing intestinal permeability. A tiered approach is often used, which involves high-throughput (but less predictive) models for primary screening followed by low throughput (but more predictive) models for secondary screening and mechanistic studies. PAMPA and cell culture-based models offer the right balance between predictability and throughput, and currently enjoy wide popularity throughout the pharmaceutical industry.

Adequate permeability is required not only for oral absorption but also for sufficient drug distribution to pharmacological target organs (e.g., tumor, liver, etc.). In addition to simple passive diffusion across lipid bilayers, numerous transporters appear to play a critical role in selective accumulation and distribution of drugs into target organs. P-gp is one of the most extensively studied transporters that have been unequivocally known to impact the ADMET characteristics of drug molecules (Kim *et al.*, 1999; Polli *et al.*, 1999; Lin, 2003; Lin and Yamazaki, 2003). It's role in influencing the pharmacokinetics of anti-cancer drugs has been extensively reviewed (Krishna and Mayer, 2000). It is a ubiquitous transporter, which is present on the apical surface of the enterocytes, canalicular membrane of hepatocytes, and on the apical surface of kidney, placenta and endothelial cells of brain membrane. Because of its strategic location, it is widely recognized that P-gp is a major determinant in disposition of a wide array of drugs in humans. The oral bioavailability of fexofenadine increased significantly when erythromycin or ketoconazole (well known inhibitor of P-gp) was co-administered in humans, suggesting P-gp as a permeability barrier at the absorption site (Simpson and Jarvis, 2000). Similarly, P-gp at the blood–brain barrier limits the entry of drugs into the brain. The biliary elimination of vincristine decreased significantly in the presence of verapamil (a known P-gp substrate/inhibitor) (Watanabe *et al.*, 1992). Therefore, the early screening of drug candidates for their potential to interact with P-gp (either as a substrate or inhibitor) is becoming necessary and critical. There are various *in vitro* and *in vivo* models used for assessing P-gp interaction (Adachi *et al.*, 2001; Polli *et al.*, 2001; Yamazaki *et al.*, 2001; Perloff *et al.*, 2003). *In vitro* assays such as ATPase activity, rhodamine-123 uptake, calcein AM uptake, cell-based bidirectional transport, radio-ligand binding along with *in vivo* models such as transgenic (knockout mice) animals are often used to assess the involvement of P-gp. The cell-based bidirectional permeability assay is the most popular method for identification of P-gp substrate in drug discovery labs (Polli *et al.*, 2001). This cell model provides the right balance of adequate throughput and functional utility. However, there are certain caveats that the user must be aware of in order to maximize the utility of the assay. In addition to P-gp, other pharmaceutically important drug transporters (e.g., MRP2, BCRP, OATP, OAT, OCT, etc.) can be examined using a plethora of models (e.g., transfected cell, vesicles, recombinant vaccinia, xenopus oocytes, etc.). A detailed discussion of these models is beyond the scope of this book chapter.

This book chapter will focus on the various techniques that are currently used by drug discovery scientists in evaluating permeability/absorption and P-gp interaction potential of drug candidates. In particular, two experimental models (PAMPA and Caco-2 cells) will be discussed in detail with special emphasis on their complimentary aspects. PAMPA is used as the primary screening tool, and it is capable of providing the structure–permeability relationship that enables a successful lead optimization. Caco-2 cell model is often used as the secondary tool to perform in-depth mechanistic studies, to delineate the various pathways of absorption and to assess the P-gp interaction potential of drug candidates.

5.2 Anatomy and Physiology of the Small Intestine

The human small intestine is approximately 2–6 m and is loosely divided into three sections: duodenum, jejunum and ileum, which comprise 5, 50, and 45% of the length. The biological and physical parameters of human intestinal tract are listed in Table 5.1 (Carr and Toner, 1984; Daugherty and Mrsny, 1999). Approximately 90% of all absorption in the gastrointestinal tract occurs in the small intestinal region. The small intestine has projections that increase the potential surface area for digestion and absorption. Macroscopic valve like folds, called circular folds, encircling the inside of the intestinal lumen is estimated to increase the surface area of the small intestine threefold. Villi increase the area 30-fold and the microvilli increase it by a factor of 600. Thus, such unique structures lead to a tremendous increase in surface area available for absorption in the small intestine.

The major role of the small intestine is the selective absorption of major nutrients and to serve as a barrier to digestive enzymes and ingested foreign substances. The epithelial cells in the intestinal region are a heterogeneous population of cells which include enterocytes or absorptive cells, goblet cells which secrete mucin, endocrine cells, paneth cells, M cells, tuft, and cup cells. The most common epithelial cells are the enterocytes or the absorptive cells. This cell is responsible for the majority of the absorption of both nutrients and drugs in the small intestine. It is polarized with distinct apical and basolateral membrane that are separated by tight junctions. Thus, the bulk of absorption takes place in the small intestine by various mechanisms such as passive diffusion (paracellular and transcellular) and carrier-mediated process (facilitated and active).

TABLE 5.1. Biological and physical characteristics of human intestinal tract

Gastrointestinal segment	Surface area (m^2)	Segment length (cm)	pH of the segment
Stomach	3.5	0.25	1–2
Duodenum	1.9	~35	4.0–5.5
Jejunum	184	~280	5.5–7.0
Ileum	276	~420	7.0–7.5
Colon and Rectum	1.3	~150	7.0–7.5

5.3 Permeability Absorption Models

5.3.1 *Physicochemical Methods*

For orally administered compounds, the systemic exposure depends on many different factors. In general, physicochemical properties such as molecular weight, p*K*a, lipophilicity, charge/ionization, solubility, gastrointestinal pH and molecular size are the major determinants of intestinal permeability. Aqueous solubility of the compound in the intestinal tract is also an important factor that dictates the dissolution characteristics of the compound eventually influencing the oral bioavailability. Physicochemical methods are attractive because of their high-throughput capacity, efficiency and reproducibility to predict passive diffusion. However, these models often lead to inaccurate prediction because of the lack of real physiological conditions that govern membrane permeability/absorption *in vivo*.

5.3.1.1 Lipophilicity (Log *P*/Log *D*)

As a measure of drug–membrane interaction, lipophilicity is one of the most important physicochemical parameters in predicting and interpreting membrane permeability (Ho *et al.*, 1977). In most of the early studies (Schanker *et al.*, 1958; Houston *et al.*, 1974; Dressman *et al.*, 1985), the oral drug absorption was demonstrated to be dependent on the lipophilicity. Historically, the octanol–water partition coefficient (Log *P*) was accepted as a surrogate to biological systems for predicting absorption. But now it is widely recognized that use of Log *P* alone for predicting absorption is an over simplification of a complex process and often leads to inaccurate estimation. The transcellular diffusion of compounds from the luminal to serosal side in the intestinal epithelia involves the partitioning of the drug from the aqueous luminal region to nonpolar lipid bilayers of the cell membrane followed by the partitioning out from the lipid layers to the aqueous serosal region. Transport across biological membranes is a complex process that includes passive as well as carrier-mediated processes for influx and efflux. Log *P* values provide indirect information of the extent of passive transcellular transport possible for various drugs. For a structurally related series of compounds in which transport is largely mediated by a passive mechanism, the relationship between permeability and lipophilicity is generally bell shaped. However, for a diverse set of compounds in which parallel transport processes may be occurring in addition to the passive component, the correlation with lipophilicity is normally lacking (Ho *et al.*, 1977).

5.3.1.2 Absorption Potential

Dressman and colleagues (Dressman *et al.*, 1985) proposed a parameter, Absorption Potential (AP), that incorporated the various basic physicochemical parameters in one single equation and was highly predictive of the extent of absorption in humans.

$$\mathrm{AP} = \mathrm{Log}\left(PF_{\mathrm{non}}\left[\frac{S_{\mathrm{o}}V_{\mathrm{L}}}{X_{\mathrm{o}}}\right]\right),$$

where AP is the absorption potential, P is the octanol–water partition coefficient for the drug, F_{non} is the fraction of drug nonionized at pH 6.5, S_o is the aqueous solubility of nonionized species at 37 °C, V_L is the luminal volume (~250 mL) and X_o is the drug dose. A relatively good correlation was demonstrated between absorption potential vs. fraction absorbed in human subjects. There was a sigmoidal relationship observed between fraction absorbed and absorption potential. However, absorption potential does not account for carrier-mediated transport and thus is limited in its utility for predicting permeabilities only for passively transported compounds.

5.3.1.3 Immobilized Artificial Membrane (IAM)

IAM is a chromatographic surface prepared by covalently immobilizing cell membrane phospholipids to solid surfaces at monolayer density. IAM chromatography column emulates the lipid environment of cell membranes (Pidgeon 1990a,b). IAM chromatography is experimentally simple and potentially capable of screening a large number of compounds. Pidgeon and colleagues have demonstrated the utility of this methodology as an accurate, cost effective and efficient predictor of permeability of test compounds. The predominant factor that regulates the passage of drugs across the gastrointestinal mucosa is their ability to passage through the lipid cell membranes. Log k' derived from the IAM column showed reasonable correlation to drug partitioning into liposomes and permeability across Caco-2 cell monolayers. Various modifications of IAM have been studied by different investigators (Beigi *et al.*, 1995; Yang *et al.*, 1996; Stewart and Chan, 1998; Krause *et al.*, 1999) and the results obtained have been shown to correlate with other parameters such as: partitioning into liposome membranes, Caco-2 cell permeability and intestinal absorption.

The IAM methodology has been used to predict not only drug intestinal absorption but also solute partitioning into liposomes, brain uptake, and human skin permeability. But, it is important to recognize that lipid composition of various cell membranes in the body differs and is not necessarily consistent even within a given cell type under different conditions. Also, the artificial membranes themselves lack the paracellular pores that form an integral part of a biological membrane architecture and the transporter proteins that are involved in the carrier-mediated transport of drugs. Thus, even though these simplified artificial membranes might be capable of predicting the absorption of a series of compounds that are passively transported across cell membrane, it has severe limitations in screening compounds that are hydrophilic and small in size (candidates for paracellular transport) or predominantly transported by carrier proteins.

5.3.2 *In Vitro* Methods

Numerous *in vitro* methods have been used in the drug selection process for assessing the intestinal absorption potential of drug candidates. *In vitro* techniques for assessment of permeability are less labor and cost intensive compared to *in vivo* animal studies. One universal issue with all the *in vitro* systems is that the effect of physiological factors such as gastric emptying rate, gastrointestinal transit rate, gastrointestinal pH, etc. cannot be incorporated in the data interpretation.

Each *in vitro* method has its distinct advantages and drawbacks. Based on the specific goal, one or more of these methods can be used as a screening tool for selecting compounds during the drug discovery process. The successful application of *in vitro* models to predict drug absorption across the intestinal mucosa depends on how closely the *in vitro* model mimics the characteristics of the *in vivo* intestinal epithelium. Although it is very difficult to develop a single *in vitro* system that can simulate all the conditions existing in the human intestine, various *in vitro* systems are used routinely as decision making tools in early drug discovery.

5.3.2.1 Animal Tissue-Based Methods

It is extremely difficult to obtain viable human tissues for permeability studies on a regular basis. Since animal intestinal tissues are also made up of essentially the same kind of endothelial cells, permeability screening for drug discovery purposes is routinely carried out using various animal species. Excised animal tissue models have been used since the 1950s to explore the mechanism of absorption of nutrients from the intestine. Evidence on the uptake of glucose against a concentration gradient (Quastel, 1961) provided impetus for mechanistic studies involving excised tissues from the intestine of animals. However, the viability of the excised tissues is difficult to maintain since the tissues are devoid of direct blood supply and need constant oxygenation. Some of the more widely used methods for absorption and permeability studies are described below.

Everted Gut Technique

The everted gut technique has been in use from as early as the 1950s (Wilson and Wiseman, 1954) in the transport study of sugars and amino acids. Over the years modifications such as use of physiologically relevant tissue culture media, means to maintain constant oxygenation and improved incubation apparatus have all contributed to increased viability of *ex vivo* tissues leading to better predictability with everted tissue studies. This model is ideal for studying the absorption mechanism of drugs since both the passive and active transport can be studied. With the recent interest in the field P-glycoprotein activity in the gut, this model may be used to evaluate the role of efflux transporters in the intestinal absorption of drugs by comparing the transport kinetics of drug in the absence and presence of P-glycoprotein inhibitors or substrates. The everted gut model has an additional analytical advantage over other *in vitro* models because the sample volume on the serosal side is relatively small and drugs accumulate faster. Some of the disadvantages are the lack of active blood and nerve supply which can lead to a rapid loss of viability. In addition everting the intestinal tissue can lead to morphological damage causing misleading results.

Ussing Chamber

Transport studies across intestinal tissues from animals are also a widely used *in vitro* method to study drug absorption. This method involves the isolation of the intestinal tissues, cutting it into strips of appropriate size, clamping it on a suitable device and then the rate of drug transport across this tissue is measured. The utility of this *in vitro* device (Ussing Chamber) was first demonstrated by Ussing and Zerahn (1951). In this method, the permeability is measured based on the appearance of drug in the serosal side rather than the disappearance of drug in the mucosal side. The unique feature of this approach is that electric resistance of the membrane can be measured during the course of the experiment. The short circuit current across the membrane, as well as the resistance across the membrane, can be monitored. These parameters are routinely used as an indicator of the viability of the intestinal tissue during the transport studies with Ussing chamber. The apparent permeability coefficient (P) is estimated using the equation (Grass and Sweetana, 1989)

$$P = \left(\frac{V}{AC_0}\right)\left(\frac{\mathrm{d}C}{\mathrm{d}T}\right),$$

where V is the volume of the receiver chamber, A is the exposed tissue surface area, C_0 is the initial concentration drug in the donor chamber, and $\mathrm{d}C/\mathrm{d}T$ is the change in drug concentration in the receiver with time.

The Ussing chamber technique is an ideal method to study regional differences on the absorption of drugs by mounting intestinal tissues from various intestinal regions (Ungell *et al.*, 1998). It is also possible to perform studies with human intestinal tissues thus providing a methodology to compare the permeability values across species. The amounts of drug required for the study are relatively small (mg quantity) and the collected samples are analytically clean which facilitates quantitative analysis. Apart from the disadvantages commonly associated with any *in vitro* studies, the drawbacks of this technique include: lack of blood and nerve supply, rapid loss of viability of the tissues during the experiment and changes in morphology and functionality of transporter proteins during the process of surgery and mounting of tissues.

Isolated Membrane Vesicles

Use of membrane vesicles isolated from numerous species (including humans) for transport characterization studies has been popular for some time now (Murer and Kinne, 1980; Hillgren *et al.*, 1995; Sinko *et al.*, 1995). Membrane vesicles prepared from intestinal tissues provide the flexibility of examining the interaction of drugs to a specific membrane of interest (e.g., brush-border membrane vs. basolateral membrane of enterocytes). Vesicles offer a unique opportunity to study the properties of drug and nutrient transport at the cellular level (brush-border as well as basolateral side). Membrane vesicle studies allow a complete manipulation of solute environment both inside and outside of the vesicle, thus making it an ideal system for mechanistic absorption studies. Studies with vesicles also facilitate isolation and identification of transporter proteins that are specifically expressed either on the brush-border or the basolateral side of the membranes.

Compared to the other conventional *in vitro* techniques (Ussing chamber and intestinal perfusions), the vesicles studies are much better suited when availability of sufficient quantities of drug becomes difficult. Drug uptake study in membrane vesicle can be performed with a small amount of drug (mg quantity), and typically multiple experiments can be performed with vesicles prepared from a single laboratory animal. A unique advantage of this method is that vesicles can be cryopreserved and used for a long duration. Thus the vesicle system offer unique experimental capabilities not possible with more integrated methodologies. However, prepared vesicles are not pure and very often contain other membrane or organelle fragments. It is practically not feasible to isolate 100% pure brush-border membrane vesicles (or basolateral vesicles) without the contamination with the other type of vesicles. During the process of isolation of vesicles often leads to damage of the transporter proteins and enzymes. Since the volume of the vesicles is very small, they typically require a sensitive analytical method.

5.3.2.2 Cell-Based Methods

Varieties of cell monolayer models that mimic *in vivo* intestinal epithelium in humans have been developed and currently enjoy widespread popularity. Unlike enterocytes, human immortalized (tumor) cells grow rapidly into confluent monolayers followed by a spontaneous differentiation providing an ideal system for transport studies. Table 5.2 lists a few of these cell models commonly used for permeability assessment in drug discovery and development.

Caco-2 Cells and Others

Caco-2 cells have been the most extensively characterized and useful cell model in the field of drug permeability and absorption for over a decade now (Arturrson,

TABLE 5.2. Cell culture models commonly used for permeability assessment

Cell Line	Species/origin	Special characteristics
Caco-2	Human colon adenocarcinoma	Most well-established cell model, differentiates and expresses some relevant efflux transporters, expression of influx transporters variable (lab-to-lab)
MDCK	Mardin-Darby canine kidney epithelial cells	Polarized cells with low intrinsic expression of ABC transporters, ideal for transfections
LLC-PK1	Pig kidney epithelial cells	Polarized cells with very low intrinsic transporter expression, ideal for transfections
2/4/A1	Rat fetal intestinal epithelial cells	Temperature-sensitive, ideal for paracellularly absorbed compounds (leakier pores)
TC-7	Caco-2 sub-clone	Similar to Caco-2
HT-29	Human colon	Contains mucus producing goblet cells
IEC-18	Rat small intestine cell line	Provides a size selective barrier for paracellularly transported compounds

1991; Artursson *et al.*, 1996; Rubas *et al.*, 1996; Hidalgo, 2001; Balimane and Chong, 2005a). Caco-2 cells, a human colon adenocarcinoma, undergo spontaneous enterocytic differentiation in culture. When they reach confluence on a semi-permeable porous filter, the cell polarity and tight junctions are well established. In the last 10–15 years, there has been a tremendous growth in the use of Caco-2 cells for mechanistic studies and as a rapid *in vitro* screening tool in support of drug discovery within the pharmaceutical industry. The predictability and utility of this model to rank order a large number of compounds in terms of absorption potential have been demonstrated conclusively by several investigators (Artursson and Karlsson, 1991; Rubas *et al.*, 1993). The recent trend in many pharmaceutical research laboratories is to completely automate Caco-2 cell permeability screen using robotics. A fully automated Caco-2 cell system allows much greater throughput in the order of 500–2,000 compounds studies per month without a proportional increase in resources. The use of 24-well monolayer (cell surface area ~0.33 cm^2) coupled with the use of LC-MS/MS significantly reduced the amount of compound (no more than 50 μg) required to perform permeability experiment in this model. A more in-depth discussion on the Caco-2 cell permeability model along with its strengths and weaknesses is discussed later in the chapter.

TC-7 is one of the sub-clones isolated from Caco-2 cells. Gres and colleagues (Gres *et al.*, 1998) have reported comprehensive comparison of TC-7 to its parental Caco-2 cells. TC-7 clone had very similar cell morphology as seen in Caco-2 cells: presence of brush-border membrane and microvilli, and formation of tight junctions. Although permeability of mannitol and PEG-4000 was identical in both cell lines, TC-7 had a significantly higher transepithelial electrical resistance (TEER) value at 21 days in culture and beyond. Permeability values of passively absorbed drugs obtained in TC-7 clone correlated equally well as in parental Caco-2 cells to the extent of absorption in humans.

2/4/A1 cells is a modified cell line reported (Tavelin *et al.*, 2003) to have been generated from fetal rat intestine. It is touted to better mimic the permeability of the human small intestine especially with regards to passive transcellular and paracellular drug transport. This immortalized cell line forms viable differentiated monolayers with tight junctions, brush-border membrane enzymes as well as the transporter proteins. Since the tight junctions in the Caco-2 cell line appear unrealistically tighter than the tight junctions in the endothelial cells in human intestine, the 2/4/A1 cells were proposed as a better model to study passively transported compounds *via* paracellular route. The TEER value in the 2/4/A1 cells reached a plateau of 25 $\Omega\,cm^2$ compared to a plateau of 234 $\Omega\,cm^2$ in the Caco-2 cells. The transport rate of poorly permeable compounds (e.g., mannitol and creatinine) in 2/4/A1 monolayers was comparable to that in the human jejunum, and was up to 300 times faster than that in the Caco-2 cell monolayers, suggesting that a cell line like 2/4/A1 will be more predictive for compounds that are absorbed *via* paracellular route.

MDCK Cells

MDCK is another cell line that is widely utilized for permeability assessment. These cells differentiate into columnar epithelial cells and form tight junctions (like Caco-2 cells) when cultured on semi-permeable membranes. The use of MDCK cell line as a model to evaluate the intestinal transport characteristics of compounds was first discussed in 1989 (Cho *et al.*, 1989). More recently, Irvine and colleagues (Irvine *et al.*, 1999) investigated the use of MDCK cells as a tool for assessing the membrane permeability properties of early drug discovery compounds. MDCK cells were grown on transwell-COL membrane culture inserts at high density and cultured for 3 days. Apparent permeability values of 55 compounds with known human absorption values were determined in MDCK cell system. For comparison purposes, the permeability of the same compounds was also determined using Caco-2 cells. The authors reported that the permeability obtained with MDCK cells correlated equally well (as with permeability determined in Caco-2 cells) to human absorption. Given the fact that Caco-2 cells are derived from human colon carcinoma cells whereas the MDCK cells are derived from dog kidney cells, it is very likely that the expression level of various transporters may be grossly different in these two cell lines. Species difference should therefore be kept in mind before using MDCK cells as a primary screening tool for permeability in early drug discovery. One major advantage of MDCK cells over Caco-2 cells is the shorter cultivation period (3 days vs. 3 weeks). A shorter cell culture time becomes a significant advantage considering reduced labor and reduced down time in case of cell contamination.

5.3.3 In Situ Methods

In situ perfusion of intestinal segments of rodents (rats or rabbits) is frequently used to study the permeability and absorption kinetics of drugs. The biggest advantage of the *in situ* system compared to the *in vitro* techniques discussed earlier is the presence of an intact blood and nerve supply in the experimental animals. Various modifications of the perfusion technique have been studied by different investigators: single pass perfusion (Komiya *et al.*, 1980), recirculating perfusion (Van Rees *et al.*, 1974), oscillating perfusion (Schurgers and DeBlaey, 1984), and the closed loop method (Doluisio *et al.*, 1969). Lennernas and colleagues (Lennernas *et al.*, 1997; Lennernas, 1998) have extended the perfusion studies to humans. This methodology is found to be highly accurate for predicting the permeability of passively transported compounds, however the use of a scaling factor has been recommended for predicting permeability of carrier-mediated compounds.

In situ experiments for studying intestinal drug absorption involve perfusion of drug solution prepared in physiological buffer through isolated cannulated intestinal segments. Absorption is assessed based on the disappearance of drug from the intestinal lumen. Hydrodynamics of the flow of buffer through the intestinal segments can influence the absorption characteristics of the drugs due to the effect of unstirred water layer. In this technique the difference in the concentration of the inlet and outlet flow is used to calculate the permeability. The presence of an intact blood supply, nerve and clearance capabilities at the site of absorption lead to an excellent experimental system which mimics the *in vivo* condition. Also, the input of compounds can be closely controlled with respect to concentration, pH, flow rate, intestinal region, etc. In the single pass perfusion approach the net effective permeability (P_{eff}) is calculated as:

$$P_{\text{eff}} = \frac{\left[-Q_{\text{in}}\text{Ln}\left(\frac{C_{\text{o}}}{C_{\text{i}}}\right)\right]}{2\pi RL},$$

where C_{i} and C_{o} are the inlet and outlet concentrations at steady state of the compound in the perfusate, Q_{in} is the inlet perfusion flow rate, and $2\pi RL$ is the mass transfer area available for absorption in the intestinal cylinder with length L and radius R.

Despite its advantages, the use of single pass perfusion method is severely limited because this method relies on the disappearance of compound from the luminal side as an indication of absorption, but the rate of decrease of concentration in the perfusate does not always represent the rate of absorption of the drug into the systemic circulation (especially for compounds undergoing presystemic or luminal metabolism). The method is also limited because of its cost factor as it requires a large number of animals to get statistically significant absorption data. Relatively high amounts of test compounds are also required to perform studies (>10 mg), which is not feasible in early drug discovery. It was also demonstrated that the surgical manipulation of intestine combined with anesthesia caused a significant change in the blood flow to intestine and had a remarkable effect on absorption rate (Uhing and Kimura, 1995).

5.3.4 *In Vivo Methods*

Extrapolation of animal permeability/absorption data to humans should be performed with caution because of potential species differences. Based on the similarity of the composition of epithelial cell membranes of the mammals, and the fact that absorption is basically an interaction between the drug and the biological membrane, permeability across the gastrointestinal tract should be similar across the species. However, there are physiological factors such as pH, GI motility, transit time, and differential distribution of enzymes and transporters that can affect the absorption leading to species variability. Excellent reviews and discussions (Dressman, 1986; Kararli, 1995; Lin, 1995) discuss species similarity and differences in ADME, with an attempt to address the question on limitations in extrapolating data from animals to humans.

Despite the fact that it is extremely resource intensive, *in vivo* evaluation of drug absorption in laboratory animals is a commonly used method to predict the extent of absorption of drug candidates in humans. Comparison of AUC (plasma concentration vs. time curve) values after intravenous, oral and intraportal (or intraperitoneal) administration can often indicate the absolute extent of *in vivo* absorption. More common practice for higher throughput *in vivo* absorption screening is that limited number of blood samples (3–4 time points up to 6 or 8 h) are collected after oral administration. As a result, a larger number of drug candidates can be ranked on the basis of drug concentration in the systemic circulation. It should be recognized that truncated pharmacokinetic studies (with reduced sampling times) are not optimal and the results obtained from such a study may not be accurate. However, the method is capable of differentiating well absorbed compounds from poorly bioavailable compounds.

5.3.5 *In Silico Methods*

Computational or virtual screening has received much attention in the last few years. *In silico* models which can accurately predict the membrane permeability of test drugs based on lipophilicity, hydrogen bonding capacity, molecular size, polar surface area, and quantum properties has the potential to specifically direct the chemical synthesis and therefore, revolutionize the drug discovery process. Such an *in silico* predictive model would minimize extremely time consuming steps of synthesis as well as experimental studies of thousands of test compounds.

Lipinski *et al.* (1997) proposed an *in silico* computational method for qualitatively predicting the developability of compounds. The "rules of five" proposed by them predicted lower permeabilities for compounds with more than five H-bond donors, ten H-bond acceptors, with molecular weight greater than 500 and *c* Log $P > 5$. Using this completely empirical model, useful permeability predictions were achieved for closely related analog series of compounds.

Quantitative structure property relationship (QSPR) has been recommended to predict human intestinal absorption without the need for actual compound synthesis. QSPR methods have been used to model physicochemical, chromatographic, and spectroscopic properties of compounds (Rubas *et al.*, 1993; Artursson and Borchardt, 1997). Several computational methods have been described to predict the intestinal absorption parameters based on factors such as polar surface area, molecular surface area, dynamic surface area, etc. But most reports involve *in silico* modeling studies performed on compounds closely related in structure thus making the model ineffective when applied to a wider structurally diverse data set. It is desirable to have predictive QSPR models that covers a diverse set of compounds with respect to their properties (e.g., physicochemical and pharmacological) as well as chemical structures.

Stenberg *et al.* (1999) compared the utility of three different predictive models for intestinal absorption. They demonstrated that molecular surface descriptors and descriptors derived from quantum mechanics were much more useful than the simple "rule of five" as the predictor of intestinal absorption. These *in silico* methods have a great potential as virtual screens in testing permeability of test drugs. The utility of polar molecular surface area (PSAd) as a predictor of intestinal absorption was demonstrated by Clark (1999). Similarly, the evaluation of the dynamic PSAd as a predictor of drug absorption was performed by Palm *et al.* (1998). PSAd of the compounds was calculated from all low energy conformations identified in molecular mechanics calculations in vacuum and in simulated chloroform and water environment. PSAd was determined to be a better predictor compared to the octanol–water partition coefficient or the experimentally obtained immobilized liposome chromatography retention time.

Wessel *et al.* (1998) proposed an *in silico* method that used the molecular structural descriptors to predict the human intestinal absorption of drugs. Topological descriptors (based on 2D information of the compound), electronic descriptors (partial atomic charge and dipole moment), geometric descriptors (surface area, volume, etc.), and hybrid descriptors (combination of molecular surface area and partial atomic charge) were used for analysis. Based on the root-mean-square errors seen in the training set as well as the study set, it was demonstrated that there was a correlation between structure and intestinal absorption. The data set in the study included 86 structurally diverse compounds and the results confirm the potential of applying QSPR methods for estimating absorption. With further improvements in the choice of the descriptors, this *in silico* method can be used as a potential virtual screen in drug discovery processes. To date, one major critical impediment to a successful *in silico* modeling is the lack of a sufficiently large data base with reliable information. Also, the *in silico* methods, even at their best are not as reliable as real experimental data for predicting the permeability and absorption characteristics of compounds. However, in spite of the limitations, there are commercially available software packages (QMPRplus and Oraspotter) that incorporate the *in silico* methodology for predicting *in vivo* human absorption numbers.

One draw-back of these modeling efforts is that most published work is based on small sets of permeability data (<100 compounds) that are collected from literature sources. Since the validity of any model depends on the data set used, a large and self-consistent data set is required for the development of global models with universal applicability. The quality and rigidity of the data set is also as important as the quantity of data. It is desirable to utilize the highest quality data to develop the models by using data generated from the same lab, using the same protocols and having appropriate controls. These *in silico* models, when carefully developed and rigorously validated, have the potential to be used in early screening set or library design. They can also play a role in lead optimization and preclinical candidate selection.

5.4 Comparison of PAMPA and Caco-2 Cells

5.4.1 Parallel Artificial Membrane Permeability Assay

PAMPA model was first introduced in 1998 (Kansy *et al.*, 1998), and since then numerous reports have been published illustrating the general applicability of this model as a high-throughput permeability screening tool (Avdeef, 2001; Di *et al.*, 2003; Ruell *et al.*, 2003; Kerns *et al.*, 2004; Balimane *et al.*, 2006). The model consists of a hydrophobic filter material coated with a mixture of lecithin/phospholipids dissolved in an inert organic solvent such as dodecane creating an artificial lipid membrane barrier that mimics the intestinal epithelium. The rate of permeation across the membrane barrier was shown to correlate well with the extent of drug absorption in humans. The use of 96-well microtiter plates coupled with the rapid analysis using a spectrophotometric plate reader makes this system a very attractive model for screening a large number of compounds and libraries. PAMPA is much less labor intensive than cell culture methods but it appears to show similar predictability. One of the main limitations of this model is that PAMPA underestimates the absorption of compounds that are actively absorbed *via* drug transporters. Despite the limitation, PAMPA may serve as an invaluable primary permeability screen during early drug discovery process because of its high-throughput capability.

5.4.1.1 PAMPA Study Protocol

A 96-well microtiter plate and a 96-well filter plate (Millipore, Bedford, MA, USA) were assembled into a "sandwich" such that each composite well was separated by a 125 μm microfilter disc (0.45 μm pores). The hydrophobic filter material of the 96-well filter plate was coated with 5 μL of the pION lipid solution and gently shaken to ensure uniform coating. Subsequently, the filter plate was placed on the microtiter plate containing 200 μL of 100 μM test compounds (dissolved in 1% DMSO/buffer), which constituted the donor well. The donor solution containing the test compound was prepared by dilution (100-fold) from a 10 mM stock solution in DMSO using the pION© buffer solution at pH 7.4 followed by filtration through a 0.20 μm polyvinylidene fluoride (PVDF) 96-well filter plate (Corning Costar, Corning, NY, USA). The receiver wells (i.e., the top of the wells) of the sandwich were hydrated with 200 μL of the specialized ionic buffer solution. The system was then incubated at room temperature for 4 h. At the end of the incubation time, samples were removed from the receiver and donor wells and analyzed by a UV-plate reader. The permeability studies were performed in triplicates (i.e., three wells per compound). The apparent permeability coefficient was estimated using the software provided by pION.

5.4.2 *Caco-2 Cells*

Caco-2 cell model has been the most popular and the most extensively characterized cell-based model in examining the permeability of drugs in both the pharmaceutical industry and academia. Caco-2 cells, a human colon adenocarcinoma, undergoes spontaneous enterocytic differentiation in culture and become polarized cells with well-established tight junctions, resembling intestinal epithelium in humans. It has also been demonstrated that the permeability of drugs across Caco-2 cell monolayers correlated very well with the extent of oral absorption in humans. In the last 10–15 years, the Caco-2 cells have been widely used as an *in vitro* tool for evaluating the permeability property of discovery compounds and for conducting in depth mechanistic studies (Chong *et al.*, 1996; Aungst *et al.*,

2000; Braun *et al.*, 2000; Horie *et al.*, 2003; Ungell, 2004; Balimane and Chong, 2005b).

5.4.2.1 Caco-2 Cell Culture

Caco-2 cells (passage #17) were obtained from the American Type Culture Collection (Rockville, MD) and cultured as described below. The cells were seeded onto 24-well polycarbonate filter membrane (HTS-Transwell® inserts, surface area: $0.33\,cm^2$) at a cell density of $80,000\,cells/cm^2$. The cells were grown in culture medium consisting of Dulbecco's modified Eagle's medium supplemented with 10% fetal bovine serum, 1% nonessential amino acids, 1% L-glutamine, 100 U/mL penicillin-G, and 100 μg/mL streptomycin. The culture medium was replaced every 2 days and the cells were maintained at 37 °C, 95% relative humidity, and 5% CO_2. Permeability studies were conducted with the monolayers cultured for approximately 21 days with the cell passage numbers between 50 and 80. Caco-2 cell monolayers with TEER values greater than $400\ \Omega\ cm^2$ were used.

5.4.2.2 Caco-2 cells Study Protocol

The transport medium used for the permeability studies was HBSS buffer containing 10 mM HEPES. Prior to all experiments, each monolayer was washed twice with buffer and TEER was measured to ensure the integrity of the monolayers. The concentration of test compounds ranged from 10 to 200 μM in this assay. The permeability studies were initiated by adding an appropriate volume of buffer containing test compound to either the apical (for apical to basolateral transport; A to B) or basolateral (for basolateral to apical transport; B to A) side of the monolayer. The monolayers were then placed in an incubator at 37 °C. Samples were taken from both the apical and basolateral compartment at the end of the incubation period (typically at 2 h) and the concentrations of test compound were analyzed by a high performance liquid chromatography method as described earlier (Chong *et al.*, 1996). Permeability coefficient (P_c) was calculated according to the following equation:

$$P_c = dA/(dt\,SC_o),$$

where dA/dt is the flux of the test compound across the monolayer (nmol/s), S is the surface area of the cell monolayer (cm^2), and C_o is the initial concentration (μM) in the donor compartment. The P_c values were expressed as nm/s. All permeability data reported in this chapter were generated at the Bristol–Myers Squibb PRI. A minor variation in the experimental procedure often resulted in a significant difference in the permeability measurement making an interlaboratory data comparison very difficult. Therefore, the study protocol of PAMPA and Caco-2 cell permeability assay is also included as a point of reference.

5.4.3 PAMPA and Caco-2 Cell: Synergies

PAMPA and Caco-2 cell models are often used in combination to evaluate the permeability properties of a large number of compounds at the early drug discovery stage. PAMPA model has been demonstrated in recent years to be an efficient, economical and high-throughput model. Caco-2 cell model, on the other hand, has been the gold standard model for over a decade for permeability screening in drug discovery phase. PAMPA captures the transcellular passive permeability across lipoidal membrane barrier without the contribution from pores or drug transporters. Caco-2 cell model is capable of incorporating not only the transcellular passive permeability but also the transporter mediated (efflux and influx) and the paracellular components of transport. Figure 5.2 demonstrates that a reasonably good prediction of the extent of absorption in humans can be obtained for ~22 marketed drugs by both Caco-2 cell and PAMPA models. Table 5.3 lists the permeability values in these two models and the fraction absorbed in humans for the validation set. These drugs were selected from different therapeutic areas and they represent diverse structures and physicochemical properties. Also, they are well known to be passively absorbed with no known major transporter involvement. It is evident that for passively absorbed drugs, both permeability models showed similar correlation. The two models also shared several other characteristics that are quite common. The dynamic range of permeability values (i.e., the fold increase in permeability value for highly absorbed drugs compared to a poorly absorbed drugs) was close to 2-orders of magnitude in both models with similar slope. The steepness of the slopes may reflect the predictability of the model for drugs that are moderately absorbed, and it appears that the two models have similar predictability for drugs with moderate permeability. The figure also demonstrates that it is very difficult to differentiate drugs in the mid range because of a large variability in the mid range of the curve. However, both models may work very well in binning the drug candidates into broad categories (high, medium, low) based on their permeability values, but might lack the sensitivity to accurately predict small differences in permeability values. The two models also demonstrate similar variability of data (20–30% coefficient of variation) and mass balance/recovery issues (many compounds had incomplete recovery).

Figure 5.3 represents the correlation observed between the two permeability models for ~100 internal research compounds from various drug discovery programs. The compounds were carefully selected from several research programs and they reflected a wide diversity of chemical space and physicochemical properties. The permeability values for these compounds were determined in both models using standard experimental conditions described above. Next, the compounds were classified into "low" and "high" permeability bins based on internal calibration data. Permeability value of ~100 nm/s was selected as a cut-off for both models. A cursory glance at the correlation figure might suggest a lack of linear correlation between the two models in terms of their permeability values. However, incorporation of the binning strategy (using 100 nm/s as the cut-off value) suggests that the two models demonstrate an acceptable agreement. Around 80% of the compounds tested were assigned to the same bin (i.e., the results were in agreement) by both the models. These compounds are represented in quadrants 2 and 3 in the figure. Almost half of those compounds were classified as "low" permeability by both methods (quadrant 3) and the other half

(a)

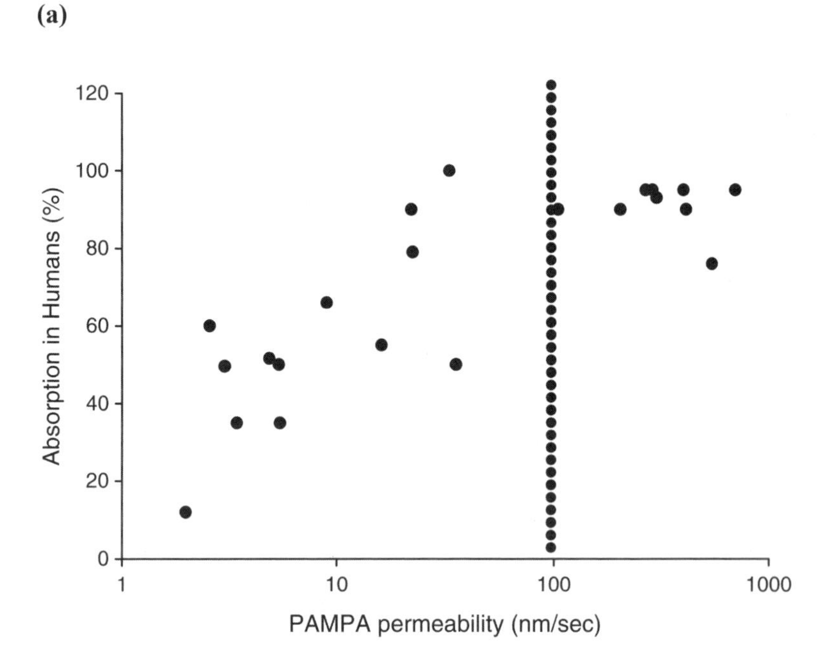

(b)

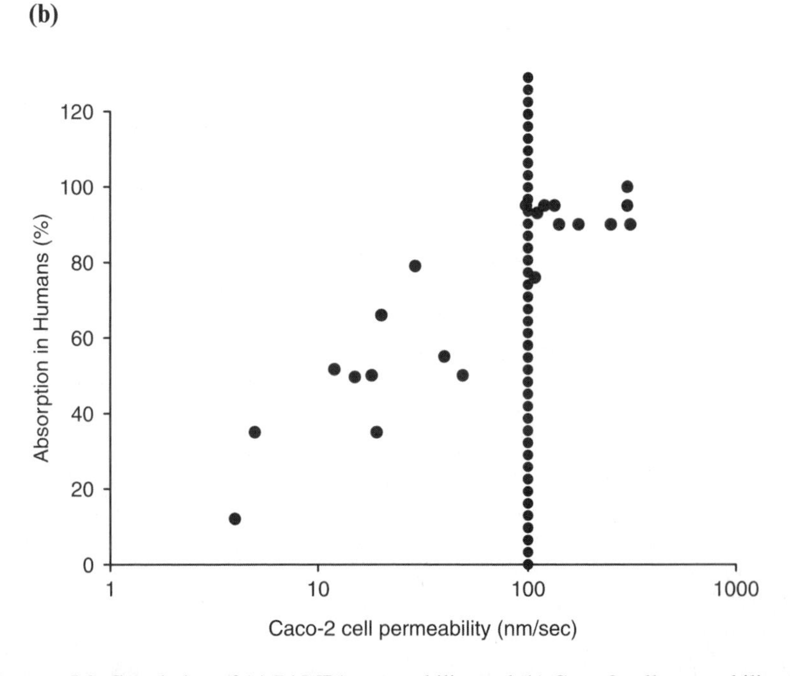

FIGURE 5.2. Correlation of (a) PAMPA permeability and (b) Caco-2 cell permeability with the extent of absorption in humans for marketed drugs. These drugs are known to be primarily absorbed *via* passive diffusive pathway. Each point is the mean of three or more repeats. The dotted line represents a cut-off of 100 nm/s

TABLE 5.3. PAMPA and Caco-2 cell permeability values of marketed drugs

Drug	PAMPA permeability (nm/s)	Caco-2 Cell permeability (nm/s)	Percentage of fraction absorbed in humans
Acebutalol	16 ± 11	40 ± 4	55
Alprenolol	299 ± 68	111 ± 30	93
Amiloride	5 ± 3	49 ± 8	50
Cimetidine	22 ± 3	29 ± 12	79
Desipramine	700 ± 170	300 ± 21	95
Dexamethasone	287 ± 11	134 ± 13	95
Etoposide	35 ± 4	18 ± 10	50
Fenaterol	3 ± 1	na	60
Furosemide	9 ± 1	20 ± 4	66
Hydralazine	105 ± 4	141 ± 16	90
Ketoconazole	542 ± 37	108 ± 18	76
Ketoprofen	22 ± 6	250 ± 26	90
Metformin	5 ± 2	12 ± 3	52
Metoprolol	266 ± 16	120 ± 10	95
Naproxen	33 ± 3	300 ± 41	100
Norfloxacin	5 ± 1	19 ± 3	35
Phenytoin	204 ± 9	310 ± 15	90
Propranolol	411 ± 110	175 ± 26	90
Sulfasalazine	2 ± 1	4 ± 1	12
Sulpiride	3 ± 1	5 ± 2	35
Terbutaline	3 ± 1	15 ± 6	50
Verapamil	399 ± 112	98 ± 19	95

na not available

were "high" permeability (quadrant 2). The compounds that fell in quadrants 1 and 4 were the disconnects between the two permeability models. The disagreement observed for ~20% of the compounds highlights the fundamental difference between the two models. PAMPA is an absorption model that captures uncontaminated transcellular diffusion across the lipid bilayers. It is completely devoid of any influx/efflux transporters or paracellular pores. On the other hand, cell tight junctions and various drug transporters (influx:PepT1 and efflux: P-gp, MRP2, BCRP, etc.) are expressed in the Caco-2 cell monolayer. The compounds shown in quadrant 4 (high PAMPA and low Caco-2 cell permeability) are likely to be substrates of efflux transporters that would limit their permeability in Caco-2 cell model. Because the PAMPA is a simple lipid bilayers, the permeability of compounds in quadrant 4 remain high. A handful of compounds from the quadrant 4 were evaluated for their P-gp substrate potential by performing bidirectional transport study in Caco-2 cells. As expected, they were all shown to be substrates with efflux ratio (ratio of B to A/A to B) higher than 3. The quadrant 1 represents compounds that have high Caco-2 cell permeability and low PAMPA permeability. Compounds in quadrant 1 are likely to permeate Caco-2 cell monolayer *via* paracellular pores and/or influx transporters. In general, a good agreement between the two permeability models for passively absorbed compounds is expected. If there is disagreement between these two models, it may provide a hint that specific influx

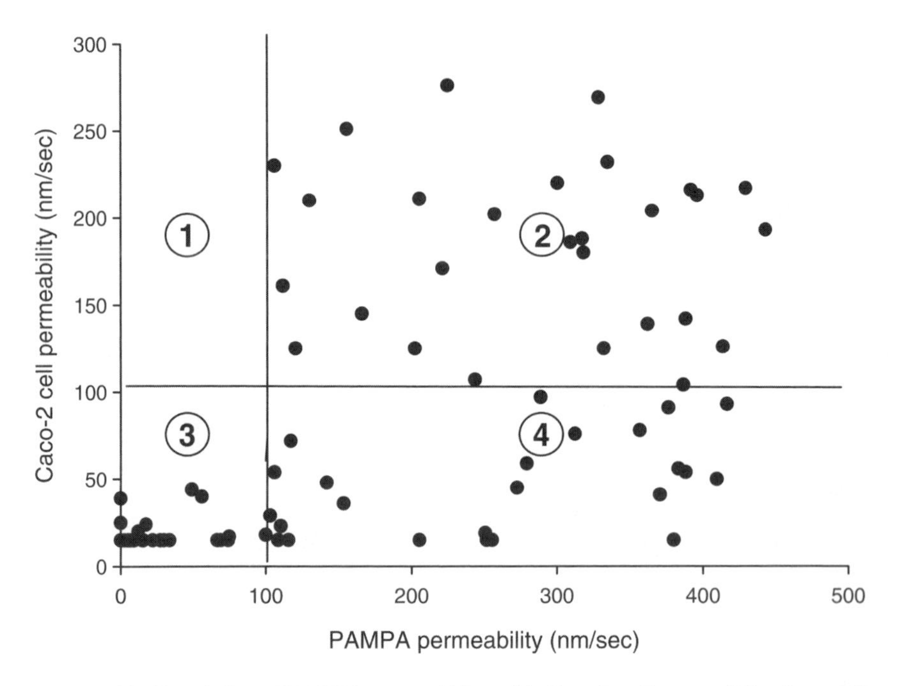

FIGURE 5.3. Correlation of PAMPA permeability with Caco-2 cell permeability for ~100 internal research compounds from the Bristol-Myers Squibb labs. Each point is the mean of three or more repeats

or efflux transporters are involved. And the follow-up studies using more mechanistic models (e.g., cell line with specific transporter expressed, xenopus oocytes, etc.) might be warranted.

A combination of PAMPA and Caco-2 cells is often used to assess the permeability of test compounds in the most cost-effective manner (Ano *et al.*, 2004; Kerns *et al.*, 2004). PAMPA model is preferred by virtue of its easy set-up and rapid operation. It is a low cost assay (cost per test is a fraction of Caco-2 cells cost) and is less resource and time intensive. Automated robotic driven PAMPA model is capable of screening thousands of compounds per week. Caco-2 cells, on the other hand, is a cost and resource intensive assay and not amenable to ultra HTS. To obtain the maximum permeability/absorption information in the least amount of time with minimal use of resources, a combination of PAMPA along with unidirectional Caco-2 cells (A to B permeability) is increasingly becoming popular. Figure 5.4a shows the Caco-2 cell permeability values obtained for a series of compounds within a chemotype from a typical research program. The majority of the compounds had "low" permeability values below 100 nm/s in the Caco-2 cell model. These compounds with low Caco-2 cell permeability would generally be assumed to be poorly absorbed in humans and may not make the initial cut to move forward (i.e., low priority). However, a combination assay (PAMPA and unidirectional Caco-2 cell) provided a better understanding of the

(a)

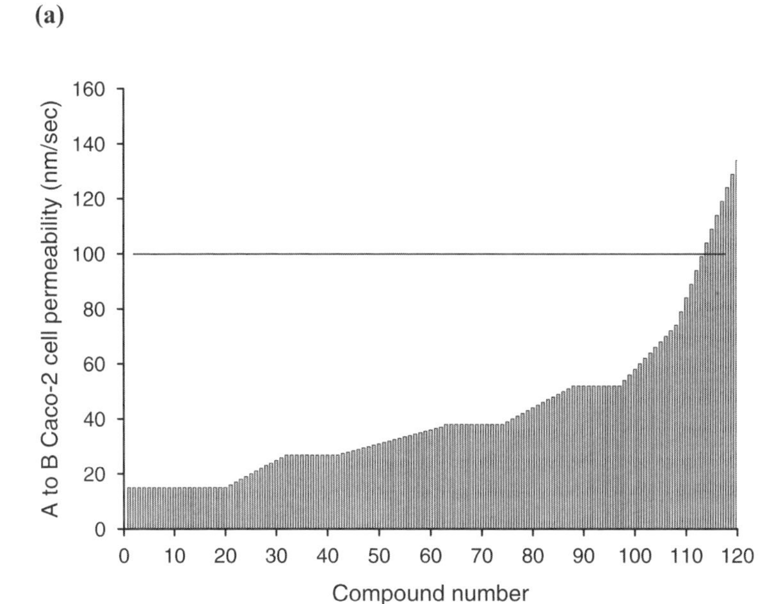

(b)

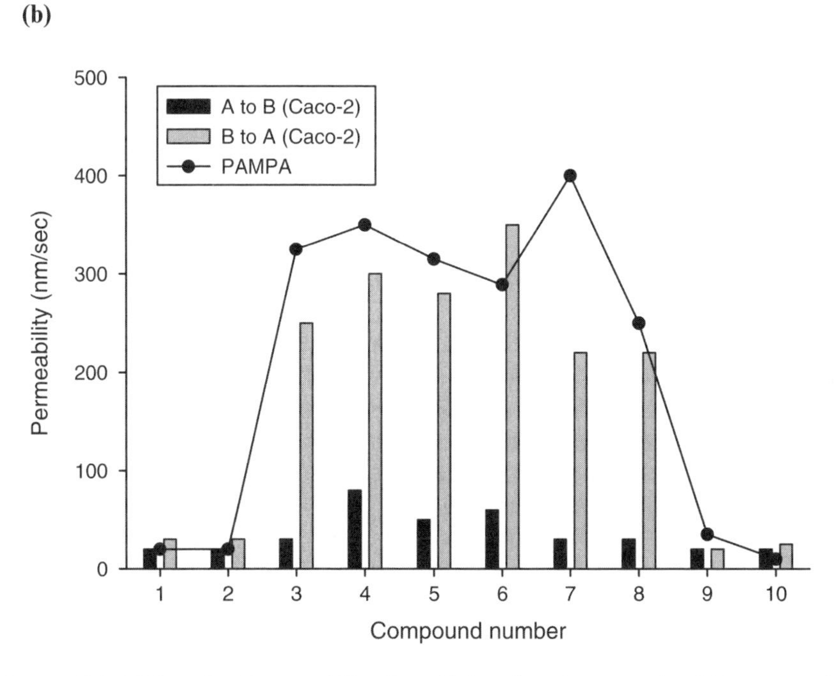

FIGURE 5.4. (a) Caco-2 cell permeability (A to B) for a list of compounds within a research program. Permeability values were obtained using Caco-2 cells cultured for ~21 days followed by transport studies at 200 μM. (b) Bi-directional Caco-2 cell permeability and PAMPA permeability values for some randomly selected compounds within the program. PAMPA study was performed at 100 μM

absorption characteristics of these compounds. A randomly selected set of compounds with Caco-2 cell permeability values lower than 100 nm/s was evaluated for their PAMPA permeability as well as their bidirectional permeability in Caco-2 cell model. The results from these two permeability models are shown in Fig. 5.4b. The combination assay demonstrated that the low A to B Caco-2 cell permeability of many compounds was due to the fact that they are substrates of P-gp (i.e., B to A permeability is much higher than A to B permeability) rather than due to low intrinsic permeability of the compound. The high PAMPA permeability coupled with low Caco-2 cell permeability would indicate an active efflux process (in Caco-2 cells) and not necessarily low intrinsic permeability of the test compound. It can be expected that these compounds may have more than adequate intestinal permeability if intestinal P-gp is saturated due to high drug concentration in the gut. In fact, many of these compounds were shown to exhibit good oral bioavailability in preclinical animal models. Despite being a substrate of P-gp, these compounds were well absorbed *in vivo*, and the high PAMPA permeability value was reflective of *in vivo* absorption.

A combination of PAMPA and unidirectional (A to B) Caco-2 cell permeability models can synergistically provide invaluable permeability/absorption assessment of test compounds. If both PAMPA and A to B Caco-2 cell permeability are low, then the compound can be deprioritized to have low intrinsic permeability and thus likely to have poor absorption in humans. However, if A to B Caco-2 cell permeability is low but PAMPA is high, then the compound is likely to be a substrate of P-gp and not necessarily a poorly permeable compound *in vivo*. At a relevant oral dose level, P-gp can be saturated to allow adequate absorption to occur in humans. A parallel combination assessment of PAMPA and A to B Caco-2 cell permeability followed by more detailed bidirectional Caco-2 cell assay might be a prudent path for research programs.

Drug absorption primarily occurs in the small intestine where the pH may vary from acidic to neutral and slightly basic (Dressman *et al.*, 1990; Russell *et al.*, 1993). In the upper small intestine where pH is likely to be more acidic, weakly acidic drugs exist primarily as unionized form and the passive transcellular pathway becomes the dominant permeation route. On the contrary, weakly basic drugs will be mostly in the form of ionized species, and consequently the passive transcellular route plays a minor role. Therefore, the apical pH used in the permeability assay becomes critical. Both PAMPA and Caco-2 cell permeability assays are typically performed at a single donor pH 6.5 to maintain adequate throughput (i.e., maximum number of compounds evaluated in minimum time frame). Single donor pH is perfectly appropriate for neutral and zwitterionic compounds where a change in pH is not expected to affect the ionization status. However, a majority of new drug candidates are either weak bases or weak acids. Therefore, the permeability of drug candidates with ionizable groups depends significantly on the experimental pH used. Table 5.4 lists a handful of drugs (acids and bases) along with their permeability values in the PAMPA model under two different experimental pH conditions. As expected, acidic drugs (e.g., ibuprofen, ketoprofen, piroxicam, etc.)

TABLE 5.4. Effect of pH on PAMPA permeability

	PAMPA (nm /s)		Percentage of FA
	pH 5.0	pH 7.4	
ACIDS			
Glipzide	562 ± 27	14[a] ± 1	95
Ibuprofen	756 ± 69	5[a] ± 1	100
Ketoprofen	567 ± 49	9[a] ± 1	90
Naproxen	500 ± 85	17[a] ± 2	100
Piroxicam	621 ± 42	21[a] ± 3	100
BASES			
Timolol	34[a] ± 2	595 ± 17	90
Hydralazine	2[a] ± 2	177 ± 44	90
Metoprolol	40[a] ± 8	512 ± 124	95

[a]Permeability value is inconsistently low despite complete absorption in humans

showed significantly higher permeability when studied at pH 5.0 than 7.4, and the basic drugs (e.g., metoprolol, timolol, etc.) had much higher permeability at pH 7.4 than 5.0. All drugs listed are almost completely absorbed in humans (>90%). But the permeability of these drugs when inappropriate pH was used is comparable to drugs that are not absorbed at all. Thus, it appears that an appropriate donor pH should be used depending on the physicochemical properties of test compounds. The degree of pH dependency is often very difficult to estimate because the pKa of a large number of drug candidates cannot be measured experimentally due to resources limitation. To minimize the occurrence of false negatives (i.e., classified as a poorly permeable when it is not), it becomes necessary to run the assay at two different pH condition (low of pH ~5 and high pH ~7.4) simulating the dynamic pH environment in the intestine to capture true intrinsic permeability. Because of a higher throughput potential, the PAMPA model can be used in the first tier pH dependent permeability study for similar chemotypes. Once an optimum pH is identified, all subsequent studies (including cell-based, tissue-based studies) can be carried out at that pH. However, it should be recognized that the permeability studies performed at multiple pH would negatively impact the throughput and cost involved.

5.4.4 PAMPA and Caco-2 Cell: Caveats

5.4.4.1 Transporter- and Paracellular-Mediated Absorption

PAMPA is a high-throughput, non-cell-based permeability model that provides estimates of the passive transcellular permeability property. The lack of any functional drug transporters and paracellular pores in PAMPA makes it an inappropriate model for compounds that are absorbed *via* transporter- and pore-mediated processes. However, the lack of transporter- and pore-mediated permeability might be an advantage of the PAMPA model. Because the PAMPA provides an

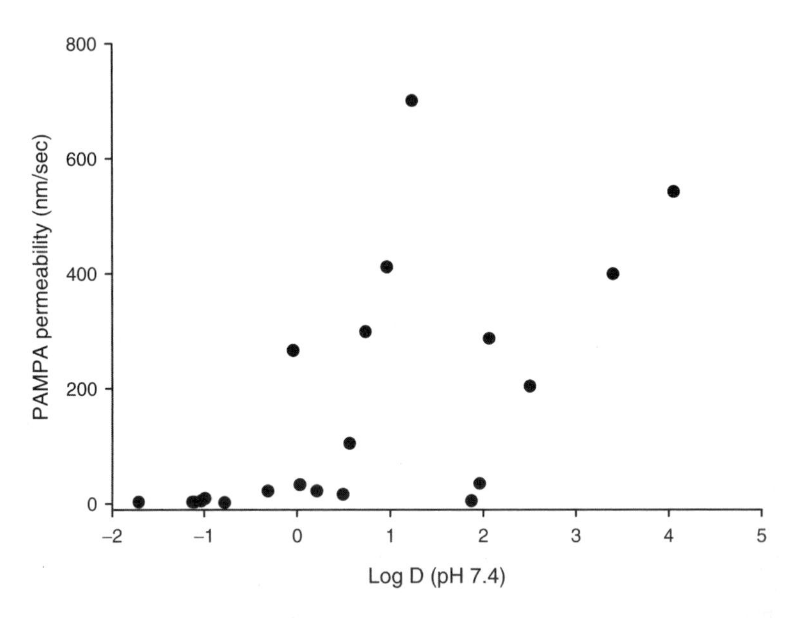

FIGURE 5.5. Correlation of PAMPA permeability with lipophilicity. Log *D* (pH 7.4) value was calculated using ACD/Log *D* module

uncontaminated transcellular passive permeability data, it could be more useful in constructing the structure–permeability relationship at the chemistry bench. Lipophilicity (most commonly expressed as Log *P* or Log *D* values) plays a major role in passive diffusion. An adequate lipophilicity is required for a permeant to travel across the phospholipid membrane. However, as shown in the Fig. 5.5, the PAMPA permeability of 22 marketed drugs (listed in Table 5.3) did not demonstrate any correlation with lipophilicity alone. As expected, several other factors (e.g., polar surface area, molecular volume/flexibility, hydrogen bonding, etc.) in addition to lipophilicity are involved in dictating the overall passive permeability of any compound. Although pharmaceutically important drug transporters (e.g., PEPT1, OCT, OAT, etc.) are functionally expressed in Caco-2 cells (Sun *et al.*, 2002; Anderle *et al.*, 2004; Behrens *et al.*, 2004), they are quantitatively under-expressed when compared to *in vivo* situation. For example, beta-lactam antibiotics (e.g., cephalexin, amoxicillin) and ACE inhibitors, that are known unequivocal substrates of dipeptide transporters, are poorly permeable across the Caco-2 cell monolayer despite the fact that they are completely absorbed *in vivo* (Chong *et al.*, 1996). This model is likely to generate false negatives with drug candidates that are transported by carrier-mediated process. Caco-2 cells have tight junctions that are significantly tighter compared to human intestine, and it does not differentiate drugs that are absorbed primarily *via* paracellular pathway. The low molecular weight hydrophilic compounds (e.g., metformin, ranitidine, atenolol, furosemide, hydrochlorothiazide, etc.) showed poor permeability (i.e., equal or less than mannitol) in Caco-2 cells despite adequate absorption (greater than 50% of dose) in humans. Therefore, models such as PAMPA and Caco-2 cells can only serve as a one-way screen such that compounds with high permeability in these models are typically well absorbed, however, compounds with low permeability cannot be ruled out as poorly absorbed compounds in humans.

5.4.4.2 Incomplete Mass-Balance Due to Nonspecific Binding

Nonspecific drug binding to plastic devices and cells during the permeability study is a common problem which often makes the data interpretation difficult. Permeability is calculated based on several factors: the amount of drugs appeared in the receiver compartment, the initial concentration in the donor compartment, and the surface area of the physical barrier (e.g., lipid bilayers and cell monolayer). When significant drug loss occurs during the incubation due to a nonspecific binding, two things occur (1) the concentration in the donor compartment (a driving force) is reduced and (2) the concentration in the receiver compartment is artifactually reduced. This will lead to an underestimation of permeability estimates, and potentially lead to false negatives. "Cacophilicty" or "membrane retention" has been used as a term to describe the drug binding to Caco-2 cell monolayer, and the significant binding results in an incomplete recovery. Incomplete recovery is particularly common with lipophilic drug candidates. A few approaches may be able to minimize the nonspecific binding. Rather than using the initial donor concentration, the final donor concentration at the termination of incubation can be used. Assumption is that the nonspecific binding occurs relatively quickly, therefore, the final donor concentration is a better estimate for the concentration gradient between two compartments. For certain drug candidates with poor recovery, the results can be significantly different depending on the calculation method used. Another approach involves addition of serum proteins (bovine serum albumin in case of Caco-2 cells) or surfactants (in case of PAMPA) to the receiver compartment to minimize nonspecific binding, therefore, improving the assay recovery and overall predictability of the model (Aungst *et al.*, 2000; Krishna *et al.*, 2001; Saha and Kou, 2002).

5.4.4.3 Inadequate Aqueous Solubility

Drug candidates at the early discovery stage are often optimized in terms of SAR around potency and pharmacological activity, and the SAR generally lead to rather lipophilic and poorly soluble candidates (solubility in aqueous buffer <0.01 mg/mL). As a result, significant percentage of new drug candidates cannot be evaluated in the permeability models due to their poor aqueous solubility. This is particularly problematic in the cell-based model because the cells do not tolerate the typical organic cosolvents (e.g., DMSO, PG, etc.) very well. Beyond a small percentage of cosolvent (>1%), the cell tight junction is easily compromised making the data interpretation difficult or impossible (Dimitrijevic *et al.*, 2000; Rege *et al.*, 2001, 2002). Consequently, for some discovery programs, the permeability data cannot be provided for a majority (>95%) of new compounds because they tend to come from a similar chemotype which may share a problematic

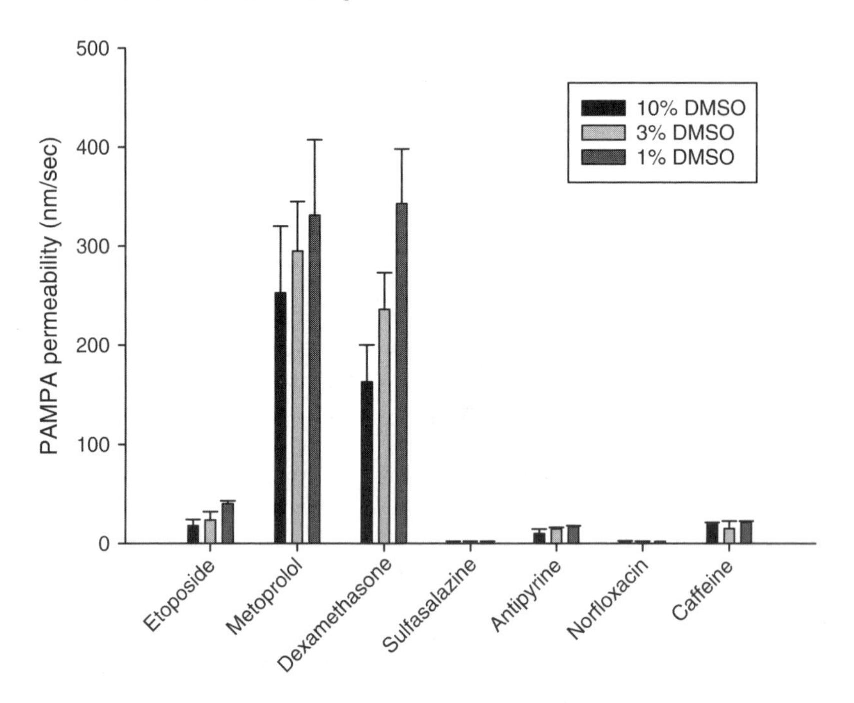

FIGURE 5.6. Effect of DMSO on the PAMPA permeability

pharmacophore in terms of aqueous solubility. PAMPA model has a significant advantage over Caco-2 cells in this regard. Figure 5.6 shows the effect of various concentrations of DMSO on the PAMPA permeability of some probe compounds (high and low P_c compounds). Contrary to a cell-based model, where higher than 1% DMSO would compromise the cell monolayer leading to unreliable results, the PAMPA permeability was consistent in the presence of DMSO up to 10%. Having higher cosolvent concentration not only increases the percentage of compounds that can be successfully studied in a permeability assay, but it also improves the mass-balance recovery. A higher cosolvent concentration is likely to minimize the physical loss (e.g., nonspecific binding to devise and membrane) during the permeability study.

5.4.4.4 Other Experimental Variability

It is well documented that the Caco-2 cell permeability of the same set of drugs obtained from different laboratories varies significantly (Walter and Kissel, 1995; Ungell, 2004; Balimane and Chong, 2005a). There are a variety of factors that can influence the outcome. Minor differences in cell culture conditions (e.g., seeding density, feeding frequency and composition of the cell media, etc.), experimental protocol (e.g., initial concentration of drugs, composition of the permeability buffer, pH and monolayer washing steps, etc.) and age of the cells (e.g., passage number, culture duration and tightness of junction) can produce dramatic differences in the permeability values. In addition, the function of drug transporters expressed in the cell-based models can fluctuate significantly with difference in culture conditions (Walter and Kissel, 1995; Anderle *et al.*, 2004). PAMPA permeability model, on the other hand, is a relatively rigid model less prone to interlaboratory variability problems. The lipid solution used to create the bilayers membrane is the only component of the model that can incorporate fluctuations in permeability value. Moreover, since PAMPA captures only the passive diffusive permeability and is devoid of any transporter proteins, other experimental factors play a negligible role.

5.5 P-gp Studies Using Caco-2 Cells

The bidirectional permeability assay, where the basolateral to apical (secretory direction, B to A) permeability is compared to the apical to basolateral (absorptive direction, A to B) permeability, is regarded as the gold standard in identifying P-gp substrates because it is functionally the most direct method of measuring efflux characteristics of drug candidates. Compounds with an efflux ratio (ratio of B to A/A to B) greater than 2–3 are typically considered as P-gp substrates. However, it is well known that efflux transporters other than P-gp (e.g., MRP2, BCRP, etc.) are also functionally expressed in the Caco-2 cells. Therefore, a simple bidirectional difference may not ascertain that the compounds being tested is indeed P-gp substrates. As a confirmatory study, a follow-up bidirectional experiment is routinely repeated in the presence of known inhibitors of P-gp, BCRP and MRP2. GF120918 at 2–4 μM is often used to selectively inhibit the P-gp-based efflux transport. However, some recent publications have reported that GF120918 interacts not only with P-gp but also with the BCRP (Maliepaard *et al.*, 2001; Woehlecke *et al.*, 2003). Similarly, MK-571 and FTC are used to selectively inhibit the MRP2 (Chen *et al.*, 1999; Dantzig *et al.*, 1999) and BCRP activity (Volk and Schneider, 2003; Zhang *et al.*, 2004), respectively. Comparison of efflux ratio in the absence and presence of these specific inhibitors can delineate the potential role of the individual efflux transporter. Because it is a cell-based assay with a physical barrier (lipid bilayer), the test compounds must have adequate cell permeability for the bidirectional P-gp assay. One major drawback of the bidirectional permeability assay is that the P-gp substrates with insufficient transcellular permeability cannot be identified. Drugs such as famotidine and ranitidine are substrates for secretory transporter proteins (Lee *et al.*, 2002) but often they fail to be detected as P-gp substrates due to their low passive permeability.

Several efflux (e.g., P-gp, MRP2, and BCRP) and influx transporters (e.g., PEPT1) are expressed in Caco-2 cells under standard cell culture conditions. However, the functional expression of drug transporters in Caco-2 cells may vary significantly depending on the passage number and minor changes in culture conditions. For example, the efflux ratio of digoxin or sulfasalazine (both suggested to be efflux transporter substrate) varied drastically with passage number. It is interesting to note that often the lack of P-gp functional expression becomes

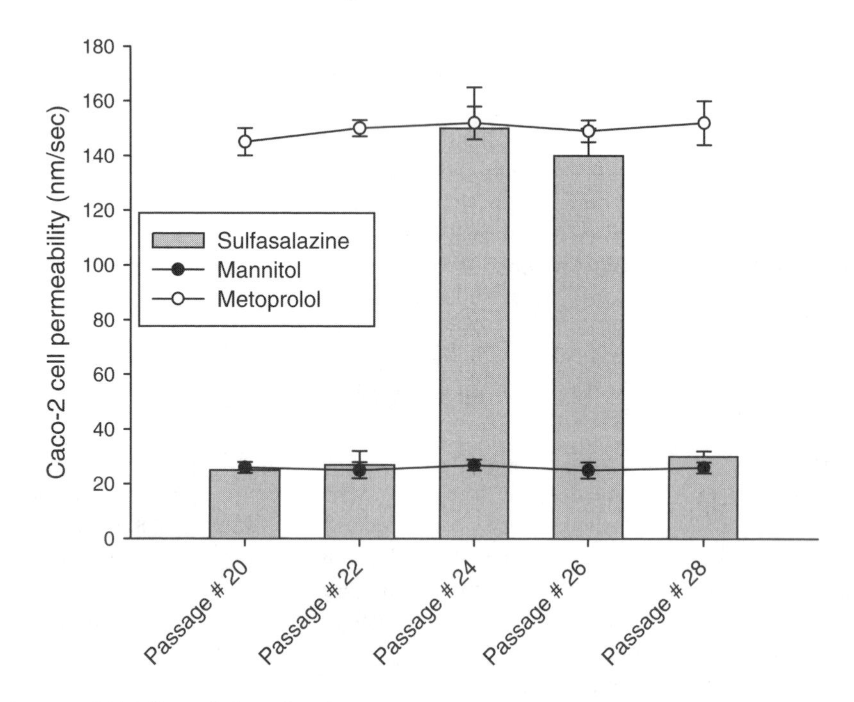

FIGURE 5.7. Effect of Caco-2 cell passage number on the permeability of sulfasalazine, mannitol, and metoprolol

problematic sporadically. There was no identifiable trend with the change in passage number. Figure 5.7 demonstrates that the levels of P-gp expression in Caco-2 cells can vary significantly depending on the cell passage number. Mannitol (a hydrophilic paracellular marker) and metoprolol (a lipophilic transcellular marker) demonstrated consistent permeability values over the different passage numbers. The reproducibility for these two compounds confirms that the cell monolayers are not compromised. However, sulfasalazine, a P-gp substrate that typically has low A to B permeability value (<30 nm/s), showed significantly higher permeability (>120 nm/s) for only certain passage numbers. The variability in the levels of expression of P-gp was directly reflected by the variability in the permeability values of sulfasalazine. Similarly, the levels of P-gp expression have also been known to fluctuate widely amongst the wells even in a single passage study. Digoxin efflux ratio can vary from low of 8 to a high of 20 in different wells of the same passage number Caco-2 cells highlighting the variability in the expression of efflux transporter.

Because of inconsistent P-gp functional expression in Caco-2 cells, the cell line which over-express efflux transporters might be a better *in vitro* tool to examine the drug–transporter interactions. The engineered cell lines selectively express the transporter of interest and facilitate interaction studies with a specific transporter in isolation. Bidirectional studies performed in these cell lines (over-expressing only one transporter) would specifically demonstrate the involvement of a single transporter without any interplay with other transporters. Follow-up studies can be conducted in the presence of selective inhibitor to further confirm the involvement of the transporter (viz: transporter phenotyping). MDCK and LLC-PK1 cell lines stably transfected with a specific efflux transporter (e.g., MDR1, mdr1, MRP2, and BCRP) may be used to tease out transporter interactions at early discovery stage. The functional expression of transporters in these cell lines appears to be more stable compared to Caco-2 cells.

5.5.1 Experimental Factors Effecting Efflux Ratio

Currently there is no universally accepted standard study protocol for conducting the cell based bidirectional permeability assay. Different laboratories perform these studies under different experimental conditions that often showed a large interlaboratory variability. Standardization of experimental conditions (e.g., pH, drug concentration and duration of incubation, etc.) can provide better consistency of the results and significantly minimize the occurrence of false negatives (i.e., unable to identify a true substrate). The use of an optimal substrate concentration is a key parameter for attaining accurate results from this model. Performance of bidirectional studies at high concentrations (50 μM or more) often lead to saturation of the efflux transporter and result in efflux ratio of unity even for well-known P-gp substrates. Figure 5.8 demonstrates the dramatic effect of using the lower (3 μM) concentration for improving the utility of this model. Classical

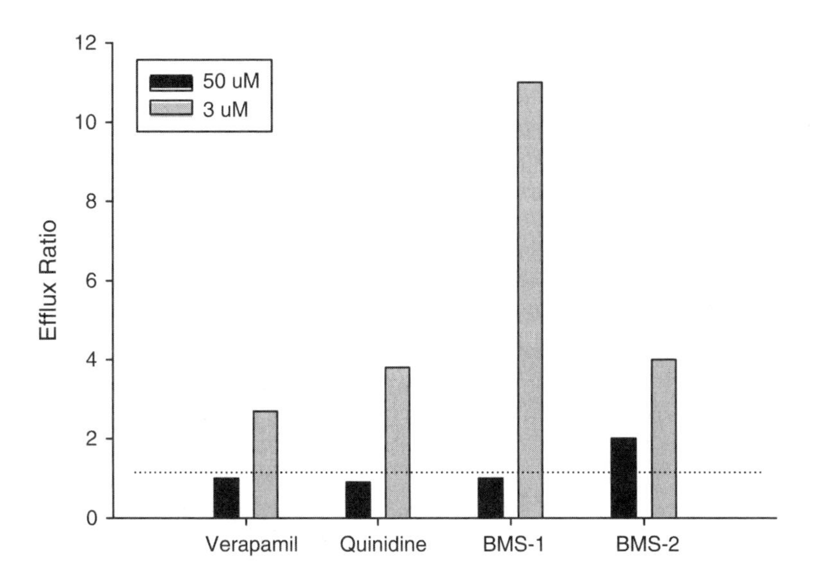

FIGURE 5.8. Effect of substrate concentration (3 μM vs. 50 μM) on the efflux ratio (ratio of B to A/A to B) in the Caco-2 cell bidirectional study. The efflux ratio was calculated using the mean data. The coefficient of variation in the permeability data (both direction) is typically less than 25% amongst replicates

P-gp substrates such as verapamil, quinidine and two internal research compounds are shown to have efflux ratio (ratio of B to A/A to B) of ~1 when studied at 50 μM. However, when they were studied at lower concentration of 3 μM, they demonstrated unequivocal efflux characteristics with efflux ratio >2. The P-gp is saturable at high substrates concentrations and thus at 50 μM, the P-gp is knocked out functionally leading to false negative data. Using lower substrate concentrations (<5 μM) could potentially minimize the saturation of the efflux transporters and can significantly improve the predictability of the assay. It should be recognized that lowering the substrate concentration imposes an analytical challenge, and one must make sure that the analytical technique is capable of measuring the sample concentrations with adequate accuracy.

The preliminary screen at a low substrate concentration is an efficient approach to evaluate a large number of compounds and maintain sufficient throughput. However, "yes or no" classification may not be very useful because it is difficult to quantify (if possible) the clinical significance of the potential role of P-gp in the drug disposition. As a follow-up study, it is recommended that a range of substrate concentrations to be tested, and the estimated Km value can be related to *in vivo* relevant drug concentrations either in the intestine (during absorption) or systemic circulation (during distribution and excretion). One of the common limitations in conducting concentration dependency experiment is the lack of adequate aqueous solubility of new drug candidates (i.e., poor solubility does not allow a wide range of concentration to be tested).

Another potential issue with cell-based models is that the integrity of tight junctions in the Caco-2 cell monolayers can be compromised (i.e., becomes leaky) when incubated with test compounds, and the extent of damage is typically concentration dependent. The permeability across the compromised monolayer is often much higher compared to the intact monolayers, and the permeability becomes artificially high in both direction. In that case, the efflux ratio often becomes unity, and the test compound is classified as a nonsubstrate even if it is a true substrate (again false negative). To detect cell damage during the incubation, one can measure TEER value before and after the study. But the change in the TEER value is not very sensitive to detect a minor cell damage. More definitive way to detect cell damage is to coincubate a paracellular route marker such as radiolabeled mannitol in the test run. It is also possible to see false positives. It appears that the basolateral membrane of Caco-2 cell monolayer is more sensitive than the apical membrane (i.e., the cell damage occurs more frequently in the B to A direction than the A to B direction). When this occurs, the B to A permeability is significantly greater than A to B permeability, and the test compound is classified as a P-gp substrate even if it is a nonsubstrate (false positives). Figure 5.9 shows that the cell damage due to BMS-X is concentration dependent because the mannitol permeability was significantly higher when incubated with 50 μM compared to 10 μM BMS-X, and it also illustrated that the only basolateral membrane was sensitive to BMS-X. When the substrate concentration was reduced to 10 μM, the B to A permeability was reduced and became similar to A to B permeability. Therefore, lowering the substrate concentration is a good way

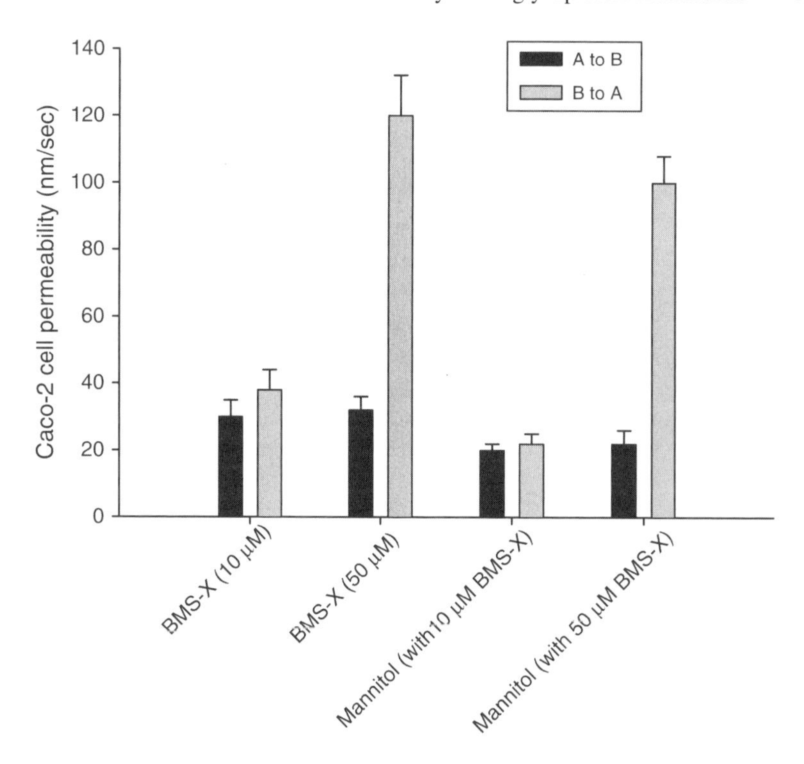

FIGURE 5.9. Effect of drug concentrations on the integrity of Caco-2 cell monolayer

to minimize the potential cell damage, and consequently the incidence of both false negatives and false positives.

Another experimental factor that needs to be fully optimized is the pH used in the bidirectional permeability studies. Optimization of the apical pH is not as straightforward as the substrate concentration. Based on the p*K*a of the test compounds, the change in apical pH can lead to dramatically different extent of ionized and unionized fractions in the apical compartment and consequently significant changes in the permeability value. Figure 5.10 demonstrates the effect apical pH on the efflux ratio observed for well-known P-gp substrates. Digoxin, a neutral P-gp substrate, had no ionization changes with pH and showed a uniform efflux ratio ~10 at all three pH. However, saquinavir, a weak base, showed a significant change in the efflux ratio where the ratio was greater than 50 at an apical pH of 5.5 as compared to the ratio of ~10 at pH 7.4. On the contrary, sulfasalazine (contains a carboxylic acid) demonstrated a higher efflux ratio at an apical pH of 7.4. The higher efflux ratio observed with saquinavir and sulfasalazine at an apical pH of 5.5 and 7.4, respectively, is primarily due to lower apical to basolateral permeabilities (i.e., smaller denominator in the ratio calculation) rather than the changes in the basolateral to apical permeability. Therefore, the apical pH needs to be optimized depending on the type of ionizable function group of the compounds of

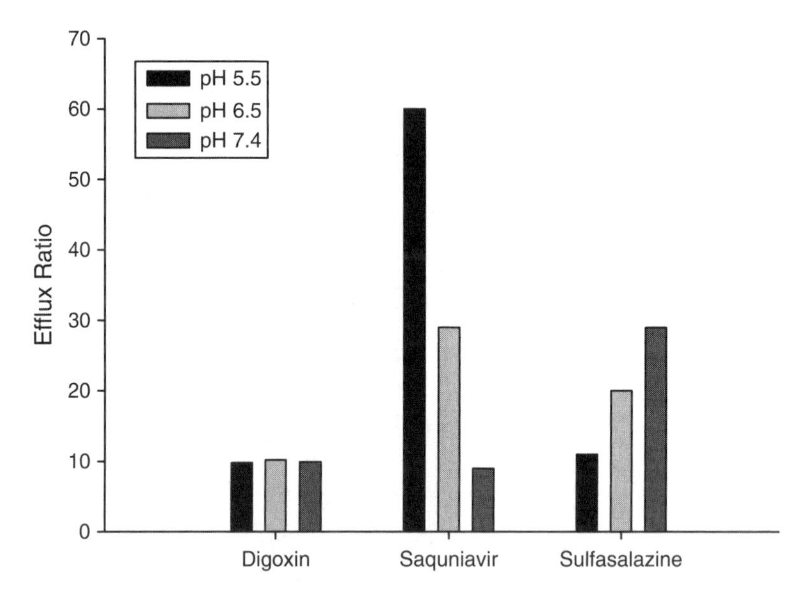

FIGURE 5.10. Effect of apical pH (5.5, 6.5 and 7.4) on the efflux ratio of acidic, basic and neutral P-gp substrates in the Caco-2 cell bidirectional study. The efflux ratio was calculated using the mean data. The coefficient of variation in the permeability data (both direction) is typically less than 25% amongst replicates

interest. Prior information on the p*K*a and acidic/basic nature of the compound will be very useful in the P-gp substrate assay. Other experimental variables such as incubation duration, composition of transport buffer, presence/absence of serum, cell-culturing conditions, etc. need to be calibrated using a set of known P-gp substrates. Factors such as culturing conditions, composition of media, cell plate architecture (12-well vs. 24-well Transwell) all have a significant effect on the expression of P-gp and the interlaboratory variability is often significantly large. Therefore, a direct comparison of the results obtained from different laboratories should be done cautiously.

5.6 Conclusions

One of the most important challenges facing the pharmaceutical industry at present is to develop high-throughput, cost effective and highly predictive screening models of drug absorption that can be used during the decision making process early in drug discovery. Although physicochemical parameters such as lipophilicity, charge, molecular weight, etc. are often used as initial indicators of absorption, they are often not reliable because of their inability to incorporate the physiologic conditions. On the other hand, the conventional experimental methods for assessing the absorption characteristics are not streamlined to keep pace with the large number of compounds synthesized by combinatorial chemistry. The available experimental models have various pros and cons and their judicious use can increase the likelihood of progressing drugs with favorable oral absorption characteristics. Physicochemical and/or *in silico* methods that are rapid and require no usage of animal tissues should be used as the primary screening models in early drug discovery stage. Subsequently, automated high-throughput *in vitro* systems (cell culture-based or animal tissue-based) can be used in selecting and optimizing the chemical leads to identify potential drug candidates. Multiple complementary screening models involving *in situ* and *in vivo* methods should also be utilized simultaneously at this stage to minimize "false-positives" and "false-negatives" from getting into the development process.

Caco-2 cells and PAMPA are two valuable research tools that are currently the method-of-choice for screening compounds for absorption and P-gp interaction potential. Despite the popularity and acceptability of PAMPA and Caco-2 cell models, it is important to recognize the caveats associated with these models to fully realize their potential. Standardization of experimental variables to develop a "uniform" methodology is an important first step prior to implementation of these models in high-throughput mode. Calibration of the models with appropriate reference probes (known marketed drugs as well as internal research compounds) is essential to maintain the validity of the model. These reference compounds should cover a broad structural and physicochemical space and should cover a range of solubility/permeability. Other probes to represent paracellular and efflux (P-gp) transport are also recommended. These reference compounds should be included as quality control set in every test run performed with unknown compounds. Like all other *in vitro* models, both PAMPA and Caco-2 cells have strengths and weaknesses. A thorough understanding of the rationale underlying the caveats (such as low recovery, solubility, etc.) associated with these models could certainly help putting the results in the proper perspectives. The results obtained from these assays should not be interpreted in isolation but should be assessed in conjunction with solubility and stability characteristics of the compounds. Finally, it is important to realize that simplistic *in vitro* models like PAMPA and Caco-2 cells are inadequate to represent the complicated absorption machinery in the human intestinal tract. Ultimately, it is the judicious use of these models in combination with *in vivo* studies that would enhance our ability to predict the drug disposition in humans and ensure that only drug candidates with a high-developability potential are moved forward into the development pipeline.

References

Adachi Y, Suzuki H and Sugiyama Y (2001) Comparative studies on in vitro methods for evaluating in vivo function of MDR1 P-gp. *Pharmaceutical Research* **18**:1660–1668.

Anderle P, Huang Y and Sadee W (2004) Intestinal membrane transport of drugs and nutrients: genomic membrane transporters using expression microarray. *European Journal of Pharmaceutical Sciences* **21**:17–24.

Ano R, Kimura Y, Shima M, Matsuno R, Ueno T and Akamatsu M (2004) Relationship between structure and high-throughput screening permeability of papetide derivatives and related compounds with artificial membranes: application to prediction of Caco-2 cell permeability. *Bioorganic & Medicinal Chemistry* **12**:257–264.

Arturrson P (1991) Cell cultures as models for drug absorption across the intestinal mucosa. *Critical Reviews in Therapeutic Drug Carrier Systems* **8**:305–330.

Artursson P and Borchardt R (1997) Intestinal drug absorption and metabolism in cell cultures: Caco-2 and beyond. *Pharmaceutical Research* **14**:1655–1658.

Artursson P and Karlsson J (1991) Correlation between oral drug absorption in humans and apparent drug permeability coefficients in human intestinal epithelia (Caco-2) cells. *Biochemical and Biophysical Research Communications* **175**:880–890.

Artursson P, Palm K and Luthman K (1996) Caco-2 monolayers in experimental and theoretical predictions of drug transport. *Advanced Drug Delivery Reviews* **22**:67–84.

Aungst B, Nguyen N, Bulgarelli J and Oates-Lenz K (2000) The influence of donor and reservoir additives on Caco-2 permeability and secretory transport of HIV protease inhibitors and other lipophilic compounds. *Pharmaceutical Research* **17**:1175–1180.

Avdeef A (2001) Physicochemical profiling (solubility, permeability and charge state). *Current Topics in Medicinal Chemistry* **1**:277–351.

Balimane PV and Chong S (2005a) Cell culture-based models for intestinal permeability: a critique. *Drug Discovery Today* **10**:335–343.

Balimane PV and Chong S (2005b) A combined cell based approach to identify P-glycoprotein substrates and inhibitors in a single assay. *International Journal of Pharmaceutics* **301**:80–88.

Balimane PV, Chong S and Morrison RA (2000) Current methodologies used for evaluation of intestinal permeability and absorption. Journal of Pharmacological and Toxicological Methods **44**:301–312.

Balimane PV, Han YH and Chong S (2006) Current industrial practices of assessing permeability and P-glycoprotein interaction. *AAPSJ* **8**:E1–13.

Behrens I, Kamm W, Dantzig A and Kissel T (2004) Variation of peptide transporter (PepT1 and HPT1) expression in Caco-2 cells as a function of cell origin. *Journal of Pharmaceutical Sciences* **93**:1743–1754.

Beigi F, Yang Q and Lundahl P (1995) Immobilized-liposome chromatographic analysis of drug partitioning into lipid bilayers. Journal of Chromatography *A*. **704**:215–321.

Braun A, Hammerle S, Suda K, Rothen-Rutishauser B, Gunthert M and Wunderli-Allenspach H (2000) Cell cultures as tools in biopharmacy. *European Journal of Pharmaceutical Sciences* **11**:S51–S60.

Carr K and Toner P (1984) Morphology of the Intestinal Mucosa, in *Pharmacology of the Intestine* (Csaky T ed) pp 1–50, Springer, Berlin Heidelberg New York.

Chen Z, Kawabe T, Ono M, Aoki S, Sumizawa T, Furukawa T, Uchiumi T, Wada M, Kuwano M and Akiyama S (1999) Effect of multidrug resistance-reversing agents on transporting activity of human canalicular multispecific organic anion transporter. *Molecular Pharmacology* **56**:1219–1228.

Cho M, Thomson D, Cramer C, Vidmar T and Scieszka J (1989) The MDCK epithelial cell monolayer as a model cellular transport barrier. *Pharmaceutical Research* **6**:71–77.

Chong S, Dando S, Soucek K and Morrison R (1996) In vitro permeability through Caco-2 cells is not quantitatively predictive of in vivo absorption for peptide-like drugs absorbed via the dipeptide transporter system. *Pharmaceutical Research* **13**:120–123.

Clark D (1999) Rapid calculation of polar molecular surface area and its application to the prediction of transport phenomena. 1. Prediction of intestinal absorption. *Journal of Pharmaceutical Sciences* **88**:807–814.

Dantzig A, Shepard R, Law K, Tabas L, Pratt S, Gillespie J, Binkley S, Kuhfeld M, Starling J and Wrighton S (1999) Selectivity of the multidrug resistance modulator, LY335979, for P-gp and effect on CYP-450 activities. *Journal of Pharmacology and Experimental Therapeutics* **290**:854–562.

Daugherty A and Mrsny R (1999) Regulation of the intestinal epithelial paracellular barrier. *Pharmaceutical Sciences and Technology Today* **2**:281–287.

Di L, Kerns EH, Fan K, McConnell OJ and Carter GT (2003) High throughput artificial membrane permeability assay for blood–brain barrier. *European Journal of Medicinal Chemistry* **38**:223–232.

Dimitrijevic D, Shaw A and Florence A (2000) Effects of some non-ionic surfactants on transepithelial permeability in Caco-2 cells. *Journal of Pharmacy and Pharmacology* **52**:157–162.

Doluisio J, Billups N, Dittert L, Sugita E and Swintosky J (1969) *Journal of Pharmaceutical Sciences* **58**:1196–1200.

Dressman J (1986) Comparison of canine and human gastrointestinal physiology. *Pharmaceutical Research* **3**:123–131.

Dressman J, Amidon G and Fleisher D (1985) Absorption potential: estimating the fraction absorbed for orally administered compounds. *Journal of Pharmaceutical Sciences* **74**:588–589.

Dressman J, Berardi R, Dermentzoglou L, Russell T, Schmaltz S, Barnett J and Jarvenpaa K (1990) Upper gastrointestinal (GI) pH in young, healthy men and women. *Pharmaceutical Research* **7**:756–761.

FDA (2004) Challenges and opportunity on the critical path to new medical products. *FDA Report.*

Grass G and Sweetana S (1989) A correlation for permeabilities of passively transported compounds in monkey and rabbit jejunum. *Pharmaceutical Research* **6**:857–862.

Gres M, Julian B, Bourrie M, Meunier V, Roques C, Berger M, Boule, Berger Y and Fabre G (1998) Correlation between oral drug absorption in humans, apparent drug permeability in TC-7 cells, a human epithelial intestinal cell line: comparison with the parental Caco-2 cell line. *Pharmaceutical Research* **15**:726–733.

Hidalgo I (2001) Assessing the absorption of new pharmaceuticals. *Current Topics in Medicinal Chemistry* **1**:385–401.

Hillgren K, Kato A and Borchardt R (1995) In vitro systems for studying intestinal drug absorption. *Medical Research Reviews* **15**:83–109.

Ho N, Park J, Morozowich W and Higuchi W (1977) Physical model approach to the design of drugs with improved intestinal absorption, in *Design of biopharmaceutical properties through prodrugs and analogues.* (Roche E ed) pp 136–277, APhA/APS, Washington, DC.

Horie K, Tang F and Borchardt R (2003) Isolation and characterization of Caco-2 subclones expressing high levels of multidrug resistance efflux transporter. *Pharmaceutical Research* **20**:161–168.

Houston J, Upshall D and Bridges J (1974) A Reevaluation of the importance of partition coefficients in the gastrointestinal absorption of nutrients. *The Journal of Pharmacology and Experimental Therapeutics* **189**:244–254.

Irvine J, Takahashi L, Lockhart K, Cheong J, Tolan J, Selick H and Grove J (1999) MDCK cells: a tool for membrane permeability screening. *Journal of Pharmaceutical Sciences* **88**:28–33.

Kansy M, Senner F and Gubernator K (1998) Physicochemical high throughput screening: parallel artificial membrane permeation assay in the description of passive absorption processes. *Journal of Medicinal Chemistry* **41**:1007–1010.

Kararli T (1995) Comparison of the gastrointestinal anatomy, physiology and biochemistry of humans and commonly used laboratory animals. *Biopharmaceutics & Drug Disposition* **16**:351–380.

Kerns E, Di L, Petusky S, Farris M, Ley R and Jupp P (2004) Combined application of parallel artificial membrane permeability assay and Caco-2 permeability assays in drug discovery. *Journal of Pharmaceutical Sciences* **93**:1440–1453.

Kim R, Wendel C, Leake B, Cvetkovic M, Fromm M, Dempsey P, Roden M, Belas F, Chaudhary A, Roden D, Wood A and Wilkinson G (1999) Interrelationship between substrates and inhibitors of human CYP3A and P-gp. *Pharmaceutical Research* **16**:408–414.

Kola I and Landis J (2004) Can pharmaceutical industry reduce attrition rates? *Nature Reviews: Drug Discovery* **3**:711–715.

Komiya I, Park J, Yamani A, Ho N and Higuchi W (1980) *International Journal of Pharmaceutics* **4**:249–262.

Krause E, Dathe M, Wieprecht T and Bienert M (1999) Noncovalent immobilized artificial membrane chromatography, an improved method for describing peptide-lipid bilayer interactions. *Journal of Chromatography* **849**:125–133.

Krishna R and Mayer LD (2000). Multidrug resistance (MDR) in cancer. Mechanisms, reversal using modulators of MDR and the role of MDR modulators in influencing the pharmacokinetics of anticancer drugs. *European Journal of Pharmaceutical Sciences* **11**(4):265–83.

Krishna G, Chen K, Lin C and Nomeir A (2001) Permeability of lipophilic compounds in drug discovery using in vitro human absorption model, Caco-2. *International Journal of Pharmaceutics* **222**:77–89.

Lee K, Brower K and Thakker D (2002) Secretory transport of ranitidine and famotidine across Caco-2 cell monolayers. *Journal of Pharmacological and Toxicological Methods* **303**:574–580.

Lennernas H (1998) Human intestinal permeability. *Journal of Pharmaceutical Sciences* **87**:403–410.

Lennernas H, Nylander S and Ungell A (1997) Jejunal permeability: a comparison between the Ussing chamber technique and the single pass perfusion in humans. *Pharmaceutical Research* **14**:667–671.

Lin JH (1995) Species similarities and differences in pharmacokinetics. *Drug Metabolism and Disposition: the Biological Fate of Chemicals* **23**:1008–1021.

Lin J (2003) Drug–drug interaction mediated by inhibition and induction of P-glycoprotein. *Advanced Drug Delivery Reviews* **55**:53–81.

Lin J and Yamazaki M (2003) Role of P-glycoprotein in pharmacokinetics. *Clinical Pharmacokinetics* **42**:59–98.

Lipinski T, Lombardo F, Dominy B and Feeney P (1997) Experimental and computational approaches to estimate solubility and permeability in drug discovery and development settings. *Advanced Drug Delivery Reviews* **23**:3–25.

Maliepaard M, van Gastelen M, Tohgo A, Hauseer F, van Waardengurg R, de Jong L, Pluim D, Beijnen J and Schellens J (2001) Circumvention of BCRP-mediated resistance to camptothecins in vitro using non-substrate drugs or the BCRP inhibitor GF120918. *Clinical Caner Research* **7**:935–941.

Murer H and Kinne R (1980) The use of isolated vesicles to study epithelial transport processes. *The Journal of Membrane Biology* **55**:81–95.

Palm K, Luthman K, Ungell AL, Strandlund G, Beigi F, Lundahl P and Artursson P (1998) Evaluation of dynamic polar molecular surface area as predictor of drug absorption: comparison with other computational and experimental predictors. *Journal of Medical Chemistry* **41**:5382–5392.

Perloff M, Stromer E, von Moltke L and Greenblatt D (2003) Rapid assessment of P-gp inhibition and induction in vitro. *Pharmaceutical Research* **20**:1177–1183.

Pidgeon C (1990a) Immobilized artificial membranes, in *US patent* p 498.

Pidgeon C (1990b) Solid phase membrane mimetics: immobilized artificial membranes. *Enzyme and Microbial Technology* **12**:149–150.

Polli J, Jerrett J, Studenberg J, Humphreys J, Dennis S, Brower K and Wooley J (1999) Role of P-gp on CNS disposition of amprenavir, an HIV protease inhibitor. *Pharmaceutical Research* **16**:1206–1212.

Polli J, Wring S, Humphreys J, Huang L, Morgan J, Webster L and Serabjit-Singh C (2001) Rational use of in vitro P-gp assays in drug discovery. *The Journal of Pharmacology and Experimental Therapeutics* **299**:620–628.

Quastel J (1961) Methods of study of Intestinal absorption and Metabolism, in *Methods in Medical Research* (Quastel J ed) pp 255–259, Year Book Medical Publishers, Chicago.

Rege B, Yu L, Hussain A and Polli J (2001) Effect of common excipients on caco-2 transport of low-permeability drugs. *Journal of Pharmaceutical Sciences* **90**:1776–1786.

Rege B, Kao J and Polli J (2002) Effect of non-ionic surfactants on membrane transport in Caco-2 cell monolayers. *European Journal of Pharmaceutical Sciences* **16**:237–246.

Rubas W, Jezyk N and Grass GM (1993) Comparison of the permeability characteristics of a human colonic epithelial (Caco-2) cell line to colon of rabbit, monkey, and dog intestine and human drug absorption. *Pharmaceutical Research* **10**:113–118.

Rubas W, Cromwell M, Shahrokh Z, Villagran J, Nguyen T, Welton M, Nguyen T and Mrsny R (1996) Flux measurements across Caco-2 monolayers may predict transport in human large intestinal tissue. *Journal of Pharmaceutical Sciences* **85**:165–169.

Ruell JA, Tsinman KL and Avdeef A (2003) PAMPA – a drug absorption in vitro model. 5. Unstirred water layer in iso-pH mapping assays and p*K*a(flux) – optimized design (pOD-PAMPA). *European Journal of Pharmaceutical Sciences* **20**:393–402.

Russell T, Berardi R, Barnett J, Dermentzoglou L, Jarvenpaa K, Schmaltz S and Dressman J (1993) Upper gastrointestinal pH in 79 healthy, elderly, north American men and women. *Pharmaceutical Research* **10**:187–196.

Saha P and Kou J (2002) Effect of bovine serum albumin on drug permeability estimation across Caco-2 monolayers. *European Journal of Pharmaceutics and Biopharmaceutics* **54**:319–324.

Schanker L, Tocco D, Brodie B and Hogben C (1958) Absorption of drugs from the rat small intestine. *The Journal of Pharmacology and Experimental Therapeutics.* **123**:81–88.

Schurgers N and DeBlaey C (1984) *International Journal of Pharmaceutics* **19**:283–295.

Simpson K and Jarvis B (2000) Fexofenadine: a review of its use in the management of seasonal allergic rhinitis and chronic idiopathic urticaria. *Drugs* **59**:301–321.

Sinko PJ, Hu P, Waclawski AP and Patel NR (1995) Oral absorption of anti-AIDS nucleoside analogues. 1. Intestinal transport of didanosine in rat and rabbit preparations. *Journal of Pharmaceutical Sciences* **84**:959–965.

Stenberg P, Luthman K, Ellens H, Lee CP, Smith PL, Lago A, Elliott JD and Artursson P (1999) Prediction of the intestinal absorption of endothelin receptor antagonists using three theoretical methods of increasing complexity. *Pharmaceutical Research* **16**:1520–1526.

Stewart BH and Chan OH (1998) Use of immobilized artificial membrane chromatography for drug transport applications. *Journal of Pharmaceutical Sciences* **87**:1471–1478.

Sun D, Lennernas H, Welage L, Barnett J, Landowski C, Foster D, Fleisher D, Lee K and Amidon G (2002) Comparison of human duodenum and Caco-2 gene expression profiles for 12,000 gene sequence tags and correlation with permeability of 26 drugs. *Pharmaceutical Research* **19**:1400–1416.

Tavelin S, Taipalensuu J, Hallbook F, Vellonen K, Moore V and Artursson P (2003) An improved cell culture model based on 2/4/A1 cell monolayers for studies of intestinal drug transport: characterization of transport routes. *Pharmaceutical Research* **20**:373–381.

Uhing M and Kimura R (1995) The effect of surgical bowel manipulation and anesthesia on intestinal glucose absorption in rats. *The Journal of Clinical Investigation* **95**:2790–2798.

Ungell A-L (2004) Caco-2 replace or refine? *Drug Discovery Today* **1**:423–430.

Ungell A, Nylander S, Bergstrand S, Sjoberg A and Lennernas H (1998) Membrane transport of drugs in different regions of the intestinal tract of the rat. *Journal of Pharmaceutical Sciences* **87**:360–366.

Ussing H and Zerahn K (1951) Active transport of sodium as a source of electric current in the short-circuited isolated frog skin. *Acta Physiologica Scandinavica* **23**:110–127.

Van Rees H, De Wolff F and Noach E (1974) *European Journal of Pharmacology* **28**:310–315.

Volk E and Schneider E (2003) Wild type BCRP is a methotrexate ployglutamate transporter. *Cancer Research* **63**:5538–5543.

Walter E and Kissel T (1995) Heterogeneity in the human intestinal cell line Caco-2 leads to differences in transepithelial transport. *European Journal of Pharmaceutical Sciences* **3**:215–230.

Watanabe T, Miyauchi S, Sawada Y, Iga T, Hanano M, Inaba M and Sugiyama Y (1992) Kinetic analysis of hepatobiliary transport of vincristine in perfused rat liver: possible roles of P-gp in biliary excretion of vincristine. *Journal of Hepatology* **16**:77–88.

Wessel M, Jurs P, Tolan J and Muskal S (1998) Prediction of human intestinal absorption of drug compounds. *Journal of Chemical Information and Computer Sciences* **38**:726–735.

Wilson T and Wiseman G (1954) The use of sacs of everted small intestine for the study of the transference of substances from the mucosal to the serosal surface. *The Journal of Physiology* **123**:116–125.

Woehlecke H, Pohl A, Alder-Berens N, Lage H and Herrmann A (2003) Enhanced exposure of phosphatidylserine in human gastric carcinoma cells overexpressing the half-size ABC transporter BCRP (ABCG2). *Biochemical Journal* **376**:489–495.

Yamazaki M, Neway W, Ohe T, Chen I, Rowe J, Hochman J, Chiba M and Lin J (2001) In vitro substrate identification studies for P-gp mediated transport: Species difference and predictability of in vivo results. *The Journal of Pharmacology and Experimental Therapeutics* **296**:723–735.

Yang C, Cai S, Liu H and Pidgeon C (1996) Immobilized artificial membranes – screens for drug membrane interactions. *Advanced Drug Delivery Reviews* **23**:229–256.

Zhang S, Yang X and Morris M (2004) Flavonoids are inhibitors of BCRP-mediated transport. *Molecular Pharmacology* **65**:1208–1216.

6
Excipients as Absorption Enhancers

Hans E. Junginger

6.1 Introduction

Excipients or auxiliary materials are used to formulate a delivery system for a drug to achieve optimal therapeutic effects. They should be able to deliver the drug at the right place at the right time and with the right dose with the optimal delivery characteristics. Basically, an excipient has to fulfill the same safety profile as a drug with the exception that it should not exert a therapeutic effect. Most excipients do comply with these requirements and do have the GRAS (generally regarded as safe, a system used by the US FDA) status when used in those amounts which are normally used to fabricate a drug delivery system.

However, various substances that are commonly used as excipients do not show complete inertness, but may have additional effects on the tissues of the absorption sites which are not intended, because the excipients are primarily used for another purpose. Many surfactants are used in a drug delivery system either as a wetting agent or as a lubricant. Others are used as a solubilizer for poorly soluble drugs. Depending on their unique structure as amphiphilic compound combining hydrophilic and lipophilic characteristics those compounds also may interact with the tissue of the mucosal surfaces and change (transiently) their structure allowing for improved absorption of especially hydrophilic and high molecular weight drugs. In order to improve the bioavailability of poorly absorbable drugs the nuisance of these surfactant compounds has turned into their primary intended action as absorption enhancers.

However, in many cases there is a direct relationship between absorption enhancing effect and toxicity of the used low molecular weight surfactants as transmucosal absorption enhancers, which excludes them for therapeutic use and hampers their commercialization especially if the poorly absorbable drug is meant for chronic use.

According to Barry (1983), an ideal absorption enhancer should have the following desirable attributes:

- The absorbing enhancing action should be immediate and unidirectional, and the duration of the effect should be specific, predictable, and suitable.

- After removal of the material from the applied membrane, the tissue should immediately fully recover its normal barrier property.
- The enhancer should show no systemic and toxic effects.
- The enhancer should not irritate or damage the applied membrane surface.
- The enhancer should be physically compatible with a wide range of drugs and pharmaceutical excipients.
- The enhancer should be applicable for chronic use.

As evident from the above points, we are currently far from having an absorption enhancer available that fulfills all the requirements. Furthermore, no knowledge is available about the long-term application of such low molecular weight absorption enhancers.

Although a lot of information is available in the literature about the efficacy and safety profile of most of the surfactants to be used as absorption enhancers, little knowledge is still available about the possible interference of these compounds with the inherent transporter systems of the cells. Such effects may become crucial when a biowaiver is granted for a drug substance that is predominantly actively absorbed *via* transporter systems and when the innovator product may use excipients that do not interfere with these transporter systems. When an excipient used in the multi source (generic) product does or when a different excipient used in the other formulation, these may account for strong differences in drug absorption.

There have been many attempts to search for new and safer absorption enhancers especially with respect to improving absorption of hydrophilic compounds with high molecular weight such as peptides, insulin, calcitonin, etc. These have become available in the last two decades in sufficient and affordable amounts due to the progress in biotechnology and which are used for chronic therapy mostly by injection. The development of suitable alternative delivery systems (for the nasal, buccal, rectal, vaginal, ocular, and peroral route) (yet) have not kept pace with the availability of endogenous peptides because of the lack of suitable absorption enhancers for this class of substances. However, it turned out rather surprisingly that special polymers, which show mucoadhesive properties, also are able to act as safe penetration enhancers for improved drug absorption of especially hydrophilic (peptide) drug substances.

This chapter aims to classify the existing types of absorption enhancers according to their mode of action, and to introduce the new categories of polymeric absorption enhancers for hydrophilic compounds.

6.2 Basic Mechanisms in Transcellular and Paracellular Transport

Mammalian cell membranes separate cells from their environment and from one another. Cell membranes of the mucosal enterocyte linings consist of phospholipid bilayers in which proteins for signal transfer are incorporated.

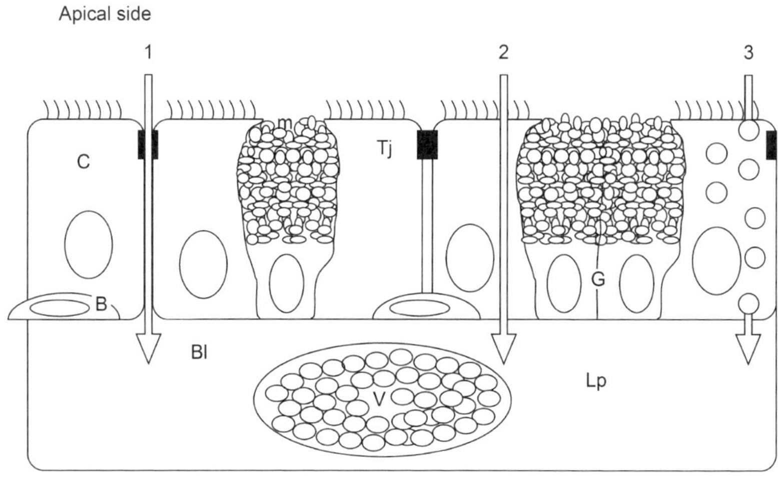

FIGURE 6.1. Transport routes across nasal respiration mucosa: (1) paracellular across tight junctions, (2) transcellular, and (3) transcytotic. Mucus secreting goblet cells (G), ciliated columnar cells (C), and tight junctions (Tj) are represented. Basal cells (B) are located on the basal lamina (Bl) adjacent to the lamina propria (Lp) with blood vessels. With permission from Junginger and Verhoef (1998)

These cell membranes are barriers to most polar compounds and also to macromolecules, but they are relatively permeable to water and small hydrophobic molecules. Methods to facilitate transport of molecules, either small or large, across epithelial cells can be categorized into two major groups: transcellular and paracellular transport (Hayashi *et al.*, 1997, Fasano, 1998) as highlighted in Fig. 6.1.

6.2.1 Transcellular Transport

As reviewed before, basic mechanisms of transepithelial transport of drugs include passive transport of small molecules, active transport of ionic and polar compound, and endocytosis and transcytosis of macromolecules (Fig. 6.1). Small and nonionic molecules usually cross cell monolayers by passive transport. The rate at which a molecule diffuses across the lipid bilayer of cell membranes depends largely on the size of the molecule and its relative lipid solubility. In general, the smaller and more lipophilic the molecule is, the more rapidly it will diffuse across the bilayer. However, cell membranes are also permeable to some small water-soluble molecules such as ions, sugars, and amino acids (Elsenhans *et al.*, 1983.)

Passive transport is the movement of a solute along its concentration gradient. The passive transcellular transport of hydrophilic compounds, including macromolecules such as peptides, can be enhanced by interaction of the absorption-enhancing materials with both the phospholipid bilayer and the integrated proteins, thereby making the membrane more fluid and thus more permeable to both lipophilic and hydrophilic compounds.

6.2.2 Paracellular Transport

Paracellular transport is the transport of molecules around or between cells (Fig. 6.1). Tight junctions or similar interconnections exist between cells. At the level of tight junctions, cell membranes are brought into extremely close apposition, but are not fused, so as to occlude the extracellular space (Fig. 6.2). Because proteolytic activity is thought to be deficient in the paracellular space (in contrast to the presence of cytosolic enzymes in the transcellular space), the investigation of paracellular transport for hydrophilic compounds in general and for peptides and proteins in particular has recently become of great interest (Fasano, 1998). However, most of the commonly used absorption enhancers improve both the transcellular and paracellular pathways, albeit in different ratios (Junginger and Verhoef, 1998).

6.2.3 Mechanisms of Action of Absorption Enhancers

Absorption promoters (penetration enhancers) can influence the mucosa in different ways:

- By acting on the mucous layer
- By acting on the membrane components
- By acting on the tight junctions

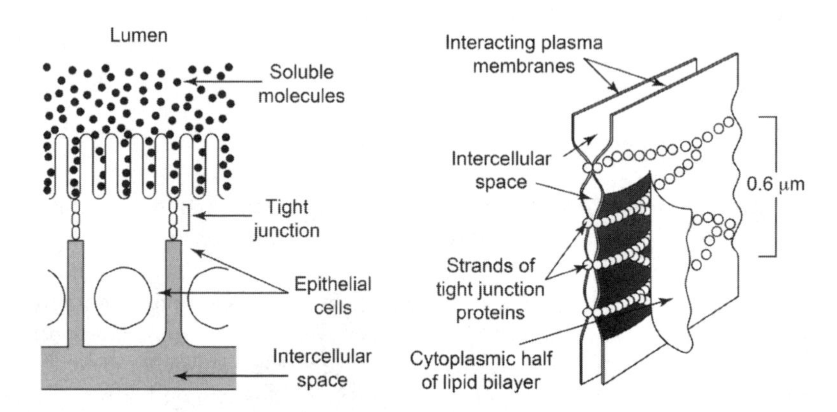

FIGURE 6.2. Tight junctions are intercellular connections that hold epithelial cells together at their apical end (left). An enlargement of a tight junction is also presented (right). With permission from Junginger and Verhoef (1998)

6.2.3.1 Action on the Mucus Layer

The mucous layer covering the cell surface of the mucosa can be seen as an unstirred layer, acting as a barrier to the diffusion of drug molecules (Marriott and Gergory, 1990). The main components of the mucous layers include water (up to 95% by weight), mucin (generally no more than 5% by weight), inorganic salts (about 1% by weight), carbohydrates, and lipids (Marriott and Gregory, 1990). Mucin represents more than 80% of the organic components of mucus (Lichtenberger, 1992) and controls the gel-like structure (Marriott and Gregory, 1990). Mucins are O-linked glycoproteins (Strous and Dekker, 1992). From a polymer science viewpoint, mucins are block copolymers with branched and unbranched blocks. Both types of the blocks have protein backbone chains, but the branched blocks have highly branched oligosaccharide chains attached to them (Peppas and Huang, 2004). Ionic surfactants have been found to be able to reduce the mucus viscosity and elasticity (Martin *et al.*, 1978). On the other side absorption of drugs across any mucosal tissue may involve interactions with the mucus gel overlying the tissue. Bhat *et al.* (1996) demonstrated that drug binding to the mucus glycoproteins is nonspecific in nature with similar types of binding forces and can reduce the amount of free drug available for absorption.

In addition, mucoadhesive polymers are thought to interfere with the mucous layer, first by covering the mucous surface and then by interpenetration of the mucous network (Lehr *et al.*, 1992c; Lehr *et al.*, 1993; Peppas and Huang, 2004). As a consequence of both mechanisms, absorption enhancers are thought to reduce the barrier function of the mucus layer and increase drug permeability. On the other side, Schipper *et al.* (1999) could show that the binding of the mucoadhesive polymer chitosan (cf. Sect. 6.3.3.3) to the epithelial cell surface and subsequent absorption enhancing effect of the hydrophilic drug atenolol were significantly reduced in mucus producing and covered cell cultures (HT29-H cultures). When the mucus layer was removed prior to the addition of chitosan, the cell surface binding and absorption-enhancing effects of the chitosans were increased. It is suggested that the only modest absorption-enhancing effect of chitosan with mucus can be overcome by increasing the local concentrations of both chitosan and drug, i.e., through formulation of the chitosan into a particulate dosage form.

6.2.3.2 Action on Membrane Components

As already mentioned, the membranes of epithelial cells contain phospholipids and proteins. The hydrophobic interaction between the acyl chains of lipid molecules results in the formation of a well-organized phospholipid bilayer. These ordered bilayers are poorly permeable to both macromolecules and highly polar low molecular weight compounds. Numerous studies have shown that absorption enhancers can increase the permeability of membranes by affecting biological membrane components such as proteins and lipids.

In an early but very important review Swenson and Curatolo (1992) reported on methods for the enhancement of the oral absorption of polar drugs, including polar peptides and proteins. The enhancers reviewed are bile salts, anionic detergents, nonionic detergents, medium chain glycerides, salicylates, acyl amino acids, acyl carnitines, lysolecithin, ethylene diamine tetraacetic acid (EDTA), and various particulate systems. Also Anderberg and Artursson (1992) studied a series of surfactants registered in solid oral drug products. The effects of anionic sodium dodecyl sulfate (SDS) and nonionic (polysorbate 80 and polyoxyl 40 hydrogenated castor oil) surfactants as well as the effects of bile salts (sodium taurocholate, sodium taurodeoxycholate, and sodium taurodihydrofusidate (STDHF)) on epithelial permeability and integrity were studied using Caco-2 cell monolayers. It was observed that all surfactants demonstrated a concentration-dependent effect on the permeability of hydrophilic markers. However, the effects of anionic surfactants were more pronounced compared to those of nonionic absorption enhancers. Altered cell morphology and cell membrane damage were observed after exposure to SDS, STDHF, and polysorbate 80. It was also shown that the absorption enhancers increased the permeability of marker molecules *via* the paracellular and transcellular routes.

The most likely mechanism by which low molecular weight absorption enhancers promote drug absorption is by solubilizing the phospholipids and membrane proteins, and thus increasing membrane permeability (Lichtenberg *et al.*, 1983). In addition, surfactants, bile salts, and fatty acids influence both the transcellular and paracellular routes of absorption (Table 6.1). As most of the examples of the absorption enhancers given also possess (to some extent) surfactant-like properties, they might mix with the phospholipid bilayer of the membrane and thus be partly absorbed. This absorption could result not only in strong membrane damage, but also in toxic side-effects due to penetration interfering with cell organelles (Junginger and Verhoef, 1998).

A special class of absorption enhancers for hydrophilic compounds are cyclodextrins, which are cyclic oligosaccharides of six, seven, or eight d-glycopyranose units, denoted as α-, β-, and γ- cyclodextrins, respectively. The three-dimensional ring structure of these compounds resembles a truncated cone, of which the internal cavity has slight hydrophobic properties and the outer surface is hydrophilic. Because of these structural features, cyclodextrins can form inclusion complexes with lipophilic drugs, thereby increasing their solubility in aqueous solutions (Uekama *et al.*, 1982; Szejtli, 1988). Just recently Mannila and coworkers (2005) studied the effects of randomly methylated β-cyclodextrin on the sublingual bioavailabililty of various cannabinoids in rabbits and found increased uptake in the systemic circulation. Furthermore, cyclodextrins (particularly methylated β-cyclodextrins) can enhance the nasal absorption of peptide drugs such as insulin, although marked interspecies differences have been reported (Verhoef *et al.*, 1994). The mechanism of action of methylated β-cyclodextrins as absorption enhancers for hydrophilic drugs is probably by transiently changing the mucosal permeability (by extraction and inclusion of membrane cholesterol) and opening of the tight junctions (Marttin *et al.*, 1997; Junginger and Verhoef, 1998).

TABLE 6.1. Classes of absorption enhancers and their mechanisms of action

Class	Examples	Mechanism	Transport ways
Surfactants	Na-laurylsulfate Polyoxyethelyne-*g*-laurylether	Phospholipid acyl chain petrubation	Transcellular↑ Paracellular↑
	Bile salts:		
	Na-deoxycholate Na-glycocholate	Reduction mucus viscosity Peptidase inhibition	
	Na-taurocholate		
Fatty acids	Oleic acid Short fatty acids	Phospholipid acyl chain petrubation	Transcellular↑ Paracellular
Cyclodextrins	α-, β- and γ-cyclodextrin Methylated β-cyclodextrins	Inclusion of membrane compounds	Transcellular↑ Paracellular↑
Chelators	EDTA	Complexation of Ca^{2+}	Transcellular↑ Paracellular↑
	Polycrylates	Opening of tight junctions	Paracellular↑
Positively charged polymers	Chitosan salts Trimethyl chitosan	Ionic interactions with negatively charged groups of glycocalix	Paracellular↑

Potential Excipient Effect on Bioavailability

The Biopharmaceutics Classification System (BCS) (Amidon *et al.*, 1995) allows waivers of *in vivo* bioequivalence for rapidly dissolving immediate-release (IR) formulations of drugs with high solubility and high permeability. One potential issue in possibly extending BCS biowaivers to low-permeability drugs is the potential for excipients to modulate the intestinal permeability of the drug. Rege *et al.* (2001) investigated the effects of common excipients on Caco-2 transport of such low-permeability drugs. The effects of nine individual excipients (lactose monohydrate, hydroxypropyl methyl cellulose (HPMC), sodium lauryl sulfate (SLS), EDTA, Tween 80, docusate sodium (dioctyl sodium sulfosuccinate), propylene glycol, poly(ethylene glycol 400), and anhydrous cherry flavor) on the Caco-2 permeability were investigated using seven low-permeability compounds that differ in their physical properties. With the exception of SLS no excipients affected Caco-2 cell monolayer integrity. SLS moderately increased the permeability of almost all the drugs. Tween 80 significantly increased the apical-to-basolateral directed permeability of furosemide, cimetidine, and hydrochlorothiazide, presumably by inhibiting their active efflux, without affecting mannitol permeability. Additionally, docusate sodium moderately increased cimetidine permeability. Other excipients did not have a significant effect on the permeability of these drugs.

In another study, Rege *et al.* (2002) investigated the transporter inhibition activity of three nonionic surfactant (Tween 80, Chromophor EL, and vitamin E

TPGS) on P-glycoprotein, the human intestinal peptide transporter, and the monocarboxylic acid transporter in Caco-2 monolayers. Additionally they evaluated the role of membrane fluidity of protein kinase C in surfactant-induced transporter inhibition. All three surfactants inhibited P-glycoprotein (P-gp). Tween 80 and Cremophor EL increased apical-to-basolateral permeability and decreased basolateral-to-apical permeability of the P-gp substrate rhodamine 123. The effect of vitamin E TPGS was equally large, but essentially only reduced the basolateral-to-apical permeability on rhodamine 123. These P-gp inhibition effects would appear to be related to these excipients' modulation of membrane fluidity, where Tween 80 and Cremophor EL fluidized cell lipid bilayers, while vitamin E TPGS rigidized the bilayers. However, among the three surfactants, only Tween 80 inhibited the peptide transporter, as measured by glycyl sarcosine permeability. A common functional feature of these three surfactants was their ability to modulate fluidity, although the results indicated that even strong membrane fluidity modulation alone was not sufficient to reduce transporter activity. Protein kinase C inhibitor failed to affect rhodamine 123 and glycyl sarcosine permeability, suggesting that protein kinase C inhibition was not the mechanism of transporter inhibition. These results suggest that surfactants, which are absorption enhancers, can inhibit multiple transporters, but that changes in membrane fluidity may not be the generalized mechanism to reduce transporter activity.

In a summary of a workshop report on BCS – Implementation Challenges and Extension Opportunities (Polli *et al.*, 2004) it is stated:

The BCS guidance indicates that excipients that are currently in US FDA-approved IR solid oral dosage forms will generally not affect the rate and extent of absorption of a highly soluble and highly permeable drug substance that is formulated in a rapidly dissolving IR product. Quantities of excipients should be consistent with intended function. The guidance further indicates that when new excipients or atypically large amounts of common excipients are used, there needs to be a documentation of the absence of an excipient effect. The FDA has recently made available to the public the Inactive Ingredients Database, which lists inactive ingredients in FDA-approved drug products. The Inactive Ingredients Database can be accessed at http://www.accessdata.fda.gov/scripts/cder/iig/index.cfm. This list is searchable by ingredient name. Each database cites for each route/dosage form containing the inactive ingredient.

Given the consensus to extend potential biowaivers to include drugs whose fraction dose absorbed is less than 85%, with perhaps some lower limit (e.g., 40%), excipient effects were discussed as a potential concern. There was no consensus on this potential concern. There was general acknowledgement that most products only employ common excipients in typical quantities. There was some level of support for the expectation that common excipients in typical quantities do not modulate permeability or gastrointestinal transit of low permeability drugs. This viewpoint was countered by the growing understanding of the role of transporters in drug permeation and disposition, including transporter-mediated drug interactions. There was consensus that dose linearity extending sufficiently above the highest dose strength is a basis to conclude that excipients in such studies do not represent a significant risk for the drug.

Potential Excipient Effect on Tight Junctions

The intercellular tight junction is one of the major barriers to the paracellular transport of macromolecules and polar compounds (Denker and Nigam, 1998). They have two physiological functions: first, they constitute the principal barrier to passive movement of fluid, electrolytes, macromolecules, and cells through the paracellular pathway (the "gate" function), and secondly, they contribute to transepithelial transport of compounds promoting epithelial cell polarity (Madara, 1987; Citi, 1992; Citi and Denisenko, 1995). Tight junction's structure and permeability can be regulated by many potential physiological factors, including the concentration of cyclic AMP (cAMP) (Duffey *et al.*, 1981), intercellular calcium concentration (Palant *et al.*, 1983), and transient mucosal loads (Madara *et al.*, 1986). Several studies have shown that one of the possible mechanisms of penetration enhancers is to loosen the tight junctions of epithelial membranes thereby increasing the paracellular transport of poorly absorbable drugs (Gonzales-Mariscal and Nava, 2005). Hence the issue of tight junction's regulation by absorption enhancers appears to be crucial for macromolecular drug absorption.

In a study by Thanou (2000), the tight junction's membrane protein occludin was visualized by immunocytochemistry staining in the presence and absence of Trimethyl Chitosan 60 (TMC60) (degree of quaternization 60%) using confocal laser scanning microscopy (CLSM). Additionally, the effects of TMC60 on cytoskeletal F-actin were determined by visualization using CLSM. The transmembrane protein occludin displayed a disrupted pattern after incubation with 1.0% (w/v) TMC60, suggesting that the interaction of TMC60 with the tight junction's protein is the major mechanism for opening of the tight junctions and subsequent increased paracellular permeability. These observations were quite similar to images of Caco-2 cells with 0.1% (w/v) chitosan, but the effect appeared to be stronger than for the reported 0.1% (w/v) chitosan. Chitosan treated cells showed a thickened pattern of occludin at the cell periphery and not a disrupted one, which might be due to the ten-fold difference in concentration or to an effect exclusively related to the quaternized derivative of chitosan, TMC60. Additionally it was observed that TMC60 provoked a redistribution of the cytoskeletal F-actin, a phenomenon that appeared to correlate well with the opening of epithelial tight junctions.

Calcium depletion by chelating agent (e.g., EDTA, EGTA) has been reported to increase paracellular permeability. These agents induce general changes in the cell physiology such as disruption of actin filaments and adherence junctions, diminished cell adhesion, and activation of protein kinases (Citi, 1992). It was proposed that EGTA provokes alterations on the tight junctions, being a consequence of its effects on Ca^{2+} dependent adhesion molecules (which are concentrated in adherence junctions), through a contraction of the junction-associated microfilament cytoskeleton (Citi and Denisenko, 1995). It has been demonstrated that serosal rather than apical Ca^{2+} levels play a more important role in this process (Collares *et al.*, 1994). Basolateral Ca^{2+} levels vary and a particular chelation enhancer cannot accomplish full depletion of the calcium ions from the adherence junction to provoke the paracellular widening. Therefore, the approach of using chelating agent as permeation enhancers leads to variable results, even in controllable *in vitro* conditions like the Caco-2 cell system (LeCluise and Sutton, 1997).

Functional polymers such as polyacrylic acid derivatives and chitosan (derivatives) appear to be a valuable alternative solution to increase exclusively the paracellular permeation and absorption of hydrophilic drugs. Being high molecular weight and hydrophilic polymers it is assumed that their intrinsic absorption and related to this their toxicity is minimal and they are not expected to show systemic adverse side effects (Junginger and Verhoef, 1998; Thanou *et al.*, 2001c). These functional polymers will be discussed in the next sections of this chapter.

6.3 Mucoadhesive Polymers as Absorption Enhancers

6.3.1 Theories of Mucoadhesion

Various theories have been worked out in the last decades that explain the mechanisms with which mucoadhesives adhere to the mucous layer. The theories of mucoadhesion are primarily based on the classical theories of metallic and polymer adhesion. Four main theories exist that describe the possible mechanisms of mucoadhesion: the *electronic*, the *adsorption*, the *wetting*, and the *diffusion* theory.

- The *electronic* theory assumes that transfer of electrons occurs between the mucus and the mucoadhesive due to differences in their electronic structures (Derjaguin *et al.*, 1977, 1994). The electron transfer between the mucus and the mucoadhesive polymer leads to the formation of a double layer of electrical charges at the interface of the mucus and the mucoadhesive with the result of attraction forces inside the double layer.
- The *adsorption* theory is based on the attraction forces between the mucus and the mucoadhesive. The attraction is achieved *via* molecular bonding caused by secondary forces such as hydrogen and van der Waals bonds (Kinloch, 1979, 1980; Gu *et al.*, 1988; Mikos and Peppas, 1989; Chickering and Mathiowitz, 1999). The resulting attractive forces are considerably larger than the forces described by the electronic theory.
- The *wetting* theory correlates the surface tension of the mucus and the mucoadhesive with the ability of the mucoadhesive to swell and spread on the mucus layer and indicates that interfacial energy plays an important role in mucoadhesion (Good and Girrfalco, 1960; Helfland and Tagami, 1972; Kaelble and Moacanin, 1977; Peppas and Buri, 1985). By calculating the interfacial energy from the individual spreading coefficients of the mucus and mucoadhesive or by calculating a combined spreading coefficient, good predictions about the mucoadhesive performance can be obtained (Lehr *et al.*, 1992a, 1993). The

wetting theory has the most impact on the mechanism of mucoadhesion since spreading of the mucoadhesive over the mucus (and vice versa) is a prerequisite for the validity of all other theories.

- The *diffusion* theory was proposed first by Voyutskii (1963) and assumes that both the mucoadhesive surface and the mucoadhesive polymer come in the first step in contact to each other. In a second step it is postulated that both the mucin polymers and the mucoadhesive polymer chains interpenetrate each other with subsequent physical entanglement and hydrogen bonds. The interpenetration has to be sufficiently deep in order to become substantial and is dependent on the molecular weight, degree of crosslinking, chain length, flexibility, and spatial conformation of the polymers (Kinloch, 1980; Park and Robinson, 1985; Mikos and Peppas, 1986; Ponchel *et al.*, 1987; Duchêne *et al.*, 1988; Peppas and Stahlin, 1996). Jabbari *et al.* (1993) and Peppas and Huang (2004) were the first to introduce the interdiffusion theory in mucoadhesion. It was proposed that in an aqueous environment the free polymers have enough mobility to diffuse. After intimate contact of the mucus and the mucoadhesive carrier, the free polymer chains, which are initially in the mucus or mucoadhesive parts, may diffuse across the interface due to a chemical potential gradient. After a period of time, the diffused chains form effective interaction sites in the interfacial region. Desai *et al.* (1992) experimentally estimated the diffusion coefficients of certain proteins in the porcine mucus on the order of 10^{-7} cm^2/s. The diffusion coefficients of free mucins were about 10^{-8} cm^2/s in mucus while aggregated mucins have diffusion coefficients of 10^{-11} to 10^{-12} cm^2/s (Bansil *et al.*, 1995). In addition to their low diffusion coefficients, the dynamics of polymer chain diffusion across the interface is rather complex (Wool, 1995). First visualization studies (plastic sections of freeze substituted samples) showed the mucoadhesive interface as an irregular borderline with many coves and invaginations, but were sharp rather than hazy. While with light microscopy mucus glycoproteins could be identified unambiguously by specific histochemical reactions, there was no evidence for intermixing using plastic sections of freeze substituted samples. Hence, the interpenetration depth at the mucus/polymer interface may not be in the micron scale but in the nanorange. (Lehr *et al.*, 1992c). Though direct observations of free chain interpenetration in the interface between mucus and mucoadhesives are not possible, recent experimental observations support the interdiffusion contribution to adhesion. Jabbari and coworkers (1993) proved mucin interpenetration at the poly(acrylic acid)/mucin interface using ATR-FTIR spectroscopy. Their results showed clearly that the concentration of mucin inside the PAA gel increases with time. Sahlin and Peppas (1996) used near-field FTIR microscopy to study the free PEG chains diffusion across PAA hydrogel. The diffusion process was confirmed and the diffusion coefficients were on the order of 10^{-8} to 10^{-9} cm^2/s.

None of these theories give a complete description of the mechanisms involved in mucoadhesion. The total phenomenon of mucoadhesion most probably is a combination of all these theories. Some investigators divide the mucoadhesion process into sequential phases, each of which is associated with a different mucoadhesion mechanism (Lee *et al.*, 2000; Solomonidou *et al.*, 2001; Dodou *et al.*, 2005): First, the polymer gets wet and swells (wetting theory). Then, noncovalent (physical) bonds are created within the mucus–polymer interface (electronic and absorption theory). Finally, the polymer and protein chains interpenetrate (diffusion theory) and entangle together to form subsequently noncovalent (physical) and covalent (chemical) bonds (electronic and adsorption theory).

It may become clear that the mechanisms of mucoadhesion are of utmost importance for the effectiveness of mucoadhesive polymers, which are intended to act as absorption enhancers for improved drug absorption. There are two important aspects to consider: first the residence time of a mucoadhesive (particulate) drug delivery system after attachment to the mucus according to the mechanisms discussed above for prolonged drug delivery, and second the interpenetration of the mucoadhesive polymers into the mucus layers covering the absorptive mucosal tissues, and their interactions with the sugar residues of the glycocalix in order to elicit a response reaction which results in the transient opening of the tight junctions and allows a paracellular transport of the hydrophilic drug molecules along this route. These aspects will be discussed in the following sections.

6.3.2 Material Properties of Mucoadhesives

Mucoadhesives are characterized by material properties that contribute to good adhesiveness according to one or more theories of mucoadhesion. Such material properties are the ability to *swell*, their ability to form *molecular bonds* with the mucus layer, and their *spatial conformation* due to the entanglement of chains. The creation of molecular bonds and the entanglement of chains result in changes in their rheological behavior of the mucoadhesive polymers. The *rheological properties* of mucoadhesives can therefore be used as an indication of the extent of molecular bonding and spatial conformation. The *cohesiveness* of mucoadhesives contributes indirectly to their adhesive ability, since it deals with the internal strength of the mucoadhesive. Dodou *et al.* (2005) have discussed and revisited these properties of mucoadhesive polymer:

- *Swelling*

The ability of mucoadhesive polymers to swell is a prerequisite for mucoadhesion since it concerns wetting, uncoiling, and spreading of the polymer over the mucus (wetting theory). This spreading process, controlled by the interfacial properties of the mucus and mucoadhesive, allows intimate contact at the mucus–mucoadhesive interface, thus governing the formation of bonds (Lehr *et al.*, 1992a, 1993). Overhydrating of the polymer, however, may result in a slippery mucilage, deteriorating mucoadhesion (Mortazavi and Smart, 1993). Furthermore, swelling is a key-parameter for the environment-sensitive drug delivery (Qiu and Perk, 2001), where controlled drug release can be obtained by a reversible volume change of an environmental-sensitive polymer with controlled swelling–deswelling properties (Gutowska *et al.*, 1997).

- *Molecular bonding*

The presence of suitable molecular groups in the mucoadhesives leads to the formation of covalent bonds (e.g., disulfide bonds), as well as noncovalent bonds (e.g., ionic, hydrogen, and van der Waals bonds) with the mucus layer. These molecular bonds contribute considerably to good adhesion, according to the electronic and the adsorption theory. The advantage of covalent bonds may be that they are stronger than the noncovalent bonds, which may result in higher mucoadhesive forces (Bernkop-Schnürch and Steiniger, 2000). However, covalent bonds require time to be created, whereas noncovalent bonds are formed immediately as soon as the mucus and the mucoadhesive polymer come into contact. The delay time that is required for covalent bonding does not play an impeding role for such drug delivery systems, in which maintaining the delivery system at a particular location for an extended period of time (about 3 h in gastrointestinal delivery) is advantageous (Lee *et al.*, 2000; Junginger *et al.*, 2002), whereas a desired longer residence time at the intestinal gut surface is most likely not possible as the turnover time of mucus is estimated to be in the order of 47–270 min in the rat (Lehr *et al.*, 1991).

- *Spatial conformation*

The interpenetration rate of the mucus–mucoadhesive chains depends on the diffusion coefficient and the chemical potential gradient of the interacting macromolecules (Huang *et al.*, 2000). The flexibility and mobility of the mucoadhesive chains as well as the expanded form of the mucoadhesive network control the effective chain length which can penetrate into the mucus (Lee *et al.*, 2000; Peppas and Huang, 2004). In this way spatial conformation is critical for the interpenetration of mucus–mucoadhesive chains.

- *Rheological properties*

The chain entanglement and the molecular bonding that occur between the mucus and the mucoadhesive lead to changes in the rheological behavior of the two materials (Rossi *et al.*, 2001). Since changes in the rheological properties reflect the degree of interaction between mucus and mucoadhesive, rheological methods constitute a common way to evaluate the strength of mucoadhesion. Mucoadhesive systems with a high elastic component showed good mucoadhesiveness (Tamburic and Craig, 1997). Moreover, a high viscosity and viscoelasticity of the mucus–mucoadhesive system indicates improved cohesiveness and resistance to deformation (Madsen *et al.*, 1998). A number of authors (Huang *et al.*, 2000; Madsen *et al.*, 1998; Rossi *et al.*, 1994; Caramella *et al.*, 1994) found experimentally that the viscosity of the mucus–mucoadhesive system can be larger as the sum of the separate viscosities. This phenomenon is called "rheological synergism." High rheological synergism indicates extensive chain entanglement (diffusion theory) and thus good mucoadhesiveness.

- *Cohesiveness*

Mucoadhesives exhibit high adhesiveness at their interface with the mucus layer, but should exhibit sufficient cohesiveness as well in order to prevent internal fracture of the (swollen) mucoadhesive polymers. Solid forms of mucoadhesives show in general satisfying cohesiveness. Another aspect of the correlation between cohesiveness and mucoadhesive ability has been pointed out by Hägerström *et al.* (2000). Together with his coworkers he investigated the mucoadhesiveness of common polymers to several kinds of mucins. The results show that a too high interaction between the polymer and the mucins led to weakening instead of strengthening of the internal gel structure, since the increased interaction at the interface disturbed the internal cohesive structure of the polymer network.

In a recent investigation, Accili and collaborators (2004) showed that the polyacrylates Carbopol 974P and Pharmacoat 606 showed different mucoadhesive properties depending on which type of mucus (sublingual, esophageal, and duodenal bovine) they were brought in contact to. The significantly different behavior of the two polymers was correlated with the desquamation layer thickness and the differential sialic acid and fucose expositions in the targeted mucosae.

A number of solid mucoadhesive dosage forms for drug delivery in the gastrointestinal tract, such as tablets, micro- and nanoparticles, granules, pellets, and capsules have been already studied *in vitro* and *in vivo* in 1992 (Duchêne and Ponchel, 1992). They showed satisfying mucoadhesive properties, although a correlation between *in vitro* and *in vivo* performance of the delivery systems cannot always be made. Until today no reliable mucoadhesive drug delivery system for intestinal application is available. This is mainly due to the fact that the mucoadhesive polymers have to first swell in order to obtain their mucoadhesive properties. Second, the particulate or single unit dose delivery systems have to get attached to the mucous linings of the gut, which is a randomly occurring process during which most of the delivery systems already come in contact with soluble mucus (fragments), amply available in the gut fluids, and which adhere also to the mucoadhesive surface of the drug delivery systems, hence deactivating the mucoadhesive properties of the drug delivery systems before they are able to reach the mucosal surface of the gut. In this case, most of the drug is delivered into the intestinal liquids rather than to the mucosal gut surface and most of the (peptide) drug is degraded and lost for drug absorption.

6.3.3 Classes of Mucoadhesive Polymers

6.3.3.1 Polyacrylates

The term polyacrylates includes synthetic, high molecular weight polymers of acrylic acid (polyacrylic acid or PAA) (Fig. 6.3a) that are also known as Carbomers. They are either linear or (weakly) crosslinked (either by allyl sucrose (Carbomers) or divinyl glycol (Polycarbophils)) polymers that are broadly applied in pharmaceutical and cosmetic industry (mostly as excipient for controlled drug release for oral dosage forms and as stabilizers for gels). Crosslinked Carbomers, manufactured by the Performance Materials Segment of the BF Goodrich company under the commercial name Carbopols and Polycarbophils (PCPs), are also used as mucoadhesive platforms for drug delivery. Carbopols and PCPs have

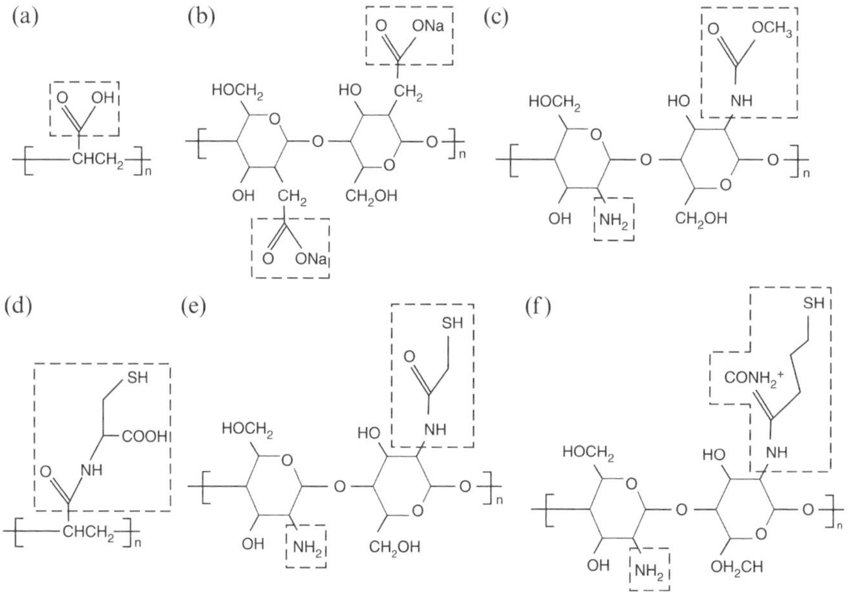

FIGURE 6.3. Structural formulas of mucoadhesives: (a) poly(acrylic acid) (PAA), (b) sodium carboxymethylcellulose (NaCMC), (c) chitosan and carboxylic methylester derivative, (d) thiolated polymer of PAA, (e) chitosan–thioglycolic acid (TGA) conjugate, and (f) chitosan-4-thio-butyl-amidine (chitosan-TBA) conjugates. The dashed boxes indicate the side groups that are responsible for mucoadhesiveness. With permission from Dodou *et al.* (2005)

received extensive review and toxicological evaluation. The PCPs and calcium PCPs are classified as category 1 GRAS materials (Goodrich, 2002). Polyacrylates interact with mucus by hydrogen and van der Waals bonds, created between the carboxylic groups of the polyacrylates and the sulfate and sialic acid residues of mucin glycoproteins (Dodou *et al.*, 2005). However, polyacrylates do possess also properties as absorption enhancers for the paracellular absorption of hydrophilic compounds as peptides and are additionally able to inhibit the activities of enzymes present in the intestinal fluid (Lueßen *et al.*, 1996a).

Polyacrylates as Absorption Enhancers

The absorption across rat intestinal tissue of the model peptide drug 9-desglycinamide, 8-L-arginine vasopressin (DGAVP) from bioadhesive formulations was studies by Lehr *et al.* (1992b) *in vitro*, in a chronically isolated intestinal loop *in situ* and after intraduodenal administration *in vivo*. Only the Polycarbophil suspensions of the drug could show significant increases of bioavailabilities in all three models, whereas a controlled release bioadhesive drug delivery system consisting of microspheres of poly(2-hydroxyethyl methacrylate) with a mucoadhesive Polycarbophil coating was practically ineffective, because its mucoadhesive coating was deactivated by soluble mucins before reaching the intestinal mucosa.

A prolongation of the absorption phase *in vitro* and in the chronically isolated loop *in situ* suggested that the polymer was able to protect the peptide from proteolytic degradation.

In another study, Lueßen and coworkers (1997) compared the absorption enhancing effects of polycarbophil, chitosan, and chitosan glutamate and found that all three mucoadhesive polymers were potent enhancers of the model peptide drug DGAVP using Caco-2 cell layers and the vertically perfused intestinal loop model of the rat. However, the observed comparable transport effect of polycarbophil in the intestinal loop model was mainly ascribed to the protection of DGAVP against proteolytic degradation in the intestinal lumen, which allows for sufficient concentration and thus transport of the peptide drug when a polycarbophil induced paracellular transport is less pronounced.

Enzyme Inhibitory Effects of Polyacrylates

Lueßen and coworkers (1996a) have studied the potency of mucoadhesive excipients to inhibit intestinal proteases. Among the different mucoadhesive polymers investigated, uniquely the poly(acrylates) polycarbophil and carbomer 934P were able to inhibit the activities of trypsin, α-chymotrypsin, carboxypeptidase A, and cytosolic leucine aminopeptidase. However, they failed to inhibit microsomal leucine aminopeptidase and pyroglutamyl aminopeptidase. Carbomer was found to be more efficient to reduce proteolytic activity than polycarbophil (most probably due to various flexibilities of the polymer chains). The authors also could demonstrate the pronounced binding properties of polycarbophil and carbomer for bivalent cations such as zinc and calcium, being a major reason for the observed inhibitory effect. These polymers were able to deprive Ca^{2+} and Zn^{2+}, respectively, from the enzyme structures, thereby inactivating their activities. Carboxypeptidase A and α-chymotrypsin activities were observed to be reversible upon addition of Zn^{2+} and Ca^{2+} ions, respectively. Table 6.2 shows some of the results of luminal enzyme inhibition by polyacrylates. However, the poly(acrylic acid) derivatives polycarbophil and carbomer showed rather weak inhibitory effects on enzymes of the intestinal brush border cell membranes responsible for DGAVP and metkephamid degradation (Lueßen *et al.*, 1996b).

In another study, Lueßen and coworkers (1996c) compared the enhanced intestinal absorption of carbomer, the neutralized carbomer–sodium salt (NaC934P), and chitosan hydrochloride for the peptide drug buserelin and concluded that the higher bioavailability with chitosan hydrochloride compared to carbomer and NaC934P is an indication that for buserelin the intestinal transmucosal transport enhancing effect of the polymer plays a more dominant role than the protection against proteases such as α-chymotrypsin.

It is emphasized again that the decrease of enzyme activity is time dependent (i.e., after 10–20 min maximum deactivation is achieved). Although enzyme inhibition by these polyacrylates is not an absorption-enhancing effect per se, especially in the intestinal peptide and protein absorption, it is a more than favorable effect to increase the amount of peptide drug absorbed (Lueßen *et al.*, 1996b, 1997).

TABLE 6.2. Effects of Polycarbophil® (PCP) and Carbomer 934P® (C934P) on protease activities (Lueßen *et al.*, 1995, 1996a)

Enzyme	Polymer	Concentration (%)	Inhibition[a]
Trypsin	PCP	0.35	+
		0.25	±
		0.15	−
	C934P	0.25	++
		0.15	+
		0.1	±
α-Chymotrypsin	PCP	0.5	±
		0.25	−
	C934P	0.5	+
		0.25	−
Carboxypeptidase A	PCP	0.1	+
		0.05	±
	C934P	0.1	++
		0.05	+
Leucine	PCP	0.5	−
Aminopeptidase M	C934P	0.5	−
Leucine	PCP	0.5	++
Aminopeptidase C		0.25	+
	C934P	0.5	++
		0.25	+
Pyroglutamyl	PCP	0.5	−
Aminopeptidase	C934P	0.5	−

[a] ++, strong inhibition; +, inhibition; ±, slightly reduced enzyme activity; −, no inhibition

Removal of endogeneous Ca^{2+} from the intestinal epithelial cells by the formation of poly(acrylic acid)-Ca^{2+} complexes loosens the cellular barrier, especially by triggering the (reversible) opening of the tight junctions. This effect has been demonstrated by Borchardt *et al.* (1996) by flux studies of mannitol and dextran fluorescently labeled fluorescein isothiocyanate (FITC-dextran, MW 4,400 Da) across intestinal Caco-2 cell monolayers after incubation with poly(acrylic acid) derivatives and by visualization studies using CLSM. It should be noted that the observed reversibility in tight junction opening was gradual only because of the high viscosity and adhesive character of the polymer solutions to the cell layer.

Carbopol 934P was also used by Thanou and collaborators (2001b) for the enhancement of the intestinal absorption of low molecular weight heparin (LMWH) in rats and pigs. LMWH is a polyanion and does not interact with polycarbophil 934P. To both animal species LMWH was administered intraduodenally and the antiXa levels were measured. Both studies showed a remarkably enhanced LMWH uptake after about 1 h and the effect for providing sufficient antithrombotic effect lasted for both animal species about 7 h showing that this polyacrylate may be a good absorption enhancer for LMWH, provided a good delivery system can be developed based on polycarbophil 934P.

6.3.3.2 Chitosan

Application, Mechanism and Safety Aspects

Chitosan (poly[β-(1-4)-2-amino-2-deoxy-D-glucopyranose]) is a cationic polysaccharide comprising copolymers of glucosamine and *N*-acetylglucosamine (Fig. 6.3c). Nowadays chitosan is available in different molecular weight (polymers 500,000–50,000 Da, oligomers 2,000 Da), viscosity grades, and degree of deacetylation (40–98%). It is next to cellulose the most abundant polysaccharide in nature. Chitosan is insoluble at neutral and alkaline pH values, whereas it forms salts with inorganic and organic acids such as glutamic acid, hydrochloric acid, lactic acid, and acetic acid. Chitosan is generally regarded as biocompatible, slowly biodegradable natural origin polymer (Hirano and Noishiki, 1985; Chandy and Sharma, 1990). It is widely used in the food industry as a food additive and as a weight loss product. Chitosans have found a number of applications as biomaterials in tissue engineering and in controlled drug release systems for various routes of delivery (Dodane and Vilivalam, 1998; Illum, 1998; Suh and Matthew, 2000). Chitosan polymers are also used as a safe excipient for a number of pharmaceutical applications (e.g., excipient in granules and tablets, gels and microspheres) (Baldrick, 2000). Chitosan has been included in the European Pharmacopoeia since 2002.

The bioadhesive properties were first described by Lehr *et al.* (1992d) demonstrating that chitosan in the swollen state is an excellent mucoadhesive at porcine intestinal mucosa and is also suitable for repeated adhesion. The authors also reported that chitosan underwent minimal swelling in artificial intestinal fluids due to its poor aqueous solubility at neutral pH values, proposing that substitution of the free-NH_2 groups with short alkyl chains would change the solubility and hence the mucoadhesion profile. The strong mucoadhesive properties of chitosan are due to the formation of hydrogen and ionic bonds between the positively charged amino groups of chitosan and the negatively charged sialic acid residues of mucin glycoproteins (Rossi *et al.*, 2000).

Illum *et al.* (1994) described that chitosan solutions at 0.5% (w/v) concentrations are highly effective at increasing the absorption of insulin across nasal mucosa in rats and sheep. The mechanism of action of chitosan was suggested to be a combination of bioadhesion and a transient widening of the tight junctions in the membrane. The influence of chitosan's degree of deacetylation and MW was also investigated on the permeability of Caco-2 cell intestinal monolayers. Schipper and coworkers (1996) studied the effect of chitosan solutions at pH 5.5 on the permeability of the nonabsorbable paracellular marker [^{14}C]mannitol and intracellular dehydrogenase activity. It was found that chitosans with a high degree of deacetylation were effective as absorption enhancers at low and high molecular weight, and also showed clear dose-dependent toxicity, whereas chitosans of low degree of deacetylation were effective at only high molecular weight and showed low toxicity. The effects of chitosans of both low and high molecular weight and degree of deacetylation were further investigated by the same authors with respect to their ability to bind at epithelial Caco-2 cell monolayers. Both chitosans appeared to bind tightly to the epithelium, inducing a redistribution of F-actin (change from a filamentous to a globular structure) and the tight junction's zonula occludens-1 protein. No intracellular uptake of chitosan could be observed. It was also shown that these effects were mediated by chitosan's cationic charges, since addition of the highly anionic heparin to the test solution inhibited the absorption enhancing effect (Schipper *et al.*, 1997).

Kerec and coworkers (2005) just recently investigated the role of Ca^{2+} on the permeability effect of chitosan on the isolated pig urinary bladder. Their results show that when calcium ions were applied together with chitosan to the luminal surface of the urinary bladder, they decrease the permeability of the model drug moxifloxacin in a concentration dependent way. These experiments show that Ca^{2+} ions are of no benefit to absorption enhancement when simultaneously given to both chitosans and polyacrylates (Lueßen *et al.*, 1996a).

Whereas for most absorption enhancers studied the cytotoxicity profile was evident, chitosan gave contradictory results regarding safety (Carreno-Gomez and Duncan, 1997). Dodane *et al.* (1999) investigated the effect of chitosan (degree of deacetylation 80%) solutions at pH 6.0–6.5 on the structure and function of Caco-2 cell monolayers. Using a series of microscopic techniques, the authors were able to show that chitosan had a transient effect on the tight junction's permeability and that viability of the cells was not affected. However, chitosan treatment slightly perturbed the plasma membrane, but this effect was reversible.

In a preliminary study, Chae and coworkers (2005) investigated the molecular weight (MW) dependent Caco-2 cell layer transport phenomena (*in vitro*) and the intestinal absorption patterns after oral administration (in rats *in vivo*) of water soluble chitosans. The absorption of chitosans was significantly influenced by its MW. As the MW increases, the absorption decreases. The absorption both *in vitro* and *in vivo* of a chitosan with a MW of 3.8 kDa was about 25 times higher in comparison to a high MW chitosan (230 kDa). On the other side, the chitosans showed concentration- and MW-dependent cytotoxic effects: the chitosan oligosaccharides (MW < 10 kDa) showed negligible cytotoxic effects on the Caco-2 cells whereas the high MW chitosans were more toxic in this experimental setting. However, the abundant use of chitosans in the food industry and the use of chitosan as excipient for peroral drug delivery systems prove that also chitosan with a high molecular weight can be regarded as safe.

Chitosan as Absorption Enhancer of Hydrophilic Macromolecular Drugs

Illum and coworkers (1994) reported at first that chitosan is able to promote the transmucosal absorption of small polar molecules as well as peptide and protein drugs across nasal epithelia. Immediately afterward Artursson and collaborators (1994) reported that chitosan can increase the paracellular permeability of [^{14}C]mannitol (a marker for paracellular routes) across Caco-2 intestinal epithelia.

Chitosan gels were first tested *in vivo* for their ability to increase the intestinal absorption by Lueßen and coworkers in 1996c. The absorption enhancement of the peptide analog buserelin was studied after intraduodenal coadministration with chitosan (pH 6.7) in rats. Chitosan substantially increased the bioavailability of the peptide (5.1%) in comparison to control (no polymer) or Carbopol 934P containing formulations. Borchardt and his team (1996) investigated chitosan glutamate solutions at pH 7.4 for their effect in increasing the paracellular permeability of [^{14}C]mannitol and fluorescently labeled dextran (MW 4,400 Da) *in vitro* in Caco-2 cells. No effect on the permeability of the monolayer could be observed, indicating that at neutral pH value chitosan is not effective as absorption enhancer. The pH dependency of chitosan's effect on epithelial permeability was further investigated by Kotzé and coworkers (1998). Two chitosan salts (hydrochloride and glutamate) were evaluated for their ability to enhance the transport of [^{14}C]mannitol across Caco-2 cell monolayers at two pH values, 6.2 and 7.4. At low pH both chitosans showed a pronounced effect on the permeability of the marker, leading to 25- (glutamate salt) and 36-fold (hydrochloride salt) enhancement. However, at pH 7.4 both chitosans failed to increase the permeability, due to their insolubility for use as absorption enhancer in more basic environment such as in the large intestine. These results made quite clear that chitosan (salts) cannot be used as absorption enhancers for *in vivo* studies when the drug should be released in the jejunum because of the insolubility and hence ineffectiveness at pH values higher than 6.5.

6.3.3.3 N,N,N,-Trimethyl Chitosan Hydrochloride (TMC)

Synthesis and Characterization of TMC

Sieval *et al.* (1998) and Kotzé and collaborators (1998) based on the method of Domard *et al.* (1986) synthesized TMC. TMC is a partially quaternized derivative of chitosan, which is prepared by reductive methylation of chitosan with methyl iodide in a strong basic environment at an elevated temperature. The degree of quaternization can be altered by increasing the number of reaction steps by repeating them or by increasing the reaction time. TMC proved to be a derivative of chitosan with superior solubility and basicity, even at low degrees of quaternization, compared to chitosan salts. This quaternized chitosan shows much higher aqueous solubility than chitosan in a much broader pH and concentration range. The reason for this improved solubility is the substitution of the primary amine with methyl groups and the prevention of hydrogen bond formation between the amine and the hydroxylic groups of the chitosan backbone.

The absolute molecular weights, radius, and polydispersity of a range of TMC polymers with different degrees of quaternization (22.1, 36.3, 48.0, and 59.2%) were determined with size exclusion chromatography and multiangle laser light scattering (MALLS). The absolute molecular weight of the TMC polymers decreased with an increase in the degree of quaternization. The respective molecular weights measured for each of the polymers were 2.02×10^5, 1.95×10^5, 1.66×10^5, and 1.43×10^5 g/mole. It should be noted that the molecular weight of the polymer chain increases during the reductive methylation process due to the addition of the methyl groups to the amino group of the repeating monomer. However, a net decrease in the absolute molecular weight is observed due to degradation of the polymer chain caused by exposure to the specific reaction condition during the synthesis (Snyman *et al.*, 2002). Polnok and coworkers (2004) investigated the influence of the methylation process on the degree of quaternization on *N*-trimethyl chitosan chloride. ^{1}H-Nuclear magnetic resonance spectra showed that the degree of quaternization was higher when using sodium hydroxide as base compared to dimethyl amino pyridine. The degrees of quaternization as well as *O*-methylation of TMC increased with the number of reaction steps.

The mucoadhesive properties of TMC with different degrees of quaternization, ranging between 22 and 49%, were investigated by the group of Snyman *et al.* (2002). TMC was found to have a lower intrinsic mucoadhesivity compared to the chitosan salts, chitosan hydrochloride and chitosan glutamate, but if compared to the reference polymer, pectin, TMC possesses superior mucoadhesive properties. The decrease in the mucoadhesion of TMC compared to the chitosan salts was explained by a change in the conformation of the TMC polymer due to interaction between the fixed positive charges on the quaternary amino group, which possibly also decreases the flexibility of the polymer backbone. The interpenetration into the mucus layer by the polymer is influenced by a decrease in flexibility resulting in a subsequent decrease in mucoadhesivity (Snyman *et al.*, 2003).

N-*Trimethyl Chitosan as Absorption Enhancer of Peptide Drugs*

TMC was first investigated for permeation enhancing properties and toxicity by Kotzé and coworkers (1997, 1999), using the Caco-2 cells as a model for intestinal epithelium. Initially a trimethylated chitosan having a degree of trimethylation of 12% (dimethylation 80%) was tested. This polymer (1.5–2.5%, w/v; pH 6.7) caused large increases in the transport rate of [^{14}C]mannitol (32- to 60-fold), fluorescently labeled dextran 4,400 (167- to 373-fold), and the peptide drug buserelin (28- to 73-fold). CLSM confirmed that TMC opens the tight junctions of intestinal epithelial cells to allow increased transport of hydrophilic compounds along the paracellular transport pathway. No intracellular transport of the fluorescent marker could be observed (Kotzé *et al.*, 1999).

Chitosan HCl and TMCs of different degrees of trimethylation were tested by Kotzé and collaborators (1999) for enhancing the permeability of [^{14}C]mannitol in Caco-2 intestinal epithelia at a pH value of 7.2. Chitosan HCl failed to increase the permeability of these monolayers and so did TMC with a degree of methylation of 12.8%. However, TMC with a degree of trimethylation of 60% increased significantly the [^{14}C]mannitol permeability across Caco-2 intestinal monolayers, indicating that a threshold value at the charge density of the polymer is necessary to trigger the opening of the tight junctions at neutral values.

TMC polymers were further investigated by Thanou *et al.* (1999, 2000a) to see if they provoke cell membrane damage on Caco-2 cell monolayers during enhancement of the transport of hydrophilic macromolecules. Using cell membrane impermeable fluorescent probes and CLSM, it was visualized that TMC polymers widen the paracellular pathways without cell damage. From such visualization studies it also appears that the mechanism of opening the tight junctions is similar to that of protonated chitosan (Thanou *et al.*, 2001c). Because of the absence of significant toxicity, TMC polymers (particularly with a high degree of trimethylation) are expected to be safe absorption enhancers for improved transmucosal delivery of peptide drugs (Thanou *et al.*, 1999).

The effects of TMC60 (degree of trimethylation 60%) polymers were subsequently studied *in vivo* in rats, using the peptide drug buserelin (pH = 6.8) and octreotide (pH = 8.2) (Thanou *et al.*, 2000b, 2000c). Buserelin formulations with or without TMC60 (pH 7.2) were compared with chitosan dispersions at neutral pH values after intraduodenal administration in rats. A remarkable increase in buserelin serum concentrations was observed after coadministration of the peptide with TMC60, whereas buserelin alone was poorly absorbed. In the presence of TMC60 buserelin was rapidly absorbed from the intestine having t_{max} at 40 min, whereas chitosan dispersions (at pH 7.2) showed a slight increase in buserelin absorption compared to the control. Chitosan did not manage to increase the buserelin concentrations to the levels achieved with TMC60. The absolute bioavailability of buserelin after coadministration with 1.0% TMC60 was 13.0%. Similar to the buserelin studies (Thanou *et al.*, 2000b), octreotide absorption after intrajejunal administration was substantially increased, resulting in peptide absolute bioavailability of 16%.

Octreotide was also administered to juvenile pigs with or without TMC60 at a pH of 7.4. The solutions were administered intrajejunally through an in-dwelling fistula that was inserted one week prior to the octreotide. Intrajejunal administration of 10 mg of octreotide, coadministered with 5 and 10% (w/v) TMC60, resulted in a 7.7- and 14.5-fold increase in octreotide absorption with absolute bioavailabilities of 13.9 ± 1.3% and 24.8 ± 1.8%, respectively (Fig. 6.4) (Thanou *et al.*, 2001a).

It is stated by the authors that a gel was obtained with the 10% (w/v) concentration of the polymer. This high concentration of the TMC60 polymer was chosen to counteract the dilution of the 20 ml administration volume by the luminal fluids and mucus of the intestinal tract and to ensure that substantial amounts of both peptide and enhancer could reach the absorptive site of the intestinal mucosa (Thanou *et al.*, 2001a). Although the results show very high bioavailabilities (also taking into account the small absorptive area which is created by only widening of the tight junctions), the impracticality of administering such high concentrations in a solid dosage form cannot be overlooked as concentrations of 1–2 g of the polymer have to be administered in an attempt to obtain the same results (Van der Merwe *et al.*, 2004a).

In order to overcome these problems a completely new and different approach has been chosen by Dorkoosh and coworkers (2002). The platform of their delivery systems consists of superporous hydrogels (SPH) and superporous hydrogels composite (SPHC). These hydrogels can swell very rapidly and have the capacity to take up between 100 and 200 times of intestinal liquid of their original volume. Arriving in the intestine those SPHs swell quickly and bring the delivery systems (small tablet in which the drug is incorporated) which is attached to the outside of the SPH platform in direct contact to the absorbing surface. TMC at the outside of the small tablet will interfere at the interface between swollen SPH and intestinal wall as a polymeric penetration enhancer widening locally the tight junctions

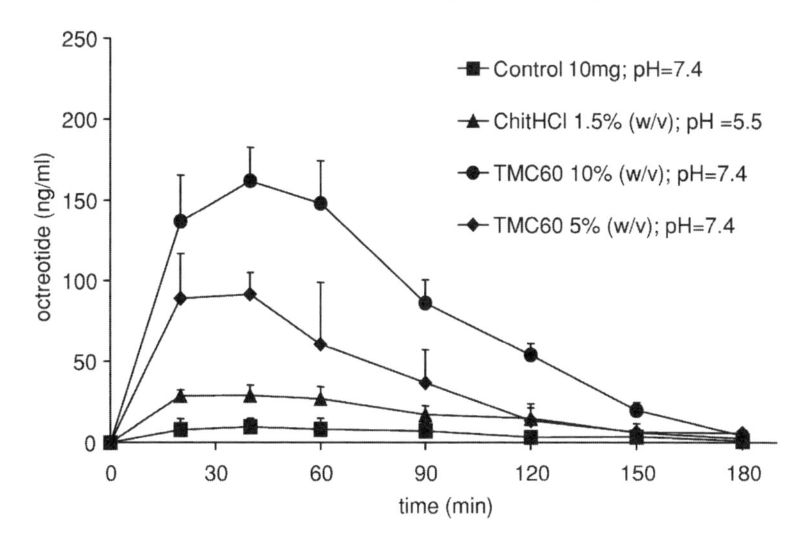

FIGURE 6.4. Plasma octreotide concentration (mean ± SE) versus time curves after intrajejunal administration in pigs (10 mg/20 ml/pig) with the polymers chitosan HCl [CS1.5, 1.5% (w/v), pH 5.5; $n = 6$] and TMC [TMC 10, 10% (w/v); pH 7.4; $n = 6$, and TMC, 5% (w/v); pH 7.4; $n = 3$] or without any polymer [OA 10, octreotide in 0.9% NaCl; pH 7.4; $n = 5$]. With permission from Thanou *et al.* (2001a)

to allow for paracellular absorption of the peptide drug. In an *in vivo* study with pigs the achieved absolute bioavailabilities of octreotide were between 8.7 ± 2.4% and 16.1 ± 3.3% depending on the type of delivery system used. The value of 16.1 ± 3.3% was achieved with TMC60 as absorption enhancer. After the peptide's release from the dosage form the SPH platforms get overhydrated and are easily broken down by the peristaltic forces of the gut. Scintigraphic studies in human have shown the good performance of these oral peptide drug delivery systems with prolonged residence times in the gut. Incorporating the SPH© delivery platforms in enteric coated gelatin capsules of size 000 lead to various stomach transit times (2–6 h in pigs) and 1.5–3 h in human volunteers (Dorkoosh *et al.*, 2004). Capsules of smaller size (00) may reduce the variability in gastric transit times.

6.3.3.4 Monocarboxymethyl Chitosan

An usual approach to increase chitosan's solubility at neutral pH values is the substitution of the primary amine. Whereas *N*-substitution with alkyl groups (i.e. $-CH_3$ groups) can increase the aqueous solubility without affecting its cationic character, substitution with moieties bearing carboxyl groups can yield polymers with polyampholytic properties (Muzzarelli *et al.*, 1982). Monocarboxymethylated chitosan (MCC) was synthesized and further evaluated as potential absorption enhancer (Thanou *et al.*, 2001d). This chitosan derivative (degree of substitution 87–90%) has polyampholytic (zwitterionic) character, which allows the formation of clear gels or solutions (dependent on the concentration of the polymer) even in the presence of polyanionic compounds like heparins at neutral and alkaline pH values, whereas it aggregates at acidic pH. Chitosan and the quaternized derivative TMC form complexes with polyanions that precipitate out of the solution. In contrast, MCC appeared to be compatible with polyanions.

Two viscosity grades MCC (high and low) were initially investigated to see if they are able to increase the permeation of LMWH (4,500 Da) across Caco-2 intestinal cell monolayers. However, the MCC concentrations necessary to open the tight junctions were several times higher than that of TMC60 at neutral pH value. Low viscosity MCC induced higher transport of LMW when compared with the high viscosity derivative. Cell viability tests at the end of the experiments showed that this type of polymer had no damaging effect on cell membranes, whereas recovery of the transepithelial electrical resistance values to initial levels indicated the functional integrity of the monolayer. The mechanism by which polyampholytic chitosans interacts with the tight junctions is not clear yet.

For *in vivo* studies, LMWH was administered intraduodenally with or without MCC to rats. Three percent (w/v) low viscosity MCC significantly increased the intestinal absorption of LMWH, reaching the therapeutic anticoagulant blood levels of LMWH for at least 5 h determined by measuring anti-Xa levels (Thanou *et al.*, 2001d).

6.3.3.5 Thiolated Polymers

Thiolated Polymers of Polyacrylates and Cellulose Derivatives

Thiolated polymers are synthesized by immobilizing thiol groups on polyacrylates or cellulose derivatives (Fig. 6.3d) by the group of Bernkop-Schnürch (Bernkop-Schnürch and Steiniger, 2000; Bernkop-Schnürch *et al.*, 2000; Leitner *et al.*, 2003a; Clausen and and Bernkop-Schnürch, 2000). The main purpose of introducing free thiol groups into polymers, which already have mucoadhesive properties, is to further strongly increase the strength of their mucoadhesiveness due to the chemical reaction of the thiol groups of the mucins and the thiol groups of the thiolated polymers by forming stable covalent disulfide bridges. With this elegant approach the mucoadhesiveness of such polymers and additionally their cohesiveness could be strongly increased. However, because of the turnover time of mucus, which has been estimated in the isolated intestinal loop of the rat by Lehr *et al.* (1991) to be in the order of 47–270 min, the residence time of these thiolated polymers and the delivery systems made of such thiolated polymer will be restricted to this time interval. A reasonable assumption is that the polymers will stick to the mucus at the time of 3 h after peroral application.

Bernkop-Schnürch and coworkers (2000; Bernkop-Schnürch *et al.*2000) linked L-cysteine covalently to PCP (Fig. 6.3d) mediated by a carbodiimide. The resulting thiolated polymers displayed 100 ± 8 and $1,280 \pm 84$ µmol thiol groups per gram, respectively. In aqueous solutions these modified polymers were capable of forming inter and/or intramolecular disulfide bonds. Due to the formation of disulfide bonds within the thiol-containing polymers, the stability of matrix tablets could be strongly improved. Whereas tablets based on the corresponding unmodified polymer disintegrated within 2 h, the swollen carrier matrices of thiolated NaCMC and PCP remained stable for 6.2 h and for more than 48 h, respectively. With the model drug rifampicin controlled release characteristics of these thiolated matrix tablets could be demonstrated. Tensile studies carried out with the unmodified and thiolated polymers at pH 3, 5, and 7, respectively, revealed that only if the polymer displays a pH value of 5, the total work of adhesion could be improved significantly due to the covalent attachment of thiol groups. The permeation enhancing effect of thiolated polycarbophil on intestinal mucosa from guinea pigs showed weak enhancement ratios (1.1–1.5) in comparison to control tests.

Thiolated Polymers of Chitosan

With the same aim as described in the last paragraph, chitosan has been chemically modified by covalent binding of sulfur containing moieties. To date, different thiolated chitosan have been synthesized: chitosan–thioglycolic acid conjugate (Fig. 6.3e), chitosan–cysteine conjugates (Bernkop-Schnürch and Hopf, 2001; Kast and Bernkop-Schnürch, 2001; Hornof *et al.*, 2003), chitosan–cystein conjugates (Bernkop-Schnürch *et al.*, 1999), and chitosan-4-thio-butyl-amide (chitosan-TBA) conjugates (Fig. 6.3f) (Bernkop-Schnürch *et al.*, 2003). These thiolated chitosans have numerous advantageous features in comparison to unmodified chitosan, such as significantly improved mucoadhesive properties and permeation enhancing properties.

The strong cohesive properties of thiolated chitosans make them highly suitable excipients for controlled drug release dosage forms (Bernkop-Schnürch *et al.*, 2003; Kast and Bernkop-Schnürch, 2001). Moreover, solutions of thiolated chitosans display *in situ* gelling properties at physiological pH values, which make them suitable for novel application systems to the eye (Bernkop-Schnürch *et al.*, 2004).

The improved mucoadhesive properties of thiolated chitosans were explained by the formation of covalent bonds between thiol groups of the polymer and cysteine-rich subdomains of glycoporteins in the mucus layer (Leitner *et al.*, 2003b). These covalent bonds are supposed to be stronger than noncovalent bonds, such as ionic interactions of chitosan with nonionic substructures as sialic acid moieties of the mucus layer. This theory was supported by the results of tensile studies with tablets of thiolated chitosan, which demonstrated a positive correlation between the degree of modification with thiol bearing moieties and the adhesive properties of the polymer (Kast and Bernkop-Schnürch, 2001; Roldo *et al.*, 2004). These findings were confirmed by another *in vitro* mucoadhesion system, where the time of adhesion of tablets on intestinal mucosa was determined. The contact time of the thiolated chitosan derivatives increased with increasing amounts of immobilized thiol groups (Kast and Bernkop-Schnürch, 2001; Bernkop-Schnürch *et al.*, 2003). With chitosan–thioglycolic acid conjugates a 5- to 10-fold increase in mucoadhesion in comparison to unmodified chitosan was achieved.

6.3.3.6 Solid Dosage Form Design Based on TMC and Thiolated Polymers and Their *In Vivo* Evaluation

As in the sections described above about TMC, this chitosan derivative has only been administered as a gel formulation or solution. However, the impracticability of administering a polymer solution intraduodenally with the peptide dispersed or dissolved in it, as well as the fact that most peptides are unstable in the presence of an aqueous milieu, has led to the need for a solid oral dosage form in which TMC can be administered together with peptide drugs. To optimally make use of the absorption enhancing properties of TMC in a solid dosage form, the polymer should be able to dissolve rapidly and then be allowed to spread over a wide area of the epithelium in the small intestine. The opening of the tight junctions is a time-dependent process and it is therefore necessary that most of the TMC should be released from the dosage form prior to the release of the peptide drug. The site at which the peptide is released should coincide with the side where the TMC is opening the paracellular pathway for maximum paracellular absorption of the peptide drug (Van der Merwe *et al.*, 2004a).

Minitablets with a diameter of 2–3 mm and granule formulation were developed as solid oral dosage form for the delivery of TMC and the peptide drug desmopressin (1-(3-mercaptopropionic acid)-8-D-arginine vasopressin monoacetate; DDAVP). Both the developed minitablet and granule formulations showed an initial burst release of TMC with a delayed release for DDAVP. Maximum release of TMC was in the order of 50% for all formulations, which is acceptable considering the high molecular weight of the polymer. Domestic pigs were used for the evaluation of the developed minitablet and granule formulations. However, the somatostatin analogue, octreotide, was used in this study and therefore the formulations were slightly adapted to give the same release profiles (Van der Merwe *et al.*, 2004b). The delivery systems were filled in 000 capsules and enteric coated and applied with a special designed applicator into the stomach of the pigs.

Statistical analysis showed no significant difference between the absolute bioavailabilities for the different formulations administered *via* the peroral route. The average bioavailabilities for the negative control, minitablet formulation, granule formulation, and TMC/octreotide solution were, respectively, $0.9 \pm 0.5\%$, $1.0 \pm 1.5\%$, $1.4 \pm 0.5\%$, and $0.5 \pm 0.2\%$. The combination of the gelatin, the enteric coating, and the sticky properties of TMC might have resulted in a delay of release of both the octreotide and the TMC, resulting in the unsatisfactory absorption enhancement with the polymer.

A similar low bioavailability with the peptide drug antide has been the result of an *in vivo* study of Bernkop-Schnürch and coworkers (2005). Antide and the permeation mediator glutathione were embedded in the thiolated polymer chitosan-4-thio-butylamidine conjugate and compressed to tablets. Because it turned out that antide was strongly degraded in the small intestine by elastase, a stomach targeted delivery system was designed. The absolute and relative bioavailability after oral application of the tablet delivery system to pigs were 1.1% and 3.2%, respectively.

In an earlier study, Guggi and coworkers (2002, 2003) developed a solid dosage for the peroral delivery of salmon calcitonin to rats. Different drug carrier matrices, comprising chitosan-4-thio-butyl-amide (chitosan-TBA) conjugate as substantial polymeric excipient and containing equal amounts of salmon calcitonin and optionally the permeation mediator reduced glutathione (Bernkop-Schnürch and Scerbe-Saiko, 1998) were developed. In order to avoid an enzymatic degradation of the peptide drug in the gastrointestinal tract chitosan inhibitor conjugates were also added. All compounds were homogenized and directly compressed to tablets. To enteric-coated tablets targeted to the small intestine, a chitosan-BBI conjugate (Bowman-Birk inhibitor) (Guggi and Bernkop-Schnürch, 2003) and a chitosan-elastatinal conjugate were added (Bernkop-Schnürch and Scerbe-Saiko, 1998). The different tablets were given orally to rats and the plasma calcium levels were monitored as a function of time. Pharmacological efficacy was calculated on the basis of the area under the reduction in plasma calcium levels of the oral matrix tablets versus i.v. injection.

No significant effects were measured when calcitonin was given as a solution orally and also when chitosan was used as the main tablet ingredient due to its insolubility in pH values above 6.5. Only with tablets containing chitosan TBA conjugate as the main tablet excipient a moderate decrease of the calcium level of more than 5% for several hours have been reached. The increased absorption of the peptide, when embedded in a thiolated chitosan matrix, occurs due to the properties of this compound: the high stability and cohesiveness can provide a sustained release of the peptide, while the mucoadhesive features should lead to a prolonged residence time of the dosage form on the site of absorption (which still has to be demonstrated). Moreover, the combination of thiolated chitosan with the permeation mediator reduced glutothione, seems to have an impact on the bioresponse of orally given calcitonin. The significantly higher pharmacological efficacy of thiolated chitosan tablets containing glutathione in comparison to corresponding tablets without glutathione indicates that glutathione contributes additionally to the drug absorption enhancing process.

As pointed out before, it is extremely difficult to develop a peptide drug delivery system for peroral application intended to trigger paracellular uptake of the drug by locally widening of the tight junctions due to the deactivation of the mucoadhesive polymers by soluble mucins in the intestinal liquids before the delivery system reaches the mucosal epithelial tissue and could exert its polymer–mucus (and even preferred directly with the epithelial tissue sialic and sulfate groups) interactions. With small animals as mice and rats, some success with small particulate delivery systems has been achieved due to the short distances between delivery system and gut wall. However, translating the efficacy of such delivery systems to the gut of the pig or human being with much wider diameter of the gut lumen has not been successfully solved so far.

6.4 Conclusions

Most of the pharmaceutical excipients, which when used in appropriate amounts for the manufacturing of common drug delivery systems, are inert and do not show interactions with absorbing tissues. Some excipients such as low molecular weight surfactants may not be as neutral in their interactions with the mucosal membranes. In most cases, their intrinsic properties to form at a defined concentration micelles is used to solubilize poorly soluble drugs. When those solubilisates come in contact with the mucosal tissues of the gut they may interact with the phospholipid bilayer of this membrane and cause some perturbations of the phospholipid bilayer structure and form also micelles with those components or solubilize membrane proteins. Dependent on their concentration, this may lead locally to toxic effects. It is well known that the absorption enhancing effect is directly related to the surfactant concentration. STDHF seemed to be a promising absorption enhancer for the nasal application of insulin; however, chronic toxicity studies with ciliated chicken membranes showed that this compound was too toxic for chronic use and was therefore withdrawn from the market.

Cyclodextrins show basically the same effect as low molecular mass permeation enhancers: they are predominantly used for the solubility increase for poorly soluble drugs such as estrogens, progesterone, testosterone, and hydrocortisone (Duchêne *et al.*, 1999). There seem to be also some species differences: whereas, e.g., β-cyclodextrins showed very promising results in the nasal absorption of insulin of rats and sheep, no significant improvement could be obtained in the human nose. It can be concluded that until today low molecular weight surfactants and also cyclodextrins play a minor role as absorption enhancer for hydrophilic drugs.

With the advent of new biotechnological techniques endogenous compounds like insulin have become available at affordable prices. However, until, today the development of alternative dosage forms (for the nasal, buccal, peroral, rectal, vaginal, and ocular route) for the administration of those class III drugs (high solubility/low permeability) according to the BCS (Amidon *et al.*, 1995) could not keep pace with this development of endogenous peptides. Multifunctional high molecular weight polymers as polyacrylates and chitosan with its various derivatives show promising properties as specific penetration enhancers for the paracellular route of absorption of hydrophilic molecules with high enhancing potency of reversibly opening of the tight junctions and practically with no toxicity when used in normal doses. However, the physical properties (poor solubility and high viscosity and easy saturation of the mucoadhesive properties by soluble mucins in the intestinal liquids) make it very difficult to develop suitable dosage forms which are able to quickly swell in the intestinal gut fluids, to develop mucoadhesive properties, and finally reach the mucous linings of the (human) gut in a still mucoadhesive form. After adhesion to the gut mucus and widening of the tight junctions the peptide drug should be released in the desired controlled manner. Hence, the development of such dosage forms is still in its infancy, but there are promising perspectives that such delivery systems will be successfully developed further on and will see the light of the market.

References

Accili, D., Menghi, G., Bonacucina, G., Di Martino, P., and Palmieri, G. F. (2004) Mucoadhesion dependence of pharmaceutical polymers on mucosa characteristics. *Eur. J. Pharm. Sci.* 22: 225–234.

Amidon, G. L., Lennernäs, H., Shah, V. P., and Crison, J. R. (1995) A theoretical basis for a biopharmaceutics drug classification. The correlation of in vitro product dissolution and in vivo bioavailability. *Pharm. Res.* 12: 413–420.

Anderberg, E. K. and Artursson, P. (1992) Epithelial transport of drugs in cell culture. VII. Effects of pharmaceutical surfactant excipients and bile acids on transepithelial permeability in monolayers of human intestinal epithelia. *J. Pharm. Sci.* 81: 879–887.

Artursson, P., Lindmark, T., Davis, S. S., and Illum, L. (1994) Effect of chitosan on the permeability of monolayers of intestinal epithelial cells (Caco-2). *Pharm. Res.* 11: 1358–1361.

Baldrick, P. (2000) Pharmaceutical excipient development: the need for preclinical guidance. *Regul. Toxicol. Pharmacol.* 32: 210–218.

Bansil, R., Stanley, E., and LaMont, J. T. (1995) Mucin biophysics. *Ann. Rev. Physiol.* 57: 635–657.

Barry, B. W. (1983) Percutaneous absorption. In: Barry, B. W. (ed.). *Dermatological Preparations*, Marcel Dekker, New York, pp. 127–233.

Bernkop-Schnürch, A. and Hopf, T. E. (2001) Synthesis and in vitro evaluation of chitosan-thioglycolic acid conjugates. *Sci. Pharm.* 69: 109–118.

Bernkop-Schnürch, A. and Scerbe-Saiko, A. (1998) Synthesis and in vitro evaluation of chitosan–EDTA–protease-inhibitor conjugates which might be useful in oral delivery of peptides and proteins. *Pharm. Res.* 15: 263–269.

Bernkop-Schnürch, A. and Steiniger, S. (2000) Synthesis and characterization of mucoadhesive thiolated polymers. *Int. J. Pharm.* 194: 239–247.

Bernkop-Schnürch, A., Brandt, U. M., and Clausen, A. E. (1999) Synthesis and in vitro evaluation of chitosan–cysteine conjugates. *Sci. Pharm.* 67: 196–206.

Bernkop-Schnürch, A., Scholler, S., and Biebel, R. G. (2000) Development of controlled drug release systems based on thiolated polymers. *J. Control. Release* 66: 39–48.

Bernkop-Schnürch, A., Hornof, M., and Zoidl, T. (2003) Thiolated polymers – thiomers: modification of chitosan with 2-iminothiolane. *Int. J. Pharm.* 260: 229–237.

Bernkop-Schnürch, A., Hornof, M., and Guggi, D. (2004) Thiolated chitosans. *Eur. J. Pharm. Biopharm.* 57: 9–17.

Bernkop-Schnürch, A., Pinter, Y., Guggi, D., Kahlbacher, H., Schöffmann, G., Schuh, M., Schmerold, I., Del Curto, M. D., D'Antonio, M., Esposito, P., and Huck, C. (2005) The use of thiolated polymers as carrier matrix in oral peptide delivery – proof of concept. *J. Control. Release* 106: 26–33.

BF Goodrich Company. (2002) Carbopol, Noveon, Nomenclature and Chemistry (product information). *Bulletins 3 and 12*, USA.

Bhat, P. G., Flanagan, D. R., and Donovan, M. D. (1996) Drug binding to gastric mucus glycoproteins. *Int. J. Pharm.* 134: 15–25.

Borchardt, G., Lueßen, H. L., de Boer, A. G., Verhoef, J. C., Lehr, C.-M., and Junginger, H. E. (1996) The potential of mucoadhesive polymers in enhancing intestinal peptide drug absorption. III. Effects of chitosan glutamate and carbomer on the epithelial tight junctions in vitro. *J. Control. Release* 39: 131–138.

Caramella, C., Bonferoni, M. C., Rossi, S., and Ferrari, F. (1994) Rheological and tensile test for the assessment of polymer–mucin interaction. *Eur. J. Pharm. Biopharm.* 40: 213–217.

Carreno-Gomez, B. and Duncan, R. (1997) Evaluation of the biological properties of soluble chitosan and chitosan microspheres. *Int. J. Pharm.* 148: 231–240.

Chae, S. Y., Jang, M.-K., and Nah, J.-W. (2005) Influence of molecular weight on oral absorption of water soluble chitosans. *J. Control. Release* 102: 383–394.

Chandy, T. and Sharma, C. P. (1990) Chitosan as a biomaterial. *Biomater. Artif. Cells Artif. Organs* 18: 1–24.

Chickering III, D. E. and Mathiowitz, E. (1999) Definitions, mechanisms, and theories of bioadhesion. In: Mathiowitz, E., Chickering, D. E., and Lehr, C.-M. (eds.). *Bioadhesive Drug Delivery Systems: Fundamentals, Novel Approaches, and Development*, Marcel Dekker, New York, pp. 1–10.

Citi, S. (1992) Protein kinase inhibitors prevent junction dissociation induced by low extracellular calcium in MDCK epithelial cells. *J. Cell Biol.* 117: 169–178.

Citi, S. and Denisenko, N. (1995) Phosphorylation of the tight junction protein cingulin and the effects of protein kinase inhibitors as activators in MDCK epithelial cells. *J. Cell Sci.* 108: 2917–2926.

Clausen, A. E. and Bernkop-Schnürch, A. (2000) In vitro evaluation of permeation-enhancig effect of thiolated polycarbophil. *J. Pharm. Sci.* 89: 1253–1261.

Collares, B. C., McEwan, G. T., Jepson, M. A., Simmons, N. L., and Hirst, B. H. (1994) Paracellular barrier and junctional protein distribution depend on basolateral extracellular Ca^{2+} in cultured epithelia. *Biochim. Biophys. Acta* 1222: 147–158.

Denker, B. M. and Nigam, S. K. (1998) Molecular structure and assembly of the tight junction. *Am. J. Physiol.* 274: F1–F9.

Derjaguin, B. V., Toporov, Y. P., Muller, V. M., and Aleinikova, I. N. (1977) On the relationship between the electrostatics and the molecular component of the adhesion of elastic particles to a solid surface. *J. Colloid Interface Sci.* 58: 528–533.

Derjaguin, B. V., Aleinikova, I. N., and Toporow, Y. P. (1994) On the role of the electrostatic forces in the adhesion of polymer particles to solid surfaces. *Prog. Surf. Sci.* 45: 119–123.

Desai, M. A., Mutlu, M., and Vadgama, P. (1992) A study of macromolecular diffusion through native porcine mucus. *Experientia* 48: 22–26.

Dodane, V. and Vilivalam, V. D. (1998) Pharmaceutical applications of chitosan. *Pharm. Sci. Technol. Today* 1: 246–253.

Dodane, V., Khan, A. M., and Merwin, J. R. (1999) Effect of chitosan on epithelial permeability and structure. *Int. J. Pharm.* 182: 21–32.

Dodou, D., Breedveld, P., and Wieringa, P. A. (2005) Mucoadhesives in the gastrointestinal tract: revisiting the literature for novel applications. *Eur. J. Pharm. Biopharm.* 60: 1–6.

Domard, A., Rinaudo, M., and Terrassin, C. (1986) New method for quaternization of chitosan. *Int. J. Biol. Macromol.* 8: 105–107.

Dorkoosh, F. A., Verhoef, J. C., Verheijden, J. H. M., Rafiee-Tehrani, M., Borchardt, G., and Junginger, H. E. (2002) Peroral absorption of octreotide in pigs formulated in delivery systems on the basis of superporous hydrogel polymers. *Pharm. Res.* 19: 1532–1536.

Dorkoosh, F. A., Stokkel, M. P. M., Blok, D., Borchardt, G., Rafiee-Tehrani, M., Verhoef, J.C., and Junginger, H. E. (2004) Feasibility study on the retention of superporous hydrogel (SPH) composite polymer in the intestinal tract of man using scintigraphy. *J. Control. Release* 99: 199–206.

Duchêne, D. and Ponchel, G. (1992) Principle and investigation of the bioadhesive mechanism of solid dosage forms. *Biomaterials* 13: 709–714.

Duchêne, D., Touchard, F., and Peppas, N. A. (1988) Pharmaceutical and medical aspects of bioadhesive systems for drug administration. *Drug Dev. Ind. Pharm.* 14: 283–318.

Duchêne, D., Wouessidjewe, D., and Ponchel, G. (1999) Cyclodextrins as carrier systems. *J. Control Release* 62: 263–268.

Duffey, M. E. Hainau, B., Ho, S., and Bentzel, C. J. (1981) Regulation of epithelial tight junction permeability by cyclic AMP. *Nature* 294: 451–453.

Elsenhans, B., Blume, R. R., Lembcke, B., and Caspary, W. F. (1983) Polycations. A new class of inhibitors for in vitro small intestinal transport of sugars and amino acids in the rat. *Biochim. Biophys. Acta (BBA) – Biomembranes* 727: 135–143.

Fasano, W. (1998) Innovative strategy for the oral delivery of drugs and peptides. *Trends Biotechnol* 16: 152–157.

Gonzales-Mariscal, L. and Nava, P. (2005) Tight junctions, from tight intercellular seals to sophisticated protein complexes involved in drug delivery, pathogen interaction and cell proliferation. *Adv. Drug Deliv. Rev.* 57: 811–814.

Good, R. J. and Girrfalco, L. A. (1960) A theory for estimation of surface and interfacial energies. III. Estimation of surface energies of solids from contact angle data. *J. Phys. Chem.* 64: 561–565.

Gu, J. M., Robinson, J. R., and Leung, S. H. S. (1988) Binding of acrylic polymers to mucin epithelial surfaces – structure–property relationship. *Crit. Rev. Ther. Drug Carrier Syst.* 5: 21–67.

Guggi, D. and Bernkop-Schnürch, A. (2003) In vitro evaluation of polymeric excipients protecting calcitonin against degradation by intestinal serine proteases. *Int. J. Pharm.* 252: 187–196.

Guggi, D., Kast, C. E., and Bernkop-Schnürch, A. (2002) In vitro evaluation of an oral calcitonin delivery system for rats based on a thiolated chitosan matrix. *Proceed. 11th Int. Pharm. Technol. Symp.* pp. 41–42.

Gutowska, A., Berk, J. S., Kwon, I. C., Bae, Y. H., Cha, Y., and Kim, S. W. (1997) Squeezing hydrogels for controlled oral drug delivery. *J. Control. Release* 48: 141–148.

Hägerström, H., Paulsson, M., and Edman, K. (2000) Evaluation of mucoadhesion for two polyelectrolyte gels in simulated physiological conditions using a rheological method. *Eur. J. Pharm. Sci.* 9: 301–309.

Hayashi, M., Tomita, M., and Awazu, S. (1997) Transcellular and paracellular contribution to transport processes in the colorectal route. *Adv. Drug Deliv. Rev.* 28: 191–204.

Helfland, E. and Tagami, Y. (1972) Theory of the interface between immiscible polymers. *J. Chem. Phys.* 57: 1812–1813.

Hirano, S. and Noishiki, Y. (1985) The blood compatibility of chitosan and N-acetylchitosans. *J. Biomed. Mater. Res.* 19: 413–417.

Hornof, M. D., Kast, C. E., and Bernkop-Schnürch, A. (2003) In vitro evaluation of the viscoelastic behavior of chitosan–thioglycolic acid conjugates. *Eur. J. Pharm. Biopharm.* 55: 185–190.

Huang,Y., Leobandung, W., Foss, A., and Peppas, N. A. (2000) Molecular aspects of muco- and bioadhesion: tethered structures and site-specific surfaces. *J. Control. Release* 65: 63–71.

Illum, L. (1998) Chitosan and its use as a pharmaceutical excipient. *Pharm. Res.* 15: 1326–1331.

Illum, L., Farraj, N. F., and Davis, S. S. (1994) Chitosan as a novel nasal delivery system for peptide drugs. *Pharm. Res.* 11: 1186–1189.

Jabbari, E., Wisniewski, N., and Peppas, N. A. (1993) Evidence of mucoadhesion by chain interpenetration at a poly(acrylic acid)/mucin interface using ATR-FTIR spectroscopy. *J. Control. Release* 26: 99–108.

Junginger, H. E. and Verhoef, J. C. (1998) Macromolecules as safe penetration enhancers for hydrophilic drugs – a fiction? *Pharm. Sci. Technol. Today* 1: 370–376.

Junginger, H. E., Thanou, M., and Verhoef, J. C. (2002) Mucoadhesive hydrogels in drug delivery. In: Swarbrick, J. (ed.). *Encyclopedia of Pharmaceutical Technology*, Marcel Dekker, New York, pp. 1848–1863.

Kaelble, D. H. and Moacanin, J. (1977) A surface energy analysis of bioadhesion. *Polymer* 18: 475–482.

Kast, C. E. and Bernkop-Schnürch, A. (2001) Thiolated polymers – thiomers: development and in vitro evaluation of chitosan-thioglycolic acid conjugates. *Biomaterials* 22: 2345–2352.

Kerec, M., Bogataj, M., Verani, P., and Mrhar, A. (2005) Permeability of pig urinary bladder wall: the effect of chitosan and the role of calcium. *Eur. J. Pharm. Sci.* 25: 113–121.

Kinloch, A. J. (1979) Interfacial fracture mechanical aspects of adhesive bonded joints – review. *J. Adhes.* 10: 193–219.

Kinloch, A. J. (1980) The science of adhesion. I. Surface and interfacial aspects. *J. Mater. Sci.* 15: 2141–2166.

Kotzé, A. F., Lueßen, H. L., de Leeuw, B. J., de Boer, A. G., Verhoef, J. C., and Junginger, H. E. (1997) *N*-trimethyl chitosan chloride as a potential absorption enhancer across mucosal surfaces: in vitro evaluation in intestinal epithelial cells (Caco-2). *Pharm. Res.* 14: 1197–1202.

Kotzé, A. F., Lueßen, H. L., de Leeuw, B. J., de Boer, A. G., Verhoef, J. C., and Junginger, H. E. (1998) Comparison of the effect of different chitosan salts and N-trimethyl chitosan chloride on the permeability of intestinal epithelial cells (Caco-2). *J. Control. Release* 51: 35–46.

Kotzé, A. F., Thanou, M. M., Lueßen, H. L., de Boer, A. G., Verhoef, J. C., and Junginger, H. E. (1999) Enhancement of paracellular drug transport with highly quaternized *N*-trimethyl chitosan chloride in neutral environment: in vitro evaluation in intestinal epithelial cells (Caco-2). *J. Pharm. Sci.* 88: 253–257.

LeCluise, E. L. and Sutton, S. C. (1997) In vitro models for selection of development candidates. Permeability studies to define mechanisms of absorption enhancement. *Adv. Drug Deliv. Rev.* 23: 163–183.

Lee, J. W., Park, J. H., and Robinson, J. R. (2000) Bioadhesive-based dosage forms: the next generation. *J. Pharm. Sci.* 89: 850–866.

Lehr, C.-M., Poelma, F. G. J., Junginger, H. E., and Tukker, J. J. (1991) An estimate of turnover time of intestinal mucus gel layer in the rat in situ loop. *Int. J. Pharm.* 70: 235–240.

Lehr, C.-M., Bouwstra, J. A., Boddé, H. E., and Junginger, H. E. (1992a) A surface energy analysis of mucoadhesion: contact angle measurements on polycarbophil and pig intestinal mucosa in physiologically relevant fluids. *Pharm. Res.* 9: 70–75.

Lehr, C.-M., Bouwstra, J. A., Kok, W., de Boer, A. G., Tukker, J. J., Verhoef, J. C., Breimer, D. D., and Junginger, H. E. (1992b) Effects of the mucoadhesive polymer polycarbophil on the intestinal absorption of a peptide drug in the rat. *J. Pharm. Pharmacol.* 44: 402–407.

Lehr, C.-M., Bouwstra, J. A., Spies, F., Onderwater, J., Noordeinde, C., Vermeij-Keers, C., van Munsteren, C. J., and Junginger, H. E (1992c) Visualization studies of the mucoadhesive interface. *J. Control. Release* 18: 249–260.

Lehr, C.-M., Bouwstra, J. A., Schacht, E. H., and Junginger, H. E. (1992d) In vitro evaluation of mucoadhesive properties of chitosan and some other natural polymers. *Int. J. Pharm.* 78: 43–48.

Lehr, C.-M., Boddé, H. E., Bouwstra, J. A., and Junginger, H. E. (1993) A surface energy analysis of mucoadhesion – the combined spreading coefficient as a new criterion for adhesion in a three phase (solid–liquid–solid) system. *Eur. J. Pharm. Sci.* 1: 19–30.

Leitner, V. M., Walker, G. F., and Bernkop-Schnürch, A. (2003a) Thiolated polymers: evidence for the formation of disulphide bonds with mucus glycoproteins. *Eur. J. Pharm. Biopharm.* 56: 207–214.

Leitner, V. M., Marschütz, M. K., and Bernkop-Schnürch, A. (2003b) Mucoadhesive and cohesive properties of poly(acrylic acid)–cysteine conjugates with regard to their molecular mass. *Eur. J. Pharm. Sci.* 18: 89–96.

Lichtenberger, L. M. (1992) The hydrophobic barrier properties of gastrointestinal mucus. *Ann. Rev. Physiol.* 57: 565–583.

Lichtenberg, D., Robson, R. J., and Dennis, E. A. (1983). Solubilization of phospholipids by detergents: structural and kinetic aspects. *Biochim. Biophys. Acta* 737: 285–304.

Lueßen, H.L., Verhoef, J.C., Borchardt, G., Lehr, C.-M., de Boer, A.G., and Junginger, H.E. (1995) Mucoadhesive polymers in peroral drug delivery II. Carbomer and Polycarbophil are potent inhibitors of the intestinal proteolytic enzyme trypsin. *Pharm. Res.* 12: 1293–1298.

Lueßen, H. L., de Leeuw, B. J., Pérard, D., Lehr, C.-M., de Boer, A. G., Verhoef, J. C., and Junginger, H. E. (1996a) Mucoadhesive polymers in peroral drug delivery. I. Influence of mucoadhesive excipients on the proteolytic activity of intestinal enzymes. *Eur. J. Pharm. Sci.* 4: 117–128.

Lueßen, H. L., Bohner, V., Pérard, D., Langguth, P., de Boer, A. G., Merkle, H. P., and Junginger, H. E. (1996b) Mucoadhesive polymers in peroral peptide drug delivery. V. Effect of poly(acrylates) on the enzymatic degradation of peptide drugs by intestinal brush border membrane vesicles. *Int. J. Pharm.* 141: 39–52.

Lueßen, H. L., de Leeuw, B. J., Langemeijer, M. W. E., de Boer, A. G., Verhoef, J. C., and Junginger, H. E. (1996c) Mucoadhesive polymers in peroral drug delivery. VI. Carbomer and chitosan improve the intestinal absorption of the peptide drug buserelin in vivo. *Pharm. Res.* 13: 1668–1672.

Lueßen, H. L., Rentel, C.-O., Kotzé, A. F., de Boer, A. G., Verhoef, J. C., and Junginger, H. E. (1997) Mucoadhesive polymers in peroral peptide drug delivery. IV. Polycarbophil and chitosan are potent enhancers of peptide transport across intestinal mucosae in vitro. *J. Control. Release* 45: 15–23.

Madara, J. L. (1987) Intestinal absorptive cell tight junctions are linked to cytoskeleton. *Am. J. Physiol.* 253: C171–C175.

Madara, J. L., Barenberg, D., and Carlson, S. (1986) Effects of cytochalasin D on occluding junctions of intestinal absorptive cells – further evidence that the cytoskeleton may influence paracellular permeability and junctional charge activity. *J. Cell Biol.* 102: 2125–2136.

Madsen, F., Eberth, K., and Smart, J. D. (1998) A rheological examination of the mucoadhesive/mucus interaction: the effect of mucoadhesive type and concentration. *J. Control. Release* 50: 167–178.

Mannila, J., Järvinen, T., Järvinen, K., Tarvainen, M., and Jarho, P. (2005) Effects of RM-β-CD on sublingual bioavailability of Δ^9-tetrahydrocannabinol in rabbits. *Eur. J. Pharm. Sci.* 26: 71–77.

Marriott, C. and Gregory, N. P. (1990). Mucus physiology and pathology. In: Lenaerts, V. and Gurny. R. (eds.). *Bioadhesive Drug Delivery systems*, CRC Press, Baco Raton, pp. 1–24.

Martin, G. P., Marriott, C., and Kellaway, I. W. (1978) Direct effect of bile salts and phospholipids on the physical properties of mucus. *Gut* 19: 1103–1107.

Marttin, E., Romeijn, S. G., Verhoef, J. C., and Merkus, F. W. H. M. (1997) Nasal absorption of dihydroergotamine from liquid and powder formulations in rabbits. *J. Pharm. Sci.* 86: 802–807.

Mikos, A. G. and Peppas, N. A. (1986) Systems for controlled release of drugs. V. Bioadhesive systems. *STP Pharma.* 2: 705–716.

Mikos, A. G. and Peppas, N. A. (1989) Measurement of the surface tension of mucin solutions. *Int. J. Pharm.* 53: 1–5.

Mortazavi, S. A. and Smart, J. D. (1993) An investigation into the role of water movement and mucus gel dehydration in mucoadhesion. *J. Control. Release* 3: 197–203.

Muzzarelli, R. A. A., Tanfani, F., Emmanueli, S., and Mariotti, S. (1982) *N*-(carboxymethylidene)-chitosans and N-(carboxymethyl)-chitosans: novel chelating polyampholytes obtained from chitosan glyoxylate. *Carbohydr. Res.* 107: 199–214.

Palant, C. E., Duffey, M. E., Mookerjee, B. K., Ho, S., and Bentzel, C. J. (1983) Ca^{2+} regulation of tight-junction permeability and structure in Necturus gallbladder. *Am. J. Physiol. Cell Physiol.* 245: C203–C212.

Park, H. and Robinson, J. R. (1985) Physico-chemical properties of water insoluble polymers important to mucin/epithelial adhesion. *J. Control Release* 2: 47–57.

Peppas, N. A. and Buri, P. A. (1985) Surface, interfacial and molecular aspects of polymer bioadhesion on soft tissues. *J. Control. Release* 2: 257–275.

Peppas, N. A. and Huang, Y. (2004). Nanoscale technology of mucoadhesive interactions. *Adv. Drug Deliv. Rev.* 56: 1675–1687.

Peppas, N. A. and Stahlin, J. J. (1996) Hydrogels as mucoadhesive and bioadhesive materials: a review. *Biomaterials* 17: 1553–1561.

Polli, J. E., Yu, L. X., et al. (2004) Summary workshop report: biopharmaceutics classification system – implementation challenges and extension opportunities. *J. Pharm. Sci.* 93: 1375–1381.

Polnok, A., Borchardt, G., Verhoef, J. C., Sarisuta, N., and Junginger, H. E. (2004) Influence on methylation process on the degree of quaternization of *N*-trimethyl chitosan chloride. *Eur. J. Pharm. Biopharm.* 57: 77–83.

Ponchel, G., Touchard, D., Duchêne, D., and Peppas, N. A. (1987) Bioadhesive analysis of controlled-release systems. I. Fracture and interpenetration analysis in poly(acrylic-acid)-containing systems. *J. Control. Release* 5: 129–141.

Qiu, Y. and Perk, K. (2001) Environment-sensitive hydrogels for drug delivery. *Adv. Drug Deliv. Rev.* 53: 321–339.

Rege, B. D., Yu, L. X., Hussain, A. S., and Polli, J. E. (2001) Effect of common excipients on Caco-2 transport of low-permeability drug. *J. Pharm. Sci.* 90: 1776–1786.

Rege, B. D., Kao, J. P. Y., and Polli, J. E. (2002) Effects of nonionic surfactants on membrane transporters in Caco-2 cell monolayers. *Eur. J. Pharm. Sci.* 16: 237–246.

Roldo, M., Hornof, M., Caliceti, P., and Bernkop-Schnürch, A. (2004) Mucoadhesive thiolated chitosans as platforms for oral controlled drug delivery: synthesis and in vitro evaluation. *Eur. J. Pharm. Biopharm.* 57: 115–121.

Rossi, S., Bonferoni, M. C., Caramella, C., and Colombo, P. (1994) A rheometric method for assessing the sucralfate-mucin interaction. *Eur. J. Pharm. Biopharm.* 40: 179–182.

Rossi, S., Ferrari, F. Bonferoni, M. C., and Caramella, C. (2000) Charaterization of chitosan hydrochlorde–mucin interactions by means of viscosimetric and turbidimetric measurements. *Eur. J. Pharm. Sci.* 10: 251–257.

Rossi, S., Ferrari, F., Bonferoni, M. C., and Caramella, C. (2001) Characterization of chitosan hydrochloride–mucin rheological interactions influence of polymer concentration and polymer weight ratio. *Eur. J. Pharm. Sci.* 12: 479–485.

Sahlin, J. J. and Peppas, N. A. (1996) An investigation of polymer diffusion in hydrogel laminates using near-field FTIR microscopy. *Macromolecules* 29: 7124–7129.

Schipper, N. G. M., Varum, K. M., and Artursson, P. (1996) Chitosans as absorption enhancers for poorly absorbable drugs. 1. Influence of molecular weight and degree of deacetylation on drug transport across human intestinal epithelial (Caco-2) cells. *Pharm. Res.* 13: 1686–1692.

Schipper, N. G. M., Olsson, S., Hoogstraate, J. A., de Boer, A. G., Varum, K. M., and Artursson, P. (1997) Chitosans as absorption enhancers for poorly absorbable drugs. 2. Mechanism of absorption enhancement. *Pharm. Res.* 14: 923–929.

Schipper, N. A., Vårum, K. M., Sternberg, P., Ocklind, G., Lennernäs, H., and Artursson, P. (1999) Chitosans as absorption enhancers of poorly absorbable drugs. 3. Influence of mucus on absorption enhancement. *Eur. J. Pharm. Sci.* 8: 335–343.

Scott Swenson, E. and Curatolo, W. J. (1992). Intestinal permeability enhancement for proteins, peptides and other polar drugs: mechanisms and potential toxicity. *Adv. Drug Deliv. Rev.* 8: 39–92.

Sieval, A. B., Thanou, M., Kotzé, A. F., Verhoef, J. C., Brussee, J., and Junginger, H. E. (1998) Preparation and NMR-characterization of highly substituted *N*-trimethyl chitosan hydrochloride. *Carbohydr. Polym.* 36: 157–165.

Snyman, D., Hamman, J. H., Kotzé, J. S., Rollings, J. E., and Kotzé, A. F. (2002) The relationship between the absolute molecular weight and the degree of quaternization of *N*-trimethyl chitosan chloride. *Carbohydr. Polym.* 50: 145–150.

Snyman, D., Hamman, J. H., and Kotzé, A. F. (2003) Evaluation of the mucoadhesive properties of *N*-trimethyl chitosan chloride. *Drug Dev. Ind. Pharm.* 29: 59–67.

Solomonidou, D., Cremer, K., Krumme, M., and Kreuter, J. (2001) Effect of carbomer concentration and degree of neutralization on the mucoadhesive properties of polymer films. *J. Biomater. Sci. Polym. Ed.* 12: 1191–1205.

Strous, G. J. and Dekker, J. (1992) Mucin-type glycoproteins. *Crit. Rev. Biochem. Mol. Biol.* 27: 57–92.

Suh, J. K. F. and Matthew, H. W. T. (2000) Application of chitosan-based polysaccharide material in cartilage tissue engineering: a review. *Biomaterials* 21: 2589–2598.

Szejtli, J. (1988) *Cyclodextrin Technology*, Kluwer Academic Publishers, Boston.

Tamburic, S. and Craig, D. Q. M. (1997) A comparison of different in vitro methods for measuring mucoadhesive performance. *Eur. J. Pharm. Biopharm.* 44: 159–167.

Thanou, M., Verhoef, J. C., Romeijn, S. G., Nagelkerke, J. F., Merkus, F. W. H. M., and Junginger, H. E. (1999) Effects of *N*-trimethyl chitosan chloride, a novel absorption enhancer, on Caco-2 intestinal epithelia and the ciliary beat frequency of chicken embryo trachea. *Int. J. Pharm.* 185: 73–82.

Thanou, M. (2000) Chitosan derivatives in drug delivery. Trimethylated and carboxymethylated chitosan as safe enhancers for the intestinal absorption of hydrophilic drugs, PhD Thesis, Leiden University, Leiden, pp. 91–108.

Thanou, M. M., Kotzé, A. F., Scharringhausen, T., Lueßen, H. L., de Boer, A. G., Verhoef, J. C., and Junginger, H. E. (2000a). Effect of degree of quaternization of N-trimethyl chitosan chloride for enhanced transport of hydrophilic compounds across intestinal Caco-2 cell monolayers. *J. Control. Release* 64: 15–25.

Thanou, M., Florea, B. I., Langemeijer, M. W. E., Verhoef, J. C., and Junginger, H. E. (2000b) *N*-trimethylated chitosan chloride (TMC) improves the intestinal permeation of the peptide drug buserelin in vitro (Caco-2 cells) and in vivo (rats). *Pharm. Res.* 17: 27–31.

Thanou, M., Verhoef, J. C., Marbach, P., and Junginger, H. E. (2000c) *N*-trimethyl chitosan chloride (TMC) ameliorates the permeability and absorption properties of the somatostatin analogue in vitro and in vivo. *J. Pharm. Sci.* 89: 951–957.

Thanou, M., Verhoef, J. C., Verheijden, J. H. M., and Junginger, H. E. (2001a) Intestinal absorption of octreotide using trimethyl chitosan chloride: studies in pigs. *Pharm. Res.* 18: 823–828.

Thanou, M., Verhoef, J. C., Nihot, M. T., Veheijden, J. H. M., and Junginger, H. E. (2001b) Enhancement of the intestinal absorption of low molecular weight heparin (LMWH) in rats and pigs using carbopol 934P. *Pharm. Res.* 18: 1638–1641.

Thanou, M., Verhoef, J. C., and Junginger, H. E. (2001c) Oral drug absorption enhancement by chitosan and its derivatives. *Adv. Drug Deliv. Rev.* 52: 117–126.

Thanou, M., Nihot, M. T., Jansen, M., Verhoef, J. C., and Junginger, H. E. (2001d) Mono-*N*-caboxymethyl chitosan (MCC), a polyampholytic chitosan derivative, enhances the intestinal absorption of low molecular weight heparin across intestinal epithelia in vitro and in vivo. *J. Pharm. Sci.* 90: 38–46.

Uekama, K., Fujinaga, T., Hirayama, F., Otagiri, M., and Yamasaki, M. (1982) Inclusion complexations of steroid hormones with cyclodextrins in water and in solid phase. *Int. J. Pharm.* 10: 1–15.

Van der Merwe, S. M., Verhoef, J. C., Verheijden, J. H. M., Kotzé, A. F., and Junginger, H. E. (2004a) Trimethylated chitosan as polymeric absorption enhancer for improved peroral delivery of peptide drugs. *Eur. J. Pharm. Biopharm.* 58: 225–235.

Van der Merwe, S. M., Verhoef, J. C., Kotzé, A. F., and Junginger, H. E. (2004b) N-Trimethyl chitosan chloride as absorption enhancer in oral peptide drug delivery. Development and characterization of minitablet and granule formulations. *Eur. J. Pharm. Biopharm.* 57: 85–91.

Verhoef, J. C., Schipper, N. G. M., Romeijn, S. G., and Merkus, F. W. H. M. (1994) The potential of cyclodextrins as absorption enhancers in nasal delivery of peptide drugs. *J. Control. Release* 29: 35–360.

Voyutskii, S. S. (1963) *Autoadhesion and Adhesion of High Polymers*, Wiley, New York.

Wool, R. P. (1995) *Polymer Interfaces: Structure and Strength*, Hanser Publishing, Munich.

7 Intestinal Transporters in Drug Absorption

Rajinder K. Bhardwaj, Dea R. Herrera-Ruiz, Yan Xu, Stephen M. Carl, Thomas J. Cook, Nicholi Vorsa, and Gregory T. Knipp

7.1 Introduction

Effective drug therapy relies on the interplay between the pharmacokinetics and pharmacodynamics (PK/PD) of the agent upon administration. During the initial stages of drug discovery, numerous studies are performed to assess the pharmacological effectiveness of new chemical entities (NCEs) to select a lead compound(s) that offers the greatest promise for therapeutic efficacy. While the ability of a drug to bind to a therapeutic target is critical to its clinical success, the ultimate effectiveness is also a function of its ability to reach the therapeutic target in sufficient concentrations to mitigate or treat the ailment. Therefore, the pharmacokinetics of any NCE must also be evaluated early in the drug discovery stages to enhance the rational selection of a lead compound from the many NCEs that are screened, based on not only biological activity but also potential *in vivo* bioavailability. Bioavailability is defined by the US FDA as "the rate and extent to which the active ingredient or active moiety is absorbed from a drug product and becomes available at the site of action" (21 CFR 320.1(a)). The overall bioavailability is largely determined by the absorption, distribution, metabolism, and excretion (ADME) of selected compounds in targeted patient populations. While ADME involves transport/permeability processes across cellular barriers in numerous tissues, we will restrict our discussion to intestinal absorption (absorptive influx) and excretion (secretory efflux).

The gastrointestinal (GI) tract varies greatly in morphological characteristics from relatively no folding in the esophagus to high degrees of folding (villi) in the small intestine (Tortora and Grabowski, 1993). The small intestinal villous epithelium is the primary mediator and barrier to GI absorption of orally administered drugs and nutrients into systemic circulation. The primary cells mediating drug absorption across the intestinal villous epithelium are the polarized columnar enterocytes, which are distinguished by the presence of apical membrane microvilli. The villous structure and the enterocyte microvilli provide a significant increase in the intestinal absorptive surface area (Tortora and Grabowski, 1993); however, it is the compound's physicochemical properties that dictate the route and extent of absorption.

Paracellular and transcellular diffusion are the two routes of GI permeation (Adson *et al.*, 1995; Knipp *et al.*, 1997; Sorensen *et al.*, 1997). Paracellular absorption occurs *via* diffusion of dissolved solute between cells through the tight junctional complex and tortuous pathway in the intercellular spacing (Adson *et al.*, 1995; Knipp *et al.*, 1997). The paracellular pathway is quite restrictive depending on the pore size and charge of the tight junctions as well as the cell barrier's porosity. There are several physicochemical characteristics of a drug that favor paracellular diffusion including charge, hydrophilicity, shape/conformation, size, and molecular weight (Adson *et al.*, 1995; Knipp *et al.*, 1997).

The transcellular route is comprised of several potential parallel pathways for drug permeation including passive transcellular diffusion, ion channels, facilitated diffusion, active transport, and endocytosis (Oh and Amidon, 1999). A more comprehensive discussion on the characteristics of each transcellular route of permeation is provided by Oh and Amidon (1999).

Passive transcellular diffusion has traditionally been viewed as the most desirable route for GI drug absorption. The degree of passive transcellular permeation of a compound is also largely dependant on those physicochemical properties mentioned above, including the degree of ionization, lipophilicity, molecular weight, and shape/conformation. In the past, the pH-partition hypothesis, first postulated in the mid- to early 1900s, was used as a model for predicting the absorption and/or disposition of a drug across biological membranes based on the lipid to aqueous partition coefficients as a function of molecular ionization. Jacobs (1940) initially linked biological permeability and accumulation to pH, demonstrating a correlation between absorption and an electrolyte's degree of dissociation. In fact, much of the early work was based on the observation that the rate of drug absorption is related to the drug's degree of dissociation in solution, where drugs that exhibit a higher lipophilic versus ionic character will diffuse much more readily across biological membranes (Hogben *et al.*, 1959). Given this clear correlation, researchers postulated that the physical barriers to drug absorption must be lipoidic in nature. The pH-partition hypothesis was then mathematically described under sink conditions based on the fraction of unionized drug in solution using Fick's first law of diffusion, assuming that aqueous boundary layer does not affect the transport process:

Fick's first law of diffusion

$$\frac{\mathrm{d}M}{\mathrm{d}t} = \frac{D_\mathrm{m} S K_\mathrm{p}}{h_\mathrm{m}} (C_\mathrm{d} - C_\mathrm{r}), \tag{7.1}$$

under sink conditions

$$\frac{\mathrm{d}M}{\mathrm{d}t} = \frac{D_\mathrm{m} S K_\mathrm{p} C_\mathrm{d}}{h_\mathrm{m}} = PSC_\mathrm{d}, \quad P = \frac{D_\mathrm{m} K_\mathrm{p}}{h_\mathrm{m}} f_\mathrm{un}, \tag{7.2}$$

where dM/dt is the flux of material per time, D_m the membrane diffusion coefficient, S the cross-sectional surface area, K_p the partition coefficient, h_m the membrane thickness, C_d the concentration of donor chamber, C_r the concentration of receiver chamber, $C_d - C_r$ the concentration gradient at time t, P the permeability, and f_{un} is the fraction unionized.

Based on the pH-partition hypothesis, an *in vitro* method was widely utilized to predict drug absorption by measuring a compound's ability to partition (log P) between a fairly immiscible lipophilic solvent (octanol) and water or buffers at different pHs (Leo *et al.*, 1971). While this method does provide adequate predictive power within series of compounds, its broad utility is often limited by observed deviations between *in vitro* and *in vivo* permeability. These deviations were observed in early literature, but a correlative explanation was often incorrectly proposed to rationalize a fit to the pH-partition hypothesis (Shore *et al.*, 1957).

In its most basic sense, under the partitioning model absolute bioavailability should increase linearly with increasing lipophilicity (log P) due to the lipophilic nature of biological membranes, where permeation is a function of diffusion. Expansion of the model realizes that highly hydrophilic compounds (log $P < 1$–2) would not be absorbed transcellularly due to their polar/ionic character, just as highly hydrophobic compounds (log $P > 4$–5) would accumulate in the interior aliphatic portion of the cellular membrane due to their lipophilicity. For example, an absorption model that is depended solely on partitioning would result in a parabolic relationship between the fraction absorbed (log 1/concentration) and lipophilicity (log P), which would only occur in unusual cases *in vivo* (Higuchi and Davis, 1970). Numerous deviations in the predictive power of log P values for assessing the permeability of a compound have been observed, and therefore, other predictive tools were advanced to better address these inconsistencies. One of the more recognized tools, "The Rule of Five" was proposed by Lipinski *et al.* (1997, 2000, 2001) to estimate the permeability of compounds *in silico* based on molecular descriptors at the early stages of drug discovery. Lipinski's Rule of Five states: ". . . .poor absorption or permeability is more likely when there are more than 5 *H-bond donors*, 10 *H-bond acceptors*, the *molecular weight is greater than 500* and the *calculated log P is greater than 5*." (Lipinski *et al.*, 1997). The rule of five has found broad utility to predict the developability and optimization of NCEs in industry. However, several confounding factors, most importantly the role of drug transporters, have acted to limit the predictability and applicability of this and many analogous techniques.

The complex nature of membrane physiology and the lack of predictive absorption methodologies is better understood when one considers the numerous roles elucidated for different drug transporters in mediating transcellular influx and efflux of xenobiotics. There are numerous classes of transporter proteins that have been identified to date, each with different, sometimes overlapping, substrate specificity, capacity and affinity, as well as specific tissue, cellular and temporal expression patterns. Transporter proteins are integral proteins that function *via* either a facilitated diffusion, or active, energy-dependent mechanisms to mediate transcellular flux of xenobiotics and nutrients (Oh and Amidon 1999). Not surprisingly, a compound's physicochemical properties greatly influence its interactions with transporters and lipophilic character (i.e., partitioning) plays a major role in determining these interactions. As such, not only is there great overlap in substrate selectivity of many transporters, but there is also variability in uptake due to lipophilicity and solubility differences and resulting membrane interactions. Much of this variability may also be due to interactions with other potential substrates present in biological fluids, stearic influences of transporter binding, possible membrane interactions, such as changes in fluidity, the rate of transport (V_{max}), or even competitive binding of substrates to other transporters. Thus researchers have developed various *in vitro* and *in vivo* models to delineate the role of individual transporter activities in mediating xenobiotic uptake (Stewart *et al.*, 1997; Ekins *et al.*, 2000, 2005; Kimura *et al.*, 2002; van de Waterbeemd, 2002; Harrison *et al.*, 2004; Kassel, 2004; Sun *et al.*, 2004).

A number of cell models have been used to evaluate the intestinal permeability of drugs, with the Caco-2 cell model, derived from colorectal adenocarcinoma, being the most widely used. Caco-2 cells exhibit much of the barrier functionality of the normal endogenous intestinal epithelium. Under the right conditions, Caco-2 cells not only grow in a tight-knit monolayer and exhibit tight junctions, but they express many of the same receptors and transporters of the intestinal epithelium. The simplicity of this model makes *in vitro* permeability and uptake measurements relatively straightforward and conducive to automation. However, despite the obvious benefits of *in vitro* drug transport studies using cell lines, there are also many disadvantages and hurdles to overcome when using cell line models. For instance, even though transepithelial electrical resistance (TEER) can be determined to verify the integrity of the monolayer, leakiness is always of concern when performing these studies and can be a source of artifacts. Additionally, cellular energy requirements dictate the endogenous expression of a multitude of transporters, many with overlapping substrate specificities. This not only provides a source of variation, but also makes it difficult to assess the transport properties due to a single transporter. Moreover, culture conditions can greatly affect the genetic regulation of a multitude of functional proteins, resulting in intra- and interlaboratory variations and error. In short, assessing drug transport function *via* various cell line models can be a useful tool, however, it also presents many technological hurdles, and the techniques employed do raise some issues of applicability to the physiologic model.

In addition to those limitations mentioned above, cell lines do not adequately address the issue of transcriptional and translational variations within any particular patient population. Single nucleotide polymorphisms (SNPs) and the regulatory mechanisms associated with these genetic variations have broadened the pharmacogenomics field quite extensively. The biochemical architecture of the human intestine, including drug transporters and their variants, must be viewed in the light of evolution of mammalian nutritional requirements. For the majority of mammalian evolution, the intestine served as the primary gatekeeper that compounds must traverse when entering the mammalian organism. As such, the *in vivo* model of drug absorption and permeation with respect to drug transporters is a function of all those transporter proteins expressed at any given time and location within the vicinity of the drug product as it moves along the intestine, working in concert to facilitate the acquisition of nutrients. While useful for permeability prediction and screening of large databases of compounds, cell culture models fall short of simulating the actual *in vivo* conditions and should be rationalized in context.

Due to the innate limitations in studying drug permeability *in vitro*, the pharmaceutical industry has been ever evolving in its quest to further understand and enhance the intestinal absorption of pharmaceuticals. Current approaches include formulation design to either exploit, or inhibit transporter function through the use of various excipients. Surfactants, for example, are well known to alter membrane fluidity, thus altering potential substrate interactions and transporter interactions. Another approach to maximize bioavailability is to tailor NCE design to improve substrate affinity for a particular transporter. A classic example utilizing this approach is the addition of a peptidic valine moiety to the antiviral acyclovir to produce valacyclovir. While the free compound does absorb well through the GI tract, the valyl addition increases the compound's transporter affinity for the oligopeptide transporter, PepT1, thus drastically increasing its oral bioavailability, as will be discussed in greater detail later.

A fundamental understanding of drug transporters is essential when analyzing a drug's PK/PD behavior after oral administration. Difficulties in characterizing the intestinal transepithelial transport of drugs underscores the need for a complete understanding of the biophysical and biochemical barriers that are present in the GI tract. This chapter aims to delineate the role of several selected transporter families likely involved in the intestinal bioavailability of orally administered drugs. These transporter families have been selected due to the breadth of literature demonstrating their respective importance in intestinal drug absorption. The chapter will highlight the interplay of the molecular and functional characteristics for each of the different transporters and summarized with respect to the impact of these characteristics on altered bioavailability and pharmacokinetics.

7.2 ATP Binding Cassette Transporters

Several members of the ATP binding cassette (ABC) transporter protein superfamily have been shown to impart multidrug resistance by virtue of their ability to efflux xenobiotics from the cytoplasm and across the cellular membrane in an energy-dependent, polarized manner. These various ABC transporter superfamily isoforms constitute a broad array of substrate specificities from endogenous fatty acid metabolites to synthetic therapeutics. A common characteristic shared by ABC transporters is the presence of nucleotide binding domain(s) (NBDs),

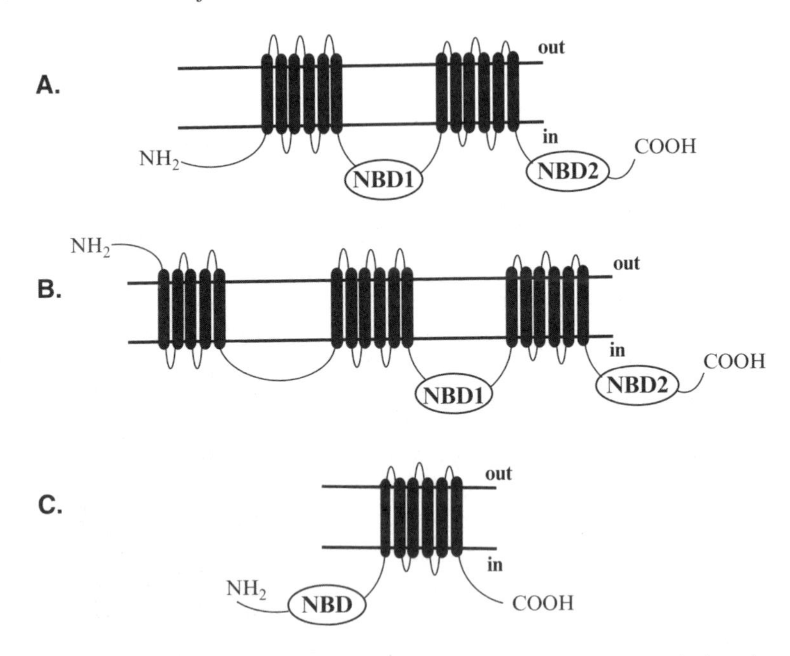

FIGURE 7.1. Transmembrane arrangement of drug transporter proteins. (A) is the schematic of P-gp (MDR1), MDR3, MRP4, MRP5, and MRP8 that have 12 TM (transmembrane) regions and two NBDs (nucleotide binding domains), (B) shows the MRP transporters (MRP1, MRP2, MRP3 and MRP6, and MRP7) that has five extra TM regions toward the N terminus. (C) depicts the BCRP that has six TM regions and one NBD also called as a half transporter

which enable these integral membrane proteins to hydrolyze ATP to drive efflux (Fig. 7.1) (Higgins, 1991).

The cloning of the human genome, coupled with various genomic and functional studies, has revealed 49 human ABC transporter isoforms that are separated into seven distinct subfamilies based on their sequence homology (ABCA to ABCG). A complete description concerning the various ABC transporter subfamilies can be found at the Web site http://www.gene.ucl.ac.uk/nomenclature/genefamily/abc.html. Among these subfamilies, the ABCB and ABCC subfamilies contain the most widely investigated transporters influencing human intestinal absorption. Specifically, P-glycoprotein (ABCB1, P-gp) and multidrug resistance-associated proteins (ABCC, MRPX) have not only been shown to be expressed along the GI tract, but due to their cellular localizations and broad substrate specificities, appear to be the primary efflux pumps determining xenobiotic absorption (Table 7.1). The ABCG2 isoform, also known as the breast cancer resistance protein (BCRP), has also been demonstrated to affect the intestinal absorption of various therapeutic agents. The molecular and functional characteristics of these isoforms on the intestinal absorption of drugs are discussed more comprehensively below.

TABLE 7.1. Classification details of the important influx and efflux transporters present in the human small intestine. The table has been summarized from the information obtained from HUGO (Human Genome Organization) gene nomenclature committee's Web site (http://www.gene.ucl.ac.uk/nomenclature/)

General name	Other names	Approved gene symbol	Gene name	Location	Sequence accession IDs
P-glycoprotein (P-gp)	CD243 GP170 ABC20	ABCB1	ATP-binding cassette, sub-family B (MDR/TAP), member 1	7q21.1	M14758 NM_000927
Sister of P-gp (sP-gp)	ABC16 PFIC-2 PGY4	ABCB11	ATP-binding cassette, sub-family B (MDR/TAP), member 11	2q24	AF091582
Multidrug resistance related proteins 1 (MRP1)	GS-X	ABCC1	ATP-binding cassette, sub-family C (CFTR/MRP), member 1	16p13.1	L05628
Multidrug resistance related proteins 2 (MRP2)	DJS cMRP	ABCC2	ATP-binding cassette, sub-family C (CFTR/MRP), member 2		U63970
Multidrug resistance related proteins 3 (MRP3)	cMOAT2 EST90757 MLP2 MOAT-D	ABCC3	ATP-binding cassette, sub-family C (CFTR/MRP), member 3	17q21	Y17151 NM_020038
Multidrug resistance related proteins 4 (MRP4)	MOAT-B	ABCC4	ATP-binding cassette, sub-family C (CFTR/MRP), member 4	13q31	U66682
Multidrug resistance related proteins 5 (MRP5)	SMRP, MOAT-C	ABCC5	ATP-binding cassette, sub-family C (CFTR/MRP), member 5	3q27	AF104942 NM_005688
Multidrug resistance related proteins 6 (MRP6)	EST349056, MLP1	ABCC6	ATP-binding cassette, sub-family C (CFTR/MRP), member 6	16p13.1	X95715
Breast cancer resistance protein	MXR ABCP	ABCG2	ATP-binding cassette, sub-family G (WHITE), member 2	4q22–q23	AF103796
Peptide transporter 1 (PEPT1)		SLC15A1	Solute carrier family 15 (oligopeptide transporter), member 1	13q33–q34	NM_005073

(continued)

TABLE 7.1. (Continued)

General name	Other names	Approved gene symbol	Gene name	Location	Sequence accession IDs
Peptide histidine transporter 1 (PHT1)	PTR4	SLC15A4	Solute carrier family 15, member 4	12q24.32	AY038999 NM_145648
Peptide histidine transporter 1 (PHT2)	hPTR3	SLC15A3	Solute carrier family 15, member 3	11q12.2	AB020598 NM_016582
Concentrative nucleoside transporter (CNT1)		SLC28A1	Solute carrier family 28 (sodium-coupled nucleoside transporter), member 1	15q25–26	U62967
Concentrative nucleoside transporter (CNT2)	SPNT1 HCNT2 HsT17153	SLC28A2	Solute carrier family 28 (sodium-coupled nucleoside transporter), member 2	15q15	U84392 NM_004212
Monocarboxylate transporter (MCT1)		SLC16A1	Solute carrier family 16 (monocarboxylic acid transporters), member 1	1p12	BC026317 NM_003051
Organic Cation transpoter (OCT1)		SLC22A1	Solute carrier family 22 (organic cation transporter), member 1	6q26	U77086
Organic anion transporting polypeptide (OATP1B1)	SLCO1B1O ATP-C LST-1	SLC21A6	Solute carrier organic anion transporter family, member 1B1	12p12	NM_006446

7.2.1 *P-Glycoprotein (P-gp; ABCB1)*

Influencing drug transport in mammalian tumor cells, P-glycoprotein (P-gp) was originally identified as a xenobiotic efflux pump by Juliano and Ling (1976) providing a mechanism for multidrug resistance. Subsequent studies confirmed that P-gp was a 170–180kDa, ATP-dependent transmembrane glycoprotein, which is formed by the posttranslational glycosylation of a 140kDa pro-P-gp protein (Kramer *et al.*, 1995). Topological analyses of P-gp showed that the protein comprised four major domains; two membrane-bound domains, each with six transmembrane segments and two cytosolic ATP binding motifs, commonly known as Walker A and B NBDs that bind and hydrolyze ATP. P-gp consists of 1,280 amino acid residues and exhibits a large degree of homology between the carboxy and amino terminal halves (Leveille-Webster and Arias, 1995). Each half consists of both a hydrophobic and hydrophilic NBD, containing approximately 300 amino acids. Xenobiotics bind to separate sites on P-gp, demonstrating that different drugs and/or the different NBDs can each independently regulate P-gp function (Leveille-Webster and Arias, 1995).

The gene responsible for encoding P-gp belongs to the ABCB family of ABC transporters, and is commonly known as the Multidrug Resistance 1 (*MDR1*) gene. There are 11 members of the ABCB family; however, the discussion here is restricted to the two commonly considered MDR gene families in humans *MDR1* (ABCB1) and *MDR2/3* (ABCB4) (Lincke *et al.*, 1991; Germann *et al.*, 1993). The human genome nomenclature details and chromosome location are provided in Table 7.1 and at the Web site http://www.gene.ucl.ac.uk/nomenclature/genefamily/abc.html. It is worth noting that the homologous *MDR3* and Bile-Salt Exporting Protein (ABCB11; sister of P-gp) are also important efflux transporters of ABCB subfamily and mainly involved in the active transport of bile salts across the hepatocyte canalicular membrane (Meier and Stieger, 2002). Both MDR3 and sister of P-gp demonstrated low expression and variable ethnicity-based expression in the human small and large intestines, which suggests that they play a minimal role in mediating oral absorption (Wang *et al.*, 2004).

In rodents, the gene responsible for the primary MDR isoforms' expression are depicted by lower case letters as *mdr1 (a and b)* and *mdr2* (Torok *et al.*, 1999). The *MDR1* gene product in humans and the *mdr1a* and *mdr1b* gene products in rodents confer resistance by effluxing xenobiotics from the cytosolic compartments in cells (Higgins, 1992). The *MDR2* and *mdr2* genes encode a protein primarily expressed in the bile canalicular membrane that is engaged in the transport of phosphatidylcholine into the liver bile canaliculi (Thiebaut *et al.*, 1987; Smit *et al.*, 1993).

7.2.1.1 The Expression of P-gp

Immunohistological studies with human small intestinal samples indicate that P-gp is localized to the apical brush-border membrane of the intestinal epithelium (Thiebaut *et al.*, 1987). Due to the localized expression of P-gp at the microvillous tip of enterocytes (Terao *et al.*, 1996), P-gp will limit the absorption of compounds by directly effluxing them back into the intestinal lumen. Interestingly, the level of P-gp expression increases from proximal to distal regions of the intestine (Mouly and Paine, 2003).

P-gp expression has also been demonstrated in kidney, adrenal gland, liver, colon, and lungs (Fojo *et al.*, 1987; Gatmaitan and Arias, 1993). Additionally, P-gp expression in endothelial cells is lining the blood–tissue barrier that includes the brain capillaries (Thiebaut *et al.*, 1989), implicating a protective functional role for P-gp through the active efflux of xenobiotics from the endothelial cytoplasm into the capillary lumen, as confirmed by Joly *et al.* (1995). P-gp is also expressed in the apical membrane of the placental syncytial trophoblasts, which faces the maternal blood compartment (Sugawara *et al.*, 1988), and thus forms a functional barrier between the maternal and fetal blood circulations. Some peripheral blood mononuclear cells, such as cytotoxic T lymphocytes and natural killer cells, also express P-gp, suggesting involvement in cell-mediated cytotoxicity. Moreover, P-gp is expressed and functions in human hematopoietic stem cells, indicating it may contribute to the established chemoresistance of these cells (Chaudhary and Roninson, 1991; Drenou *et al.*, 1993). Low-level P-gp expression is also found in prostate, skin, spleen, heart, skeletal muscle, stomach, and ovary (Fojo *et al.*, 1987; Gatmaitan and Arias, 1993).

7.2.1.2 The Regulation of P-gp Expression

Expectedly, P-gp expression can be modulated by various factors, such as xenobiotics, environmental stress, differentiating agents, and hormones under cell culture conditions. In the rat liver, P-gp expression was increased after acute treatment by chemical carcinogens including 2-acetylaminofluorene and aflatoxin B1 (Burt and Thorgeirsson, 1988), suggesting xenobiotic-mediated transcriptional induction. Furthermore, cholestasis or carbon tetrachloride intoxication has also resulted in increased P-gp expression in the liver of rodents and nonhuman primates (Schrenk *et al.*, 1993). In addition to anticancer drugs, other xenobiotics, including protein kinase C agonists and chemical carcinogens, have been demonstrated to induce *in vitro* P-gp expression in several human carcinoma cell lines (Chaudhary and Roninson, 1993). Steroid hormones have also been shown to increase P-gp expression levels, as estradiol treatment has resulted in increased efflux of rhodamine 123 in rat pituitary cells expressing P-gp (Jancis *et al.*, 1993). Moreover, concomitant rifampin therapy (600 mg/day for 10 days, p.o.) has also resulted in significant reduction in the area under the plasma concentration time curve (AUC) of oral digoxin (single-dose, 1 mg oral and 1 mg intravenous), a known P-gp substrate. Furthermore, rifampin treatment resulted in a threefold increase in intestinal P-gp levels that correlated with a decrease in the AUC of orally administered digoxin (Greiner *et al.*, 1999). Differentiating agents such as retinoic acid and sodium butyrate also upregulate P-gp expression in human neuroblastoma and colon carcinoma cells, respectively (Bates *et al.*, 1989; Mickley *et al.*, 1989). Environmental stresses, such as heat, shock, arsenite, and cadmium chloride treatment have also been shown to affect *in vitro* P-gp expression (Chin *et al.*, 1990).

7.2.1.3 P-gp Mediated Drug Transport

To explain the mechanism by which P-gp actively effluxes xenobiotics, two hypotheses have been postulated: the "hydrophobic vacuum cleaner" (HVC) and the "flippase model" (FM). The HVC model suggests P-gp clears substrates before they enter the cytoplasm (Higgins and Gottesman, 1992; Gottesman and Pastan, 1993). As such P-gp forms a hydrophilic pathway, and the drugs are transported from the cytosol to the extracellular media through the middle of a pore. Alternatively, the FM proposes that P-gp interacts with the xenobiotics as they enter through the lipid membrane and "flips" the drug from the inner leaflet to the outer leaflet and back into the extracellular media. Evidence also supports the presence of at least two allosterically coupled drug-binding sites (Ferry *et al.*, 1992; Martin *et al.*, 1997), although it is not clear if these sites facilitate drug efflux by separate mechanisms, it does suggest that they may convey broader P-gp substrate affinity.

Cornwell *et al.* (1986) demonstrated the role of P-gp in the specific and saturable binding of vinblastine in membrane vesicles from highly multidrug-resistant human KB carcinoma cell lines. The binding of vinblastine to P-gp is competitively inhibited by vincristine and daunorubicin, suggesting these compounds share the same binding site. Competitive binding studies using colchicine and actinomycin D revealed a lack of competition for the vinblastine-binding site, further supporting the findings of Ferry *et al.* (1992) and Martin *et al.* (1997) that P-gp has multiple drug binding domains (Cornwell *et al.*, 1986; Akiyama *et al.*, 1988). Shapiro *et al.* (1999) demonstrated that progesterone was not effluxed by P-gp, although it was shown to bind P-gp and block the efflux of other substrates. This study further supports the possibility of additional potential binding sites, or high affinity of endogenous progesterone to one of the two binding sites suggested above. There is still confusion about whether these studies have identified the same two binding sites or, in fact there are additional binding sites present in P-gp. As such, considerable research is required to elucidate the functional mechanisms of P-gp active efflux, as well as determine the structural moieties responsible for imparting such broad substrate specificity.

7.2.1.4 The Substrate Specificity of P-gp

Numerous studies have demonstrated that P-gp possesses broad substrate specificity, with a preference for hydrophobic, amphipathic molecules containing a planar ring system ranging in size from 200 to 1,900 Da. P-gp is also involved in the transport of neutral compounds such as digoxin and cyclosporine A, negatively charged carboxyl groups such as those found on atorvastine and fexofenadine, and hydrophilic drugs such as methotrexate (Sharom, 1997). The degree of hydrogen bonding and partitioning into the lipid membrane has been determined to be a rate-limiting step for substrate interactions with P-gp (Seelig and Landwojtowicz, 2000). A representative list of substrates/inhibitors is listed in Table 7.2 and includes anticancer agents, antibiotics, antivirals, calcium channel blockers, and immunosuppressive agents.

TABLE 7.2. Partial list of compounds that have shown to interact with P-glycoprotein

Group	List of drugs
Anticancer drugs	Doxorubicin, daunorubicin, vinblastine, vincristine, actinomycin D, paclitaxel, teniposide, etoposide
Immunosuppressive	Cyclosporin A, FK506, valinomycin, gramicidine
Lipid lowering agents	Lovastatin, atorvastatin, pravastatin, simvastatin
Anti-Histaminic	Fexofenadine, terfenadine
Antidiarrheal agents	Loperamide, antiemetics, domperidone, ondansetron
Antibiotic	Ivermectin, itraconazole, dactinomycin, ketoconazole, erythromycin, valinomycin, grepafloxacin, actinomycin D
HIV protease inhibitors	Ritonavir, amprenavir, saquinavir, indinavir, nelfinavir
Steroids	Aldosterone, hydrocortisone, cortisol, corticosterone, dexamethasone, dopamine antagonist, domperidone
Cardiac drugs	Digoxin, digitoxin, quinidine
Analgesics	Morphine, asimadoline, fentanyl
Beta-Adrenoceptor antagonists	Bunitrolol, carvedilol, celiprolol, talinolol
Food/Herbal Constitutents	Piperine, quercetin, naringin, curcumin, bergamottin, kaempferol, rutin
Pharmaceutical excipients	Cremophor EL, Tween 20, Tween 80, Nonidet P-40, Acacia, polyethylene glycols, Triton-X 100, pluronic block copolymers, Brij 30 & 35, solutol HS 15, poloxamers, 1-[(3-cholamidopropyl)dimethyla amino]-1-propanesulfonate (CHAPS)

The table has been modified from the information reported by Fisher *et al.* (1996), Sikic (1997), Zuylen *et al.* (2000), and Kunta and Sinko (2004)

It is important to note that the substrate specificity of P-gp may also vary across populations due to genetic polymorphisms. For example, Kurata *et al.* (2002) demonstrated a significant difference in oral digoxin bioavailability between two allelically diverged MDR1 populations, which resulted in population specific absorption and/or distribution outcomes. A recent comprehensive review of the clinical implications of P-gp polymorphisms is suggested for additional information on this subject (Ieiri *et al.*, 2004).

While the observed absorption of numerous drugs may be affected by P-gp affinity, there are certain drugs that can also be used to inhibit P-gp activity (Krishna and Mayer, 2000). The P-gp inhibitors include several calcium channel blockers, immunosuppressive agents, and other well-characterized compounds such as SDZ, PSC 833, LY335979, and GF120918 (Table 7.2) (Hyafil *et al.*, 1993; Schinkel and Jonker, 2003). Drug–drug interactions resulting from concomitant administration of P-gp substrates and/or inhibitors may result in an increase in the absorption of one or more of these agents, potentially leading to toxic side effects by virtue of increased plasma levels that rise above the minimum toxic concentration. There are numerous clinical examples of these effects being observed and reported in the literature. For example, a significant therapeutic increase in the plasma levels of digoxin was observed upon coadministration with valspodar (PSC833) in healthy patients. The increase in AUC of this narrow therapeutic index compound leads to the conclusion that the digoxin dose should be decreased by 50% and clinical toxicity of digoxin should be monitored upon concomitant administration of these two agents (Kovarik *et al.*, 1999).

In contrast, drug–drug interactions can also result in a lowering of the plasma levels of a therapeutic agent, where the coadministered drug induces P-gp expression that subsequently results in decreased plasma levels due to reduced absorption and increased intestinal clearance. One compelling example of this phenomenon was observed in a study with rifampin and fexofenadine (Hamman *et al.*, 2001), where rifampin induced the intestinal expression of P-gp resulting in increased oral clearance and reduced bioavailability of fexofenadine.

Additionally, common pharmaceutical excipients such as hydrophilic cyclodextrin (2,6-di-*O*-methyl-β-cyclodextrin) (Arima *et al.*, 2001), cosolvents (poly(ethylene)glycol (PEG) 400), and surfactants (Tween 80, Cremophor EL) (Nerurkar *et al.*, 1996; 1997; Hugger *et al.*, 2002) have also been shown to inhibit P-gp activity. Given the obvious adverse effects this may have on drug bioavailability and overall therapy in general, excipient selection should be an important factor to be considered in rational formulation design. Surfactants or cosolvents have also been shown to indirectly influence P-gp by inducing changes in cellular membrane fluidity. Changes in membrane fluidity can alter the microenvironment of the apically oriented TM domain, and subsequently alter substrate recognition, binding, and efflux by P-gp (Ferte, 2002).

Dietary constituents and phytochemicals also affect drug absorption and disposition through drug–nutrient P-gp interactions, potentially impacting the clinical pharmacokinetics of therapeutic agents (Walter-Sack and Klotz, 1996). Recently, it has been estimated in the US that approximately 30–50% of the population used a dietary supplement that can alter the absorption of drugs through mechanisms like competitive inhibition or transporter induction (Kauffman *et al.*, 2002). The flavonoids, particularly the flavonols, flavanones, flavones as well as coumarins, and other ingredients present in fruits, vegetables, and herbs have been found to modulate the activity of P-gp function and may cause detrimental effects on drug pharmacokinetics (Izzo, 2005). One of the most recognized interactions is the ingestion of grapefruit juice and/or St John's Wort which results in the change of the pharmacokinetic profiles of cyclosporine A and digoxin due to inhibition or induction of P-gp mediated transcellular intestinal epithelial absorption (Bailey *et al.*, 1998; Dürr *et al.*, 2000). In a representative case, the oral coadministration of grapefruit juice with talinolol (10 mg/kg) resulted in an increased talinolol C_{max}, AUC and a reduced t_{max} without significantly affecting the terminal talinolol half-life (Spahn-Langguth and Langguth, 2001). This study indicated that grapefruit juice acted to inhibit intestinal P-gp mediated efflux and resulted in enhanced talinolol bioavailability. In a separate study, 1 g of black pepper administered in soup along with phenytoin significantly increased the human plasma levels of phenytoin (Velpandian *et al.*, 2001). Piperine, an active constituent of black pepper, is an inhibitor of P-gp and CYP3A4 (Bhardwaj *et al.*, 2002). Orange juice components, such as methoxyflavones, increase vinblastine uptake by Caco-2 cells, possibly by interacting with P-gp (Honda *et al.*, 2004). Therefore concomitant intake of herbal extracts or fruits/foods and nutraceuticals may modulate the pharmacokinetic profile of the therapeutic index of drugs, in particular for those agents exhibiting a narrow therapeutic index, resulting in an altered clinical response.

7.2.2 Multidrug Resistance-Associated Protein Family (MRP; ABCC)

The human *MRP* gene encodes an MRP polypeptide with an apparent mass of 170 kDa, which is posttranslationally converted to a 190 kDa form by the addition of N-linked complex oligosaccharides (Almquist *et al.*, 1995). To date, nine members within the MRP family have been identified, delineated as MRP1 to MRP6 (ABCC1 to ABCC6) and MRP7 to MRP9 (ABCC10 to ABCC12) (Belinsky *et al.*, 1998; Bera *et al.*, 2001, 2002; Kubota *et al.*, 2001; Kruh and Belinsky, 2003). Membrane topology of MRP1 identified three transmembrane spanning domains (two are of 6 and the third is of 5 helices) (Fig. 7.1) (Bakos *et al.*, 1996; Hipfner *et al.*, 1997). Additionally, MRP1 has two NBDs and two intracellular loops with the first linker segment located between two transmembrane domains (TMD) whereas the second is between the transmembrane and a nucleotide binding domains. MRP4 (ABCC4), MRP5 (ABCC5), MRP8 (ABCC11), and MRP9 (ABCC12) lack the five helices of the third membrane spanning domain, but possess intracellular loops whereas MRP2 (ABCC2), MRP3 (ABCC3), MRP6 (ABCC6), and MRP7 (ABCC10) resemble MRP1 (Bera *et al.*, 2001, 2002; Hopper *et al.*, 2001; Yabuuchi *et al.*, 2001; Kruh and Belinsky, 2003). The human genome nomenclature details and chromosomal localization of important intestinal MRP transporters are listed in Table 7.1.

7.2.2.1 The Expression of MRPs

MRP1 was first cloned by Cole *et al.* (1992) from a multidrug-resistant human lung cancer cell line. MRP1 is highly expressed in the intestine, where it is localized in the basolateral membrane of intestinal epithelial cells and is involved in the absorptive efflux of its substrates into blood (Peng *et al.*, 1999). MRP1 expression has also been demonstrated in brain, liver, lung, kidney, and testis (Flens *et al.*, 1996; Zhang *et al.*, 2000; Cherrington *et al.*, 2002).

Comparatively, MRP2 was first cloned from human liver as a canalicular multispecific organic anion transporter (cMOAT) (Paulusma *et al.*, 1996) and was found to be expressed on the apical membrane of the intestine, liver, and kidney tubules (Schaub *et al.*, 1997; Fromm *et al.*, 2000; Scheffer *et al.*, 2002). Furthermore, MRP2 expression was localized in the brush-border membrane of villi in rabbit small intestine with expression decreasing from the villous tip to the crypt (Van Aubel *et al.*, 2000). Moreover, MRP2 expression was shown to vary substantially along the human GI tract, with higher expression levels found in small intestine and minimal expression observed in colonic segments (Zimmermann *et al.*, 2005).

MRP3 expression is also evident in the basolateral membrane of the small intestine, as well as the liver, colon, lung, spleen, and kidney (Kool *et al.*, 1997;

Scheffer *et al.*, 2002). Rost *et al.* (2002) identified MRP3 mRNA expression throughout the rat intestine by RT-PCR, with lower expression observed in the duodenum and jejunum and markedly increased expression observed in the ileum and colon. MRP4 is expressed in several tissues including jejunum, kidney, brain, lung, and gall bladder (Kool *et al.*, 1997; Zhang *et al.*, 2000; Taipalensuu *et al.*, 2001; Van Aubel *et al.*, 2002). MRP5 expression is found to be in colon, liver, kidney, skeletal muscle, and brain (Kool *et al.*, 1997; McAleer *et al.*, 1999; Zhang *et al.*, 2000), while MRP6 is highly expressed in the kidney and liver, with low expression in several other tissues, including duodenum, colon, brain, and salivary gland (Zhang *et al.*, 2000). The functional activity and expression of MRP7, MRP8, and MRP9 have not been as well characterized, although they are gaining increasing attention due to their involvement in conveying multidrug resistance (Bera *et al.* 2002; Kruh and Belinsky, 2003; Chen *et al.*, 2003, 2005; Hopper-Borge *et al.*, 2004).

The comparative mRNA expression of MRPs 1–5 and P-gp studied in ten healthy human intestinal tract biopsy samples (Zimmermann *et al.*, 2005) demonstrated a rank order expression of MRP3 $\gg$ MDR1 > MRP2 > MRP5 > MRP4 > MRP1 and MDR1 > MRP3 $\gg$ MRP1 in the duodenum and the terminal ileum, respectively. The comparative rank order of mRNA expression remains consistent throughout the colon (ascending, transverse, descending, and sigmoid colon) with MRP3 $\gg$ MDR1 > MRP4 $\cong$ MRP5 > MRP1 $\gg$ MRP2. Total RNA analysis by Taipalensuu *et al.* (2001) showed a ranking of BCRP $\cong$ MRP2 > MDR1 $\cong$ MRP3 $\cong$ MRP6 $\cong$ MRP5 $\cong$ MRP1 > MRP4 > MDR3, as determined by quantitative polymerase chain reaction in jejunal biopsies from 13 healthy human subjects. Direct comparison between the expression of BCRP, MDR1, MDR3, and MRP1–6 found that BCRP exhibited a 100-fold lower transcript level in Caco-2 cells compared with jejunum (Taipalensuu *et al.*, 2001). Furthermore, based upon the ratio of MRP:18S rRNA expression, the rank order for MRP expression in Caco-2 cells was MRP2 $\geq$ MRP6 > MRP4 $\geq$ MRP3 > MRP1 = MRP5 (Prime-Chapman *et al.*, 2004).

Our laboratory also observed a variable intestinal expression of P-gp, MDR3, S-P-gp, LRP, MRP1–4 in tissue slides and intestinal protein lysates obtained from normal adult small and large intestinal epithelium of Chinese and Caucasian donors (Wang *et al.*, 2004). Furthermore, the expressions of P-gp, MDR3, LRP, MRP1, and MRP2 were found to be higher in the small intestine of Chinese when compared to Caucasian samples, suggesting distinct expression profiles of these transporters (Wang *et al.*, 2004). Interestingly, there has been concern over the variable expression of these isoforms in traditional cell screening models (e.g., Caco-2 cells), as compared with normal physiological expression observed in patients. Given the widespread use of these screening models, these differences in transporter expression may not properly translate to the actual physiological condition resulting in selection of lead candidates based on erroneous data. Finally, immunofluorescent staining of Caco-2 cells revealed the apical and basolateral localizations of MRP2 and MRP3, respectively, whereas MRP1 was not observed on either membrane (Prime-Chapman *et al.*, 2004).

7.2.2.2 The Regulation of MRP Isoform Expression

Nuclear receptors that regulate transcription of MRP2 proteins include pregnane X receptor (PXR, NR1I2), farnesoid X receptor (FXR, NR1H4), and the constitutive androstane receptor (CAR, NR1I3) (Tanaka *et al.*, 1999; Kast *et al.*, 2002). Data indicate that steroid-based compounds, such as bile acids, cholesterol, hormones, as well as other drugs, including rifampin, dexamethasone, and phenobarbital, all affect the above nuclear receptors. For example, the mRNA expression of MRP2 was up regulated in the presence of rifampin in the rhesus monkey liver, human small intestine, and primary cultures of human hepatocytes (Kauffmann *et al.*, 1998; Fromm *et al.*, 2000; Dussault *et al.*, 2001). Moreover, rats treated with known PXR ligands, such as the antiglucocorticoid/antiprogestin RU486 and the antifungal clotrimazole, showed upregulation of *MRP2* gene expression, suggesting a role of PXR ligands in regulation of MRP2 expression (Courtois *et al.*, 1999). Naturally occurring chenodeoxycholic acid, synthetic GW4064, and known FXR ligands have induced MRP2 mRNA expression in both human and rat hepatocytes. Finally, the CAR agonist, phenobarbital also induced MRP2 mRNA expression (Kast *et al.*, 2002).

Intestinal MRP3 mRNA has been shown to increase in a dose- and time-dependent manner in Caco-2 cells after treatment with a series of bile salts, chenodeoxycholic acid, taurochenodeoxycholic acid, taurocholic acid, and taurolithocholic acid (Inokuchi *et al.*, 2001). Additionally, the glucocorticoids, dexamethasone, and hydrocortisone have also been shown to transcriptionally upregulate human MRP3 mRNA in a non-small-cell lung cancer cell line, which was correlated to an increase in MRP3 protein (Pulaski *et al.*, 2005).

Recently, the involvement of various transcriptionally mediated pathways aryl hydrocarbon receptor (AhR), PXR, CAR, peroxisome proliferator-activated receptor α (PPARα), and nuclear factor-E2-related factor 2 (Nrf2) were studied on the induction of MRPs (Maher *et al.*, 2005). The mRNA expression of MRP2, 3, 5, and 6 were shown to be induced using AhR ligands [2,3,7,8-tetrachlorodibenzo-*p*-dioxin (TCDD), polychlorinated biphenyl 126 (PCB126), and β-naphthoflavone], while the CAR activator 1,4-bis[2-(3,5-dichloropyridyloxy)]benzene (TCPOBOP) induced MRP2, 3, 4, 6, and 7 mRNA expression. In addition, Nrf2 activators (butylated hydroxyanisole, oltipraz, and ethoxyquin) induced MRP2–6, further supporting the relevance of transcription factors with respect to MRP regulation (Maher *et al.*, 2005).

7.2.2.3 The Substrate Specificity of MRP's

The substrate specificity and function of MRP family members have been extensively reviewed by Kruh and Belinsky (2003), and will only be briefly discussed here. MRP1 is involved in the transport of anionic drugs (e.g., methotrexate), drug or metabolite conjugates (glutathione, glucuronate, or sulfate), leukotriene C4 (LTC4), 2,4-dinitrophenyl-*S*-glutathione(DNP-SG), bilirubin glucuronides, estradiol-17-glucuronide, and dianionic bile salts (Leier *et al.*, 1994; Jedlitschky

TABLE 7.3. Partial list of substrates that have shown to interact with Multidrug Resistance-Associated protein

Transporter	Drugs/xenobiotics	Physiological substrates
MRP1	Vinca alkaloids, epipodophyllotoxins, anthracyclines, camptothecin, MTX	LTC4, estrogen glucuronide, leukotriene E4, S-Glutathionyl prostaglandin A2, bilirubin (monoglucuronosyl, bisglucuronosyl)
MRP2	Vinca alkaloids, anthracylines, vincristine, etoposide camptothecin, MTX, ochratoxin A, *p*-Aminohippurate	Bilirubin glucuronide DNP-SG, S-Glutathionyl ethacrynic acid, 17β-glucuronosyl estradiol
MRP3	Etoposide, MTX	Glycocholic acid
MRP4	6-MP, MTX, PMEA	Cyclic nucleotides, DHEAS
MRP5	6-MP, PMEA	Cyclic nucleotides
MRP6	Anthracycline, etoposide	?
MRP7	?	17β-estradiol 17-(β-D-glucuronide)?
MRP8	5-FU, ddC, PMEA	Cyclic nucleotides
MRP9	?	?

LTC4, cysteinyl leukotriene; MTX, methotrexate; 6-MP, 6-mercaptopurine; PMEA, 9-(2-phosphonylmethoxyethyl)adenine; 5-FU, 50-fluorouracil; ddC, zalcitabine; DHEAS, dehydroepiandrosterone sulfate; DNP-SG, S-glutathionyl 2,4-dinitrobenzene. Table modified from the information reported by Konig *et al.* (1999) and Kruh and Belinsky (2003)

et al., 1996; Nies *et al.*, 1998; Hooijberg *et al.*, 1999). Representative substrates of MRP1 are listed in Table 7.3 and include anthracyclines, vinca alkaloids, pipodophyllotoxins, and camptothecins (Cole *et al.*, 1994; Konig *et al.*, 1999; Kruh and Belinsky, 2003). Several compounds are also known to be either direct or indirect inhibitors of MRP1. Verapamil and its analogs, in the presence of reduced glutathione (GSH) or its nonreducing *S*-methyl derivative, inhibited LTC_4 transport in membrane vesicles prepared from MRP1-transfected cells. Verapamil itself was not transported by MRP1 in either intact cells or membrane vesicles, suggesting that verapamil modulates MRP1 activity through enhancing MRP1 mediated GSH transport (Loe *et al.*, 2000a,b). Leslie *et al.* (2003) demonstrated a similar effect on MRP1 mediated GSH transport in the presence of bioflavonoids, possibly through an interaction(s) with one of the several multiple flavonoid binding sites (Trompier *et al.*, 2003).

Similarly, MRP2 transports leukotrienes C_4, D_4, and E_4 and various glutathione conjugates, including oxidized glutathione, 2,4-dinitrophenyl-*S*-glutathione, bromosulfophthalein glutathione, as well as those conjugates of heavy metals including arsenic and cadmium (Jedlitschky *et al.*, 1997; Madon *et al.*, 1997; Suzuki and Sugiyama, 1998, 1999). The glucuronide conjugates of bilirubin, estradiol, acetaminophen, grepafloxacin, triiodo-L-thyronine, and SN-38 have also been demonstrated to be MRP2 substrates (Suzuki and Sugiyama, 1998, 1999). Additionally, MRP2 also transports glucuronide and sulfate conjugates of several bile salts, a range of unconjugated organic anions such as methotrexate, reduced folates, bromosulfophthalein, irinotecan and its metabolite SN-38, pravastatin, ceftriaxone, temocaprilat, ampicillin as well as Fluo-3 and *p*-aminohippurate (Konig *et al.*, 1999; Suzuki and Sugiyama, 1998, 1999; Kusuhara and Sugiyama, 2002). Keppler *et al.* (1997) have suggested the transport efficiency (V_{max}/K_m) of MRP2 substrates in the rank order of leukotriene C4 > leukotrieneD4 > *S*-(2,4-dinitrophenyl)-glutathione > monoglucuronosyl bilirubin > estradiol-17β-D-glucuronide > taurolithocholate sulfate > oxidized glutathione.

The overlapping substrate specificity of MRP2 with P-gp, coupled with their intestinal and cellular colocalization to the apical membrane, suggests a concerted function between these two transporters that would comprise a significant barrier to the intestinal absorption of many xenobiotics. Interestingly, grepafloxacin uptake was observed to be directly influenced by the combined effect of P-gp and MRP2. However, the secretory efflux of grepafloxacin was shown to be predominantly a function of MRP2 activity as opposed to P-gp, both *in vitro* and *in vivo*, suggesting that in spite of some overlapping substrate specificity, MRP members may also act independently as a functional barrier to bioavailability (Naruhashi *et al.*, 2002).

Recently, MRP2 has also been shown to be involved in the efflux of a tobacco-specific carcinogen, 4-(methylnitrosamino)-1-(3-pyridyl)-1-butanol, and the food carcinogen PhIP, reducing their oral bioavailabilities, thus protecting against food-derived carcinogenesis (Dietrich *et al.*, 2001; Leslie *et al.*, 2001). In addition, bioavailability of fungal toxin ochratoxin A and the tea flavonoid epicatechin were inhibited by MK571, an antagonist of MRP2, suggesting these as MRP2 substrates (Leier *et al.*, 2000; Vaidyanathan and Walle, 2001).

Similar to both MRP1 and MRP2, MRP3 also has the capacity to transport organic anionic drugs and glucuronate-conjugated drugs; however, MRP3 exhibits a reduced capacity for GSH conjugates (Hirohashi *et al.*, 1999), as well as a wide range of bile salts such as glycocholate, taurolithocholate-3-sulfate, and taurochenodeoxycholate-3-sulfate (Hirohashi *et al.*, 2000; Zeng *et al.*, 2000; Zelcer *et al.*, 2001). In addition MRP3, in contrast to MRP1 and MRP2, does not require glutathione for mediating the transport of natural products (Zelcer *et al.*, 2001).

Substrates for MRP4 include folic acid, folinic acid (leucovorin), and methotrexate (Chen *et al.*, 2002), as well as cAMP, cGMP, estradiol-17β-D-glucuronide, bile acids (Chen *et al.*, 2001; Lai and Tan, 2002; Van Aubel *et al.*, 2002; Zelcer *et al.*, 2003), and thiopurines (Wielinga *et al.* 2002). Recent literature also indicates that MRP4 is involved in the efflux of camptothecins, as shown in human MRP4 stably transfected HepG2 cells (Tian *et al.*, 2005). MRP5 has an affinity for nucleotide-based substrates including anticancer thiopurine and thioguanine drugs, as well as the anti-HIV drug 9-(2-phosphonylmethoxyethyl)adenine (Wijnholds *et al.*, 2000; Wielinga *et al.* 2002; Reid *et al.*, 2003). Interestingly, organic anion such as benzbromarone and sulfinpyrazone inhibit MRP5, suggesting an affinity for anionic phosphate/phosphonate moieties (Wijnholds *et al.*, 2000). With respect to the remaining MRP family members, MRP6 was shown to be involved in the transport of different natural cytotoxic agents such as etoposide, doxorubicin, and cisplatin in MRP6-transfected Chinese hamster ovary (CHO) cells (Belinsky *et al.* 2002), while MRP7 is also a lipophilic anion efflux transporter with an affinity for docotaxel and 17β-estradiol-(17β-D-glucuronide) (Chen *et al.*, 2003; Hopper-Borge *et al.*, 2004). MRP8 has been demonstrated to mediate the efflux of cyclic nucleotides (Guo *et al.*, 2003).

7.2.3 Breast Cancer Resistance Protein (BCRP; ABCG2)

The ABC transporter BCRP was first cloned from a doxorubicin-resistant MCF7 breast cancer cell line (MCF-7/AdrVp) by Doyle *et al.* (1998). Subsequently, other groups cloned the BCRP cDNA sequence from other sources and designated the gene either *MXR* (mitoxantrone resistance protein) or *ABCP* (placental ABC protein) (Allikmets *et al.*, 1998; Maliepaard *et al.*, 1999). Since structural and sequence homology revealed that BCRP belongs to the ABCG gene subfamily, the Human Genome Nomenclature Committee conferred the official designation as ABCG2 (Table 7.1). BCRP consists of six putative TMD involved in drug binding and efflux, as well as a single amino-terminal cytosolic NBD that functions as an ABC involved in ATP hydrolysis. BCRP has been suggested to be a half-transporter that may function as a homodimer or tetramer bridged by disulfide bonds (Xu *et al.*, 2004).

Study of BCRP tissue localization shows that it is expressed in the small intestine, colon, liver, placental syncytiotrophoblasts, and ovary (Maliepaard *et al.*, 2001). Further, a monoclonal antibody directed against BCRP used for immunohistochemical staining revealed prominent expression at the apical membrane of the small intestinal, colonic epithelia, and in hepatocyte canalicular membranes (Scheffer *et al.*, 2000; Maliepaard *et al.*, 2001). Comparatively, BCRP mRNA expression was found to be higher than all other ABC transporters measured (MDR1, MDR3, MRP1–6) in human jejunal mucosa biopsies obtained from 13 healthy volunteers (Taipalensuu *et al.*, 2001). Moreover, with villin-normalized data expression levels between Caco-2 cells versus jejunal samples varied less than 2.5-fold for ABC transporter genes, whereas Caco-2 cells exhibited a 100-fold lower transcript copy number in contrast to the human jejunum for BCRP (Taipalensuu *et al.*, 2001). Furthermore, later passage (p56) Caco-2 cells showed higher expression levels of BCRP in comparison to earlier passage (p33) cells, suggesting culturing conditions or epigenetic factors also influence BCRP expression (Taipalensuu *et al.*, 2001). Immunofluorescence studies demonstrated the polarized apical membrane expression of BCRP in Caco-2 cells (Xia *et al.*, 2005).

BCRP mRNA expression quantified by quantitative real-time PCR along the GI tract of 14 healthy subjects demonstrated expression to be maximal in the duodenum and continuously decreasing toward the rectum (terminal ileum 93.7%, ascending colon 75.8%, transverse colon 66.6%, descending colon 62.8%, and sigmoid colon 50.1% compared to duodenum, respectively) (Gutmann *et al.*, 2005).

Interestingly, gender-based variation was also observed for BCRP expression, although the biological relevance of these data has not yet been elucidated.

BCRP transports estrone 3-sulfate and 17β-estradiol-3-sulfate, while the parent steroids are not substrates (Imai *et al.*, 2003). BCRP also transports the active glucuronide conjugate metabolite of SN-38 (SN-38-glucuronide) with an approximately sevenfold lower affinity than the parent SN-38 itself, suggesting the involvement of BCRP in transporting glucuronide conjugated compounds (Nakatomi *et al.*, 2001). BCRP also mediates the secretion of other clinically important camptothecans, such as topotecan. Jonker *et al.* (2000) observed a more than sixfold decrease in topotecan plasma concentrations upon concomitant administration of GF120918 (an inhibitor of both P-gp and BCRP), in P-gp deficient mice. This study shows that BCRP mediates apically directed drug transport, which would reduce drug bioavailability. As such, the coadministration of an effective BCRP inhibitor may increase the oral bioavailability of topotecan. Interestingly, fumitremorgin C, an extract of *Aspergillus fumigatus*, selectively inhibits BCRP with no overlapping affinity for P-gp, as demonstrated by reversing resistance mediated by MCF-7 cells transfected with the BCRP gene (Rabindran *et al.*, 1998; Ozvegy *et al.*, 2001). These observations suggest that the oral absorption of therapeutic compounds specifically designed to be substrates of BCRP would be enhanced. A more comprehensive summary of various BCRP substrates can be found in Table 7.4.

TABLE 7.4. Partial list of compounds shown to interact with breast cancer resistance proteins

Substrates	Inhibitors
Anti-tumor drugs	GF120918
Mitoxantrone	XR9576
Methotrexate,	VX-710
Camptothecins (SN-38, topotecan)	Diethylstilbestrol
CI1033	Taxanes
NB 506	Falvonoids
J-107088	Imatinib
Flavopiridol	Gefitinib
Endogenous substrates	Fumitremorgin C
Estrone 3-sulfate	Tamoxifen
17 β-estradiol sulfate	HIV protease Inhibitors
17β-Estradiol 17-(β-D-glucuronide	Novobiocin
Folic acid	UCN-01
Protoporphyrin IX	
Fluorescent dyes	
BODIPY-Prazosin	
Hoechst 33342	
BBR 3390	

The table has been modified from the information published by Sarkadi *et al.* (2004), Leslie *et al.* (2005) and Staud and Pavek (2005)

7.3 Solute Carrier Transporters

The solute carrier (SLC) transporter superfamily is comprised of integral membrane proteins that mediate substrate transport across cell surface or cellular organelle, such as the golgi apparatus or synaptic vesicle membranes, by either a facilitated diffusion or an active transport mechanism. The function of SLC transporters may also be coupled to the cotransport of a counterion down its electrochemical gradient (e.g., Na^+, H^+, and Cl^-). The members of the SLC superfamily are involved in the transport of a wide range of substrates including amino acids, peptides, sugars, vitamins, bile acids, neurotransmitters, and xenobiotics. There are several reviews on the SLC family members identified across species that the reader is directed for more information (Saier *et al.*, 1999; Saier, 2000; Hediger *et al.*, 2004).

The Human Genome Organization (HUGO) Nomenclature Committee Database provides a comprehensive list of SLC transporter genes which are subdivided into 43 distinct subfamilies consisting of 319 separate isoforms (Hediger *et al.*, 2004; see http://www.gene.ucl.ac.uk/nomenclature/). The SLC superfamily includes many pharmacokinetically important transporters such as proton dependent oligopeptide transporters (SLC15A family), organic cation transporters (SLC22A family), organic anion transporting polypeptides (SLC21A family), nucleoside transporters (SLC28, 29A family), and the monocarboxylate transporters (SLC16A family). While many other SLC subfamilies exist, these subfamilies are thought to significantly influence the bioavailability of their respective substrates and are relevant to the observed PK/PD profiles for many existing drugs and are increasingly being utilized for NCE screening in the drug discovery phase. The classifications and molecular characteristics of several important intestinal SLC transporters are described below.

7.3.1 Proton/Oligopeptide Transporters (POT; SLC15A)

With the elucidation of the human genome, and advances in molecular biology and cloning, it should be of no surprise that peptidomimetic drugs are increasingly utilized as therapeutic agents for the treatment of numerous disorders. Peptide-like agents have a broad range of clinical applications in the treatment of many disorders including AIDS, hypertension, and cancer. The currently known peptide transporters include the Peptide Transporters 1 and 2, PepT1 (SLC15A1) and PepT2 (SLC15A2); the Peptide/Histidine Transporters 1 and 2, PHT1 (SLC15A4) and PHT2 (SLC15A3); and the Intestinal Peptide Transporter PT1 (CDH17). PT1 is the only protein identified which is not classified as a member of the proton oligopeptide transporter (SLC15) family, also known as the proton-coupled oligopeptide transporter (POT) superfamily (Fei *et al.*, 1998; Meredith and Boyd, 2000). In fact, PT1 is considered a member of the cadherin family. PepT2 expression has not been shown in the GI tract and thus it will not be discussed in this chapter.

Several comprehensive reviews can be found describing the common characteristics of the oligopeptide transporter proteins (Graul and Sadée, 1997; Nussberger *et al.*, 1997a; Yang *et al.*, 1999; Meredith and Boyd, 2000; Rubio-Aliaga and Daniel, 2002; Herrera-Ruiz and Knipp, 2003). Members of the POT superfamily are predicted to contain 12 predicted transmembrane α-helical spans, with a majority of the proteins having intracellulary localized N- and C-termini (Covitz *et al.*, 1998; Lee, 2000; Herrera-Ruiz and Knipp, 2003). Two characteristic protein signatures of the POT family members have been identified, known as the PTR2 family signatures (Steiner *et al.*, 1995): (1) [GA] – [GAS] – [LIVMFYWA] – [LIVM] – [GAS] – D – x – [LIVMFYWT] – [LIVMFYW] – G – x(3) – [TAV] – [IV] – x(3) – [GSTAV] – x – [LIVMF] – x(3) – [GA] and (2) [FYT] – x(2) – [LMFY] – [FYV] – [LIVMFYWA] – x – [IVG] – N – [LIVMAG] – G – [GSA] – [LIMF], and a third consensus proposed has been proposed by Fei *et al.* (1998) (GTGGIKPXV). Saier *et al.* (1999) have proposed three different signature sequences associated with the POT superfamily based on their phylogenetic analysis.

The cloned human PepT1 cDNA sequence encodes a 708 amino acid protein with an estimated molecular weight of 79 kDa, and an isoelectrical point of 8.6 (Liang *et al.*, 1995). PepT1 has shown expression in several animal species (Fei *et al.*, 1994, 2000; Saito *et al.*, 1995; Chen *et al.*, 1999; Pan *et al.*, 2001; Klang *et al.*, 2005; Van *et al.*, 2005), each exhibiting high homology with other species. PepT1 protein expression has been demonstrated in the human small intestine (Liang *et al.*, 1995; Herrera-Ruiz *et al.*, 2001) and was localized on the apical plasma membrane of enterocytes in rats (Ogihara *et al.*, 1999). Other studies have demonstrated that PepT1 isoforms are localized intracellularly in lysosomes (Zhou *et al.*, 2000 Sun *et al.*, 2001). Apical expression of PepT1 has been established in both prenatal and mature animals (Shen *et al.*, 2001; Rome *et al.*, 2002). Furthermore, PepT1 cellular localization has been demonstrated to vary with the stage of animal development. Hussain *et al.* (2002) revealed that PepT1 is exclusively expressed in the apical membrane of enterocytes from both prenatal and mature animals; however, immunolocalization studies showed that immediately after birth, PepT1 was also expressed intracellularly in the basal cytoplasm, as well as the basolateral membrane of the intestinal epithelium.

Recently, two putative human peptide/histidine (hPHT) transporters have been identified with expression observed in several human tissues (Botka *et al.*, 2000; Knipp and Herrera-Ruiz, 2004). The hPHT1 mRNA sequence is approximately 2.7 kb long, encoding a translated 577 amino acid protein with an estimated molecular weight of 62 kDa, and a predicted pI of 9.2. Four N-linked glycosylation sites were predicted, along with several protein phosphorylation sites (Herrera-Ruiz and Knipp, 2003). Human PHT2 has not widely been studied, and little is known about its biological significance. PHT2 was first isolated from the human placenta and has an open reading frame of 1.7 kb, encoding a protein with 581 amino acids with an estimated molecular mass of 64.6 kDa. Only the rat isoform of the PHT2 protein has been partially evaluated (Sakata *et al.*, 2001). Three N-linked glycosylation sites on the rPHT2 protein are predicted and protein phosphorylation sites (PKA and PKC) were identified (Herrera-Ruiz and Knipp, 2003). Orthologous expression in rat and mouse has been reported for both PHT1 and PHT2 (Yamashita *et al.*, 1997; Botka *et al.*, 2000; Sakata *et al.*, 2001). Both hPHT1 and hPHT2 have shown expression along the entire GI tract, especially in the small intestine and colon (Herrera-Ruiz *et al.*, 2001), an important difference in relation with hPepT1. Expression of hPHT1 has been shown in the plasma membrane of intestinal tissue segments (Bhardwaj *et al.*, 2005a). Studies have demonstrated that PHT1 and PHT2 are expressed in the human and rat GI tracts and in Caco-2 cells (Herrera-Ruiz *et al.*, 2001). Intracellular localization rPHT2 expression has also been demonstrated intracellular localization in lysosomes, autophagosomes, and vacuoles of HEK-293T and baby hamster kidney (BHK) cells (Sakata *et al.*, 2001).

The human Intestinal Peptide Transporter 1 (HPT1) has a cDNA coding region of 2.5 kb long, encoding a 120 kDa protein comprising 832 amino acids. While HPT1 is related to the cadherin family of proteins (Dantzig *et al.*, 1994), it has demonstrated peptide and cephalosporin transport activity (Yang, 1998). Since it is not a POT family member, it will not be discussed further.

7.3.1.1 Peptide Transporter Mediated Transport

Transepithelial peptide transporters use a proton gradient and membrane potential to provide the necessary driving force for substrate translocation (Daniel, 1996; Adibi, 1997; Nussberger *et al.*, 1997a,b). The required proton gradient is generated through the activity of an electroneutral proton/cation exchanger, the Na^+/H^+ antiporter (Meredith and Boyd, 2000; Theis *et al.*, 2001). Peptide or peptidomimetic molecule uptake is commensurate with the translocation of a proton into epithelial cells, making substrate uptake strongly dependent on the extracellular pH, where a pH of 4.5–6.5, depending on the net charge of the substrate, is optimal for transport activity (Temple *et al.*, 1995, 1996; Amasheh *et al.*, 1997; Balimane and Sinko, 2000; Kottra *et al.*, 2002). The optimal extracellular pH for both PepT1 and PHT1 has been estimated to be 6.0 and 5.0, respectively. In addition, this model suggests that peptides which are not appreciably degraded intracellularly are transported out of the cells by an as-yet unidentified basolateral peptide transporters which have lower affinities than the PepT-like transporters (Terada *et al.*, 1999, 2000b; Irie *et al.*, 2001, 2004).

It has been established that all PepT1 substrates share the same substrate-binding site due to the fact that uptake strictly conforms to the Michaelis–Menten equation and exhibits the competitive inhibition regardless of substrate charge (Wenzel *et al.*, 1996; Mackenzie *et al.* 1996; Sawada *et al.*, 1999). Proton coupling occurs in the H^+-binding site of PepT1, where a H^+ is bound prior to anionic or neutral substrate uptake but is not required for cationic substrates (Nussberger *et al.*, 1997b). This model assumes that anionic substrates cannot access the substrate-binding site lacking a H^+, and that the protonated substrate-binding site can accept only negatively charged substrates. Recently, a computational model of the H+-coupled substrate transport of neutral and charged molecules has been published (Irie *et al.*, 2005). The authors established a PepT1 mechanistic model demonstrating the normally observed bell shaped uptake versus pH shaped curves for different charged substrates based on two novel main hypotheses: (1) H^+ binds to not only the H^+-binding site, but also the substrate-binding site; and (2) H^+ at the substrate-binding site inhibits the interaction of neutral and cationic substrates, but is necessary for that of anionic substrates.

Studies have shown the importance of certain amino acid residues in the transport activity of PepT-like proteins (Fei *et al.*, 1997; Bolger *et al.*, 1998; Yeung *et al.*, 1998; Chen *et al.*, 2000; Meredith, 2004). Studies utilizing PepT1/PepT2 chimeras suggest that the first half of the transporters (first 400 residues) contain substrate-binding domain segments as well as defining other functional properties (Döring *et al.*, 1996, 2002; Terada *et al.*, 2000a; Kulkarni, 2003a,b). The significance of the PepT chimeras with respect to these transporters is unknown since similar studies have not been conducted for PHT1/PHT2 peptide transporters.

7.3.1.2 The Substrate Specificity of Peptide Transporters

The PepT-Like Transporters

Studies elucidating the substrate specificity of PepT1 indicate that this protein transports almost all possible dipeptides, tripeptides, as well as numerous peptidomimetics and that their respective sizes or molecular weights were not significant factors (Leibach and Ganapathy, 1996). Analysis of the binding and transport characteristics of PepT1 have led to the development of several molecular models attempting to establish a PepT-substrate template (Swaan and Tukker, 1997; Bailey *et al.*, 1999; Zhang *et al.*, 2002a,b; Gebauer *et al.*, 2003; Biegel *et al.*, 2005). Brandsch *et al.* (2004) summarized the substrate structural characteristics required for high PepT1 affinity (<0.5 mM) which include (a) L-amino acids, (b) the presence of an acidic or hydrophobic function at the C-terminus, (c) the presence of a weakly basic group in α-position at the N-terminus, (d) exhibiting a ketomethylene moiety or acid amide bond, and in the case of having a peptide bond (e) to present it in a trans configuration. PepT1 has also been shown to mediate the transport of a variety of drugs (Table 7.5) with differing degrees of affinity and capacity, depending on their chemical structure. Generally, PepT1 has a high transport capacity, which makes it highly attractive as a drug target.

The Peptide/Histidine Transporters

Little information is known concerning the function of the PHT transporters. The rat peptide/histidine transporters (rPHT1 and rPHT2) have both shown high affinity for histidine and the ability to transport dipeptides (Yamashita *et al.*, 1997; Sakata *et al.*, 2001). Rat PHT1 expression in *Xenopus oocytes* revealed a high-affinity, proton-dependent histidine uptake that was inhibited by several di- and tripeptides, but not other amino acids. Rat PHT1 also shows high affinity for the β-ala-his dipeptide, carnosine (Yamashita *et al.*, 1997). Rat PHT2 reconstitution in liposomes modeling the lysosomal environment showed proton-dependent transport activity with histidine and dipeptides, but not with amino acids (Sakata *et al.*, 2001). However, studies in our laboratory suggest that hPHT1

TABLE 7.5. Therapeutic compounds shown to interact with PepT1

Substrates	Inhibitors
Cephalosporins	
Cefaclor	Latamoxef
Cefadroxil	
Cefamandole	
Cefatrizine	
Cefepine	
Cefixime	
Cefodizime	
Cefpirome	
Cefroxadine	
Ceftibuten	
Cefotaxime	
Ceftriaxone	
Cephalexin	
Cephradine	
Loracarbef	
Moxalactam	
Penicillins	
Benzylpenicillin	Carbenicillin
Cloxacillin	
Cyclacillin	
Dicloxacillin	
Metampicillin	
Oxacillin	
Phenoxymethyl-penicillin	
Propicillin	
ACE inhibitors	
Benazepril	Benazeprilat
Captopril	Enalaprilat
Enalapril	Fosinoprilat
Fosinopril	Quinalaprilat
Quinalapril	
Antivirals	
Valaciclovir	
Valganciclovir	
Others	
Bestatin	Glibenclamide
α-methyldopa-phenylalanine	Nateglinide
Pro-Phe-alendronate	Lys[Z(NO_2)]-Pro
Renin inhibitors	
Thrombin inhibitors	
Thyrotropin-releasing hormone	

mediates the transport of not only carnosine and L-histidine, but also valacyclovir in a proton-dependent, sodium-independent manner (Bhardwaj *et al.*, 2005a). Nevertheless, there is still much to elucidate concerning the affinity and substrate specificity of these peptide transporters.

7.3.1.3 The Regulation of Peptide Transporters

Information concerning the regulation mechanisms of peptide transporters is limited to PepT1. Earlier studies have shown that PepT1 transport activity changes as a response to diet regimens (Thamotharan *et al.*, 1998, 1999a; Walker *et al.*, 1998; Shiraga *et al.*, 1999; Ihara *et al.*, 2000, Adibi, 2003). PepT1 mRNA expression levels significantly increased in Caco-2 cells cultured in a dipeptide supplemented medium (Thamotharan *et al.*, 1998; Walker *et al.*, 1998). Other studies have demonstrated that malnourishment upregulates PepT1 expression in rat intestines (Ogihara *et al.*, 1999, Thamotharan *et al.*, 1999a; Ihara *et al.*, 2000), suggestive of transcriptional regulation.

Shiraga *et al.* (1999) performed a characterization of the rPepT1 promoter to evaluate how factors including the diet participate in the transcriptional activation of the rPepT1 gene. They suggested that the AP-1 binding site (TGACTCAG, nt −295) and the AARE-like element-binding site (CATGGTG, nt −277) regions were associated with dietary protein content regulation of rPepT1. Deletion analysis of the hPEPT1 promoter region in Caco-2 cells suggested that the region spanning −172 to −35 bp was essential for basal transcriptional activity, demonstrating the significant role of the Sp1 nuclear transcription factor in the basal transcriptional regulation of hPepT1 (Shimakura *et al.*, 2005).

The regulatory effects of secondary messengers on PepT activity are still controversial. Activation of PKC is suggested to be responsible for the downregulation of Gly-Sar uptake into Caco-2 cells (Brandsch *et al.*, 1994). Other investigators have also illustrated that transepithelial peptide transport is inhibited by PKC activation, or by an increase in intracellular Ca^{2+} (Wenzel *et al.*, 1999, 2002). Additionally, di- and tripeptide transport activity has been modified by agents that interfere with intracellular cAMP levels (Muller *et al.*, 1996; Berlioz *et al.*, 1999, 2000).

The effect of hormones on peptide transport has not been comprehensively analyzed; however, several summary reviews are available (Meredith and Boyd, 2000; Gangopadhyay *et al.*, 2002; Adibi, 2003). Several studies have demonstrated an effect of insulin (Thamotharan *et al.*, 1999b) and leptin (Buyse *et al.*, 2001) on PepT1 activity in Caco-2 cells. Dipeptide uptake into Caco-2 cells is stimulated by insulin (5 nM) treatment, with an observed apparent increase in the transporter capacity (V_{max} increased twofold) with no alteration in the K_m and hPepT1 mRNA levels (Thamotharan *et al.*, 1999b). These studies suggest that the effect in transport capacity was due to an increase in the insertion of PepT1 protein in the plasma membrane from a preformed cytoplasmic pool. Similar observations were obtained in Caco-2 cells treated with leptin (Buyse *et al.*, 2001).

Ashida *et al.* (2002) reported the effect of thyroid hormone 3, 5, 3′-L-triiodothyronine (T_3) on the expression and transport activity of PepT1 in Caco-2 cells suggesting that the changes in dipeptide uptake were associated with the inhibition of the transcription of PepT1 mRNA and/or with a change in the mRNA stability, but the precise mechanism of the T_3 effect on [^{14}C]glycylsarcosine transport was not clearly elucidated. Furthermore, *in vivo* studies in euthyroid and hyperthyroid rats have demonstrated the effect of thyroid hormone on the activity and expression of PEPT1 in the small intestine (Ashida *et al.*, 2004). The [^{14}C]glycylsarcosine uptake by everted small intestinal preparations was significantly decreased in hyperthyroid rat. The mean portal vein concentrations after intrajejunal administration of [^{14}C]glycylsarcosine were also decreased in hyperthyroid rats. Moreover, hyperthyroidism caused a significant decrease in the expression of PEPT1 mRNA and protein in the small intestine. These results show the relevancy of hormonal regulation in the expression and activity of peptide transporters, providing useful information for protein nutrition and drug treatment in patients with hyperthyroidism.

Several studies have demonstrated the capability of the small intestine to compensate for possible nutritional deficiencies caused by tissue injury or resection (Tanaka *et al.*, 1998; Merlin *et al.*, 1998, 2001; Takahashi *et al.*, 2001; Ziegler *et al.*, 2002). Studies performed in sections of Short-bowel syndrome patients have revealed that hPepT1 mRNA and protein expression were up regulated in the colon mucosa, suggesting hPepT1 expression maybe an adaptation process in response to gut mucosal damage (Ziegler *et al.*, 2002). Furthermore, Merlin *et al.* (1998, 2001) have suggested that the aberrant PepT1 expression under chronic disease states implicates PepT1 function in intestinal inflammatory processes.

It has been recognized that genetic polymorphism of membrane transporters may affect their function in different ways. SNPs localized on exons might modify the intrinsic activity by changing the affinity to substrates (K_m) and/or the protein translocation ability or capacity (V_{max}). Furthermore, the capacity of the transporter can also be altered by changes in the protein expression level or impaired subcellular sorting of the protein to appropriate domains of the plasma membrane (Ishikawa *et al.*, 2004).

Peptide transporters polymorphisms have been scarcely studied. The only functional report testing polymorphisms and their impact on the transport of substrates was recently published by Zhang *et al.* (2004). In this study, nine nonsynonymous SNPs were identified in a population of 44 individuals of different ethnicities. The hPepT1 wild-type sequence as well as the hPepT1 sequences containing the individual SNPs were amplified and transiently transfected into HeLa cells, where the transport kinetics of [14C]Gly-Sar was analyzed. Western blot and immunocytochemical analyses were performed to establish the amount and expression of the hPepT1 protein in the cells. Their results showed that only the nonsynonymous P586L SNP modified the hPepT1 activity by decreasing in tenfold the transport capacity (V_{max}) of the protein. The researches demonstrated that this decrease on transport capacity was due to a lower level of expression of hPepT1 and not because of intrinsic changes in transport function. Overall, these data suggest that Pro^{586} may have an important effect on hPepT1 translation, degradation, and/or membrane insertion.

Ishikawa *et al.* (2004) have published the identification of other potentially relevant nonsynonymous SNPs found in the Japanese population; however, no functional analysis has been conducted so far. The known nonsynonymous SNPs for hPepT1 can be observed in Table 7.6.

TABLE 7.6. Nonsynonymous polymorphisms in the PepT1 gene (reference NM_005073)

NCBI SNP ID	dbSNP allele	Effect on protein	Functional relevancy
Not registered	C/T	P586L	Reduced transport capacity (Zhang *et al.*, 2004)
rs8187830	C/T	P537S	Unknown
rs2274827	C/T[a]	R459C	None (Zhang *et al.*, 2004)
rs8187838	C/A	T451N	None (Zhang *et al.*, 2004)
rs2274828	G/A[a]	V450I	None (Zhang *et al.*, 2004)
rs4646227	G/C[a]	G419A	None (Zhang *et al.*, 2004)
Not registered	G/C	V416L	None (Zhang *et al.*, 2004)
rs1782674	G/A	D383N	Unknown

[a]SNPs found in Japanese population (Ishikawa *et al.*, 2004)

7.3.2 Organic Anion Transporters (OAT, SLC22A; OATP, SLCO)

Conventional pH-partition hypothesis theory states that ionic agents generally exhibit low passive membrane permeability, resulting in their poor bioavailability. However, there is evidence demonstrating that the intestinal absorption of numerous ionic agents are mediated by the organic anion (OA) or organic cation (OC) transporter systems (Katsura and Inui, 2003; Sai and Tsuji, 2004; Steffansen *et al.*, 2004), thereby overcoming the passive membrane barriers and significantly increasing the intestinal absorption of these compounds (Table 7.7). The expression and function of a variety of these transporters have been investigated; however, their ability to mediate transport of ionic drugs across the intestinal epithelium is still poorly understood. This section will briefly describe the most recent studies involving the role of the OA (Sect. 7.3.2) and OC (Sect. 7.3) families in the mediation of intestinal absorption. Several comprehensive review articles on these transporter families are suggested for additional information (Hagenbuch and Meier, 2003; Koepsell *et al.*, 2003; Tirona and Kim, 2002; van Montfoort *et al.*, 2003; Jonker and Schinkel, 2004; Koepsell and Endou, 2004; Miyazak *et al.*, 2004; You, 2004).

7.3.2.1 OAT (SLC22A)

According to the pH-partition hypothesis, most anionic drugs are expected to traverse the intestinal epithelium by passive diffusion in the nonionized state, due to the presence of an acidic microclimate around the intestinal epithelial cells. However, involvement of specific anion transporters in the intestinal absorption of anionic compounds has also been suggested (Katsura and Inui, 2003; Mizuno *et al.*, 2003; Kunta and Sinko, 2004; Sai and Tsuji, 2004; Steffansen *et al.*, 2004). The organic anion transporters are classified into several categories: organic anion transporters (OATs), organic anion transporting polypeptides (rodents: Oatps; human: OATPs), and multiple drug resistance-associated proteins (MRPs) (Hagenbuch and Meier, 2003; van Montfoort *et al.*, 2003; Miyazak *et al.*, 2004; Koepsell and Endou, 2004).

TABLE 7.7. Organic cation and anion transporters in human instestine

Name	Transport mode	Endogenous substrates	Xenobiotics/drugs
Organic cation transporters			
hOCT1 (SLC22A2[a])	F[b]	Prostaglandin E_2, F_2	TEA[c], MPP^+, *N*-methylquinine, *N*-methylquinidine, tributylmethylammonium, Acyclovir, ganciclovir
hOCT2 (SLC22A2)	F	Choline, histamine, dopamine, serotonin, noradrenaline, agmatine, Prostaglandin E_2, F_2	TEA, MPP^+, *N*-methylnicotinamide, cimetidine amantadine, memantine
hOCT3 (SLC22A3)	F	Serotonin, adrenaline, noradrenaline, agmatine	MPP^+, cimetidine
Organic cation/cartine transporters			
hOCTN2 (SLC22A5)	C/Na^+, carnitine, F (for OC)	Acetyl-L-carnitine, L-carnitine, D-carnitine	TEA, quinidine, pyrilamine, verapamil
Organic anion transporting polypeptides			
OATP-B (SLC21A9)	ND	Prostaglandin E_2	Estrone sulfate, BSP, fexofenadine, pravastatin, temocaprilat
OATP-D (SLC21A11)	ND	Prostaglandin E_2, estrone sulfate	Benzylpenicillin
OATP-E (SLC21A12)	ND	Taurocholate, thyroid hormones, prostaglandin, estrone sulfate,	

[a] Solute carrier family gene symbol
[b] C, cotransporter; E, exchanger; F, facilitated transporter
[c] TEA, tetraethylammonium; MPP^+, 1-methyl-4-phenylpyridium; PAH, *p*-aminohippuric acid; BSP, bromosulfophthalein, DHEAS, dehydroepiandrosterone sulfate

To date, five structurally related isoforms (OAT1–5) have been identified in the OAT family (Miyazak *et al.*, 2004; You, 2004). Most OAT isoforms are predominantly expressed in the kidney and have important functions in renal clearance of relevant substrates (Miyazak *et al.*, 2004), although rat OAT2 was expressed at much higher level in liver compared to kidney (Sekine *et al.*, 1998). In contrast to the liver and kidney, OATs are expressed to a lesser extent, in brain, muscle, eye, and placenta (Miyazak *et al.*, 2004). The distribution patterns of OAT family members might be one of the important determinants influencing the substrate's pharmacokinetics. OAT family members share some common topology characteristics including twelve α-helix TMD; one large hydrophilic extracellular loop between TMD 1 and 2 carrying several potential glycosylation sites; and a large intracellular loop containing multiple potential phosphorylation sites between TMD 6 and 7 (You, 2004).

OATs are polyspecific transporters that are capable of interacting with a wide range of clinically significant organic anion drugs such as nonsteroidal anti-inflammatory drugs (NSAIDs), β-lactam antibiotics, antiviral drugs, diuretics, antitumor drugs, and angiotensin-converting enzyme inhibitors (Koepsell and Endou, 2004; You, 2004). While OAT isoforms have broad substrate specificity, members of the OAT family have not been identified in the human intestine. In fact, the intestinal expression of OAT members is quite limited, with only one report demonstrating the presence of OAT2 mRNA in mouse fetal intestine (Pavlova *et al.*, 2000). In contrast to the abundance of members of the MRP family identified in the intestine, the role of the OAT family in the intestinal absorption of drugs seems to be negligible and will not be discussed further.

7.3.2.2 OATP (SLCO)

The related OATP/Oatp isoforms are part of a rapidly expanding family of mammalian transporters that mediate the transmembrane transport of a wide range of amphipathic endogenous and exogenous organic compounds (Hagenbuch and Meier, 2003; Tirona and Kim, 2002; van Montfoort *et al.*, 2003). The OATP/Oatp genes were previously classified within the solute carrier family 21A (SLC21A) and were given various trivial names (Hagenbuch and Meier, 2004; Mikkaichi *et al.*, 2004). However, this classification does not provide a clear and species independent identification of genes. Therefore, in agreement with the HUGO Gene Nomenclature Committee (HGNC), all OATP/Oatp isoforms are currently classified within the OATP/SLCO superfamily based on their putative phylogenetic relationships and the chronology of identification (Hagenbuch and Meier, 2004; Mikkaichi *et al.*, 2004; http://www.bioparadigms.org/slc/SLC21.htm). Of the 52 members of the OATP/SLCO superfamily, 36 isoforms have been identified across the human, rat, and mouse genomes. The OATP/SLCO isoforms are identified within six out of 13 subfamilies (OATP1–OATP6) having different structural features as compared with OATs. While possessing twelve TMD, OATP/SLCO isoforms contain a large extracellular domain between TMD 9 and 10 (extracellular loop 5) and have multiple glycosylation sites in extracellular loop 2 and 5. In addition, there is an OATP superfamily signature (D-X-RW-(I,V)-GAWW-X-G-(F,L)-L.) at the border between extracellular loop 3 and TMD 6 (Hagenbuch and Meier, 2003).

The Substrate Specificity of OATPs/Oatps

OATP/Oatp family members mediate transmembrane transport of a wide range of organic compounds (Table 7.7) (Hagenbuch and Meier, 2003; Tirona and Kim, 2002; Koepsell and Endou, 2004; Sai and Tsuji, 2004) including (1) organic anions, such as bromosulfophthalein (BSP), bile salts, bilirubin, prostaglandins, and estrogen-conjugates; (2) neural steroids and steroid conjugates; (3) lipophilic organic cations, e.g., rocuronium; and (4) organic dyes, thyroid hormones, and anionic oligopeptides. Various pharmaceutically relevant compounds such as digoxin, pravastatin, methotrexate, temocaprilat, benzylpenicillin, fexofenadine, (D-Pen2, D-Pen5)-enkephalin (DPDPE), as well as some NSAIDs are also substrates of OATP/Oatp isoforms (Hagenbuch and Meier, 2003; Tirona and Kim, 2002; Koepsell and Endou, 2004; Sai and Tsuji, 2004) (Table 7.7). The clinical significance of the OATP/Oatp family is readily apparent given the wide range of pharmaceutical substrates for its members. The tissue expression patterns and cellular localization of OATP/Oatp isoforms (see below), make these transporters attractive targets to enhance drug bioavailability.

OATP/Oatp Isoform Mediated Transport

The mechanism driving OATP/Oatp-mediated transport has been investigated for several isoforms. OATP/Oatp members mediate the transport of organic anions and other compounds in a Na^+-independent manner (Hagenbuch and Meier, 2003; Tirona and Kim, 2002; van Montfoort *et al.*, 2003). Bidirectional transmembrane transport of BSP was observed in rat Oatp1 transfected cells (Shi *et al.*, 1995), as well as rat Oatp1 mediated taurocholate/HCO_3 exchange (Satlin *et al.*, 1997) suggesting an anion exchange mechanism. Evidence implicates a role of glutathione (GSH) in rat Oatp1 and Oatp2 (Li *et al.*, 1998, 2000) substrate transport. Li *et al.* (1998) demonstrated that rat Oatp1 mediated uptake of taurocholate and leukotriene C4 (LTC4) was significantly inhibited by high extracellular GSH concentrations, yet stimulated by high intracellular GSH. Additionally, GSH efflux across rat Oatp1 expressing oocytes was increased and further enhanced in the presence of extracellular Oatp1 substrates taurocholate or BSP (Li *et al.*, 1998). Increased GSH efflux in Oatp2 expressing *Xenopus oocytes* was similarly observed (Li *et al.*, 2000). Oatp2 mediated taurocholate transport was upregulated by high intracellular GSH, although changes of the extracellular GSH concentration had no effect (Li *et al.*, 2000). Therefore, physiological GSH efflux down its electrochemical gradient could be an important driving force for rat Oatp1 and Oatp2 mediated transport.

A proton-coupled transport mechanism has also been suggested for OATP/Oatp isoforms (Kobayashi *et al.*, 2003) based on the uptake of estrone-3-sulfate and pravastatin in OATP-B-transfected HEK 293 cells. It was demonstrated that uptake of both compounds were pH-dependent, with higher uptake at pH 5.5 in contrast to pH 7.4 (Kobayashi *et al.*, 2003). Moreover, an increase in V_{max} was observed with decrease of pH from 7.4 to 5.0, whereas the change in K_m was negligible (Nozawa *et al.*, 2004). This is analogous to the proton-dependent mechanism of uptake observed with POT family members discussed above. Additional work is required to identify the mechanism(s) by which each OATP/Oatp isoform mediates substrate uptake.

The Expression of OATPs/Oatps

The tissue distribution of OATPs/Oatp isoforms has been extensively studied. Several isoforms demonstrate tissue-specific expression patterns of OATP-C (Konig *et al.*, 2000a) and OATP-8 (Konig *et al.*, 2000b), which are exclusively expressed at the basolateral membrane of the hepatocytes. In contrast, several OATP/Oatp family members (e.g., OATP-B, OATP-D, and OATP-E) have a fairly broad pattern of tissue expression including the blood–brain barrier (BBB), lung, heart, kidney, placenta, and intestine (Hagenbuch and Meier, 2003; Kim, 2003; van Montfoort *et al.*, 2003).

Rat Oatp3 mRNA levels were similar down the length of the small intestine through the RNAse protection assay (Walters *et al.*, 2000). Immunofluorescence studies further localized Oatp3 to the apical brush-border membrane of rat jejunal enterocytes (Walters *et al.*, 2000) (Fig. 7.2). In contrast, expression of the human analogue OATP-A in the human small intestine has not been shown in the human small intestine (Tamai *et al.*, 2000a), whereas evidence suggests the expression of OATP-B, OATP-D, and OATP-E isoforms, at least at the mRNA level (Tamai *et al.*, 2000a). In a subsequent study, Kobayashi *et al.* (2003) identified the OATP-B isoform in the apical membrane of human intestinal epithelial cells by immunocytochemical analysis (Fig. 7.2). OATP/Oatp isoforms might play an important

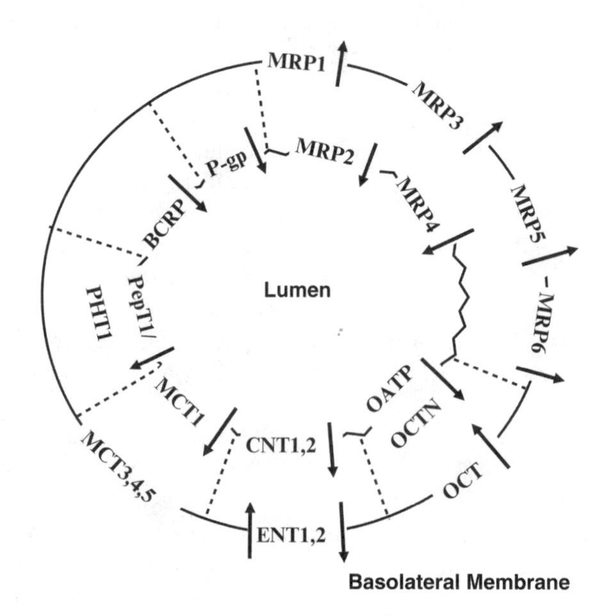

FIGURE 7.2. Localization of intestinal drug transporters P-glycoprotein (P-gp), breast cancer resistance protein (BCRP), peptide transporter (PepT1), peptide histidine transporters (PHT1), monocarboxylate transporters (MCT), concentrative nucleoside transporter (CNT), equilibrative nucleoside transporter (ENT), organic anion transporting protein (OATP), multiple resistance protein (MRP) on the apical and basolateral membranes of the intestinal epithelial cells surrounding a hypothetical lumen

role in the intestinal absorption of bile acids and other anionic drugs as well, considering that rat Oatp3 and human OATP-B mediate apical uptake of taurocholate, estrone-3-sulfate, and/or pravastatin in stably transfected cells (Walters *et al.*, 2000; Kobayashi *et al.*, 2003; Nozawa *et al.*, 2004).

The Regulation of OATPs/Oatps

The regulation of OATP/Oatp isoform expression and the functional kinetics of each transporter can occur at both the transcriptional and posttranscriptional levels (Hagenbuch and Meier, 2003). Several studies have investigated the transcriptional regulation of OATP/Oatp isoforms and various control elements were identified (Rausch-Derra *et al.*, 2001; Staudinger *et al.*, 2001; Guo *et al.*, 2002b). Increased Oatp2 mRNA and protein expression were observed from livers of rats treated with pregnenolone-16α-carbonitrile (PCN) (Rausch-Derra *et al.*, 2001). Pregnane X receptors (PXR) were suggested to play a major role in this PCN induction. Oatp2 expression was strongly induced by PCN treatment in the $PXR^{+/+}$ mice, but not in the $PXR^{-/-}$ mice (Staudinger *et al.*, 2001). Furthermore, several PXR response elements have been identified on the rat Oatp2 promoter (Guo *et al.*, 2002b). Since PXR is one of the key regulators of cytochrome P450 3A (CYP3A) (Luo *et al.*, 2004), concomitant PXR-dependent upregulation of OATP/Oatp and CYP3A in response to stimuli represents an important mechanism in the hepatic detoxification of both bile salts and xenobiotics (Staudinger *et al.*, 2001, Xie *et al.*, 2001). Hepatic expression of the human OATP-C gene may be dependent on a liver-enriched transcriptional factor, hepatocyte nuclear factor 1α (HNF-1α) (Jung *et al.*, 2001). Coexpression of HNF-1α stimulated OATP-C promoter activity by 30-fold in HepG2 and 49-fold in HeLa cells. Similarly, promoters of human OATP8 and mouse Oatp4 were also responsive to HNF-1α (Jung *et al.*, 2001). In addition, OATP8 mRNA levels were induced by ligand of FXR/BAR (Farnesoid X receptor/Bile acid receptor), but not PXR or LXR (Liver X receptor) (Jung *et al.*, 2002), suggesting a FXR-mediated OATP8 gene regulation. An inverted hexa-nucleotide repeat motif (IR-1 element) in the promoter region of OATP-8 was suggested to be the bile acid response element (Jung *et al.*, 2002), where targeted mutation abolished the inducibility of OATP8 (Jung *et al.*, 2002).

The functional regulation of rat Oatp1 and Oatp2 isoforms was also determined to occur at the posttranslational level. Glavy *et al.* (2000) demonstrated that serine phosphorylation of rat Oatp1 reduces uptake of BSP by 85% in the presence of extracellular ATP in cultured rat hepatocytes. The activation of protein kinase C (PKC), but not PKA, significantly suppressed estrone-3-sulfate uptake in Oatp1 expressing oocytes in a concentration- and time-dependent manner, while pretreatment with specific PKC inhibitor partially reversed this suppression (Guo and Klaassen, 2001). Similarly, PKC activators suppressed Oatp2 mediated digoxin transport and the downregulation effect was completely abolished by PKC inhibitors (Guo and Klaassen, 2001), demonstrating that Oatp2 is also regulated at the protein level by PKC. It is clear that increased attention needs to be focused on studying the mechanistic role of phosphorylation pathways in regulating OATP/Oatp isoform function.

Ontogenic expression patterns of OATP/Oatp isoforms have also been investigated. It appears that the rat Oatp family members follow a similar temporal pattern with regard to the developmental regulation of their mRNAs. For example, low hepatic expression of rat Oatp2 (Guo *et al.*, 2002a) and Oatp4 (Li *et al.*, 2002) was observed in newborn rats, with a gradual increase shown during postnatal development. Expression of rat Oatp1 mRNA and protein in the choroid plexus were not observed until 15 days postnatal, and were at adult levels by 30 days (Angeletti *et al.*, 1998). Rat Oatp5 expression in the kidney could not be found during the first 3 weeks after birth (Choudhuri *et al.*, 2001). Unlike rat Oatp isoforms, mouse hepatic rPGT (Oatp2a1) was expressed at adult levels at birth, while renal Oatp1, Oatp5, and Oatp-D were expressed at lower level at birth versus at 6 weeks of age (Cheng *et al.*, 2005). Other renal mouse Oatp isoforms followed a similar age-dependent expression pattern to their rat orthologues while mouse hepatic mRNA expression of Oatp1, Oatp2, Oatp4, and Oatp5 elevated gradually after birth and reached observed maximum adult levels by 6 weeks of age (Cheng *et al.*, 2005).

These data suggest that developmental changes influence the OATP/Oatp family and can significantly influence the substrate pharmacokinetic and pharmacodynamic profiles, especially those in the liver and kidney, the two major organs in drug detoxification. For example, newborn rats are more sensitive to ouabain (cardiac glycoside) toxicity due to lower Oatp2 expression in the liver (Guo *et al.*, 2002a), which results in less hepatic uptake and higher blood and tissue levels resulting in ouabain toxicity (Klaassen, 1972; Guo *et al.*, 2002a). Interestingly, pregnenalone-16α-carbonitrile (PCN) protects newborns from cardiac glycoside toxicity by dramatically inducing hepatic sinusoidal Oatp2 mRNA and protein levels in neonatal rats (Guo *et al.*, 2002a) and thus increasing hepatic ouabain clearance. The elucidation of similarities and differences in OATP/Oatp expression among rat, mouse, and human will aid in extrapolation of rodent pharmacokinetic data to humans.

Gender-specific expression of OATP/Oatp isoforms was also reported. In one study, the protein expression of Oatp2 in the female rat liver was significantly lower than those observed in male rats, while levels of Oatp1 and Oatp4 were comparable (Rost *et al.*, 2005). Rost *et al.* (2005) further demonstrated that the protein expression of both Oatp1 and Oatp4 was dramatically downregulated after DHEA (dehydroepiandrosterone) treatment in both male and female rats, while Oatp2 expression was only downregulated in male rats. Higher renal expression of Oatp1 in female rats was also inferred from increased urinary excretion of estradiol-17β-D-glucuronide when contrasted with male rats (Gotoh *et al.*, 2002). It has also been demonstrated that renal Oatp1 expression is stimulated by testosterone and inhibited by estrogens (Lu *et al.*, 1996), while hepatic Oatp1 expression is not influenced (Simon *et al.*, 1999). These findings suggest that sex hormones may play a role in the regulation of OATP/Oatp isoforms, although further work is required to elucidate the effects in the human intestine.

A number of allelic polymorphisms have been identified for each human OATP isoform (Tirona and Kim, 2002). Mutations may manifest in the promoter region influencing expression or sites impacting tertiary structure and/or substrate interacting regions. Since they may play a critical role in the pharmacokinetics of anionic drugs and other compounds, mutations in these OATP isoforms might alter the handling of certain drugs in the human body and thus enhance (or decrease, depending on the particular situation) toxicity and therapeutic efficacy. For example, SNPs in OATP-C from a population of African- and European Americans have variable pharmacokinetic attributes (Tirona *et al.*, 2001). *In vitro* assessment of 16 OATP-C alleles revealed that several variants exhibit markedly reduced OATP-C mediated uptake. Additionally, alterations in transport were associated with SNPs imparting amino acid changes within the TMD and also with those modifying extracellular loop 5. Genetic polymorphisms of OATP-B have also been studied in a Japanese population (Nozawa *et al.*, 2002). A SNP of OATP-B, S486F, had an allelic frequency of as high as 10%. This OATP-B variant demonstrated dramatic decreased activity in *in vitro* transport assay. Given the allelic differentiation arising from SNPs among populations, SNPs in human OATPs represent a heretofore unrecognized factor influencing drug disposition. Expression and functional characterization of allelic variants as to the pharmacokinetics of substrates requires further investigation.

These data demonstrate that genetic, age, and gender dependent variability exists in the expression of OATP/Oatp isoforms. The pharmacogenomic differences may result in significant pharmacokinetic and pharmacodynamic outcomes with administered OATP/Oatp substrates. This is an especially important factor when one considers those substrates that have a narrow therapeutic index. This also ignores the potential confounding idiosyncratic events that may arise from environmental/xenobiotic exposure.

7.3.3 Organic Cation Transporters (OCT, OCTN; SLC22A)

A significant number of the current therapeutic agents including antihistamines, skeletal muscle relaxants, antiarrhymics, and β-adreno receptor blocking agents, as well as endogenous bioactive amines (e.g., catecholamines, dopamine, histamine, and choline) are organic cations (OC). Based on the fact that many of these organic cations are polar and positively charged at physiological pH, membrane bound transporters are required to enhance their intestinal uptake and absorption in an acidic environment. The OC transporter family has been identified as one of the main classes of transporters that act in this fashion. It should be noted that most of the literature dealing with the function of OC transporters have been conducted in the liver and kidney (Katsura and Inui, 2003; Koepsell *et al.*, 2003; Koepsell and Endou, 2004; Sai and Tsuji, 2004; Steffansen *et al.*, 2004), while few discuss the role of OC transporters in the GI tract. The results pertaining to the expression and functional significance of OC transporters in the GI tract are summarized here.

Organic cation transporter 1 (OCT1) was the first member of the OC family identified and was cloned from rat kidney by Grundemann *et al.* (1994). Subsequently, other organic cation transporters (OCT2–3), as well as the more distantly related carnitine and organic cation transporters (OCTN1–3) have been cloned and characterized (Koepsell *et al.*, 2003; Koepsell and Endou, 2004; You, 2004). Although they share some common structural features, OCTs and OCTNs are considered distinct subfamilies within the OCT family with each member having been isolated from multiple species. Similar to OATs, the OCT and OCTN isoforms represent 12 α-helix TMD protein, which contain a large glycosylated extracellular loop between TMD 1 and 2, and a large intracellular loop carrying phosphorylation sites between TMD 6 and 7 (You, 2004). Since OCT and OCTN family members have sequence homology to the OAT family (Koepsell and Endou, 2004; Miyazak *et al.*, 2004; You, 2004), OATs, OCTs and OCTNs, together with other uncharacterized, yet homologous orphan transporters (e.g. BOCT, brain-type organic cation transporter; ORCTL, organic cation transporter like; UST, unknown solute transporter; etc.) comprise a transporter superfamily, referred to as the organic ion transporter family SLC22A (Miyazak *et al.*, 2004; http://www.bioparadigms.org/slc/SLC22.htm).

7.3.3.1 The Substrate Specificity of Organic Cation Transporters

OCT1–3 are polyspecific transporters capable of transporting various organic cations (Table 7.7), including model compounds such as tetraethylammonium (TEA) and *N*-methylquinine, as well as other xenobiotics including 1-methyl-4-phenylpyridium (MPP^+), acyclovir and ganciclovir, metformin and phenformin, memantine, as well as quinidine (Koepsell *et al.*, 2003; Koepsell and Endou, 2004; You, 2004). Endogenous substrates of the OCTs include the monoamine neurotransmitters acetylcholine, dopamine, serotonin, histamine, choline, and physiological compounds such as creatinine, guanidine, and thiamine (Koepsell *et al.*, 2003; Koepsell and Endou, 2004; You, 2004). Although organic cations are clearly the preferred ligands of the OCTs, several uncharged or anionic compounds are known to be substrates of these transporters (Table 7.7). For example, hOCT2 is partially responsible for the transport of cimetidine, a weak base (Barendt and Wright, 2002), while both hOCT1 and hOCT2 mediate the transport of anionic prostaglandins (Kimura *et al.*, 2002). Although the substrate and inhibitor specificities of OCT1–3 overlap extensively, there are distinct differences in affinity and maximal transport rates among different OCT isoforms and species, which have been summarized by Koepsell *et al.* (2003). For example, the IC50s of hOCT2 (16 μM) and hOCT3 (14 μM) to desipramine, an antidepressant, were one order of magnitude higher than that of hOCT3 (5.4 μM) (Gorboulev *et al.*, 1997; Zhang *et al.*, 1998; Wu *et al.*, 2000b). In contrast, IC50 of hOCT2 (>100 μM) to α Blocker Prazosin was much higher when compared to hOCT1 (1.8 μM) and hOCT3 (13 μM) (Hayer-Zillgen *et al.*, 2002).

Members of the OCTN subfamily have differential abilities to interact with a variety of organic cation drugs, as well as carnitine (Koepsell and Endou, 2004; You, 2004). For example, TEA is a substrate for rat, mouse, and human OCTN1 (Tamai *et al.*, 1997, Wu *et al.*, 2000a) and OCTN2 (Wu *et al.*, 1998, 1999; Tamai *et al.*, 1998; Friedrich *et al.*, 2003), but not for OCTN3, at least for mouse isoform (Tamai *et al.*, 2000b). Rat OCTN1 and hOCTN1 exhibit a very low affinity for carnitine (Tamai *et al.*, 1997; Wu *et al.*, 2000a), but mOCTN1 mediates significant transport of carnitine (Tamai *et al.*, 2000b), suggesting a species difference of substrate specificity. In addition, OCTN2 from all species (Wu *et al.*, 1998; Tamai *et al.*, 2000b; Friedrich *et al.*, 2003), and mOCTN3 (Tamai *et al.*, 2000b) exhibit medium and very high affinity for carnitine, respectively.

7.3.3.2 Organic Cation Transporter Mediated Transport

Several common transport properties have been identified for all OCTs and independent from subtype or species. OCTs translocate organic cations and other compounds in an electrogenic manner, which has been shown for the rat isoforms rOCT1, rOCT2, and rOCT3 (Busch *et al.*, 1996; Nagel *et al.*, 1997; Kekuda *et al.*, 1998; Okuda *et al.*, 1999), and for the human transporters hOCT1 and hOCT2 (Busch *et al.*, 1998; Gorboulev *et al.*, 1997). In addition, OCTs medicated transport is independent from Na^+ and H^+ ions (Busch *et al.*, 1996; Gorboulev *et al.*, 1997; Kekuda *et al.*, 1998). Driving force for substrate transport is provided by the substrate concentration gradient and the membrane potential (Busch *et al.*, 1996, 1998; Gorboulev *et al.*, 1997; Kekuda *et al.*, 1998; Okuda *et al.*, 1999). As such, OCTs are able to translocate substrates across the plasma membrane in either direction (Busch *et al.*, 1996, 1998; Nagel *et al.*, 1997; Kekuda *et al.*, 1998).

In contrast to OCTs, OCTN mediated transport mechanism depends largely on the isoform and substrate tested. Human and rat OCTN1 work as H^+/organic cation antiporters that mediate transport of tetraethylammonium (TEA) and other organic cations (Tamai *et al.*, 1997; Wu *et al.*, 1998). However, mOCTN1 mediate carnitine transport in a Na^+/dependent manner (Tamai *et al.*, 2000b). OCTN2 is a Na^+/carnitine cotransporter with a high affinity for carnitine (Tamai *et al.*, 1998, 2001; Wu *et al.*, 1999; Wagner *et al.*, 2000). It can also function alternatively as a polyspecific cation uniporter in a Na^+-independent manner (Tamai *et al.*, 1998, 2001; Wu *et al.*, 1999; Wagner *et al.*, 2000). In the presence of Na^+, hOCTN2 could transport short-chain acyl esters of carnitine (Ohashi *et al.*, 1999) as well as zwitterions, e.g., cephaloridine (Ganapathy *et al.*, 2000). Furthermore, OCTN2 mediated transport is electrogenic and pH-dependent (Wagner *et al.*, 2000). OCTN3 was found only in mice (Tamai *et al.*, 2000b). In contrast to OCTN2 from different species, mOCTN3 transports carnitine independently from Na^+ and demonstrates less affinity for organic cation compared to OCTN1 and OCTN2 (Tamai *et al.*, 2000b).

7.3.3.3 The Expression of Organic Cation Transporters

The tissue distribution and membrane localization of the OC family of proteins have been studied using different approaches. In each case, it has been demonstrated that the respective isoforms have different tissue expression patterns that can vary with species (Koepsell *et al.*, 2003; You, 2004). In general, OCT isoforms are mainly expressed in the liver or kidney, and may also be found, to a less extent, in the heart, skeletal muscle, placenta, and small intestine. hOCT1 is mainly expressed in the liver, whereas hOCT2 is mainly found in the kidney

(Gorboulev *et al.*, 1997). hOCTN1 and hOCTN2 are both abundantly expressed in the kidney, skeletal muscle, placenta, prostate, and heart (Tamai *et al.*, 1997, 1998; Wu *et al.*, 1999), with hOCTN2 also being expressed at low level in the liver (Tamai *et al.*, 1998). Although hOCTN1 is expressed strongly in the kidney (Tamai *et al.*, 1997), rOCTN1 is present principally in the liver (Wu *et al.*, 2000a). For a more detailed description of the tissue distribution of OCT and OCTN isoforms, the reader is directed to the reviews by Koepsell *et al.* (2003) and You (2004). The following sections will focus on the expression of OCTs and OCTNs in the GI tract.

Tissue expression of OCT1 isoforms indicates that rat OCT1 (Grundemann *et al.*, 1994, Zhang *et al.*, 1997a) and OCT3 (Kekuda *et al.*, 1998) mRNA are expressed in the small intestine at a relatively low and high level, respectively. With respect to the human variants, hOCT1 is expressed to a much lower extent in the human small intestine (Zhang *et al.*, 1997b), while hOCT3 expression in the human small intestine has not been confirmed. Although rOCT2 is predominantly expressed in the rat kidney (Okuda *et al.*, 1996), hOCT2 is also expressed in the human small intestine, as detected by RT-PCR (Gorboulev *et al.*, 1997). A recent study using $Oct1^{-/-}$ mice further suggests the basolateral localization of mOCT1 in the mouse intestine (Fig. 7.2) (Jonker *et al.*, 2001; Jonker and Schinkel, 2004). However, direct immunolocalization data are not currently available.

The mRNA expression of the rOCTN1 (Wu *et al.*, 2000a) and hOCTN2 (Tamai *et al.*, 1998; Wu *et al.*, 1998) were also detected in intestinal enterocytes, the latter one being very weak. To date, no further studies on the intestinal protein expression of OCTNs have been reported.

7.3.3.4 The Regulation of Organic Cation Transporters

The expression and function of OCT family members have been suggested to be regulated *via* subtype, species, and tissue-specific parameters. The short-term regulation of basolateral and apical OCTs in different experimental systems is well documented (Ciarimboli and Schlatter, 2005). For rOCT1, hOCT1, hOCT2, and hOCT3, regulation has been associated with phosphorylation/dephosphorylation of the transporter, which can result in changes in the substrate affinity (Ciarimboli and Schlatter, 2005). For example, after stable transfection in HEK293 cells, rOCT1-mediated organic cation transport was stimulated by protein kinase C (PKC), PKA, and endogenous tyrosine kinase activation (Mehrens *et al.*, 2000). Furthermore, it was determined that at least one of these phosphorylation sites is PKC dependent. PKC mediated phosphorylation in these sites leads to a conformational change at the substrate binding site, and thus results in alteration of the substrate transport including TEA and quinine (Mehrens *et al.*, 2000). Therefore, the potential phosphorylation sites may play an important role in the regulation of transporter activity, especially the S286 residue, which is conserved in almost all OCTs (Ciarimboli and Schlatter, 2005).

A gender dependent difference was observed when it was shown that renal expression of rOCT2 was significantly increased in male versus female rats (Urakami *et al.*, 1999, 2000). Furthermore, these studies demonstrated that testosterone increased rOCT2, while estradiol resulted in a moderate decrease in the expression of rOCT2 (Urakami *et al.*, 1999, 2000), suggesting the role of sex hormone regulation. In contrast, there was no gender difference observed in the renal expression of rOCT1 or rOCT3. In a separate ontogenic study, it was determined that rat renal OCT1, OCTN1, and OCTN2 mRNA levels increased gradually from infants to adults (Slitt *et al.*, 2002). Since OCT1 and OCTN1/2 mediate the renal clearance of organic cations, substrates of these isoforms may be excreted more slowly in infants and children in contrast to adults (Slitt *et al.*, 2002). Similarly, hOCTN1 mRNA was strongly expressed in the fetus when contrasted to the adult liver (Tamai *et al.*, 1997), which indicates that the potential increased fetal hepatic toxicity. Clearly, the current data suggest that the gender- and age-dependent effects on the expression of OCT isoforms need to be considered when evaluating the overall pharmacokinetics and potential toxicity of OCT substrates that have a narrow therapeutic index.

Similar to the OATP/Oatp family, SNPs have been identified for several isoforms of human OCT and OCTN family (Koepsell and Endou, 2004). Some are associated with decreased transporter activity. In a population of 57 Caucasians, 25 SNPs within the hOCT1 gene were detected and further analyzed (Kerb *et al.*, 2002). Eight SNPs resulted in single amino acid substitutions. Out of these, three SNPs (Arg61Cys, Cys88Arg, and Gly401Ser) affected *in vitro* organic cation transport. Uptake of MPP^+ by Arg61Cys variant, the most frequent mutant (16%), was dramatically decreased (50%) compared to wild-type hOCT1. A follow-up study by Shu *et al.* (2003) screened 15 protein-altering variants of hOCT1 in *Xenopus oocytes*. It was demonstrated that predicted mutations occurring in the evolutionarily conserved regions of the gene had a more significant impact on the hOCT1 function. In a separate investigation on a heterogeneous collection of 247 patients, 28 hOCT2 variants were identified with eight (four occurring at a frequency $\geq$1% in ethnic populations) causing a nonsynonomous amino acid change (Leabman *et al.*, 2002). Interestingly, the pharmacogenomic analysis of hOCT3 revealed no polymorphisms that resulted in an amino acid composition change (Lazar *et al.*, 2003). The net impact of these OCT isoform variants in mediating the intestinal absorption of xenobiotics has yet to be determined.

OCTs polymorphism may also contribute to the toxicity of drugs. Influence of OCT1 on a severe drug side effect, lactic acidosis, was observed in $Oct1^{-/-}$ mice treated with the biguanide analog, phenformin and metformin. Biguanides are substrates of OCTs, and can accumulate in the plasma due to reduced hepatic clearance caused by OCT transport (Wang *et al.*, 2003; Jonker and Schinkel, 2004). Biguanides exhibit their antidiabetic effects partially through the inhibition of the mitochondrial respiration, leading to reduced glucose, which subsequently leads to an accumulation of plasma lactic acid (Owen *et al.*, 2000; Wang *et al.*, 2003). Deletion of the mOCT1 isoform was demonstrated to decrease the excretion of biguanides and promote inhibition of the mitochondrial respiration, i.e., lactic acidosis (Wang *et al.*, 2003; Jonker and Schinkel, 2004). It is hypothesized that hOCT1 is largely responsible for the potentially fatal lactic acidosis, and may contribute to the removal of phenformin from the market (Wang *et al.*,

2003; Jonker and Schinkel, 2004; Koepsell, 2004). Other OCT knockout studies are well reviewed by Koepsell (2004) and Jonker and Schinkel (2004).

Defect mutations in hOCTN1 and hOCTN2 have also been demonstrated (Koepsell and Endou, 2004). Kawasaki *et al.* (2004) identified two SNPs in hOCTN1 of the Japanese population. One of them almost completely abrogated the TEA transport activity in stable transfected HEK293 cells. This result suggests that mutations in hOCTN1 might affect its physiological function and/or the pharmacological characteristics of its substrates. For example, decreased renal secretion of organic cations mediated by hOCTN1 might increase the nephrotoxic potential of relevant substrates.

Mutations in hOCTN2 lead to a recessive hereditary disorder called "primary systemic carnitine deficiency (SCD)" (Nezu *et al.*, 1999; Lahjouji *et al.*, 2001). This potentially lethal disease is characterized by progressive infantile-onset cardiomyopathy, skeletal myopathy, hypoketotic hypoglycemic encephalopathy, and extremely low plasma and tissue carnitine concentrations (Tein *et al.*, 1990; Stanley *et al.*, 1991; Pons *et al.*, 1997). Nonsense or missense mutations in hOCTN2 have been demonstrated to cause low or nonfunctional carnitine transporters. Therefore, defect of carnitine reabsorption in kidney and carnitine uptake in cardiac muscles and other organs leads to systemic carnitine depletion, which lead to the inhibition of β-oxidation of fatty acids (Tein, 2003). Patients with SCD are treated by oral administration of carnitine. Due to the potential drug–drug interaction, mediation with carnitine drugs interacting with hOCTN2 in patients with partial defects of hOCTN2 is suggested to be avoided or supplemented with higher dose of carnitine (Koepsell and Endou, 2004).

Mutations in hOCTNs might increase susceptibility to Crohn's disease. Peltekova *et al.* (2004) first identified that a missense mutation in hOCTN1 gene and a transversion in the promoter region of hOCTN2, which form haplotype related with an increase in the prevalence of Crohn's disease. The resulting amino acid change in hOCTN1 reduced its affinity and capacity for transporting carnitine. It, however, enhanced the affinity and uptake of other xenobiotics (Peltekova *et al.*, 2004). The hOCTN2 promoter mutation occurred in the heat shock transcription factor binding element region and altered transcription factor binding affinity (Peltekova *et al.*, 2004). Therefore, these variants alter transcription and transporter functions of the OCTNs and interact with variants in another gene associated with Crohn's disease, CARD15, to increase risk of Crohn's disease. The increased risk observed in this patient population was later confirmed by Torok *et al.* (2005).

7.3.4 Nucleoside Transporters (CNT, SLC28A; ENT, SLC29A)

Nucleosides are the ribosylated precursors of purine and pyrimidine nucleotides, and in addition to their biological importance with respect to cellular energy and signal transduction in the form of their phosphorylated analogs (e.g., ATP and cAMP, respectively), their importance to cellular function and physiology is profound. For instance, adenosine alone has been shown to exhibit cardiac and vascular effects (Ely and Berne, 1992), act as a neuromodulator (Dunwiddie, 1985;

Phillis and Wu, 1981), inhibit lipolysis in fat cells (Fain and Malbon, 1979), and act as an anti-inflammatory (Cronstein, 1994; Griffiths and Jarvis, 1996). In accordance with their overall physiological importance, the cellular transport of nucleosides is mediated by two distinct families of transporter proteins: the high-affinity, concentrative nucleoside transporters (CNT; SLC28) and the low affinity, equilibrative nucleoside transporters (ENT; SLC29). These transporter families are distinguished not only by their structural features, but also by their different transport mechanisms. In short, the SLC28 family is sodium dependent and works through an active symport mechanism, while the SLC29 family functions by a facilitated diffusion mechanism. On the cellular level, it is thought that ENTs and CNTs asymmetrically localize between the apical and basolateral membranes to mediate the vectorial transepithelial flux of nucleosides (Lai *et al.*, 2002; Kong *et al.*, 2004). However, given the broad and overlapping substrate specificities of each individual transporter family, coupled with their differential regulation, it has been surmised that these transporters may not only be key modulators of intracellular nucleoside availability, but may also fulfill various metabolic needs through selective and complementary activation (Pastor-Anglada *et al.*, 2001). Regardless, given the widespread tissue localization, in concert with those items mentioned above, nucleoside transporters should be attractive pharmaceutical targets for therapeutic nucleoside analogs.

7.3.4.1 The Molecular and Structural Characteristics of Nucleoside Transporters

CNT (SLC28A)

The human SLC28 family consists of three subtypes of sodium-dependent, concentrative nucleoside transporters, hCNT1 (SLC28A1), hCNT2 (SLC28A2; also termed SPNT for sodium-dependent purine nucleoside transporter), and hCNT3 (SLC28A3), although five functionally distinct sodium dependent concentrative nucleoside activities have been reported, as N1 (system *cif*), N2 (sytem *cit*), N3 (sytem *cib*), N4, and N5 (Griffiths and Jarvis, 1996). In summary, N1 transport has been shown to be selective for purine nucleosides, as well as the nucleobase, uridine; N2 transport is characterized by pyrimidine selectivity and adenosine; N3 transport exhibits broad substrate specificity, transporting both purine and pyrimidines alike; N4 transport is characterized by pyrimidine selectivity, guanosine and adenosine; and N5 is characterized by transport of formycin B and cladribine (Griffiths and Jarvis, 1996).

The current topological model for the SLC28 family is based on rCNT1, which is hypothesized to be comprised of 13 putative TMD, although a 15 transmembrane domain model cannot be ruled out (Hamilton *et al.*, 2001). SLC28A1 is chromosomally localized to 15q25–26 and encodes a protein, CNT1, of approximately 649 amino acids in length that exhibits approximately 83% identity with its rat homologue (Ritzel *et al.*, 1997). SLC28A2 is chromosomally localized to 15q13–14, and encodes a protein of approximately 658 amino acids with a molecular weight of approximately 72 kDa and exhibits several putative protein kinase C phosphorylation sites (Wang *et al.*, 1997). hCNT3 is approximately 48 and 47%

identical to hCNT1 and hCNT2, respectively (Ritzel *et al.*, 2001). The SLC28A3 gene is localized to 9q22.2 and encodes a protein that is approximately 691 amino acids (Ritzel *et al.*, 2001).

ENT (SLC29A)

The human SLC29 transporter family contains four members, hENT1 (SLC29A1), hENT2 (SLC29A2), hENT3 (SLC29A3), and hENT4 (SLC29A4) that are primarily distinguished from the concentrative nucleoside transporters by their facilitated diffusion transport mechanism. Early research distinguished SLC29 family members by their sensitivity to inhibition by nitrobenzylthioinosine (NBMPR), as either *es*, equilibrative sensitive, or *ei*, equilibrative insensitive; however, the usefulness of this terminology for identification has been superceded by the cloning and subsequent characterization of both hENT1 and hENT2, which are responsible for these transport mechanisms (Griffiths and Jarvis, 1996; Kong *et al.*, 2004).

Structurally, SLC29 family members are characterized by 11 putative TMD, with an intracellular amine terminus and extracellular carboxy terminus. The gene encoding hENT1 is localized to chromosome 6p21.1–2 (Coe *et al.*, 1997), and encodes a protein that is 456 amino acids long (Griffiths *et al.*, 1997a). Both the rat and mouse homologues have also been cloned and exhibit approximately 78% identity with hENT1 (Yao *et al.*, 1997; Kiss *et al.*, 2000). Similarly, the genes encoding hENT2 and hENT3 have been identified and are chromosomally localized to 11q13 and 10q22.1, respectively (Griffiths *et al.*, 1997b; Williams *et al.*, 1997; Hyde *et al.*, 2001; Clark *et al.*, 2003). Additionally, hENT2 has been shown to be 456 amino acids long, sharing approximately 46% identity with hENT1 (Griffiths *et al.*, 1997b), while hENT3 is 475 amino acids long and exhibits approximately 30–33% identity with mouse, rat, and human ENT1 and ENT2 isoforms (Hyde *et al.*, 2001). Although it has been confirmed as a nucleoside transporter capable of low affinity adenosine transport, hENT4 is a 530 amino acid protein that shows only 18% identity to hENT1 (Acimovic and Coe, 2002; Baldwin *et al.*, 2004). The gene encoding hENT4 is localized to chromosome 7p22.1 (Acimovic and Coe, 2002; Strausberg *et al.*, 2002; Baldwin *et al.*, 2004). Interestingly, hENT4 has recently been shown to be a low affinity monoamine transporter and renamed plasma membrane monoamine transporter, or PMAT (Engel *et al.*, 2004). As such, ENT4 will not be discussed further.

Both hENT1 and hENT2 exhibit glycosylation sites in the extracellular loops between TMD one and two, on Asn residue 48 (Yao *et al.*, 1997; Crawford *et al.*, 1998; Ward *et al.*, 2003). In the case of hENT1, glycosylation is not required for transport activity, but may affect the binding affinity to transport inhibitors, such as NBMPR (Vickers *et al.*, 1999; Ward *et al.*, 2003). HENT2 also contains an additional glycosylation site on Asn57, which most likely functions to target the protein to the plasma membrane (Ward *et al.*, 2003). Such structural assessments have not yet been conducted for hENT3, although it does differ from both hENT1 and hENT2 in possessing a long, hydrophilic amine terminus region preceding transmembrane domain one that contains dileucine motifs, which are responsible for its intracellular localization (Baldwin *et al.*, 2005).

7.3.4.2 The Substrate Specificities of Nucleoside Transporters

CNT (SLC28A)

Vijayalakshmi and Belt (1988) first showed the differing substrate specificities of CNT1 and CNT2 by their observations in mouse intestinal epithelium of two classes of sodium dependent concentrative nucleoside transport, which was dependent on purine/pyrimidine species. It is now known that CNT1 transports primarily pyrimidine nucleosides and the purine adenosine (N2 transport), while CNT2 transports primarily purine nucleosides and uridine (N1 transport) (Huang *et al.*, 1994; Ritzel *et al.*, 1997; Wang *et al.*, 1997). In contrast, CNT3 has been shown to be more broadly selective, transporting both purine and pyrimidine nucleosides (N3 transport) (Wu *et al.*, 1992). Interestingly, while CNT1 is specific for pyrimidines, it does exhibit specificity for adenosine in a high-affinity, low-capacity manner (Ritzel *et al.*, 2001). Furthermore, while both CNT1 and CNT2 employ a 1:1 Na^+:nucleoside coupling ratio, CNT3 requires a 2:1 ratio (Plagemann and Aran, 1990; Ritzel *et al.*, 2001).

Given their respective substrate specificities, SLC28 family members exhibit transport activity for a wide range of pharmaceutically relevant compounds. CNT1 has been shown to exhibit high transport affinity for the antiviral nucleoside analogs zidovudine, lamivudine, and zalcitabine, and the cytotoxic cytidine analogs cytarabine and gemcitabine used for treatment of a wide spectrum of tumors, while CNT2 has been shown to transport the antiviral compounds didanosine (ddI) and ribavirin (Gray *et al.*, 2004). With its broader substrate specificity, CNT3 transports a number of anticancer nucleoside analogs including cladrabine, gemcitabine, 5-fluorouridine, fludarabine, and zebularine (Ritzel *et al.*, 2001).

Substrate and cation recognition sites for CNT transporters are both located extracellularly on the carboxy half of the proteins, on TMD 7, 8, and 9 (Wang and Giacomini, 1999; Loewen *et al.*, 1999). Changing serine 319 of CNT1 to glycine has been shown to alter the substrate selectivity to include purines, while the adjacent glutamine residue was shown to be important in modulating the apparent affinity for nucleosides (Wang and Giacomini, 1999). Moreover, changing serine 353 to threonine changed CNT1 into a transporter that was highly selective for uridine (Loewen *et al.*, 1999). Recently, Lai *et al.* (2005) have demonstrated that G476 is important for correct membrane targeting, folding, and/or intracellular processing of hCNT1 and that F316H mutation confers guanine sensitivity. These researchers speculated that the naturally occurring F316H mutation in hCNT1 is responsible for one of the two CNT activities for which a transporter has not been identified (N4) (Griffiths and Jarvis, 1996; Lai *et al.*, 2005). Although such selectivity studies have not been performed for CNT2 per se, Chang *et al.* (2004) did explore the structural requirements necessary for purine and pyrimidine transport by hCNT1, hCNT2, and hENT1. Their computer modeling studies, which explore the relationships between hydroxylation position, substrate selectivity, and transporter inhibition, could prove useful for rational drug design of future nucleoside analogs for both cancer and antiviral treatment.

ENT (SLC29A)

Kinetic studies assessing the transport affinities of both hENT1 and hENT2, the first SLC29 family members identified, have been quite extensive. Earlier research identified hENT1 as the *es* transport system, while hENT2 was identified as the *ei* system, where the *es* system bound NMBPR with high affinity ($K_d = 1–10\,nM$) and the *ei* system was not affected at nanomolar concentrations, but was affected at higher concentrations (>10 μM) (Kong *et al.*, 2004). In fact, since ENT transport is mediated *via* a bidirectional facilitated diffusion mechanism, which is dependant upon substrate concentration gradient, the difference in binding NMBPR was used quite successfully for the early biochemical characterization of these two transporters.

The substrate affinities for endogenous nucleosides and some therapeutic nucleoside analogs have been described for hENT1 and hENT2 and recently for hENT3. Both hENT1 and hENT2 have been shown to be broadly selective for both purine and pyrimidine nucleosides; however, hENT2 exhibits 7.7- and 19.3-fold lower affinities for guanosine and cytidine, respectively, but a fourfold higher affinity for inosine (Kong *et al.*, 2004). Further, hENT2 transports nucleobases, whereas hENT1 does not (Yao *et al.*, 2002). In terms of therapeutic nucleoside analogs, hENT1 has been shown to interact with a number of compounds widely used in the treatment of cancer, such as cladribine, gemcitabine, fludarabine, cytarabine, tiazofurin, and benzamide riboside (Griffiths *et al.*, 1997a; Mackey *et al.*, 1998; 1999; Vickers *et al.*, 2002; Damaraju *et al.*, 2005). However, hENT1 only poorly transports the antivirals 2′, 3′-dideoxycytidine (ddC) and 2′, 3′-dideoxyinosine (ddI) and does not transport 3′-azido-3′-deoxythymidine (AZT), suggesting an important role of the 3′-hydroxyl group on the ribose moiety for substrate recognition (Yao *et al.*, 2001a). Moreover, Vickers *et al.* (2002) recently demonstrated the importance of the 3′-hydroxyl group with respect to uridine analogs for both hENT1 and hENT2. Nonetheless, hENT2 demonstrated a much broader substrate selectivity, transporting nucleobases (which lack the ribose moiety), a number of the above compounds, as well as ddC, ddI, and AZT (Yao *et al.*, 2001a; Vickers *et al.*, 2002). However the selectivity of hENT2 for cytidine and its analogs is comparatively lower than that of hENT1, suggesting a lack of tolerance for a 4′-amino moiety on the base (Vickers *et al.*, 2002). Both transporters tolerated halogen substitution at the 5′ position on the base, as well as the 2′ and 5′ positions on the ribose moiety of uridine, suggesting these positions are not essential for uridine-like substrate recognition (Vickers *et al.*, 2002).

Interestingly, recent evidence indicates the antidiabetic compound troglitazone, but not the related thiazolidinediones pioglitazone and ciglitazone, inhibits hENT1 transport of adenosine and uridine *via* a competitive inhibition mechanism (Leung *et al.*, 2005b). Troglitazone had minimal inhibitory effect on hENT2 in this study (Leung *et al.*, 2005b). Interestingly, inhibition of ENT1 in vascular smooth muscle is suggested to increase extracellular adenosine, which causes vasodilation and inhibits vascular smooth muscle cell proliferation (Rubin *et al.*, 2000; Kim *et al.*, 2004; Masaki *et al.*, 2004). In concert, these data suggest a novel therapeutic approach to modulate extracellular adenosine concentrations. To our knowledge, the transport characteristics of other diabetic drugs *via* nucleoside transporters have not been studied.

Similar to both hENT1 and hENT2, hENT3 has demonstrated broad selectivity for nucleosides (Baldwin *et al.*, 2005). Additionally, hENT3 has also been shown to transport the nucleobase adenine; however, in contrast to hENT2 it does not transport hypoxanthine (Baldwin *et al.*, 2005). Interestingly, the transport activity of hENT3 was relatively insensitive to many of the classical ENT inhibitors, such as NBMPR, dipyridamole, and dilazep although transport was found to be highly pH dependent, indicative of its intracellular localization (Baldwin *et al.*, 2005). HENT3 has also demonstrated good transport efficiency for a wide range of pharmaceutical purine and pyrimidine analogs, including cladribine, cordycepin, tubercidin, zebularine, 5-fluoro-2′-deoxyuridine, ddC, ddI, and AZT (Baldwin *et al.*, 2005). Other compounds, such as gemcitabine and gancyclovir were also substrates for hENT3, although with a much lower transport efficiency (Baldwin *et al.*, 2005). Especially intriguing is the high transport efficiency of AZT, especially compared with hENT1, which does not transport AZT, and hENT2, which has been shown to transport AZT, but at one third its capacity of uridine (Yao *et al.*, 2001a). Comparatively, hENT3 exhibits AZT selectivity comparable to adenosine and inosine, which are higher than that for thymidine and uridine, indicating that the 3′-hydroxy group is not important for substrate recognition (Baldwin *et al.*, 2005). Given the high prevalence of 3′-hydroxyl moieties on many antiviral compounds, it is unfortunate that studies exploring structure–function relationships for hENT3 have not been performed to date. Such studies comparative to either hENT1, or especially hENT2 could prove critical to elucidate the mechanism by which these compounds enter the cell to elicit their cytotoxic effects.

Although limited, some literature does explore the structure–function relationships of both hENT1 and hENT2. Using chimeric constructs, researchers have shown the primary sites of substrate binding for hENT1 reside in TMD three through six, while TMD one to two confer a secondary contribution (Sundaram *et al.*, 2001). Molecularly, studies have shown that a single substitution of M33I was sufficient to alter sensitivity to a number of substrates (Visser *et al.*, 2002), while Gly179 is important to NBMPR binding (SenGupta *et al.*, 2002). With respect to ENT2, chimeric studies using the rat homologues of ENT1 and ENT2 have shown that TMD five and six form at least part of the translocation pathway (Yao *et al.*, 2002). Modification of the corresponding Cys140 (transmembrane domain four) inhibits transport, suggesting that this residue lies within, or is adjacent to the substrate translocation pathway (Yao *et al.*, 2001b). However, additional studies are required to fully characterize and elucidate the structure–function relationships of each of these transporters.

7.3.4.3 The Expression of Nucleoside Transporters

Tissue expression of CNT1 has been shown in epithelial tissues including small intestine, kidney, and liver. In contrast, CNT2 has been shown to be widely distributed in kidney, liver, heart, brain, placenta, pancreas, skeletal muscle, colon, rectum, and throughout the small intestine, while CNT3 has been shown in pancreas, trachea, bone marrow, and mammary gland, with lower levels in intestine, lung, placenta, prostrate, testis, and liver (Gray *et al.*, 2004). Interestingly, only CNT1 and CNT2 were widely expressed on the brush-border membranes of enterocytes along the length of both the fetal and adult small intestines, as determined *via* nucleoside uptake by intestinal brush-border membrane vesicles (Ngo *et al.*, 2001). Using double transfected MDCK cells, which simultaneously expressed YFP-tagged CNT1 and ENT1, Lai *et al.* (2002) confirmed the apical cellular localization of CNT1. While the rat homologue of CNT2 has been shown to be apically localized in polarized MDCK and LLC-PK cells, in liver parenchymal cells it is highly expressed in the basolateral (sinusoidal) membrane, suggesting tissue-specific sorting (Mangravite *et al.*, 2001; Duflot *et al.*, 2002). Duflot *et al.* (2002) also demonstrated the apical localization of CNT1 in liver parenchymal cells, suggesting transcytotic membrane insertion. The subcellular localization of CNT3 has not been reported as of preparation of this chapter.

Tissue distribution studies of hENT1, hENT2, and hENT3 suggest these transporters are ubiquitously expressed, at least on the mRNA level, although there does exist appreciable intertissue and interindividual expression differences (Griffiths and Jarvis, 1996; Crawford *et al.*, 1998; Pennycooke *et al.*, 2001; Jennings *et al.*, 2001; Baldwin *et al.*, 2005). On the cellular level, it is thought that ENTs and CNTs asymmetrically distribute between the apical and basolateral membranes to mediate the vectorial transepithelial flux of nucleosides (Lai *et al.*, 2002; Kong *et al.*, 2004). However, no studies directly examine ENT1, or ENT2 protein localization in intestinal epithelial tissues. Nevertheless, hENT1 has been shown to be predominantly basolaterally localized in YFP-tagged ENT1 transfected MDCK cells, with some apical localization (Lai *et al.*, 2002; Mangravite *et al.*, 2003), whereas hENT2 exhibits only basolateral localization (Mangravite *et al.*, 2003). Interestingly, some functional data indicate that ENT1 may also exhibit some intracellular localization (Pisoni and Thoene, 1989; Mani *et al.*, 1998; Jimenez *et al.*, 2000), although further studies are required to determine the functional relevance of intracellular nucleoside transport due to ENT1 (Kong *et al.*, 2004). HENT3 is now known to be an intracellular transporter that colocalized with lysosomal markers, but not with golgi, early endosomes, or endoplasmic reticulum (Baldwin *et al.*, 2005). As expected, truncation of the N-terminus or mutation of its dileucine motif caused relocation of the transporter to the cell surface (Baldwin *et al.*, 2005). Given its lysosomal localization, it should be of no surprise that its transport kinetics were optimum at pH 5.5 (Baldwin *et al.*, 2005).

7.3.4.4 The Regulation of Nucleoside Transporters

Due to the inherent importance of nucleosides to overall cellular and tissue function, it is of no surprise that the regulation mechanisms of both CNTs and ENTs are quite complex and difficult to effectively study. In short, expression of both CNT1 and CNT2 has been shown to be tissue specific, dependent on cell cycle, certain hormones and cytokines, as well as the presence of substrate (Gomez-Angelats *et al.*, 1996; Del Santo *et al.*, 1998; Soler *et al.*, 1998, 2001; Valdes *et al.*, 2000, 2002). However, with the exception of nutritional regulation studies (Valdes *et al.*, 2000), virtually all of the regulatory studies have been conducted in hepatocytes, or immune cells. Given tissue-specific regulation dependence, one should question the applicability of these data to the intestinal model.

In short, not much is known concerning either the transcriptional, or translational regulation of either the SLC28, or SLC29 families. Regulation of hENT1 has been shown to be dependent on deoxynucleotide levels in cultured cancer cells (Pressacco *et al.*, 1995). Moreover, hENT1 expression has also been demonstrated to be a function of cell cycle, with expression doubling between G_1 and G_2-M phases (Pressacco *et al.*, 1995). Cultured cells also exhibited an upregulation of hENT1 due to phorbol ester treatment, which appears to be due to either PKC δ, or PKC ε; however, this remains to be elucidated (Coe *et al.*, 2002). Interestingly, glucose treatment has been shown to upregulate ENT1 protein activity, as well as its protein and mRNA expression in human aortic smooth muscle cells (Leung *et al.*, 2005a). The functional significance of this finding is still uknown; however, it has been shown that ENT1 expression is modulated due to diabetes (Pawelczyk *et al.*, 2003), and that it is not transcriptionally controlled *via* peroxisome proliferator-activated receptor gamma (PPARγ) (Leung *et al.*, 2005b). Still less is known concerning the mechanisms regulating the expression of hENT2 and hENT3.

7.3.5 Monocarboxylate Transporters (MCT; SLC16A)

Many endogenous and exogenous short chain anionic compounds, such as lactic acid, pyruvate, acetoacetate, β-hydroxybutyrate, acetate, propionate, and butyrate, are all good substrates for a family of proton-coupled transporter proteins, termed the monocarboxylate transporter family (MCT; SLC16) (Price *et al.*, 1998). Although investigation of MCT activity was studied biochemically in earlier literature, it was not until the first identified member of the MCT family, MCT1, was cloned in CHO cells (Kim *et al.*, 1992), and later functionally expressed in a breast tumor cell line (Garcia *et al.*, 1994), that the molecular characterization of this transporter family began (Enerson and Drewes, 2003). To date 14 members of the MCT family have been identified, each having unique tissue distribution (Halestrap and Price, 1999; Halestrap and Meredith, 2004). Given the relative size of this transporter family, this assessment will comprise only those MCT members relevant to the intestinal transport of pharmaceutical or therapeutic compounds. Additionally, given that the molecular and functional characterizations of members in this transporter family are still in their infancy, one is expected to consult the literature for a more comprehensive review (Price *et al.*, 1998; Enerson and Drewes, 2003; Halestrap and Meredith, 2004). Finally, given the convoluted nomenclature for this transporter family, representative transporters will be identified by their MCT name, as opposed to SLC or another designation.

7.3.5.1 Molecular and Structural Characteristics of Monocarboxylate Transporters

Kyte–Doolittle hydropathy plots indicate that MCT isoforms exhibit 12 putative TMD, with both the amine and carboxy termini located intracellularly (Price *et al.*, 1998; Halestrap and Price, 1999). This has been confirmed experimentally for MCT1 in rat erythrocytes (Poole *et al.*, 1996). Transporter topology varies with respect to the length of the carboxy-terminus after transmembrane domain 12 and a large intracellular loop located between TMD six and seven (Price *et al.*, 1998); however, these structures are consistent among the MCT isoforms. MCT family members also exhibit two highly conserved motifs: **[D/E]G[G/S][W/F][G/A]W** which traverses into transmembrane domain one and **YfXK[R/K][R/L]**XLAX**[G/A]**XAXA**G** leading into transmembrane domain five, where residues shown in bold are conserved in all of the sequences, residues in square brackets indicate alternative amino acids, residues that are in normal type are the consensus amino acid at that position, and "X" represents any amino acid (Halestrap and Price, 1999). Site directed mutagenesis studies have been conducted for a number of residues, which have resulted in various changes in substrate specificities; however, most interesting is that mutagenesis of the highly conserved Arg313 in MCT1 results in reduction in affinity for lactate (Rahman *et al.*, 1999). It has been surmised that this positively charged Arg binds the carboxy anion of monocarboxylates, much like lactate dehydrogenase, allowing for transport (Poole and Halestrap, 1993; Carpenter and Halestrap, 1994; Halestrap and Price, 1999).

Seemingly another defining factor of MCT family members is the presence of ancillary proteins that are required for proper intracellular trafficking, cell surface expression, and/or function. Studies indicate MCT1, MCT3, and MCT4 require the presence of CD147 [also known as OX-47, extracellular matrix metalloproteinase inducer (EMM-PRIN), HT7 or basigin], or the related protein GP70 (Embigen), both widely distributed cell surface glycoproteins, for proper cell surface expression and function (Philp *et al.*, 2003). These glycoproteins exhibit a single transmembrane domain, two immunoglobulin-like domains in the extracellular region, and a short carboxy terminus cytoplasmic tail (Schuster *et al.*, 1996). In contrast, MCT2 does not interact with CD147 specifically; however, it does appear to require an additional protein for proper expression at the cell surface (Kirk *et al.*, 2000). The associations of MCT1 and MCT4 with CD147 have been confirmed by carboxy terminus tagging with fluorescent proteins (Kirk *et al.*, 2000; Zhao *et al.*, 2001). Lack of CD147 coexpression with MCT1 and MCT4 results in accumulation in the endoplasmic reticulum, or golgi apparatus (Kirk *et al.*, 2000). Studies exploring the coexpression of CD147 or GP70 with other MCT isoforms have not been conducted.

7.3.5.2 The Substrate Specificity of Monocarboxylate Transporters

Functional evidence for the proton-dependent symport of monocarboxylates *via* various MCT isoforms is well established (Enerson and Drewes, 2003).

Comparatively, MCT1, MCT2, and MCT4 exhibit higher affinities for pyruvate than lactate, although MCT4 exhibits low affinity for both substrates (Enerson and Drewes, 2003). The difference in affinities from pyruvate to lactate indicates a preference for 2-oxoacids over 2-hydroxy acids (Enerson and Drewes, 2003). Interestingly, the affinity for MCT2 for pyruvate is extremely high compared to MCT1 (Lin *et al.*, 1998). Other monocarboxylates known to be substrates for MCTs include butyrate, acetate, propionate, etc. With respect to therapeutic compounds, it has been surmised that MCT isoforms may play a role in the intestinal absorption of some β-lactam antibiotics (Li *et al.*, 1999), penicillins (Itoh *et al.*, 1998), nonsteroidal anti-inflammatory drugs (Emoto *et al.*, 2002), valproic acid (Utoguchi and Audus, 2000; Hosoya *et al.*, 2001), atorvastatin (Wu *et al.*, 2000c), and nateglinide (Okamura *et al.*, 2002).

One isoform in the MCT family, MCT10 (TAT1; SLC16A10), is an aromatic amino acid transporter that exhibits proton independent transport of substrates, although it does exhibit 30% identity to other MCTs (Kim *et al.*, 2001a). The high sequence conservation (49%) between MCT10 and MCT8 suggests MCT8 may not function as a monocarboxylate transporter either (Friesema *et al.*, 2005). Indeed, when expressed in *Xenopus* oocytes, MCT8 transports thyroid hormones T3 and T4 in a sodium and proton independent manner (Enerson and Drewes, 2003). Neither lactate nor aromatic amino acids appear to be substrates for MCT8 (Enerson and Drewes, 2003). Although the substrate affinities of MCT5–8 have not yet been identified, their similarity to MCT10 may also suggest an alternate substrate affinity, differing from MCT1–4 (Enerson and Drewes, 2003).

7.3.5.3 The Expression of Monocarboxylate Transporters

Although each MCT isoform exhibits its own unique tissue distribution, of the 14 different MCTs currently identified, MCTs 1, 3–7 were shown to be expressed in the GI tract by Northern blotting (Price *et al.*, 1998). However, at the time of that publication only seven MCT isoforms had been established and the intestinal expression of the remaining MCT isoforms has not yet been investigated. These northern blotting experiments also indicate that MCT2 is not intestinally localized, which has been supported by a recent publication exploring the intestinal distribution of a number of MCT isoforms (Gill *et al.*, 2005). This literature suggests that MCT1, 4, and 5 are the predominant intestinally expressed isoforms, although MCT3 does exhibit some minimal expression (Gill *et al.*, 2005). This study further suggests that MCT1 expression increases along the length of the human intestine, with a predominant expression in the distal colon, followed by proximal colon, ileum, and jejunum (Gill *et al.*, 2005). Results were similar for MCT4 and MCT5, where expression increased moving down the GI tract; however, neither MCT4 nor MCT5 exhibited expression in human ileum (Gill *et al.*, 2005). In contrast to previous data suggesting MCT3 is only expressed in the retinal epithelium (Yoon *et al.*, 1997; Philp *et al.*, 1998) when antibody specific to MCT3 was used at a very low dilution of 1:50, expression was observed in human GI tract. However, in contrast to MCT1, MCT4, and MCT5, expression was higher in ileum as compared to colonic regions (Gill *et al.*, 2005). Expression of MCT6 was not observed in the human GI tract (Gill *et al.*, 2005). Gill *et al.* (2005) also showed differences in the preferential membrane localization of these transporters, suggesting basolateral localization of MCT3, MCT4, and MCT5, while MCT1 is suggested as being apically localized (Gill *et al.*, 2005). The basolateral expression of MCT10 has also been shown in human intestinal epithelial cells (Kim *et al.*, 2001a,b).

7.3.5.4 The Regulation of Monocarboxylate Transporters

With the exception of some limited studies conducted using MCT1, regulation mechanisms of the various intestinally expressed MCT isoforms has not been reported. Previous studies have shown that MCT1 is upregulated in response to excess butyrate substrate, which is reflected functionally as an increase in butyrate transport in cultured colonic epithelial cells (AA/C1) (Cuff *et al.*, 2002). Moreover, it has been reported that MCT1 expression is decreased markedly during colon carcinogenesis, indicating a role of MCT1 in cancer prevention *via* active butyrate influx (Lambert *et al.*, 2002). It has been proposed that the decrease in MCT1 expression during colon carcinogenesis may decrease the intracellular availability of butyrate required to regulate expression of genes associated with the processes maintaining tissue homeostasis within the colonic mucosa (Cuff *et al.*, 2005). In fact, studies suggest that the ability of butyrate to induce cell-cycle arrest and differentiation is dependent on the abundance and functionality of MCT1 (Cuff *et al.*, 2005). Furthermore, downregulation of MCT1 using RNAi in various colonic cell lines resulted in consistent inhibition of butyrate influx thus inhibiting its ability to modulate various indicators of carcinogenesis, namely IAP, a marker of differentiation, p21, a cell cycle inhibitor, and CD1, a positive regulator of cell cycle progression (Cuff *et al.*, 2005). This study also concluded that MCT1 inhibition did not affect those genes associated with apoptosis, bcl-x_L and bak (Cuff *et al.*, 2005).

Whereas it is known that alteration in physiological state changes MCT expression, the underlying cellular and molecular mechanisms are poorly understood (Enerson and Drewes, 2003). However, studies of MCT regulation in other tissue do suggest an upregulation of MCT expression under hypoxia, as well as in the presence of exogenous vascular endothelial growth factor, which is known to be mediated by the hypoxia inducible factor HIF-1 (Enerson and Drewes, 2003). Moreover, mRNA expression of MCT1 in cultured macrophages is upregulated by exposure to lipopolysaccharides, tumor necrosis factor-α, and nitric oxide; however, only lipopolysaccharide and tumor necrosis factor-α treatment related to an increase in MCT1 protein (Hahn *et al.*, 2000). These results suggest the presence of multiple signaling pathways that may converge to regulate MCT1 expression (Enerson and Drewes, 2003). Additionally, Leino *et al.* (2001) recently demonstrated upregulation of MCT1 and GLUT1 in rat brain after diet-induced ketosis. Results indicated an eightfold increase in MCT1 expression in brain endothelial cells after 4 weeks under ketonemic conditions (Leino *et al.*, 2001). Moreover, an increase in MCT1 levels were shown throughout the rat brain, especially in the cerebellum, indicating protein upregulation in response to dietary and possibly pathological stresses (Leino *et al.*, 2001).

Although the transport of monocarboxylates by MCTs is important with respect to several disease pathologies, such as cancer and ischemic stroke, given their widespread tissue expression (heart, brain, intestine, liver, etc.) and their substrate's impact on cellular energy and overall tissue and organism function, it is not expected that pharmaceutical intervention *via* MCTs is likely in the near future. This is not particularly surprising given the incomplete functional characterization and even identification of these transporters. However, the relative importance of MCTs to proper physiological function necessitates further research into the pathophysiology associated with this transporter family, especially considering the recent finding of the role of MCT8 dysfunction in causing severe X-linked psychomotor retardation, termed Allan–Herndon–Dudley syndrome (Dumitrescu *et al.*, 2004; Friesema *et al.*, 2004).

7.4 Impact of Intestinal Transporters on Bioavailability

The importance of membrane transporter proteins has been well established with respect to the bioavailability of many therapeutic compounds. Notwithstanding their physiologic importance (e.g., fulfillment of nutrient requirements and waste trafficking), membrane transporter proteins can be particularly important due to their substrate specificity and kinetics and provide attractive therapeutic targets for pharmaceutical intervention. However, tissue and temporal expression patterns, as well as cellular localization, are all extremely important in determining the relevance of a particular transporter with respect to the contribution of transporter activity on the bioavailability of exogenous substrates. The cellular permeation of a compound is a function of a multitude of parameters and virtually unpredictable without adequate study. Moreover, given the obvious physiological relevance of many of the endogenous transporter protein substrates, it is also essential to explore the subtle interplay between different transporter proteins to elucidate the underlying mechanisms of transitional trafficking veiled by the net transcellular flux. In short, the interplay of various transporter protein mechanisms along with transport by parallel pathways can significantly impact on the overall bioavailability of a compound.

P-gp, possibly the most studied transporter protein, is expressed on the apical surface of intestinal epithelial mucosa where its primary function is to promote the removal of toxic xenobiotics, a side-effect of which is the active efflux of various pharmaceutical compounds. The role of P-gp in restricting intestinal absorption was clearly determined by observing two- to fivefold greater plasma concentrations of the HIV-1 protease inhibitors indinavir, nelfinavir, and saquinavir in mdr1a(−/−) mice, as compared to wild-type mice (Kim *et al.*, 1998). Additionally, Chiou *et al.* (2000) have also reported the absolute bioavailability of tacrolimus increased from 22% in normal mice to 72% in P-gp knockout mice.

In lieu of functional knockout studies, a number of SNP have also been identified in the human *MDR1* gene. A number of these SNPs have been associated with changes in P-gp expression, thereby affecting the pharmacokinetic profiles of a number of substrates (Hoffmeyer *et al.*, 2000; Kim *et al.*, 2001; Kurata *et al.*, 2002). For example, Kurata *et al.* (2002) demonstrated the role of human *MDR1* gene polymorphism on the bioavailability of digoxin, a known P-gp substrate. Previously, Hoffmeyer *et al.* (2000) showed that homozygosity for a polymorphism in exon 26 (C3435T) of the *MDR1* gene was observed in 24% of a study sample population ($n = 188$) with these patients exhibiting significantly lower duodenal MDR1 expression as well as significantly increased plasma levels of digoxin. Furthermore, differences in the expression of P-gp, sP-gp, MDR3, MRP's (1–5), and lung resistance-associated protein (LRP) in the Caucasian and Chinese intestines have also been reported, suggesting variations in drug bioavailability due to ethnicity (Wang *et al.*, 2004). Genetic polymorphism is not limited to the *MDR1* gene, but has also been shown in both *MRP1* and *MRP2* in a Japanese population (Moriya *et al.*, 2002). Clearly, there are multiple factors that influence inter-individual variability in drug absorption and disposition, and determining the precise contribution of genetic polymorphism to this variability in oral absorption will continue to be an interesting and challenging area in the future.

While genetic variations and/or environmental factors can obviously affect bioavailability, it is the interplay of various transporter proteins and other metabolic (e.g., cytochrome P-450s) and nonmetabolic (e.g., transduction factors) proteins that are more likely to influence individual therapy. One such example is the underlying interplay between P-gp and CYP3A4, a major drug-metabolizing enzyme (Watkins, 1997). As mentioned above, P-gp is expressed on the apical surface of intestinal epithelial mucosa where its primary function is to promote the removal of toxic xenobiotics, *via* active efflux into the intestinal lumen. CYP3A4, in contrast, is intracellularly localized and is a broad affinity, metabolizing enzyme. The coexpression of these two proteins results in an increase in drug metabolism through an increase in exposure to not only CYP3A4, but also any proteases, drug metabolizing enzymes, and the natural sink condition of the intestines. Therefore, the repeated exposure to both the harsh intestinal conditions, as well as the drug metabolizing enzyme, results in a net decrease in plasma concentrations, thus limiting overall bioavailability (Fig. 7.3) (Watkins, 1997; Benet *et al.*, 2001).

In contrast to the efflux transporter proteins where substrate activity is a negative attribute of potential drug candidates, influx transporters (i.e., peptide transporters) provide a unique opportunity to specifically design and develop therapeutic agents with significantly higher molecular recognition for potentially increasing intestinal drug absorption and bioavailability. This strategy has been utilized in the development of prodrugs exhibiting high affinity for PepT1, as observed in the synthesis of L-Val-acyclovir (valacyclovir) prodrug of acyclovir (Friedrichsen *et al.*, 2002; Bhardwaj *et al.*, 2005b), although affinity for other transporters in addition to PepT1 cannot be ruled out (Phan *et al.*, 2003; Landowski *et al.*, 2003). Acyclovir has an oral bioavailability between 15 and 30% (Fletcher and Bean, 1985), while

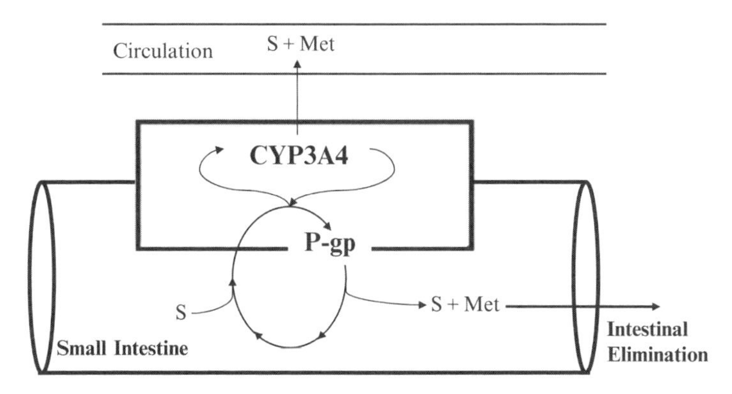

FIGURE 7.3. Interplay of P-gp and CYP3A4 affecting the absorption of substrates (S) and potential metabolites (Met) from a representative intestinal epithelium cell. Substrate enters the enterocyte through any number of mechanisms and can either enter the circulation, be metabolized by CYP3A4, or effluxed into the intestinal lumen *via* P-gp. The effluxed substrate may then be either reabsorbed into the enterocyte where the process may then be repeated, or excreted through intestinal elimination. The reabsorbed substrate may then undergo additional metabolism, be absorbed into the circulation, or be effluxed again. (Modified from Benet and Cummins (2001))

Soul-Lawton *et al.* (1995) estimated a 54.2% mean bioavailability of acyclovir in healthy volunteers after a single dose of valacyclovir. Other studies have demonstrated three to fivefold increases in bioavailability for valacyclovir as compared to acyclovir (Weller *et al.* 1993; Steingrimsdottir *et al.*, 2000). A more detailed review of the clinical advantages and pharmacokinetics of valacyclovir were elaborated by MacDougall and Guglielmo (2004). It is also important to mention one study by Kimberlin *et al.* (1998) where preliminary pharmacokinetic data for acyclovir were contrasted with valacyclovir therapy in pregnant women receiving herpes simplex virus suppressive therapy. Valacyclovir therapy resulted in an increased acyclovir bioavailability and increased acyclovir concentrations in the amniotic fluid with no evidence of a preferential accumulation in the fetus (1.7 mean maternal/umbilical vein plasma ratios for valacyclovir to 1.3 for acyclovir therapy). Other prodrugs, particularly angiotensin-converting enzyme inhibitors (e.g., captopril), have also shown to be PepT substrates, with important consequences in their pharmacokinetics (Sugawara *et al.*, 2000; Shu *et al.*, 2001).

Another strategy to increase drug bioavailability *via* peptide transporters is by stimulating PepT-like transport by increasing the proton driving force. Nozawa *et al.* (2003) used a proton-releasing polymer, Eudragit L100–55, to increase the luminal/intracellular proton gradient to stimulate PepT1 transport. Incremental increases in intestinal transport activity were seen. Nielsen *et al.* (2002), Kunta and Sinko (2004), and Steffansen *et al.* (2004, 2005) have reviewed additional examples of therapeutic applications of the PepT1 transport pathway.

Organic anion transporter proteins have also been shown to influence drug absorption of therapeutic compounds. For instance, the intestinal absorption of the

H_1-histamine receptor antagonist fexofenadine is primarily mediated by OATP-B not P-gp (Dresser *et al.*, 2002). Inhibition of OATP-B, by ingestion of grape fruit juice, decreased the fexofenadine area under the plasma concentration–time curve (AUC), the peak plasma drug concentration (C_{max}), and the urinary excretion values to 30–40% of those compared with the compound ingested with water in human. Since grape fruit juice is a more potent inhibitor of OATPs compared to P-gp, the reduced oral bioavailability of fexofenadine is suggested to result from limited influx process *via* OATP-B (Dresser *et al.*, 2002). Furthermore, the drug may be cycled between the enterocyte and the intestinal lumen, which may also result in increased metabolism and/or hydrolysis, as mentioned above. However, this model has not been studied specifically for this compound.

Further evidence suggests that OATP-B may be involved in the intestinal uptake of pravastatin, as mentioned in Sect. 7.3.2. These data suggest OATP-B not only mediates absorption of anionic compounds, but that its activity may be optimum at the acidic surface microclimate of the small intestine. In addition, infusion of pravastatin directly into the stomach resulted in maximal absorption in the duodenum and enhanced the bioavailability of pravastatin (Triscari *et al.*, 1995). This observation correlates well with the expression patterns of OATP-B in human enterocytes, where OATP-B was immunohistochemically localized at the apical membrane of intestinal epithelial cells (Kobayashi *et al.*, 2003). Therefore, determinations of the intestinal expression patterns of other important OATP/Oatp family members could have critical clinical relevance. Unfortunately, limited *in vivo* data exist to ascertain the full functional significance of this transporter family with respect to oral drug absorption.

It is clear that OCT family members function primarily in the elimination of cationic drugs and other xenobiotics in the kidney and the liver; however, their role with respect to intestinal transport is poorly understood. Studies suggest that OCTs, especially OCT1, mediate the basolateral uptake of organic cations from the blood to the intestine and may play an important role in the secretion of compounds to the small intestinal lumen. Jonker *et al.* (2001) demonstrated that the intestinal secretion of TEA was reduced about twofold after i.v. administration in Oct1(−/−) knockout mice. Wang *et al.* (2002) further reported decreased metformin distribution to the mouse small intestine due to OCT1 knockout. Additionally, the role of OCTs in the active intestinal elimination of drugs is supported in part by the finding that in the presence of cephalexin, a known OCT substrate (Karlsson *et al.*, 1993), plasma AUC of ciprofloxacin increased, while intestinal clearances decreased in rats (Dautrey *et al.*, 1999). The role of other efflux membrane transporters may also contribute to this mechanism and cannot be ruled out.

Unfortunately, targeting OCTs to increase oral bioavailability of drug substrates does not seem to be an attractive approach since these substrates are excreted across the sinusoidal membrane into the hepatocyte and further across the canalicular membrane to the bile by an as yet unidentified H^+-organic cation exchanger and/or by P-gp. In contrast, targeting OCTNs could be an efficient method to increase drug bioavailability, by taking advantage of both the transporter's tissue and cellular localizations. OCTNs are localized to the apical membranes of intestinal and renal cells. As such, the transporter can effectively increase drug bioavailability *via* the absorptive intestinal OCTNs, while simultaneously maintaining blood concentrations by renal resorption through OCTNs. In one such example, the peak concentration and AUC of sulpiride were decreased by the concomitant administration of substrates or inhibitors of OCTN1 and OCTN2 in rats (Watanabe *et al.*, 2004).

Similar to the synthesis of valacyclovir, additional strategies have been developed to exploit the activity and selectivity of specific transporters. One such example is the development of anticancer agent prodrugs, which has been developed to improve not only their physicochemical properties, but also to promote their selectivity and thus reduce undesirable toxicity effects. Floxuridine and gemcitabine are anticancer agents where studies have shown the feasibility of achieving enhanced transport and selective antiproliferative action of amino acid ester prodrugs in cell systems overexpressing PepT1 (Vig *et al.*, 2003; Landowski *et al.*, 2005; Song *et al.*, 2005). However, one should question the relevance of PepT1 transport for these particular compounds in normal physiologic systems in light of the fact that they are also potential substrates for other transporter families, such as the nucleoside transporters. However, the broad substrate specificity, tissue expression, and physiologic relevance of their substrates limit the utility of both nucleoside and monocarboxylate transporters as effective drug targets (see Sects. 7.3.4 and 7.3.5). This example illustrates one of the many problems inherent to the study of drug transport in that overlapping specificity of many of these transporter systems confounds the applicability of transport data to physiologic systems. Technology hurdles also limit the feasibility of studying multiple transporter systems simultaneously, thereby further limiting our understanding of the relevance of a particular transporter to overall transcellular flux.

Drug transporters play an important role in intestinal drug absorption and secretion, and can be major determinants of oral bioavailability. Transporters exhibit affinity for an extraordinary range of compounds and provide great insight for advancing the field of rational drug design. However, understanding the limitations associated with transporter research will aid future scientists in understanding the subtle interactions of each of the distinct transporter families, and may help elucidate their overlapping specificities as well as provide additional therapeutic targets. By understanding the substrate specificity, transport mechanism, and expression profile of transporters, efficient intestinal absorption may be made feasible by strategies including appropriately modifying either the structural recognition elements of NCEs or through rational formulation design to tailor delivery to windows of optimized drug delivery. In addition, complete understanding of the mechanisms of intestinal absorption of various drugs and the underlying regulation of drug transporters could help pharmaceutical scientists to predict the intra- and interindividual variability inherent to the study of oral bioavailability.

References

Acimovic, Y., and Coe, I.R. (2002). Molecular evolution of the equilibrative nucleoside transporter family: identification of novel family members in prokaryotes and eukaryotes. *Mol. Biol. Evol.* 19(12):2199–2210.

Adibi, S.A. (1997). The oligopeptide transporter (PepT1) in human intestine: biology and function. *Gastroenterology* 113:322–340.

Adibi, S.A. (2003). Regulation of expression of the intestinal oligopeptide transporter (Pept-1) in health and disease. *Am. J. Physiol. Gastrointest. Liver Physiol.* 285(5):G779–G788.

Adson, A., Burton, P.S., Raub, T.J., Barsuhn, C.L., Audus, K.L., and Ho, N.F. (1995). Passive diffusion of weak organic electrolytes across Caco-2 cell monolayers: uncoupling the contributions of hydrodynamic, transcellular, and paracellular barriers. *J. Pharm. Sci.* 84(10):1197–1204.

Akiyama, S.I., Cornwell, M.M., Kuwano, M., Pastan, I., and Gottesman, M.M. (1988). Most drugs that reverse multidrug resistance also inhibit photoaffinity labeling of P-glycoprotein by a vinblastine analog. *Mol. Pharmacol.* 33:144–147.

Allikmets, R., Schriml, L.M., Hutchinson, A., Romano-Spica, V., and Dean, M. (1998). A human placenta-specific ATP-binding cassette gene (ABCP) on chromosome 4q22 that is involved in multidrug resistance. *Cancer Res.* 58(23):5337–5339.

Almquist, K.C., Loe, D.W., Hipfner, D.R., Mackie, J.E., Cole, S.P., and Deeley, R.G. (1995). Characterization of the 190 kDa multidrug resistance protein (MRP) in drug-selected and transfected human tumor cells. *Cancer Res.* 55:102–110.

Amasheh, S., Wenzel, U., Boll, M., Dom, D., Weber, W., and Clauss, W.D.H. (1997). Transport of charged dipeptides by the intestinal H+/peptide symporter PepT1 expressed in Xenopus laevis oocytes. *J. Membr. Biol.* 155:247–256.

Angeletti, R.H., Bergwerk, A.J., Novikoff, P.M., and Wolkoff, A.W. (1998). Dichotomous development of the organic anion transport protein in liver and choroid plexus. *Am. J. Physiol.* 275:C882–C887.

Arima, H., Yunomae, K., Hirayama, F., and Uekama, K. (2001). Contribution of P-glycoprotein to the enhancing effects of dimethyl-a-cyclodextrin on oral bioavailability of tacrolimus. *J. Pharmacol. Exp. Ther.* 297:547–555.

Ashida, K., Katsura, T., Motohashi, H., Saito, H., and Inui, K.I. (2002). Thyroid hormone regulates the activity and expression of the peptide transporter PepT1 in Caco-2 cells. *Am. J. Physiol. Gastrointest. Liver Physiol.* 282:G617–G623.

Ashida, K., Katsura, T., Saito, H., and Inui, K. (2004). Decreased activity and expression of intestinal oligopeptide transporter PEPT1 in rats with hyperthyroidism in vivo. *Pharm. Res.* 21(6):969–975.

Bailey, D.G., Malcolm, J., Arnold, O., and Spence, J.D. (1998). Grapefruit juice-drug interactions. *Br. J. Clin. Pharmacol.* 46:101–110.

Bailey, P.D., Boyd, C.A., Bronk, J.R., Collier, I.D., Meredith, D., Morgan, K.M., and Temple, C.S. (1999). How to make drugs orally active: a substrate model for the peptide transporter PepT1. *Angew Chem. Int. Ed. Engl.* 39:505–508.

Bakos, E., Hegedus, T., Hollo, Z., Welker, E., Tusnady, G.E., Zaman, G.J., Flens, M.J., Varadi, A., and Sarkadi, B. (1996). Membrane topology and glycosylation of the human multidrug resistance-associated protein. *J. Biol. Chem.* 271(21):12322–12326.

Baldwin, S.A., Beal, P.R., Yao, S.Y.M., King, A.E., Cass, C.E., and Young, J.D. (2004). The equilibrative nucleoside transporter family, SLC29. *Plugers Arch. Eur. J. Physiol.* 447:735–743.

Baldwin, S.A., Yao, S.Y., Hyde, R.J., Ng, A.M., Foppolo, S., Barnes, K., Ritzel, M.W., Cass, C.E., and Young, J.D. (2005). Functional characterization of novel human and mouse equilibrative nucleoside transporters (hENT3 and mENT3) located in intracellular membranes. *J. Biol. Chem.* 280(16):15880–15887.

Balimane, P.V., and Sinko, P.J. (2000). Effect of ionization on the variable uptake of valacyclovir via the human intestinal peptide transporter (hPepT1) in CHO cells. *Biopharm. Drug Dispos.* 21:165–174.

Barendt, W.M., and Wright, S.H. (2002). The human organic cation transporter (hOCT2) recognizes the degree of substrate ionization. *Biol. Chem.* 277:22491–22496.

Bates, S.E., Mickley, L.A., Chen, Y.N., Richert, N., Rudick, J., Biedler, J.L., and Fojo, A.T. (1989). Expression of a drug resistance gene in human neuroblastoma cellines: modulation by retinoic acid-induced differentiation. *Mol. Cell. Biol.* 9:4337–4344.

Belinsky, M.G., Bain, L.J., Balsara, B.B., Testa, J.R., and Kruh, G.D. (1998). *J. Natl Cancer. Inst.* 90:1735–1741.

Belinsky, M.G., Chen, Z.S., Shchaveleva, I., Zeng, H., and Kruh, G.D. (2002). Characterization of the drug resistance and transport properties of multidrug resistance protein 6 (MRP6, ABCC6). *Cancer Res.* 62(21):6172–6177.

Benet, L.Z., and Cummins, C.L. (2001). The drug efflux–metabolism alliance: biochemical aspects. *Adv. Drug Deliv. Rev.* 50(Suppl. 1): S3–S11.

Bera, T.K., Lee, S., Salvatore, G., Lee, B., and Pastan, I. (2001). MRP8, a new member of ABC transporter superfamily, identified by EST database mining and gene prediction program, is highly expressed in breast cancer. *Mol. Med.* 7(8):509–516.

Bera, T.K., Iavarone, C., Kumar, V., Lee, S., Lee, B., and Pastan, I. (2002). MRP9, an unusual truncated member of the ABC transporter superfamily, is highly expressed in breast cancer. *Proc. Natl. Acad. Sci. USA* 99(10):6997–7002.

Berlioz, F., Julien, S., Tsocas, A., Chariot, J., Carbon, C., Farinotti, R., and Roze, C. (1999). Neural modulation of cephalexin intestinal absorption through the di- and tripeptide brush border transporter of rat jejunum in vivo. *J. Pharmacol. Exp. Ther.* 288:1037–1044.

Berlioz, F., Maoret, J.J., Paris, H., Laburthe, M., Farinotti, R., and Roze, C. (2000). Alpha(2)-adrenergic receptors stimulate oligopeptide transport in a human intestinal cell line. *J. Pharmacol. Exp. Ther.* 294:466–472.

Bhardwaj, R.K., Glaeser, H., Becquemont, L., Klotz, U., Gupta, S.K., and Fromm, M.F. (2002). Piperine, a major constituent of black pepper, inhibits human P-glycoprotein and CYP3A4. *J. Pharmacol. Exp. Ther.* 302:645–650.

Bhardwaj, R.K., Herrera-Ruiz, D., Eltoukhy, N., Saad, M., and Knipp, G.T. (2005a). The functional evaluation of human peptide/histidine transporter 1 (hPHT1) in transiently transfected COS-7 cells. *Eur. J. Pharm. Sci.* (Epub ahead of print; DOI:10.1016/j.ejps.2005.09.014).

Bhardwaj, R.K., Herrera-Ruiz, D., Sinko, P.J., Gudmundsson, O.S., and Knipp, G. (2005b). Delineation of HPEPT1 meditated uptake and transport of substrates with varying transporter affinities utilizing stably transfected HPEPT1/MDCK clones and Caco-2 cells. *J. Pharmacol. Exp. Ther.* 314(3):1093–1100.

Biegel, A., Gebauer, S., Hartrodt, B., Brandsch, M., Neubert, K., and Thondorf, I. (2005). Three-dimensional quantitative structure-activity relationship analyses of beta-lactam antibiotics and tripeptides as substrates of the mammalian H+/peptide cotransporter PEPT1. *J. Med. Chem.* 48(13):4410–4419.

Bolger, M.B., Haworth, I.S., Yeung, A.K., Ann, D., von Grafenstein, H., Hamm-Alvarez, S., Okamoto, C.T., Kim, K.J., Basu, S.K., Wu, S., and Lee, V.H. (1998). Structure, function, and molecular modeling approaches to the study of the intestinal dipeptides transporter PepT1. *J. Pharm. Sci.* 87:1286–1291.

Botka, C.W., Witting, T.W., Graul, R.C., Nielsen, C.U., Sadée, W., Higaki, K., and Amidon, G.L. (2000). Human proton/oligopeptide transporter (POT) genes: identification of putative human genes using bioinformatics. *AAPS Pharm Sci* 2:Article 16.

Brandsch, M., Miyamoto, Y., Ganapathy, V., and Leibach, F.H. (1994). Expression and protein kinase C-dependent regulation of peptide/H+ co-transport system in the Caco-2 human colon carcinoma cell line. *Biochem. J.* 299:253–260.

Brandsch, M., Knutter, I., and Leibach, F.H. (2004). The intestinal H+/peptide symporter PEPT1: structure-affinity relationships. *Eur. J. Pharm. Sci.* 21(1):53–60.

Burt, R.K., and Thorgeirsson, S.S. (1988). Coinduction of MDR-1 multidrug resistance and cytochrome P-450 genes in rat liver by xenobiotics. *J. Natl Cancer Inst.* 80:1383–1386.

Busch, A.E., Quester, S., Ulzheimer, J.C., Waldegger, S., Gorboulev, V., Arndt, P., Lang, F., and Koepsell, H. (1996). Electrogenic properties and substrate specificity of the polyspecific rat cation transporter rOCT1. *J. Biol. Chem.* 271:32599–32604.

Busch, A.E., Karbach, U., Miska, D., Gorboulev, V., Akhoundova, A., Volk, C., Arndt, P., Ulzheimer, J.C., Sonders, M.S., Baumann, C., Waldegger, S., Lang, F., and Koepsell, H. (1998). Human neurons express the polyspecific cation transporter hOCT2, which translocates monoamine neurotransmitters, amantadine, and memantine. *Mol. Pharmacol.* 54:342–352.

Buyse, M., Berlioz, F., Guilmeau, S., Tsocas, A., Voisin, T., Péranzi, G., Merlin, D., Laburthe, M., Lewin, M.J.M., Rozé, C., and Bado, A. (2001). PepT1-mediated epithelial transport of dipeptides and cephalexin is enhanced by luminal leptin in the small intestine. *J. Clin. Invest.* 108:1483–1494.

Carpenter, L., and Halestrap, A.P. (1994). The kinetics, substrate and inhibitor specificity of the lactate transporter of Ehrlich-Lettre tumour cells studied with the intracellular pH indicator BCECF. *Biochem. J.* 304:751–760.

Chang, C., Swaan, P.W., Ngo, L.Y., Lum, P.Y., Patil, S.D., and Unadkat, J.D. (2004). Molecular requirements of the human nucleoside transporters hCNT1, hCNT2, and hENT1. *Mol. Pharmacol.* 65(3):558–570.

Chaudhary, P.M., and Roninson, I.B. (1991). Expression and activity of P-glycoprotein, a multidrug efflux pump, in human hematopoietic stem cells. *Cell* 66(1):85–94.

Chaudhary, P.M., and Roninson, I.B. (1993). Induction of multidrug resistance in human cells by transient exposure to different chemotherapeutic drugs. *J. Natl Cancer. Inst.* 85(8):632–639.

Chen, H., Wong, E.A., and Webb, K.E. (1999). Tissue distribution of a peptide transporter mRNA in sheep, dairy cows, pigs, and chickens. *J. Anim. Sci.* 77:1277–1283.

Chen, X.-Z., Steel, A., and Hediger, M.A. (2000). Functional roles of histidine and tyrosine residues in the H+-peptide transporter PepT1. *Biochem. Biophys. Res. Commun.* 272:726–730.

Chen, Z.S., Lee, K., and Kruh, G.D. (2001). Transport of cyclic nucleotides and estradiol 17-beta-D-glucuronide by multidrug resistance protein 4. Resistance to 6-mercaptopurine and 6-thioguanine. *J. Biol. Chem.* 276(36):33747–33754.

Chen, Z.S., Lee, K., Walther, S., Raftogianis, R.B., Kuwano, M., Zeng, H., and Kruh, G.D. (2002). Analysis of methotrexate and folate transport by multidrug resistance protein 4 (ABCC4): MRP4 is a component of the methotrexate efflux system. *Cancer. Res.* 62(11):3144–3150.

Chen, Z.S., Hopper-Borge, E., Belinsky, M.G., Shchaveleva, I., Kotova, E., and Kruh, G.D. (2003). Characterization of the transport properties of human multidrug resistance protein 7 (MRP7, ABCC10). *Mol. Pharmacol.* 63(2):351–358.

Chen, Z.S., Guo, Y., Belinsky, M.G., Kotova, E., and Kruh, G.D. (2005). Transport of bile acids, sulfated steroids, estradiol 17-beta-D-glucuronide, and leukotriene C4 by human multidrug resistance protein 8 (ABCC11). *Mol. Pharmacol.* 67(2):545–557.

Cheng, X., Maher, J., Chen, C., and Klaassen, C.D. (2005). Tissue distribution and ontogeny of mouse organic anion transporting polypeptides (Oatps). *Drug Metab. Dispos.* 33:1062–1073.

Cherrington, N.J., Hartley, D.P., Li, N., Johnson, D.R., and Klaassen, C.D. (2002). Organ distribution of multidrug resistance proteins 1, 2, and 3 (Mrp1, 2, and 3) mRNA and hepatic induction of Mrp3 by constitutive androstane receptor activators in rats. *J. Pharmacol. Exp. Ther.* 300(1):97–104.

Chin, K.V., Tanaka, S., Darlington, G., Pastan, I., and Gottesman, M.M. (1990). Heat shock and arsenite increase expression of the multidrug resistance (MDR1) gene in human renal carcinoma cells. *J. Biol. Chem.* 265:221–226.

Chiou, W.L., Chung, S.M., and Wu, T.C. (2000). Apparent lack of effect of P-glycoprotein on the gastrointestinal absorption of a substrate, tacrolimus, in normal mice. *Pharm. Res.* 17:205–208.

Choudhuri, S., Ogura, K., and Klaassen, C.D. (2001). Cloning, expression, and ontogeny of mouse organic anion-transporting polypeptide-5, a kidney-specific organic anion transporter. *Biochem. Biophys. Res. Commun.* 280:92–98.

Ciarimboli, G., and Schlatter, E. (2005). Regulation of organic cation transport. *Pflugers Arch.* 449:423–441.

Clark, H.F., Gurney, A.L., Abaya, E., Baker, K., Baldwin, D., Brush, J., Chen, J., Chow, B., Chui, C., Crowley, C., Currell, B., Deuel, B., Dowd, P., Eaton, D., Foster, J., Grimaldi, C., Gu, Q., Hass, P.E., Heldens, S., Huang, A., Kim, H.S., Klimowski, L., Jin, Y., Johnson, S., Lee, J., Lewis, L., Liao, D., Mark, M., Robbie, E., Sanchez, C., Schoenfeld, J., Seshagiri, S., Simmons, L., Singh, J., Smith, V., Stinson, J., Vagts, A., Vandlen, R., Watanabe, C., Wieand, D., Woods, K., Xie, M.-H., Yansura, D., Yi, S., Yu, G., Yuan, J., Zhang, M., Zhang, Z., Goddard, A., Wood, W.I., and Godowski, P. (2003). The secreted protein discovery initiative (SPDI), a large-scale effort to identify novel human secreted and transmembrane proteins: a bioinformatics assessment. *Genome Res.* 13(10):2265–2270.

Cole, S.P., Bhardwaj, G., Gerlach, J.H., Mackie, J.E., Grant, C.E., Almquist, K.C., Stewart, A.J., Kurz, E.U., Duncan, A.M., and Deeley, R.G. (1992). Overexpression of a transporter gene in a multidrug-resistant human lung cancer cell line. *Science* 258(5088):1650–1654.

Cole, S.P., Sparks, K.E., Fraser, K., Loe, D.W., Grant, C.E., Wilson, G.M., and Deeley, R.G. (1994). Pharmacological characterization of multidrug resistant MRP-transfected human tumor cells. *Cancer Res.* 54(22):5902–5910.

Coe, I.R., Griffiths, M., Young, J.D., Baldwin, S.A., and Cass, C.E. (1997). Assignment of the human equilibrative nucleoside transporter (hENT1) to 6p21.1–p21.2. *Genomics* 45(2):459–460.

Coe, I., Zhang, Y., McKenzie, T., and Naydenova, Z. (2002). PKC regulation of the human equilibrative nucleoside transporter, hENT1. *FEBS Lett.* 517:201–205.

Cornwell, M.M., Safa, A.R., Felsted, R.L., Gottesman, M.M., and Pastan, I. (1986). Membrane vesicles from multidrug-resistant human cancer cells contain a specific 150- to 170-kDa protein detected by photoaffinity labeling. *Proc. Natl Acad. Sci. USA* 83:3847–3850.

Courtois, A., Payen, L., Guillouzo, A., and Fardel, O. (1999). Up-regulation of multidrug resistance-associated protein 2 (MRP2) expression in rat hepatocytes by dexamethasone. *FEBS Lett.* 459(3):381–385.

Covitz, Y.K.-M., Amidon, G.L., and Sadée, W. (1998). Membrane topology of the human dipeptides transporter, hPepT1, determined by epitope insertions. *Biochemistry* 37:15214–15221.

Crawford, C.R., Patel, D.H., Naeve, C., and Belt, J.A. (1998). Cloning of the human equilibrative, nitrobenzylmercaptopurine riboside (NBMPR)-insensitive nucleoside transporter ei by functional expression in a transport-deficient cell line. *J. Biol. Chem.* 273:5288–5293.

Cronstein, B.N. (1994). Adenosine, an endogenous anti-inflammatory agent. *J. Appl. Physiol.* 76:5–13.

Cuff, M.A., Lambert, D.W., and Shirazi-Beechey, S.P. (2002). Substrate-induced regulation of the human colonic monocarboxylate transporter, MCT1. *J. Physiol.* 539:361–371.

Cuff, M., Dyer, J., Jones, M., and Shirazi-Beechey, S. (2005). The human colonic monocarboxylate transporter isoform 1: its potential importance to colonic tissue homeostasis. *Gastroenterology* 128:676–686.

Damaraju, V.L., Visser, F., Zhang, J., Mowles, D., Ng, A.M., Young, J.D., Jayaram, H.N., and Cass, C.E. (2005). Role of human nucleoside transporters in the cellular uptake of two inhibitors of IMP dehydrogenase, tiazofurin and benzamide riboside. *Mol. Pharmacol.* 67(1):273–279.

Daniel, H. (1996). Function and molecular structure of brush border membrane peptide/H+ symporters. *J. Memb. Biol.* 154:197–203.

Dantzig, A.H., Hoskins, J.A., Tabas, L.B., Bright, S., Shepard, R.L., Jenkins, I.L., Duckworth, D.C., Sportsman, J.R., Mackensen, D., Rosteck, P.R., Jr., and Skatrud, P.L. (1994). Association of intestinal peptide transport with a protein related to the cadherin superfamily. *Science* 264:430–433.

Dautrey, S., Felice, K., Petiet, A., Lacour, B., Carbon, C., and Farinotti, R. (1999). Active intestinal elimination of ciprofloxacin in rats: modulation by different substrates. *Br. J. Pharmacol.* 127:1728–1734.

Del Santo, B., Valdes, R., Mata, J., Felipe, A., Casado, F.J., and Pastor-Anglada, M. (1998). Differential expression and regulation of nucleoside transport systems in rat liver parenchymal and hepatoma cells. *Hepatology* 28:1504–1511.

Dietrich, C.G., de Waart, D.R., Ottenhoff, R., Schoots, I.G., and Elferink, R.P. (2001). Increased bioavailability of the food-derived carcinogen 2-amino-1-methyl-6-phenylimidazo[4,5-b]pyridine in MRP2-deficient rats. *Mol. Pharmacol.* 59(5):974–980.

Döring, F., Dorn, D., Bachfisher, U., Amasheh, S., Herget, M., and Daniel, H. (1996). Functional analysis of a chimeric mammalian peptide transporter derived from the intestinal and renal isoforms. *J. Physiol.* 497:773–779.

Döring, F., Martini, C., Walter, J., and Daniel, H. (2002). Importance of a small N-terminal region in mammalian peptide transporters for substrate affinity and function. *J. Membr. Biol.* 186:55–62.

Doyle, L.A., Yang, W., Abruzzo, L.V., Krogmann, T., Gao, Y., Rishi, A.K., and Ross, D.D. (1998). A multidrug resistance transporter from human MCF-7 breast cancer cells. *Proc. Natl Acad. Sci. USA* 95:15665–15670.

Drenou, B., Fardel, O., Amiot, L., and Fauchet, R. (1993). Detection of P glycoprotein activity on normal and leukemic CD34+ cells. *Leuk Res.* 17(12):1031–1035.

Dresser, G.K., Bailey, D.G., Leake, B.F., Schwarz, U.I., Dawson, P.A., Freeman, D.J., and Kim, R.B. (2002). Fruit juices inhibit organic anion transporting polypeptide-mediated drug uptake to decrease the oral availability of fexofenadine. *Clin. Pharmacol. Ther.* 71:11–20.

Duflot, S., Calvo, M., Casado, F.J., Enrich, C., and Pastor-Anglada, M. (2002). Concentrative nucleoside transporter (rCNT1) is targeted to the apical membrane through the hepatic transcytotic pathway. *Exp. Cell Res.* 281(1):77–85.

Dumitrescu, A.M., Liao, X.H., Best, T.B., Brockmann, K., and Refetoff, S. (2004). A novel syndrome combining thyroid and neurological abnormalities is associated with mutations in a monocarboxylate transporter gene. *Am. J. Hum. Genet.* 74:168–175.

Dunwiddie, T.V. (1985). The physiological role of adenosine in the central nervous system. *Int. Rev. Neurobiol.* 27:63–139.

Dürr, D., Stieger, B., Kullak-Ublick, G.A., Rentsch, K.M., Steinert, H.C., Meier, P.J., and Fattinger, K. (2000). St John's Wort induces intestinal P-glycoprotein/MDR1 and intestinal and hepatic CYP3A4. *Clin. Pharmacol. Ther.* 68:598–604.

Dussault, I., Lin, M., Hollister, K., Wang, E.H., Synold, T.W., and Forman, B.M. (2001). Peptide mimetic HIV protease inhibitors are ligands for the orphan receptor SXR. *J. Biol. Chem.* 276(36):33309–33312.

Ekins, S., Waller, C.L., Swaan, P.W., Cruciani, G., Wrighton, S.A., and Wikel, J.H. (2000). Progress in predicting human ADME parameters in silico. *J. Pharmacol. Toxicol. Methods* 44(1):251–272.

Ekins, S., Nikolsky, Y., and Nikolskaya, T. (2005). Techniques: application of systems biology to absorption, distribution, metabolism, excretion and toxicity. *Trends Pharmacol. Sci.* 26(4):202–209.

Engel, K., Zhou, M., and Wang, J. (2004). Identification and characterization of a novel monoamine transporter in the human brain. *J. Biol. Chem.* 279(48):50042–50049.

Ely, S.W., and Berne, R.M. (1992). Protective effects of adenosine in myocardial ischemia. *Circulation* 85:893–904.

Emoto, A., Ushigome, F., Koyabu, N., Kajiya, H., Okabe, K., Satoh, S., Tsukimori, K., Nakano, H., Ohtani, H., and Sawada, Y. (2002). H(+)-linked transport of salicylic acid, an NSAID, in the human trophoblast cell line BeWo. *Am. J. Physiol.* 282:C1064–C1075.

Enerson, B.E., and Drewes, L.R. (2003). Molecular features, regulation and function of monocarboxylate transporters: implications for drug delivery. *J. Pharm. Sci.* 92:1531–1544.

Fain, J.N., and Malbon, C.C. (1979). Regulation of adenylate cyclase by adenosine. *Mol. Cell. Biochem.* 25:143–169.

Fei, Y.J., Kanai, Y., Nussberger, S., Ganapathy, V., Leibach, F.H., Romero, M.F., Singh, S.K., Boron, W.F., and Hediger, M.A. (1994). Expression cloning of a mammalian proton-coupled oligopeptide transporter. *Nature* 368:563–566.

Fei, Y.J., Liu, W., Prasad, P.D., Kekuda, R., Oblak, T.G., Ganapathy, V., and Leibach, F.H. (1997). Identification of the histidyl residue obligatory for the catalytic activity of the human H+/peptide cotransporters PepT1 and PepT2. *Biochemistry* 14:452–460.

Fei, Y.J., Ganapathy, V., and Leibach, F.H. (1998). Molecular and structural features of the proton-coupled oligopeptide transporter superfamily. *Prog. Nucleic. Acid. Res. Mol. Biol.* 58:239–261.

Fei, Y.J., Sugawara, M., Liu, J.C., Li, H.W., Ganapathy, V., Ganapathy, M.E., and Leibach, F.H. (2000). cDNA structure, genomic organization, and promoter analysis of the mouse intestinal peptide transporter PepT1. *Biochim. Biophys. Acta* 1492: 145–154.

Ferry, D.R., Russell, M.A., and Cullen, M.H. (1992). P-glycoprotein possesses a 1,4-dihydropyridine-selective drug acceptor site which is allosterically coupled to a vinca-alkaloid-selective binding site. *Biochem. Biophys. Res. Commun.* 188:440–445.

Ferte, J. (2002). Analysis of tangled relationships between P-glycoprotein mediated multidrug resistance and the lipid phase of the cell membrane. *Eur. J. Biochem.* 267:277–294.
Fisher, G.A., Lum, B.L., Hausdorff, J., and Sikic, B.I. (1996). Pharmacological considerations in the modulation of multidrug resistance. *Eur. J. Cancer* 32A(6):1082–1088.
Flens, M.J., Zaman, G.J., Van Der Valk, P., Izquierdo, M.A., Schroeijers, A.B., Scheffer, G.L., Van Der Groep, P., De Haas, M., Meijer, C.J., and Scheper, R.J. (1996). Tissue distribution of the multidrug resistance protein. *Am. J. Pathol.* 148:1237–1247.
Fletcher, C., and Bean, B. (1985). Evaluation of oral aciclovir therapy. *Drug Intell. Clin. Pharm.* 19:518–524.
Fojo, A.T., Ueda, K., Slamon, D.J., Poplack, D.G., Gottesman, M.M., and Pastan, I. (1987). Expression of a multidrug-resistance gene in human tumors and tissues. *Proc. Natl Acad. Sci. USA* 84(1):265–269.
Friedrich, A., Prasad, P.D., Freyer, D., Ganapathy, V., and Brust, P. (2003). Molecular cloning and functional characterization of the OCTN2 transporter at the RBE4 cells, an in vitro model of the blood-brain barrier. *Brain Res.* 968:69–79.
Friedrichsen, G.M., Chen, W., Begtrup, M., Lee, C.P., Smith, P.L., and Borchardt, R.T. (2002). Synthesis of analogs of L-valacyclovir and determination of their substrate activity for the oligopeptide transporter in Caco-2 cells. *Eur. J. Pharm. Sci.* 16(1–2):1–13.
Friesema, E.C., Grueters, A., Biebermann, H., Krude, H., von Moers, A., Reeser, M., Barrett, T.G., Mancilla, E.E., Svensson, J., Kester, M.H., Kuiper, G.G., Balkassmi, S., Uitterlinden, A.G., Koehrle, J., Rodien, P., Halestrap, A.P., and Visser, T.J. (2004). Association between mutations in a thyroid hormone transporter and severe X-linked psychomotor retardation. *Lancet* 364:1435–1437.
Friesema, E.C., Jansen, J., Milici, C., and Visser, T.J. (2005). Thyroid hormone transporters. *Vitam. Horm.* 70:137–167.
Fromm, M.F., Kauffman, H.M., Fritz, P., Burk, O., Kroemer, H.K., Warzok, R.W., Eichelbaum, M., Siegmund, W., and Schrenk, D. (2000). The effect of rifampin treatment on intestinal expression of human MRP transporters. *Am. J. Pathol.* 157:1575–1580.
Ganapathy, M.E., Huang, W., Rajan, D.P., Carter, A.L., Sugawara, M., Iseki, K., Leibach, F.H., and Ganapathy, V. (2000). Beta-lactam antibiotics as substrates for OCTN2, an organic cation/carnitine transporter. *J. Biol. Chem.* 275:1699–1707.
Gangopadhyay, A., Thamotharan, M., and Adibi, S.A. (2002). Regulation of oligopeptide transporter (Pept-1) in experimental diabetes. *Am. J. Physiol. Gastrointest. Liver Physiol.* 283(1):G133–G138.
Garcia, C.K., Goldstein, J.L., Pathak, R.K., Anderson, R.G., and Brown, M.S. (1994). Molecular characterization of a membrane transporter for lactate, pyruvate, and other monocarboxylates: implications for the Cori cycle. *Cell* 76:865–873.
Gatmaitan, Z.C., and Arias, I.M. (1993). Structure and function of P-glycoprotein in normal liver and small intestine. *Adv. Pharmacol.* 24:77–97.
Gebauer, S., Knutter, I., Hartrodt, B., Brandsch, M., Neubert, K., and Thondorf, I. (2003). Three-dimensional quantitative structure-activity relationship analyses of peptide substrates of the mammalian H+/peptide cotransporter PEPT1. *J. Med. Chem.* 46(26):5725–5734.
Germann, U.A., Pastan, I., and Gottesman, M.M. (1993). P-glycoproteins: mediators of multidrug resistance. *Semin. Cell. Biol.* 4(1):63–76.
Gill, R.K., Saksena, S., Alrefai, W.A., Sarwar, Z., Goldstein, J.L., Carroll, R.E., Ramaswamy, K., and Dudeja, P.K. (2005). Expression and membrane localization of MCT isoforms along the length of the human intestine. *Am. J. Physiol. Cell Physiol.* 289(4):C846–C852.

Glavy, J.S., Wu, S.M., Wang, P.J., Orr, G.A., and Wolkoff, A.W. (2000). Down-regulation by extracellular ATP of rat hepatocyte organic anion transport is mediated by serine phosphorylation of Oatp1. *J. Biol. Chem.* 275:1479–1484.

Gomez-Angelats, M., Del Santo, B., Mercader, J., Ferrer-Martinez, A., Felipe, A., Casado, J., and Pastor-Anglada, M. (1996). Hormonal regulation of concentrative nucleoside transport in liver parenchymal cells. *Biochem. J.* 313:915–920.

Gorboulev, V., Ulzheimer, J.C., Akhoundova, A., Ulzheimer-Teuber, I., Karbach, U., Quester, S., Baumann, C., Lang, F., Busch, A.E., and Koepsell, H. (1997). Cloning and characterization of two human polyspecific organic cation transporters. *DNA Cell Biol.* 16:871–881.

Gotoh, Y., Kato, Y., Stieger, B., Meier, P.J., and Sugiyama, Y. (2002). Gender difference in the Oatp1-mediated tubular reabsorption of estradiol 17beta-D-glucuronide in rats. *Am. J. Physiol. Endocrinol. Metab.* 282:E1245–E1254.

Gottesman, M.M., and Pastan, I. (1993). Biochemistry of multidrug resistance mediated by the multidrug transporter. *Annu. Rev. Biochem.* 62:385–427.

Graul, R.C., and Sadée, W. (1997). Sequence alignments of the H(+)-dependent oligopeptide transporter family PTR: Inferences on structure and function of the intestinal PEPT1 transporter. *Pharm. Res.* 14:388–400.

Gray, J.H., Owen, R.P., and Giacomini, K.M. (2004). The concentrative nucleoside transporter family, SLC28. Pflugers Arch. *Eur. J. Physiol.* 447:728–734.

Greiner, B., Eichelbaum, M., Fritz, P., Kreichgauer, H.P., von Richter, O., Zundler, J., and Kroemer, H.K. (1999). The role of intestinal P-glycoprotein in the interaction of digoxin and rifampin. *J. Clin. Invest.* 104(2):147–153.

Griffiths, D.A., and Jarvis, S.M. (1996). Nucleoside and nucleobase transport systems of mammalian cells. *Biochim. Biophys. Acta* 1286:153–181.

Griffiths, M., Beaumont, N., Yao, S.Y.M., Sundaram, M., Boumah, C.E., Davies, A., Kwong, F.Y.P., Coe, I.R., Cass, C.E., Young, J.D., Baldwin, S.A. (1997a). Cloning of a human nucleoside transporter implicated in the cellular uptake of adenosine and chemotherapeutic drugs. *Nat. Med.* 3(1):89–93.

Griffiths, M., Yao, S.Y., Abidi, F., Phillips, S.E., Cass, C.E., Young, J.D., and Baldwin, S.A. (1997b). Molecular cloning and characterization of a nitrobenzylthioinosine-insensitive (ei) equilibrative nucleoside transporter from human placenta. *Biochem. J.* 328(3):739–743.

Grundemann, D., Gorboulev, V., Gambaryan, S., Veyhl, M., and Koepsell, H. (1994). Drug excretion mediated by a new prototype of polyspecific transporter. *Nature* 372:549–552.

Guo, G.L., and Klaassen, C.D. (2001). Protein kinase C suppresses rat organic anion transporting polypeptide 1- and 2-mediated uptake. *J. Pharmacol. Exp. Ther.* 299:551–557

Guo, G.L., Johnson, D.R., and Klaassen, C.D. (2002a). Postnatal expression and induction by pregnenolone-16α-carbonitrile of the organic anion-transporting polypeptide 2 in rat liver. *Drug Metab. Dispos.* 30:283–288.

Guo, G.L., Staudinger, J., Ogura, K., and Klaassen, C.D. (2002b). Induction of rat organic anion transporting polypeptide 2 by pregnenolone-16alpha-carbonitrile is via interaction with pregnane X receptor. *Mol. Pharmacol.* 61:832–839.

Guo, Y., Kotova, E., Chen, Z.S., Lee, K., Hopper-Borge, E., Belinsky, M.G., and Kruh, G.D. (2003). MRP8, ATP-binding cassette C11 (ABCC11), is a cyclic nucleotide efflux pump and a resistance factor for fluoropyrimidines 2-3-dideoxycytidine and 9-(2-phosphonylmethoxyethyl)-adenine. *J. Biol. Chem.* 278:29509–29514.

Gutmann, H., Hruz, P., Zimmermann, C., Beglinger, C., and Drewe, J. (2005). Distribution of breast cancer resistance protein (BCRP/ABCG2) mRNA expression along the human GI tract. *Biochem. Pharmacol.* 70(5):695–699.

Hagenbuch, B., and Meier, P.J. (2003). The superfamily of organic anion transporting polypeptides. *Biochim. Biophys. Acta* 1609:1–18.

Hagenbuch, B., and Meier, P.J. (2004). Organic anion transporting polypeptides of the OATP/SLC21 family: phylogenetic classification as OATP/SLCO superfamily, new nomenclature and molecular/functional properties. *Pflugers. Arch.* 447:653–665.

Hahn, E.L., Halestrap, A.P., and Gamelli, R.L. (2000). Expression of the lactate transporter MCT1 in macrophages. *Shock* 13:253–260.

Halestrap, A.P., and Meredith, D. (2004). The SLC16 gene family-from monocarboxylate transporters (MCTs) to aromatic amino acid transporters and beyond. *Pflugers Arch. Eur. J. Physiol.* 447:619–628.

Halestrap, A.P., and Price, N.T. (1999). The proton-linked monocarboxylate transporter (MCT) family: structure, function and regulation. *Biochem. J.* 343(Pt 2):281–299.

Hamilton, S.R., Yao, S.Y., Ingram, J.C., Hadden, D.A., Ritzel, M.W., Gallagher, M.P., Henderson, P.J., Cass, C.E., Young, J.D., and Baldwin, S.A. (2001). Subcellular distribution and membrane topology of the mammalian concentrative Na+-nucleoside cotransporter rCNT1. *J. Biol. Chem.* 276(30):27981–27988.

Hamman, M.A., Bruce, M.A., Haehner-Daniels, B.D., and Hall, S.D. (2001). The effect of rifampin administration on the disposition of fexofenadine. *Clin. Pharmacol. Ther.* 69(3):114–121.

Harrison, A.P., Erlwanger, K.H., Elbrond, V.S., Andersen, N.K., and Unmack, M.A. (2004). Gastrointestinal-tract models and techniques for use in safety pharmacology. *J. Pharmacol. Toxicol. Methods* 49(3):187–199.

Hayer-Zillgen, M., Bruss, M., and Bonisch, H. (2002). Expression and pharmacological profile of the human organic cation transporters hOCT1, hOCT2 and hOCT3. *Br. J. Pharmacol.* 136:829–836.

Hediger, M.A., Romero, M.F., Peng, J.B., Rolfs, A., Takanaga, H., and Bruford, E.A. (2004). The ABCs of solute carriers: physiological, pathological and therapeutic implications of human membrane transport proteins Introduction. *Pflugers Arch.* 447(5):465–468.

Herrera-Ruiz, D., and Knipp, G.T. (2003). Current Perspectives on Established and Putative Mammalian Oligopeptide Transporters. *J. Pharm. Sci.* 92(4):691–714.

Herrera-Ruiz, D., Wang, Q., Gudmundsson, O.S., Cook, T.J., Smith, R.L., Faria, T.N., and Knipp, G.T. (2001). Spatial expression patterns of peptide transporters in the human and rat gastrointestinal tracts, Caco-2 in vitro cell culture, and multiple human tissues. *AAPS Pharm Sci* 3(1):Article 9 (DOI: 10.1208/ps030109).

Higgins, C.F. (1991). Molecular basis of multidrug resistance mediated by P-glycoprotein. *Curr. Opin. Biotechnol.* 2(2):278–281.

Higgins, C.F., and Gottesman, M.M. (1992). Is the multidrug transporter a flippase? *Trends Biochem. Sci.* 17(1):18–21.

Higuchi, T., and Davis, S.S. (1970). Thermodynamic analysis of structure-activity relationships of drugs. Prediction of optimal structure. *J. Pharm. Sci.* 59(10):1376–1383.

Hipfner, D.R., Almquist, K.C., Leslie, E.M., Gerlach, J.H., Grant, C.E., Deeley, R.G., and Cole, S.P. (1997). Membrane topology of the multidrug resistance protein (MRP), A study of glycosylation-site mutants reveals an extracytosolic NH2 terminus. *J. Biol. Chem.* 272(38):23623–23630.

Hirohashi, T., Suzuki, H., and Sugiyama, Y. (1999). Characterization of the transport properties of cloned rat multidrug resistance-associated protein 3 (MRP3). *J. Biol. Chem.* 274:15181–15185.

Hirohashi, T., Suzuki, H., Takikawa, H., and Sugiyama, Y. (2000). ATP-dependent transport of bile salts by rat multidrug resistance-associated protein 3 (Mrp3). *J. Biol. Chem.* 275:2905–2910.

Hoffmeyer, S., Burk, O., von Richter, O., Arnold, H.P., Brockmoller, J., Johne, A., Cascorbi, I., Gerloff, T., Roots, I., Eichelbaum, M., and Brinkmann, U. (2000). Functional polymorphisms of the human multidrug-resistance gene: multiple sequence variations and correlation of one allele with P-glycoprotein expression and activity in vivo. *Proc. Natl Acad. Sci. USA* 97(7):3473–3478.

Hogben, C.A.M., Tocco, D.J., Brodie, B.B., and Schanker, L.S. (1959). On the mechanism of intestinal absorption of drugs. *J. Pharmacol. Exp. Ther.* 125:275–282.

Honda, Y., Ushigome, F., Koyabu, N., Morimoto, S., Shoyama, Y., Uchiumi, T., Kuwano, M., Ohtani, H., and Sawada, Y. (2004). Effects of grapefruit juice and orange juice components on P-glycoprotein and MRP2 mediated drug efflux. *Br. J. Pharmacol.* 143(7):856–864.

Hooijberg, J.H., Broxterman, H.J., Kool, M., Assaraf, Y.G., Peters, G.J., Noordhuis, P., Scheper, R.J., Borst, P., Pinedo, H.M., and Jansen, G. (1999). Antifolate resistance mediated by the multidrug resistance proteins MRP1 and MRP2. *Cancer. Res.* 59:2532–2535.

Hopper, E., Belinsky, M.G., Zeng, H., Tosolini, A., Testa, J.R., and Kruh, G.D. (2001). Analysis of the structure and expression pattern of MRP7 (ABCC10), a new member of the MRP subfamily. *Cancer Lett.* 162:181–191.

Hopper-Borge, E., Chen, Z.S., Shchaveleva, I., Belinsky, M.G., and Kruh, G.D. (2004). Analysis of the drug resistance profile of multidrug resistance protein 7 (ABCC10): resistance to docetaxel. *Cancer Res.* 64(14):4927–4930.

Hosoya, K., Kondo, T., Tomi, M., Takanaga, H., Ohtsuki, S., and Tersaki, T. (2001). MCT1-mediated transport of L-lactic acid at the inner blood-retinal barrier: a possible route for delivery of monocarboxylic acid drugs to the retina. *Pharm. Res.* 18:1669–1676.

Huang, Q.Q., Yao, S.Y., Ritzel, M.W., Paterson, A.R., Cass, C.E., and Young, J.D. (1994). Cloning and functional expression of a complementary DNA encoding a mammalian nucleoside transport protein. *J. Biol. Chem.* 269:17757–17760.

Hugger, E.D., Novak, B.L., Burton, P.S., Audus, K.L., and Borchardt, R.T. (2002). A comparison of commonly used polyethoxylated pharmaceutical excipients on their ability to inhibit P-glycoprotein activity in vitro. *J. Pharm. Sci.* 91:1991–2002.

Hussain, I., Kellett, G.L., Affleck, J., Shepherd, E.J., and Boyd, C.A.R. (2002). Expression and cellular distribution during development of the peptide transporter (PepT1) in the small intestinal epithelium of the rat. *Cell. Tissue Res.* 307:139–142.

Hyafil, F.C., Vergely, P., and Vigneaud, T. (1993). Grand-Perret: in vitro and in vivo reversal of multidrug resistance by GF120918, an acridonecarboxamide derivative. *Cancer Res.* 53:4595–4602.

Hyde, R.J., Cass, C.E., Young, J.D., and Baldwin, S.A. (2001). The ENT family of eukaryote nucleoside and nucleobase transporter: recent advances in the investigation of structure/function relationships and the identification of novel isoforms. *Mol. Membr. Biol.* 18(1):53–63.

Ieiri, I., Takane, H., and Otsubo, K. (2004). The MDR1 (ABCB1) gene polymorphism and its clinical implications. *Clin. Pharmacokinet.* 43(9):553–576.

Ihara, T., Tsujikawa, T., Fujiyama, Y., and Bamba, T. (2000). Regulation of PepT1 peptide transporter expression in the rat small intestine under malnourished conditions. *Digestion* 61:59–67.

Imai, Y., Asada, S., Tsukahara, S., Ishikawa, E., Tsuruo, T., and Sugimoto, Y. (2003). Breast cancer resistance protein exports sulfated estrogens but not free estrogens. *Mol. Pharmacol.* 64:610–618.

Inokuchi, A., Hinoshita, E., Iwamoto, Y., Kohno, K., Kuwano, M., and Uchiumi, T. (2001). Enhanced expression of the human multidrug resistance protein 3 by bile salt in human enterocytes. A transcriptional control of a plausible bile acid transporter. *J. Biol. Chem.* 276(50):46822–46829.

Irie, M., Terada, T., Sawada, K., Saito, H., and Inui, K.-I. (2001). Recognition and transport characteristics of nonpeptidic compounds by basolateral peptide transporter in Caco-2 cells. *J. Pharmacol. Exp. Ther.* 298:711–717.

Irie, M., Terada, T., Okuda, M., and Inui, K.-I. (2004). Efflux properties of basolateral peptide transporter in human intestinal cell line Caco-2. *Pflugers Arch.* 449(2):186–194.

Irie, M., Terada, T., Katsura, T., Matsuoka, S., and Inui, K.I. (2005). Computational modelling of H+-coupled peptide transport via human PEPT1. *J. Physiol.* 565(Pt 2):429–439.

Ishikawa, T., Tsuji, A., Inui, K., Sai, Y., Anzai, N., Wada, M., Endou, H., and Sumino, Y. (2004). The genetic polymorphism of drug transporters: functional analysis approaches. *Pharmacogenomics* 5(1):67–99.

Itoh, T., Tanno, M., Li, Y.-H., and Yamada, H. (1998). Transport of phenethicillin into rat intestinal brush border membrane vesicles: role of the monocarboxylic acid transport system. *Int. J. Pharm.* 172:103–112.

Izzo, A.A. (2005). Herb-drug interactions: an overview of the clinical evidence. *Fundam. Clin. Pharmacol.* 19(1):1–16.

Jacobs, M.H. (1940). Some aspects of cell permeability to weak electrolytes. *Cold Spring Harbor Symp. Quant. Biol.* 8:30–39.

Jancis, E.M., Carbone, R., Loechner, K.J., and Dannies, P.S. (1993). Estradiol induction of rhodamine 123 efflux and the multidrug resistance pump in rat pituitary cells. *Mol. Pharmacol.* 43:51–56.

Jedlitschky, G., Leier, I., Buchholz, U., Barnouin, K., Kurz, G., and Keppler, D. (1996). Transport of glutathione, glucuronate, and sulfate conjugates by the MRP gene-encoded conjugate export pump. *Cancer Res.* 56:988–994.

Jedlitschky, G., Leier, I., Buchholz, U., Hummel-Eisenbeiss, J., Burchell, B., and Keppler, D. (1997). ATP-dependent transport of bilirubin glucuronides by the multidrug resistance protein MRP1 and its hepatocyte canalicular isoform MRP2. *Biochem. J.* 327(Pt 1):305–310.

Jennings, L.L., Hao, C., Cabrita, M.A., Vickers, M.F., Baldwin, S.A., Young, J.D., and Cass, C.E. (2001). Distinct regional distribution of human equilibrative nucleoside transporter proteins 1 and 2 (hENT1 and hENT2) in the central nervous system. *Neuropharmacology* 40:722–731.

Jimenez, A., Pubill, D., Pallas, M., Camins, A., Llado, S., Camarasa, J., and Escubedo, E. (2000). Further characterization of an adenosine transport system in the mitochondrial fraction of rat testis. *Eur. J. Pharmacol.* 298:31–39.

Joly, B., Fardel, O., Cecchelli, R., Chesne, C., Puzzo, C., and Guillouzo, A. (1995). Selective drug transport and P-glycoprotein activity in an in-vitro blood brain barrier model. *Tox. In Vitro* 9:357–364.

Jonker, J.W., and Schinkel, A.H. (2004). Pharmacological and physiological functions of the polyspecific organic cation transporters: OCT1, 2, and 3 (SLC22A1-3). *J. Pharmacol. Exp. Ther.* 308(1):2–9.

Jonker, J.W., Smit, J.W., Brinkhuis, R.F., Maliepaard, M., Beijnen, J.H., Schellens, J.H., and Schinkel, A.H. (2000). Role of breast cancer resistance protein in the bioavailability and fetal penetration of topotecan. *J. Natl Cancer. Inst.* 92(20):1651–1656.

Jonker, J.W., Wagenaar, E., Mol, C.A., Buitelaar, M., Koepsell, H., Smit, J.W., and Schinkel, A.H. (2001). Reduced hepatic uptake and intestinal excretion of organic cations in mice with a targeted disruption of the organic cation transporter 1 (Oct1 [Slc22a1]) gene. *Mol. Cell Biol.* 21:5471–5477.

Juliano, R.T., and Ling, V.A. (1976). Surface glycoprotein modulating drug permeability in Chinese hamster ovary cell mutants. *Biochim. Biophys. Acta* 455:152–162.

Jung, D., Hagenbuch, B., Gresh, L., Pontoglio, M., Meier, P.J., and Kullak-Ublick, G.A. (2001). Characterization of the human OATP-C (SLC21A6) gene promoter and regulation of liver-specific OATP genes by hepatocyte nuclear factor 1 alpha. *J. Biol. Chem.* 276:37206–37214.

Jung, D., Podvinec, M., Meyer, U.A., Mangelsdorf, D.J., Fried, M., Meier, P.J., and Kullak-Ublick, G.A. (2002). Human organic anion transporting polypeptide 8 promoter is transactivated by the farnesoid X receptor/bile acid receptor. *Gastroenterology* 122:1954–1966.

Karlsson, J., Kuo, S.M., Ziemniak, J., and Artursson, P. (1993). Transport of celiprolol across human intestinal epithelial (Caco-2) cells: mediation of secretion by multiple transporters including P-glycoprotein. *Br. J. Pharmacol.* 110:1009–1016.

Kassel, D.B. (2004). Applications of high-throughput ADME in drug discovery. *Curr. Opin. Chem. Biol.* 8(3):339–345.

Kast, H.R., Goodwin, B., Tarr, P.T., Jones, S.A., Anisfeld, A.M., Stoltz, C.M., Tontonoz, P., Kliewer, S., Willson, T.M., and Edwards, P.A. (2002). Regulation of multidrug resistance-associated protein 2 (ABCC2) by the nuclear receptors pregnane X receptor, farnesoid X-activated receptor and constitutive androstane receptor. *J. Biol. Chem.* 277:2908–2915.

Katsura, T., and Inui, K. (2003). Intestinal absorption of drugs mediated by drug transporters: mechanisms and regulation. *Drug Metab. Pharmacokinet.* 18:1–15.

Kauffman, D.W., Kelly, J.P., Rosenberg, L., Anderson, T.E., and Mitchell, A.A. (2002). Recent patterns of medication use in the ambulatory adult population of the United States. *JAMA* 287:337–344.

Kauffmann, H.M., Keppler, D., Gant, T.W., and Schrenk, D. (1998). Induction of hepatic mrp2 (cmrp/cmoat) gene expression in nonhuman primates treated with rifampicin or tamoxifen. *Arch. Toxicol.* 72(12):763–768.

Kawasaki, Y., Kato, Y., Sai, Y., and Tsuji, A. (2004). Functional characterization of human organic cation transporter OCTN1 single nucleotide polymorphisms in the Japanese population. *J. Pharm. Sci.* 93:2920–2926.

Kekuda, R., Prasad, P.D., Wu, X., Wang, H., Fei, Y.J., Leibach, F.H., and Ganapathy, V. (1998). Cloning and functional characterization of a potential-sensitive, polyspecific organic cation transporter (OCT3) most abundantly expressed in placenta. *J. Biol. Chem.* 273:15971–15979.

Keppler, D., Leier, I., and Jedlitschky, G. (1997). Transport of glutathione conjugates and glucuronides by the multidrug resistance proteins MRP1 and MRP2. *Biol. Chem.* 378(8):787–791.

Kerb, R., Brinkmann, U., Chatskaia, N., Gorbunov, D., Gorboulev, V., Mornhinweg, E., Keil, A., Eichelbaum, M., and Koepsell, H. (2002). Identification of genetic variations of the human organic cation transporter hOCT1 and their functional consequences. *Pharmacogenetics* 12:591–595.

Kim, R.B. (2003). Organic anion-transporting polypeptide (OATP) transporter family and drug disposition. *Eur. J. Clin. Invest.* 33(Suppl 2):1–5.

Kim, C.M., Goldstein, J.L., and Brown, M.S. (1992). cDNA cloning of MEV, a mutant protein that facilitates cellular uptake of mevalonate, and identification of the point mutation responsible for its gain of function. *J. Biol. Chem.* 267:23113–23121.

Kim, R.B., Fromm, M.F., Wandel, C., Leake, B., Wood, A.J., Roden, D.M., and Wilkinson, G.R. (1998). The drug transporter P-glycoprotein limits oral absorption and brain entry of HIV-1 protease inhibitors. *J. Clin. Invest.* 101(2):289–294.

Kim, D.K., Kanai, Y., Chairoungdua, A., Matsuo, H., Cha, S.H., and Endou, H. (2001a). Expression cloning of a Na+-independent aromatic amino acid transporter with structural similarity to H+/monocarboxylate transporters. *J. Biol. Chem.* 276:17221–17228.

Kim, D.K., Kanai, Y., Matsuo, H., Kim, J.Y., Chairoungdua, A., Kobayashi, Y., Enomoto, A., Cha, S.H., Goya, T., and Endou, H. (2001b). The human T-type amino acid transporter-1: characterization, gene organization, and chromosomal location. *Genomics* 79:95–103.

Kim, S.J., Masaki, T., Leypoldt, J.K., Kamerath, C.D., Mohammad, S.F., and Cheung, A.K. (2004). Arterial and venous smooth-muscle cells differ in their responses to antiproliferative drugs. *J. Lab. Clin. Med.* 144(3):156–162.

Kimberlin, D.F., Weller, S., Whitley, R.J., Andrews, W.W., Hauth, J.C., Lakeman, F., and Miller, G. (1998). Pharmacokinetics of oral valacyclovir and acyclovir in late pregnancy. *Am. J. Obstet. Gynecol.* 179(4):846–851.

Kimura, T., and Higaki, K. (2002). Gastrointestinal transit and drug absorption. *Biol. Pharm. Bull.* 25(2):149–164.

Kimura, H., Takeda, M., Narikawa, S., Enomoto, A., Ichida, K., and Endou, H. (2002). Human organic anion transporters and human organic cation transporters mediate renal transport of prostaglandins. *J. Pharmacol. Exp. Ther.* 301:293–298.

Kirk, P., Wilson, M.C., Heddle, C., Brown, M.H., Barclay, A.N., and Halestrap, A.P. (2000). CD147 is tightly associated with lactate transporters MCT1 and MCT4 and facilitates their cell surface expression. *EMBO J.* 19:3896–3904.

Kiss, A., Farah, K., Kim, J., Garriock, R.J., Drysdale, T.A., and Hammond, J.R. (2000). Molecular cloning and functional characterization of inhibitor-sensitive (mENT1) and inhibitor-resistant (mENT2) equilibrative nucleoside transporters from mouse brain. *Biochem. J.* 352(2):363–372.

Klaassen, C.D. (1972). Immaturity of the newborn rat's hepatic excretory function for ouabain. *J. Pharmacol. Exp. Ther.* 183:520–526.

Klang, J.E., Burnworth, L.A., Pan, Y.X., Webb, K.E. Jr., and Wong, E.A. (2005). Functional characterization of a cloned pig intestinal peptide transporter (pPepT1). *J. Anim. Sci.* 83(1):172–181.

Knipp, G.T., and Herrera-Ruiz, D. (2004). Filing date: 5/31/01, Issued: 1/27/04. US Patent no. 6,683,169: "Nucleic acid encoding the human peptide histidine transporter 1 and methods of use thereof."

Knipp, G.T., Ho, N.F., Barsuhn, C.L., and Borchardt, R.T. (1997). Paracellular diffusion in Caco-2 cell monolayers: effect of perturbation on the transport of hydrophilic compounds that vary in charge and size. *J. Pharm. Sci.* 86(10):1105–1110.

Kobayashi, D., Nozawa, T., Imai, K., Nezu, J., Tsuji, A., and Tamai, I. (2003). Involvement of human organic anion transporting polypeptide OATP-B (SLC21A9) in pH-dependent transport across intestinal apical membrane. *Pharmacol. Exp. Ther.* 306:703–708.

Koepsell, H. (2004). Polyspecific organic cation transporters: their functions and interactions with drugs. *Trends. Pharmacol. Sci.* 25:375–381.

Koepsell, H., and Endou, H. (2004). The SLC22 drug transporter family. *Pflugers. Arch.* 447:666–676.

Koepsell, H., Schmitt, B.M., and Gorboulev, V. (2003). Organic cation transporters. *Rev. Physiol. Biochem. Pharmacol.* 150:36–90.

Kong, W., Engel, K., and Wang, J. (2004). Mammalian Nucleoside Transporters. *Curr. Drug Metab.* 5:63–84.

Konig, J., Nies, A.T., Cui, Y., Leier, I., and Keppler, D. (1999). Conjugate export pumps of the multidrug resistance protein (MRP) family: localization, substrate specificity, and MRP2-mediated drug resistance. *Biochim. Biophys. Acta* 1461:377–394.

Konig, J., Cui, Y., Nies, A.T., and Keppler, D. (2000a). A novel human organic anion transporting polypeptide localized to the basolateral hepatocyte membrane. *Am. J. Physiol. Gastrointest. Liver Physiol.* 278:G156–G164.

Konig, J., Cui, Y., Nies, A.T., and Keppler, D. (2000b). Localization and genomic organization of a new hepatocellular organic anion transporting polypeptide. *J. Biol. Chem.* 275:23161–23168.

Kool, M., De Haas, M., Scheffer, G.L., Scheper, R.J., Van Eijk, M.J., Juijn, J.A., Baas, F., and Borst, P. (1997). Analysis of expression of cMOAT (MRP2), MRP3, MRP4, and MRP5, homologues of the multidrug resistance-associated protein gene (MRP1), in human cancer cell lines. *Cancer. Res.* 57:3537–3547.

Kottra, G., Stamfort, A., and Daniel, H. (2002). PEPT1 as a paradigm for membrane carriers that mediate electrogenic bidirectional transport of anionic, cationic and neutral substrates. *J. Biol. Chem.* 277(36):32683–32691.

Kovarik, J.M., Rigaudy, L., Guerret, M., Gerbeau, C., and Rost, K.L. (1999). Longitudinal assessment of a P-glycoprotein-mediated drug interaction of valspodar on digoxin. *Clin. Pharmacol. Ther.* 66(4):391–400.

Kramer, R., Weber, T.K., and Arceci, R. (1995). Inhibition of N-linked glycosylation of P-glycoproptein by tunicamycin results in a reduced multidrug resistance phenotype. *Br. J. Cancer.* 71:670–676.

Krishna, R., and Mayer, L.D. (2000). Multidrug resistance (MDR) in cancer. Mechanisms, reversal using modulators of MDR and the role of MDR modulators in influencing the pharmacokinetics of anticancer drugs. *Eur J Pharm Sci.* 11(4):265–83.

Kruh, G.D., and Belinsky, M.G. (2003). The MRP family of drug efflux pumps. *Oncogene* 22:7537–7352.

Kubota, T., Furukawa, T., Tanino, H., Suto, A., Otan, Y., Watanabe, M., Ikeda, T., and Kitajima, M. (2001). Resistant mechanisms of anthracyclines–pirarubicin might partly break through the P-glycoprotein-mediated drug-resistance of human breast cancer tissues. *Breast Cancer* 8(4):333–338.

Kulkarni, A.A., Haworth, I.S., and Lee, V.H. (2003a). Transmembrane segment 5 of the dipeptide transporter hPepT1 forms a part of the substrate translocation pathway. *Biochem. Biophys. Res. Commun.* 306(1):177–185.

Kulkarni, A.A., Haworth, I.S., Uchiyama, T., and Lee, V.H. (2003b). Analysis of transmembrane segment 7 of the dipeptide transporter hPepT1 by cysteine-scanning mutagenesis. *J. Biol. Chem.* 278(51):51833–51840.

Kunta, J.R., and Sinko, P.J. (2004). Intestinal drug transporters: in vivo function and clinical importance. *Curr. Drug Metab.* 5:109–124.

Kurata, Y., Ieiri, I., Kimura, M., Morita, T., Irie, S., Urae, A., Ohdo, S., Ohtani, H., Sawada, Y., Higuchi, S., and Otsubo, K. (2002). Role of human MDR1 gene polymorphism in bioavailability and interaction of digoxin, a substrate of P-glycoprotein. *Clin. Pharmacol. Ther.* 72(2):209–219.

Kusuhara, H., and Sugiyama, Y. (2002). Role of transporters in the tissue-selective distribution and elimination of drugs: transporters in the liver, small intestine, brain and kidney. *J. Control Release* 78(1–3):43–54.

Lahjouji, K., Mitchell, G.A., and Qureshi, I.A. (2001). Carnitine transport by organic cation transporters and systemic carnitine deficiency. *Mol. Genet. Metab.* 73:287–297.

Lai, L., and Tan, T.M. (2002). Role of glutathione in the multidrug resistance protein 4 (MRP4/ABCC4)-mediated efflux of cAMP and resistance to purine analogues. *Biochem. J.* 361(Pt 3):497–503.

Lai, Y., Bakken, A.H., and Unadkat, J.D. (2002). Simultaneous expression of hCNT1-CFP and hENT1-YFP in Madin-Darby canine kidney cells. Localization and vectorial transport studies. *J. Biol. Chem.* 277:37711–37717.

Lai, Y., Lee, E.W., Ton, C.C., Vijay, S., Zhang, H., and Unadkat, J.D. (2005). Conserved residues F316 and G476 in the concentrative nucleoside transporter 1 (hCNT1) affect guanosine sensitivity and membrane expression, respectively. *Am. J. Physiol. – Cell Physiol.* 288(1):C39–C45.

Lambert, D.W., Wood, I.S., Ellis, A., and Shirazi-Beechey, S.P. (2002). Molecular changes in the expression of human colonic nutrient transporters during the transition from normality to malignancy. *Br. J. Cancer* 86:1262–1269.

Landowski, C., Sun, D., Foster, D., Menon, S.S., Barnett, J.L., Welage, L.S., Ramachandran, C., and Amidon, G.L. (2003). Gene expression in the human intestine and correlation with oral valaciclovir pharmacokinetic parameters. *J. Pharmacol. Exp. Ther.* 306:778–786.

Landowski, C.P., Vig, B.S., Song, X., and Amidon, G.L. (2005). Targeted delivery to PEPT1-overexpressing cells: acidic, basic, and secondary floxuridine amino acid ester prodrugs. *Mol. Cancer Ther.* 4(4):659–667.

Lazar, A., Grundemann, D., Berkels, R., Taubert, D., Zimmermann, T., and Schomig, E. (2003). Genetic variability of the extraneuronal monoamine transporter EMT (SLC22A3). *J. Hum. Genet.* 48:226–230.

Leabman, M.K., Huang, C.C., Kawamoto, M., Johns, S.J., Stryke, D., Ferrin, D.E., De Young, J., Taylor, T., Clark, A.G., Herskowitz, I., and Giacomini, K.M. (2002) Polymorphisms in a human kidney xenobiotic transporter, OCT2, exhibit altered function. *Pharmacogenetics.* 12(5):395–405.

Lee, V.H.L. (2000). Membrane transporters. *Eur. J. Pharm. Sci.* 11:S41–S50.

Leibach, F.H., and Ganapathy, V. (1996). Peptide transport in the intestine and the kidney. *Annu. Rev. Nutr.* 16:99–119.

Leier, I., Jedlitschky, G., Buchholz, U., Cole, S.P., Deeley, R.G., and Keppler, D. (1994). The MRP gene encodes an ATP-dependent export pump for leukotriene C4 and structurally related conjugates. *J. Biol. Chem.* 269:27807–27810.

Leier, I., Hummel-Eisenbeiss, J., Cui, Y., and Keppler, D. (2000). ATP-dependent para-aminohippurate transport by apical multidrug resistance protein MRP2. *Kidney Int.* 57(4):1636–1642.

Leino, R.L., Gerhart, D.Z., Duelli, R., Enerson, B.E., and Drewes, L.R. (2001). Diet-induced ketosis increases monocarboxylate transporter (MCT1) levels in rat brain. *Neurochem. Int.* 38(6):519–527.

Leo, A., Hansch, C., and Elkins, D. (1971). Partition coefficients and their uses. *Chem. Rev.* 71(6):525–616.

Leslie, E.M., Ito, K., Upadhyaya, P., Hecht, S.S., Deeley, R.G., and Cole, S.P. (2001). Transport of the beta-O-glucuronide conjugate of the tobacco-specific carcinogen 4-(methylnitrosamino)-1-(3-pyridyl)-1-butanol (NNAL) by the multidrug resistance protein 1 (MRP1). Requirement for glutathione or a non-sulfur-containing analog. *J. Biol. Chem.* 276(30):27846–27854.

Leslie, E.M., Deeley, R.G., and Cole, S.P. (2003). Bioflavonoid stimulation of glutathione transport by the 190-kDa multidrug resistance protein 1 (MRP1). *Drug Metab. Dispos.* 31:11–15.

Leslie, E.M., Deeley, R.G., and Cole, S.P. (2005). Multidrug resistance proteins: role of P-glycoprotein, MRP1, MRP2, and BCRP (ABCG2) in tissue defense. *Toxicol. Appl. Pharmacol.* 204(3):216–237.

Leung, G.P., Man, R.Y., and Tse, C.M. (2005a). d-Glucose upregulates adenosine transport in cultured human aortic smooth muscle cells. *Am. J. Physiol. Heart Circ. Physiol.* 288:H2756–H2762.

Leung, G.P., Man, R.Y., and Tse, C.M. (2005b). Effect of thiazolidinediones on equilibrative nucleoside transporter-1 in human aortic smooth muscle cells. *Biochem. Pharmacol.* 70(3):355–362.

Leveille-Webster, C.R., and Arias, I.M. (1995). The biology of the P-glycoproteins. *J. Membr. Biol.* 143(2):89–102.

Li, L., Lee, T.K., Meier, P.J., and Ballatori, N. (1998). Identification of glutathione as a driving force and leukotriene C4 as a substrate for oatp1, the hepatic sinusoidal organic solute transporter. *J. Biol. Chem.* 273:16184–16191.

Li, Y.H., Ito, K., Tsuda, Y., Kohda, R., Yamada, H., and Itoh, T. (1999). Mechanism of intestinal absorption of an orally active beta-lactam prodrug: uptake and transport of carindacillin in Caco-2 cells. *J. Pharmacol. Exp. Ther.* 290:958–964.

Li, L., Meier, P.J., and Ballatori, N. (2000). Oatp2 mediates bidirectional organic solute transport: a role for intracellular glutathione. *Mol. Pharmacol.* 58:335–340.

Li, N., Hartley, D.P., Cherrington, N.J., and Klaassen, C.D. (2002). Tissue expression, ontogeny, and inducibility of rat organic anion transporting polypeptide 4. *J. Pharmacol. Exp. Ther.* 301:551–560.

Liang, R., Fei, Y.J., Prasad, P.D., Ramamoorthy, S., Han, H., Yang-Feng, T.L., Hediger, M.A., Ganapathy, V., and Leibach, F.H. (1995). Human intestinal H+/peptide cotransporter. Cloning, functional expression, and chromosomal localization. *J. Biol. Chem.* 270:6456–6463.

Lin, R.Y., Vera, J.C., Chaganti, R.S., and Golde, D.W. (1998). Human monocarboxylate transporter 2 (MCT2) is a high affinity pyruvate transporter. *J. Biol. Chem.* 273: 28959–28965.

Lincke, C.R., Smit, J.J., van der Velde-Koerts, T., and Borst, P. (1991). Structure of the human MDR3 gene and physical mapping of the human MDR locus. *J. Biol. Chem.* 266(8):5303–5310.

Lipinski, C.A. (2000). Drug-like properties and the causes of poor solubility and poor permeability. *J. Pharmacol. Toxicol. Methods* 44(1):235–249.

Lipinski, C.A., Lombardo, F., Dominy, B.W., and Feeney, P.J. (1997). Experimental and computational approaches to estimate solubility and permeability in drug discovery and development settings. *Adv. Drug Deliv. Rev*. 23:3–25.

Lipinski, C.A., Lombardo, F., Dominy, B.W., and Feeney, P.J. (2001). Experimental and computational approaches to estimate solubility and permeability in drug discovery and development settings. *Adv. Drug Deliv. Rev*. 46(1–3):3–26.

Loe, D.W., Deeley, R.G., and Cole, S.P. (2000a). Verapamil stimulates glutathione transport by the 190-kDa multidrug resistance protein 1 (MRP1). *J. Pharmacol. Exp. Ther.* 293:530–538.

Loe, D.W., Oleschuk, C.J., Deeley, R.G., and Cole, S.P. (2000b). Structure–activity studies of verapamil analogs that modulate transport of leukotriene C(4) and reduced glutathione by multidrug resistance protein MRP1. *Biochem. Biophys. Res. Commun.* 275(3):795–803.

Loewen, S.K., Ng, A.M., Yao, S.Y., Cass, C.E., Baldwin, S.A., and Young, J.D. (1999). Identification of amino acid residues responsible for the pyrimidine and purine nucleoside specificities of human concentrative Na+nucleoside cotransporters hCNT1 and hCNT2. *J. Biol. Chem.* 274:24475–24484.

Lu, R., Kanai, N., Bao, Y., Wolkoff, A.W., and Schuster, V.L. (1996). Regulation of renal oatp mRNA expression by testosterone. *Am. J. Physiol.* 270:F332–F337.

Luo, G., Guenthner, T., Gan, L.S., and Humphreys, W.G. (2004). CYP3A4 induction by xenobiotics: biochemistry, experimental methods and impact on drug discovery and development. *Curr. Drug Metab.* 5:483–505.

MacDougall, C., and Guglielmo, B.J. (2004). Pharmacokinetics of valaciclovir. *J. Antimicrob. Chemother.* 53(6):899–901.

Mackenzie, B., Loo, D.D., Fei, Y., Liu, W.J., Ganapathy, V., Leibach, F.H., and Wright, E.M. (1996). Mechanisms of the human intestinal H+-coupled oligopeptide transporter hPEPT1. *J. Biol. Chem.* 271(10):5430–5437.

Mackey, J.R., Mani, R.S., Selner, M., Mowles, D., Young, J.D., Belt, J.A., Crawford, C.R., and Cass, C.E. (1998). Functional nucleoside transporters are required for gemcitabine influx and manifestation of toxicity in cancer cell lines. *Cancer Res.* 58(19):4349–4357.

Mackey, J.R., Yao, S.Y., Smith, K.M., Karpinski, E., Baldwin, S.A., Cass, C.E., and Young, J.D. (1999). Gemcitabine transport in Xenopus oocytes expressing recombinant plasma membrane mammalian nucleoside transporters. *J. Natl Cancer Inst.* 91(21):1876–1881.

Madon, J., Eckhardt, U., Gerloff, T., Stieger, B., and Meier, P.J. (1997). Functional expression of the rat liver canalicular isoform of the multidrug resistance-associated protein. *FEBS Lett.* 406:75–78.

Maher, J.M., Cheng, X., Slitt, A.L., Dieter, M.Z., and Klaassen, C.D. (2005). Induction of the multidrug resistance-associated protein family of transporters by chemical activators of receptor-mediated pathways in mouse liver. *Drug Metab. Dispos.* 33(7):956–962.

Maliepaard, M., van Gastelen, M.A., de Jong, L.A., Pluim, D., van Waardenburg, R.C., Ruevekamp-Helmers, M.C., Floot, B.G., and Schellens, J.H. (1999). Overexpression of the BCRP/MXR/ABCP gene in a topotecan-selected ovarian tumor cell line. *Cancer Res.* 59(18):4559–4563.

Maliepaard, M., Scheffer, G.L., Faneyte, I.F., van Gastelen, M.A., Pijnenborg, A.C., Schinkel, A.H., van De Vijver, M.J., Scheper, R.J., and Schellens, J.H. (2001). Subcellular localization and distribution of the breast cancer resistance protein transporter in normal human tissues. *Cancer Res.* 61(8):3458–3464.

Mangravite, L.M., Lipschutz, J.H., Mostov, K.E., and Giacomini, K.M. (2001). Localization of GFP-tagged concentrative nucleoside transporters in a renal polarized epithelial cell line. *Am. J. Physiol. Renal Physiol.* 280(5):F879–F885.

Mangravite, L.M., Xiao, G., and Giacomini, K.M. (2003). Localization of human equilibrative nucleoside transporters, hENT1 and hENT2, in renal epithelial cells. *Am. J. Physiol. Renal. Physiol.* 284:F902–F910.

Mani, R.S., Hammond, J.R., Marjan, J.M., Graham, K.A., Young, J.D., Baldwin, S.A., and Cass, C.E. (1998). Demonstration of equilibrative nucleoside transporters (hENT1 and hENT2) in nuclear envelopes of cultured human choriocarcinoma (BeWo) cells by functional reconstitution in proteoliposomes. *J. Biol. Chem.* 273:30818–30825.

Martin, C., Berridge, G., Higgins, C.F., and Callaghan, R. (1997). The multi-drug resistance reversal agent SR33557 and modulation of vinca alkaloid binding to P-glycoprotein by an allosteric interaction. *Br. J. Pharmacol.* 122:765–771.

Masaki, T., Kamerath, C.D., Kim, S.J., Leypoldt, J.K., Mohammad, S.F., and Cheung, A.K. (2004). In vitro pharmacological inhibition of human vascular smooth muscle cell proliferation for the prevention of hemodialysis vascular access stenosis. *Blood Purif.* 22(3):307–312.

McAleer, M.A., Breen, M.A., White, N.L., and Matthews, N. (1999). pABC11 (also known as MOAT-C and MRP5), a member of the ABC family of proteins, has anion transporter activity but does not confer multidrug resistance when overexpressed in human embryonic kidney 293 cells. *J. Biol. Chem.* 274:23541–23548.

Mehrens, T., Lelleck, S., Cetinkaya, I., Knollmann, M., Hohage, H., Gorboulev, V., Boknik, P., Koepsell, H., and Schlatter, E. (2000). The affinity of the organic cation transporter rOCT1 is increased by protein kinase C-dependent phosphorylation. *J. Am. Soc. Nephrol.* 11:1216–1224.

Meier, P.J., and Stieger, B. (2002). Bile salt transporters. *Annu. Rev. Physiol.* 64:635–661.

Meredith, D. (2004). Site-directed mutation of arginine 282 to glutamate uncouples the movement of peptides and protons by the rabbit proton–peptide cotransporter PepT1. *J. Biol. Chem.* 279(16):15795–15798.

Meredith, D., and Boyd, C.A.R. (2000). Structure and function of eukaryotic peptide transporters. *Cell Mol. Life Sci.* 57:754–778.

Merlin, D., Steel, A., Gewirtz, A.T., Si-Tahar, M., Hediger, M.A., and Madara, J.L. (1998). hPepT1-mediated epithelial transport of bacteria-derived chemotactic peptides enhances neutrophil-epithelial interactions. *J. Clin. Invest.* 102:2011–2018.

Merlin, D., Si-Tahar, M., Sitaraman, S.V., Eastburn, K., Williams, I., Liu, X., Hediger, M.A., and Madara, J.L. (2001). Colonic epithelial hPepT1 expression occurs in inflammatory bowel disease: transport of bacterial peptides influences expression of MHC class 1 molecules. *Gastroenterology* 120:1666–1679.

Mickley, A.L., Bates, S.S., Richert, N.D., Currier, S., Tanaka, S., Foss, F., Rosen, N., and Fojo, A.T. (1989). Modulation of the expression of a multidrug resistance gene (mdr-1/P-glycoprotein) by differentiating agents. *J. Biol. Chem.* 264:18031–18040.

Mikkaichi, T., Suzuki, T., Tanemoto, M., Ito, S., and Abe, T. (2004). The organic anion transporter (OATP) family. *Drug Metab. Pharmacokinet.* 19(3):171–179.

Mizuno, N., Niwa, T., Yotsumoto, Y., and Sugiyama, Y. (2003). Impact of drug transporter studies on drug discovery and development. *Pharmacol. Rev.* 55:425–461.

van Montfoort, J.E., Hagenbuch, B., Groothuis, G.M., Koepsell, H., Meier, P.J., and Meijer, D.K. (2003). Drug uptake systems in liver and kidney. *Curr. Drug Metab.* 4:185–211.

Moriya, Y., Nakamura, T., Horinouchi, M., Sakaeda, T., Tamura, T., Aoyama, N., Shirakawa, T., Gotoh, A., Fujimoto, S., Matsuo, M., Kasuga, M., and Okumura, K. (2002). Effects of polymorphisms of MDR1, MRP1, and MRP2 genes on their mRNA expression levels in duodenal enterocytes of healthy Japanese subjects. *Biol. Pharm. Bull.* 25(10):1356–1359.

Mouly, S., and Paine, M.F. (2003). P-glycoprotein increases from proximal to distal regions of human small intestine. *Pharm. Res.* 20(10):1595–1599.

Muller, U., Brandsch, M., Prasad, P.D., Fei, Y.-J., Ganapathy, V., and Leibach, F.H. (1996). Inhibition of the H+/peptide cotransporter in the human intestinal cell line Caco-2 by cyclic AMP. *Biochem. Biophys. Res. Commun.* 218:461–465.

Nagel, G., Volk, C., Friedrich, T., Ulzheimer, J.C., Bamberg, E., and Koepsell, H. (1997). A reevaluation of substrate specificity of the rat cation transporter rOCT1. *J. Biol. Chem.* 272:31953–31956.

Nakatomi, I., Yoshikawa, M., Oka, M., Ikegami, Y., Hayasaka, S., Sano, K., Shiozawa, K., Kawabata, S., Soda, H., Ishikawa, T., Tanabe, S., and Kohno, S. (2001). Transport of 7-ethyl-10-hydroxycamptothecin (SN-38) by breast cancer resistance protein ABCG2 in human lung cancer cells. *Biochem. Biophys. Res. Commun.* 288:827–832.

Naruhashi, K., Tamai, I., Inoue, N., Muraoka, H., Sai, Y., Suzuki, N., and Tsuji, A. (2002). Involvement of multidrug resistance-associated protein 2 in intestinal secretion of grepafloxacin in rats. *Antimicrob. Agents Chemother.* 46:344–349.

Nerurkar, M.M., Burton, P.S., and Borchardt, R.T. (1996). The use of surfactants to enhance the permeability of peptides through Caco-2 cells by inhibition of an apically polarized efflux system. *Pharm. Res.* 13:528–534.

Nerurkar, M.M., Ho, N.F., Burton, P.S., Vidmar, T.J., and Borchardt, R.T. (1997). Mechanistic roles of neutral surfactants on concurrent polarized and passive membrane transport of a model peptide in Caco-2 cells. *J. Pharm. Sci.* 86:813–821.

Nezu, J., Tamai, I., Oku, A., Ohashi, R., Yabuuchi, H., Hashimoto, N., Nikaido, H., Sai, Y., Koizumi, A., Shoji, Y., Takada, G., Matsuishi, T., Yoshino, M., Kato, H., Ohura, T., Tsujimoto, G., Hayakawa, J., Shimane, M., and Tsuji, A. (1999). Primary systemic carnitine deficiency is caused by mutations in a gene encoding sodium ion-dependent carnitine transporter. *Nat. Genet.* 21:91–94.

Ngo, L.Y., Patil, S.D., and Unadkat, J.D. (2001). Ontogenic and longitudinal activity of Na+-nucleoside transporters in the small intestine. *Am. J. Physiol.* 280:G475–G481.

Nielsen, C.U., Brodin, B., Jorgensen, F., Frokjaer, S., and Steffansen, B. (2002). Human peptide transporters: therapeutic applications. *Expert Opin.* 12:1329–1350.

Nies, A.T., Cantz, T., Brom, M., Leier, I., and Keppler, D. (1998). Expression of the apical conjugate export pump, Mrp2, in the polarized hepatoma cell line, WIF-B. *Hepatology* 28(5):1332–1340.

Nozawa, T., Nakajima, M., Tamai, I., Noda, K., Nezu, J., Sai, Y., Tsuji, A., and Yokoi, T. (2002). Genetic polymorphisms of human organic anion transporters OATP-C (SLC21A6) and OATP-B (SLC21A9): allele frequencies in the Japanese population and functional analysis. *Pharmacol. Exp. Ther.* 302:804–813.

Nozawa, T., Toyobuku, H., Kobayashi, D., Kuruma, K., Tsuji, A., and Tamai, I. (2003). Enhanced intestinal absorption of drugs by activation of peptide transporter PEPT1 using proton-releasing polymer. *J. Pharm. Sci.* 92(11):2208–2216.

Nozawa, T., Imai, K., Nezu, J., Tsuji, A., and Tamai, I. (2004). Functional characterization of pH-sensitive organic anion transporting polypeptide OATP-B in human. *J. Pharmacol. Exp. Ther.* 308:438–445.

Nussberger, S., Steel, A., and Hediger, M. (1997a). Structure and pharmacology of proton-linked peptide transporters. *J. Control Rel.* 46:31–38.

Nussberger, S., Steel, A., Trott, D., Romero, M., Boron, W.F., and Hediger, M. (1997b). Symmetry of H+ binding to the intra- and extracellular side of the H+-coupled oligopeptide cotransporter PepT1. *J. Biol. Chem.* 272:7777–7785.

Ogihara, H., Suzuki, T., Nagamachi, Y., Inui, K.-I., and Takata, K. (1999). Peptide transporter in the rat small intestine: ultrastructural localization and the effect of starvation and administration of amino acids. *Histochem. J.* 31:169–174.

Oh, D.-M., and Amidon, G.L. (1999). Overview of membrane transport. In *Membrane Transporters as Drug Targets*. Edited by G.L. Amidon and W. Sadee, Kluwer Academic/Plenum Publishers, New York, NY, pp. 1–27.

Ohashi, R., Tamai, I., Yabuuchi, H., Nezu, J.I., Oku, A., Sai, Y., Shimane, M., and Tsuji, A. (1999). Na(+)-dependent carnitine transport by organic cation transporter (OCTN2): its pharmacological and toxicological relevance. *J. Pharmacol. Exp. Ther.* 291:778–784.

Okamura, A., Emoto, A., Koyabu, N., Ohtani, H., and Sawada, Y. (2002). Transport and uptake of nateglinide in Caco-2 cells and its inhibitory effect on human monocarboxylate transporter MCT1. *Br. J. Pharmacol.* 137:391–399.

Okuda, M., Saito, H., Urakami, Y., Takano, M., and Inui, K. (1996). cDNA cloning and functional expression of a novel rat kidney organic cation transporter, OCT2. *Biochem. Biophys. Res. Commun.* 224:500–507.

Okuda, M., Urakami, Y., Saito, H., and Inui, K.I. (1999). Molecular mechanisms of organic cation transport in OCT2-expressing Xenopus oocytes. *Biochim. Biophys. Acta* 1417:224–231.

Owen, M.R., Doran, E., and Halestrap, A.P. (2000). Evidence that metformin exerts its anti-diabetic effects through inhibition of complex 1 of the mitochondrial respiratory chain. *Biochem. J.* 348:607–614.

Ozvegy, C., Litman, T., Szakacs, G., Nagy, Z., Bates, S., Varadi, A., and Sarkadi, B. (2001). Functional characterization of the human multidrug transporter, ABCG2, expressed in insect cells. *Biochem. Biophys. Res. Commun.* 285(1):111–117.

Pan, Y.X., Wong, E.A., Bloomquist, J.R., and Webb, K.E., Jr. (2001). Expression of a cloned ovine gastrointestinal peptide transporter (oPepT1) in Xenopus oocytes induces uptake of oligopeptides in vitro. *J. Nutr.* 131:1264–1270.

Pastor-Anglada, M., Casado, F.J., Valdes, R., Mata, J., Garcia-Manteiga, J., and Molina, M. (2001). Complex regulation of nucleoside transporter expression in epithelial and immune system cells. *Mol. Memb. Bio.* 18:81–85.

Paulusma, C.C., Bosma, P.J., Zaman, G.J., Bakker, C.T., Otter, M., Scheffer, G.L., Scheper, R.J., Borst, P., and Oude Elferink, R.P. (1996). Congenital jaundice in rats with a mutation in a multidrug resistance-associated protein gene. *Science* 271:1126–1128.

Pavlova, A., Sakurai, H., Leclercq, B., Beier, D.R., Yu, A.S., and Nigam, S.K. (2000). Developmentally regulated expression of organic ion transporters NKT (OAT1), OCT1, NLT (OAT2), and Roct. *Am. J. Physiol. Renal Physiol.* 278:F635–F643.

Pawelczyk, T., Podgorska, M., and Sakowicz, M. (2003). The effect of insulin on expression level of nucleoside transporters in diabetic rats. *Mol. Pharmacol.* 63(1):81–88.

Peltekova, V.D., Wintle, R.F., Rubin, L.A., Amos, C.I., Huang, Q., Gu, X., Newman, B., Van Oene, M., Cescon, D., Greenberg, G., Griffiths, A.M., St George-Hyslop, P.H., and Siminovitch, K.A. (2004). Functional variants of OCTN cation transporter genes are associated with Crohn disease. *Nat. Genet.* 36:471–475.

Peng, K.C., Cluzeaud, F., Bens, M., Van Huyen, J.P., Wioland, M.A., Lacave, R., and Vandewalle, A. (1999). Tissue and cell distribution of the multidrug resistance-associated protein (MRP) in mouse intestine and kidney. *J. Histochem. Cytochem.* 47:757–768.

Pennycooke, M., Chaudary, N., Shuralyova, I., Zhang, Y., and Coe, I.R. (2001). Differential expression of human nucleoside transporters in normal and tumor tissue. *Biochem. Biophys. Res. Commun.* 280:951–959.

Phan, D., Chin-Hong, P., Lin, E., Anderle, P., Sadee, W., and Guglielmo, B.J. (2003). Intra- and interindividual variabilities of valaciclovir oral bioavailability and effect of coadministration of an hPEPT1 inhibitor. *Antimicrob. Agents Chemother.* 47:2351–2353.

Phillis, J.W., and Wu, P.H. (1981). The role of adenosine and its nucleotides in central synaptic transmission. *Prog. Neurobiol.* 16:187–239.

Philp, N.J., Yoon, H., and Grollman, E.F. (1998). Monocaroxylate transporter (MCT1) is located in the apical membrane and MCT3 in the basal membrane of rat RPE. *Am. J. Physiol.* 274:R1824–R1828.

Philp, N.J., Ochrietor, J.D., Rudoy, C., Muramatsu, T., and Linser, P.J. (2003). Loss of MCT1, MCT3 and MCT4 expression in the retinal pigment epithelium and neural retina of the 5A11/basigin-null mouse. *Invest. Ophthalmol. Vis. Sci.* 44(3):1305–1311.

Pisoni, R.L., and Thoene, J.G. (1989). Detection and characterization of a nucleoside transport system in human fibroblast lysosomes. *J. Biol. Chem.* 264:4850–4856.

Plagemann, P.G., and Aran, J.M. (1990). Characterization of Na+-dependent, active nucleoside transport in rat and mouse peritoneal macrophages, a mouse macrophage cell line and normal rat kidney cells. *Biochim. Biophys. Acta* 1028:289–298.

Pons, R., Carrozzo, R., Tein, I., Walker, W.F., Addonizio, L.J., Rhead, W., Miranda, A.F., Dimauro, S., and De Vivo, D.C. (1997). Deficient muscle carnitine transport in primary carnitine deficiency. *Pediatr. Res.* 42:583–587.

Poole, R.C., and Halestrap, A.P. (1993). Transport of lactate and other monocarboxylates across mammalian plasma membranes. *Am. J. Physiol.* 264:C761–C782.

Poole, R.C., Sansom, C.E., and Halestrap, A.P. (1996). Studies of the membrane topology of the rat erythrocyte H+/lactate cotransporter (MCT1). *Biochem. J.* 320:817–824.

Pressacco, J., Wiley, J.S., Jamieson, G.P., Erlichman, C., and Hedley, D.W. (1995). Modulation of the equilibrative nucleoside transporter by inhibitors of DNA-synthesis. *Br. J. Cancer* 72:939–942.

Price, N.T., Jackson, V.N., and Halestrap, A.P. (1998). Cloning and sequencing of four new mammalian monocarboxylate transporter (MCT) homologues confirms the existence of a transporter family with an ancient past. *Biochem. J.* 329(Pt 2):321–328.

Prime-Chapman, H.M., Fearn, R.A., Cooper, A.E., Moore, V., and Hirst, B.H. (2004). Differential multidrug resistance-associated protein 1 through 6 isoform expression and function in human intestinal epithelial Caco-2 cells. *J. Pharmacol. Exp. Ther.* 311(2):476–484.

Pulaski, L., Kania, K., Ratajewski, M., Uchiumi, T., Kuwano, M., and Bartosz, G. (2005). Differential regulation of the human MRP2 and MRP3 gene expression by glucocorticoids. *J. Steroid Biochem. Mol. Biol.* 96(3–4):229–234.

Rabindran, S.K., He, H., Singh, M., Brown, E., Collins, K.I., Annable, T., and Greenberger, L.M. (1998). Reversal of a novel multidrug resistance mechanism in human colon carcinoma cells by fumitremorgin C. *Cancer Res.* 58(24):5850–5858.

Rahman, B., Schneider, H.P., Broer, A., Beitmer, J.W., and Broer, S. (1999). Helix 8 and helix 10 are involved in substrate recognition in the rat monocarboxylate transporter MCT1. *Biochemistry* 38:11577–11584.

Rausch-Derra, L.C., Hartley, D.P., Meier, P.J., and Klaassen, C.D. (2001). Differential effects of microsomal enzyme-inducing chemicals on the hepatic expression of rat organic anion transporters, Oatp1 and Oatp2. *Hepatology* 33:1469–1478.

Reid, G., Wielinga, P., Zelcer, N., De Haas, M., Van Deemter, L., Wijnholds, J., Balzarini, J., and Borst, P. (2003). Characterization of the transport of nucleoside analog drugs by the human multidrug resistance proteins MRP4 and MRP5. *Mol. Pharmacol.* 63(5):1094–1103.

Ritzel, M.W., Yao, S.Y., Huang, M.Y., Elliott, J.F., Cass, C.E., and Young, J.D. (1997). Molecular cloning and functional expression of cDNAs encoding a human Na+-nucleoside cotransporter (hCNT1). *Am. J. Physiol.* 272:C707–C714.

Ritzel, M.W., Ng, A.M., Yao, S.Y., Graham, K., Loewen, S.K., Smith, K.M., Ritzel, R.G., Mowles, D.A., Carpenter, P., Chen, X.Z., Karpinski, E., Hyde, R.J., Baldwin, S.A., Cass, C.E., and Young, J.D. (2001). Molecular identification and characterization of novel human and mouse concentrative Na+-nucleoside cotransporter proteins (hCNT3 and mCNT3) broadly selective for purine and pyrimidine nucleosides (system cib). *J. Biol. Chem.* 276:2914–2927.

Rome, S., Barbot, L., Windsor, E., Kapel, N., Tricottet, V., Huneau, J.F., Reynes, M., Gobert, J.G., and Tomé, D. (2002). The regionalization of PepT1, NBAT and EAAC1 transporters in the small intestine of rats are uncharged from birth to adulthood. *J. Nutr.* 132:1009–1011.

Rost, D., Mahner, S., Sugiyama, Y., and Stremmel, W. (2002). Expression and localization of the multidrug resistance-associated protein 3 in rat small and large intestine. *Am. J. Physiol. Gastrointest. Liver Physiol.* 282(4):G720–G726.

Rost, D., Kopplow, K., Gehrke, S., Mueller, S., Friess, H., Ittrich, C., Mayer, D., and Stiehl, A. (2005). Gender-specific expression of liver organic anion transporters in rat. *Eur. J. Clin. Invest.* 35:635–643.

Rubin, L.J., Johnson, L.R., Dodam, J.R., Dhalla, A.K., Magliola, L., Laughlin, M.H., and Jones, A.W. (2000). Selective transport of adenosine into porcine coronary smooth muscle. *Am. J. Physiol. Heart Circ. Physiol.* 279(3):H1397–H1410.

Rubio-Aliaga, I., and Daniel, H. (2002). Mammalian peptide transporters as targets for drug delivery. *Trends Pharm. Sci.* 23(9):434–440.

Sai, Y., and Tsuji, A. (2004). Transporter-mediated drug delivery: recent progress and experimental approaches. *Drug Discov. Today* 9:712–720.

Saier, M.H., Jr. (2000). A functional-phylogenetic classification system for transmembrane solute transporters. *Microbiol. Mol. Biol. Rev.* 64(2):354–411.

Saier, M.H., Jr., Eng, B.H., Fard, S., Garg, J., Haggerty, D.A., Hutchinson, W.J., Jack, D.L., Lai, E.C., Liu, H.J., Nusinew, D.P., Omar, A.M., Pao, S.S., Paulsen, I.T., Quan, J.A., Sliwinski, M., Tseng, T.T., Wachi, S., and Young, G.B. (1999). Phylogenetic characteristics of novel transport protein families revealed by genome analyses. *Biochim. Biophys. Acta* 1422:1–56.

Saito, H., Okuda, M., Terada, T., Sasaki, S., and INRI, K.-I. (1995). Cloning and characterization of a rat H+/peptide cotransporter mediating absorption of beta-lactam antibiotics in the intestine and kidney. *J. Pharmacol. Exp. Ther.* 275:1631–1637.

Sakata, K., Yamashita, T., Maeda, M., Moriyama, Y., Shimada, S., and Tohyama, M. (2001). Cloning of a lymphatic peptide/histidine transporter. *Biochem. J.* 356:53–60.

Sarkadi, B., Ozvegy-Laczka, C., Nemet, K., and Varadi, A. (2004). ABCG2 – a transporter for all seasons. *FEBS Lett.* 567(1):116–120.

Satlin, L.M., Amin, V., and Wolkoff, A.W. (1997). Organic anion transporting polypeptide mediates organic anion/HCO3-exchange. *J. Biol. Chem.* 272:26340–26345.

Sawada, K., Terada, T., Saito, H., Hashimoto, Y., and Inui, K. (1999). Effects of glibenclamide on glycylsarcosine transport by the rat peptide transporters PEPT1 and PEPT2. *Br. J. Pharmacol.* 128(6):1159–1164.

Schaub, T.P., Kartenbeck, J., König, J., Vogel, O., Witzgall, R., Kriz, W., and Keppler, D. (1997). Expression of the conjugate export pump encoded by the mrp2 gene in the apical membrane of kidney proximal tubules. *J. Am. Soc. Nephrol.* 8:1213–1221.

Scheffer, G.L., Maliepaard, M., Pijnenborg, A.C., van Gastelen, M.A., de Jong, M.C., Schroeijers, A.B., van der Kolk, D.M., Allen, J.D., Ross, D.D., van der Valk, P., Dalton, W.S., Schellens, J.H., and Scheper, R.J. (2000). Breast cancer resistance protein is localized at the plasma membrane in mitoxantrone- and topotecan-resistant cell lines. *Cancer Res.* 60(10):2589–2593.

Scheffer, G.L., Kool, M., De Haas, M., De Vree, J.M., Pijnenborg, A.C., Bosman, D.K., Elferink, R.P., Van Der Valk, P., Borst, P., and Scheper, R.J. (2002). Tissue distribution and induction of human multidrug resistant protein 3. *Lab. Invest.* 82:193–201.

Schinkel, A.H., and Jonker, J.W. (2003). Mammalian drug efflux transporters of the ATP binding cassette (ABC) family: an overview. *Adv. Drug Deliv. Rev.* 55:3–29.

Schrenk, D., Gant, T.W., Preisegger, K.H., Silverman, J.A., Marino, P.A., and Thorgeirsson, S.S. (1993). Induction of multidrug resistance gene expression during cholestasis in rats and nonhuman primates. *Hepatology* 17:854–860.

Schuster, V.L., Lu, R., Kanai, N., Bao, Y., Rosenberg, S., Prie, D., Ronco, P., and Jennings, M.L. (1996). Cloning of the rabbit homologue of mouse 'basigin' and rat 'OX-47': kidney cell type-specific expression, and regulation in collecting duct cells. *Biochim. Biophys. Acta* 1311(1):13–19.

Seelig, A., and Landwojtowicz, E. (2000). Structure-activity relationship of P-glycoprotein substrates and modifiers. *Eur. J.Pharm. Sci.* 12:31–40.

Sekine, T., Cha, S.H., Tsuda, M., Apiwattanakul, N., Nakajima, N., Kanai, Y., and Endou, H. (1998). Identification of multispecific organic anion transporter 2 expressed predominantly in the liver. *FEBS Lett.* 429:179–182.

SenGupta, D.J., Lum, P.Y., Lai, Y., Shubochkina, E., Bakken, A.H., Schneider, G., and Unadkat, J.D. (2002). A single glycine mutation in the equilibrative nucleoside transporter gene, hENT1, alters nucleoside transport activity and sensitivity to nitrobenzylthioinosine. *Biochemistry* 41:1512–1519.

Shapiro, A.B., Fox, K., Lam, P., and Ling, V. (1999). Stimulation of P-glycoprotein mediated drug transport by prazosin and progesterone. Evidence for a third drug-binding site. *Eur. J. Biochem.* 259:841–850.

Sharom, F.J. (1997). The P-glycoprotein efflux pump: how does it transport drugs? *J. Membr. Biol.* 160:161–175.

Shen, H., Smith, D.E., and Brosius, F.C., III. (2001). Developmental expression of PepT1 and PepT2 in rat small intestine, colon, and kidney. *Pediatr. Res.* 49:789–795.

Shi, X., Bai, S., Ford, A.C., Burk, R.D., Jacquemin, E., Hagenbuch, B., Meier, P.J., and Wolkoff, A.W. (1995). Stable inducible expression of a functional rat liver organic anion transport protein in HeLa cells. *J. Biol. Chem.* 270:25591–25595.

Shimakura, J., Terada, T., Katsura, T., and Inui, K.I. (2005). Characterization of the Human Peptide Transporter PEPT1 promoter: Sp1 Functions as a Basal Transcriptional Regulator of Human PEPT1. *Am. J. Physiol. Gastrointest. Liver Physiol.* 289(3):G471–G477.

Shiraga, T., Miyamoto, K., Tanaka, H., Yamamoto, H., Taketani, Y., Morita, K., Tamai, I., Tsuji, A., and Takeda, E. (1999). Cellular and molecular mechanism of dietary regulation on rat intestinal H+/peptide transporter PepT1. *Gastroenterology* 116:354–362.

Shore, P.A., Brodie, B.B., and Hogben, C.A.M. (1957). The gastric secretion of drugs: a pH partition hypothesis. *J. Pharmacol. Exp. Ther.* 119:361–369.
Shu, C., Shen, H., Hopfer, U., and Smith, D.E. (2001). Mechanism of intestinal absorption and renal reabsorption of an orally active ace inhibitor: uptake and transport of fosinopril in cell cultures. *Drug Metab. Dispos.* 29(10):1307–1315.
Shu, Y., Leabman, M.K., Feng, B., Mangravite, L.M., Huang, C.C., Stryke, D., Kawamoto, M., Johns, S.J., DeYoung, J., Carlson, E., Ferrin, T.E., Herskowitz, I., and Giacomini, K.M. (2003). Evolutionary conservation predicts function of variants of the human organic cation transporter, OCT1. *Proc. Natl Acad. Sci. USA* 100:5902–5907.
Sikic, B.I. (1997). Pharmacologic approaches to reversing multidrug resistance. *Semin Hematol.* 34(4 Suppl 5):40–47.
Simon, F.R., Fortune, J., Iwahashi, M., Bowman, S., Wolkoff, A., and Sutherland, E. (1999). Characterization of the mechanisms involved in the gender differences in hepatic taurocholate uptake. *Am. J. Physiol.* 276:G556–G565.
Slitt, A.L., Cherrington, N.J., Hartley, D.P., Leazer, T.M., and Klaassen, C.D. (2002). Tissue distribution and renal developmental changes in rat organic cation transporter mRNA levels. *Drug Metab. Dispos.* 30:212–219.
Smit, J.J., Schinkel, A.H., Oude, R.P., Groen, A.K., Wagenaar, E., van, D.L., Mol, C.A., Ottenhoff, R., van der Lugt, N.M., and van Roon, M.A. (1993). Homozygous disruption of the murine mdr2 P-glycoprotein gene leads to a complete absence of phospholipid from bile and to liver disease. *Cell* 75(3):451–462.
Soler, C., Felipe, A., Mata, J.F., Casado, F.J., Celada, A., and Pastor-Anglada, M. (1998). Regulation of nucleoside transport by lipopolysaccharide, phobol esters, and tumor necrosis factor-alpha in human B-lymphocytes. *J. Biol. Chem.* 273:26939–26945.
Soler, A.P., Gilliard, G., Xiong, Y., Knudsen, K.A., Martin, J.L., DeSuarez, G.B., Mota Gamboa, J.D., Mosca, W., and Zoppi, L.B. (2001). Overexpression of neural cell adhesion molecule in Chagas' myocarditis. *Hum. Pathol.* 32:149–155.
Song, X., Lorenzi, P.L., Landowski, C.P., Vig, B.S., Hilfinger, J.M., and Amidon, G.L. (2005). Amino acid ester prodrugs of the anticancer agent gemcitabine: synthesis, bioconversion, metabolic bioevasion, and hPEPT1-mediated transport. *Mol. Pharm.* 2(2):157–167.
Sorensen, M., Steenberg, B., Knipp, G.T., Wang, W., Steffansen, B., Frokjaer, S., and Borchardt, R.T. (1997). The effect of beta-turn structure on the permeation of peptides across monolayers of bovine brain microvessel endothelial cells. *Pharm. Res.* 14(10):1341–1348.
Soul-Lawton, J., Seaber, E., On, N., Wootton, R., Rolan, P., and Posner, J. (1995). Absolute bioavailability and metabolic disposition of valaciclovir, the L-valyl ester of aciclovir, following oral administration to humans. *Antimicrob. Agents Chemother.* 39:2759–2764.
Spahn-Langguth, H., and Langguth, P. (2001). Grapefruit juice enhances intestinal absorption of the P-glycoprotein substrate talinolol. *Eur. J. Pharm. Sci.* 12(4):361–367.
Stanley, C.A., DeLeeuw, S., Coates, P.M., Vianey-Liaud, C., Divry, P., Bonnefont, J.P., Saudubray, J.M., Haymond, M., Trefz, F.K., Breningstall, G.N., Wappner, R.S., Byrd, D.J., Sansaricq, C., Tein, I., Grover, W., Valle, D., Rutledge, S.L., and Treem, W.R. (1991). Chronic cardiomyopathy and weakness or acute coma. *Ann. Neurol.* 30(5):709–716.
Staud, F., and Pavek, P. (2005). Breast cancer resistance protein (BCRP/ABCG2). *Int. J. Biochem. Cell. Biol.* 37(4):720–725.
Staudinger, J.L., Goodwin, B., Jones, S.A., Hawkins-Brown, D., MacKenzie, K.I., LaTour, A., Liu, Y., Klaassen, C.D., Brown, K.K., Reinhard, J., Willson, T.M., Koller,

B.H., and Kliewe, S.A. (2001). The nuclear receptor PXR is a lithocholic acid sensor that protects against liver toxicity. *Proc. Natl Acad. Sci. USA* 98:3369–3374.

Steffansen, B., Nielsen, C.U., Brodin, B., Eriksson, A.H., Andersen, R., and Frokjaer, S. (2004). Intestinal solute carriers: an overview of trends and strategies for improving oral drug absorption. *Eur. J. Pharm. Sci.* 21:3–16.

Steffansen, B., Nielsen, C.U., and Frokjaer, S. (2005). Delivery aspects of small peptides and substrates for peptide transporters. *Eur. J. Pharm. Biopharm.* 60(2):241–245.

Steiner, H.Y., Naider, F., and Becker, J.M. (1995). The PTR family: a new group of peptide transporters. *Mol. Microbiol.* 16:825–834.

Steingrimsdottir, H., Gruber, A., Palm, C., Grimfors, G., Kalin, M., and Eksborg, S. (2000). Bioavailability of aciclovir after oral administration of aciclovir and its prodrug valaciclovir to patients with leukopenia after chemotherapy. *Antimicrob. Agents Chemother.* 44(1):207–209.

Stewart, B.H., Chan, O.H., Jezyk, N., and Fleisher, D. (1997). Discrimination between drug candidates using models for evaluation of intestinal absorption. *Adv. Drug Deliv. Rev.* 23(1):27–45.

Strausberg, R.L., Feingold, E.A., Grouse, L.H., Derge, J.G., Klausner, R.D., Collins, F.S., Wagner, L., Shenmen, C.M., Schuler, G.D., Altschul, S.F., Zeeberg, B., Buetow, K.H., Schaefer, C.F., Bhat, N.K., Hopkins, R.F., Jordan, H., Moore, T., Max, S.I., Wang, J., Hsieh, F., Diatchenko, L., Marusina, K., Farmer, A.A., Rubin, G.M., Hong, L., Stapleton, M., Soares, M.B., Bonaldo, M.F., Casavant, T.L., Scheetz, T.E., Brownstein, M.J., Usdin, T.B., Toshiyuki, S., Carninci, P., Prange, C., Raha, S.S., Loquellano, N.A., Peters, G.J., Abramson, R.D., Mullahy, S.J., Bosak, S.A., McEwan, P.J., McKernan, K.J., Malek, J.A., Gunaratne, P.H., Richards, S., Worley, K.C., Hale, S., Garcia, A.M., Gay, L.J., Hulyk, S.W., Villalon, D.K., Muzny, D.M., Sodergren, .J., Lu, X., Gibbs, R.A., Fahey, J., Helton, E., Ketteman, M., Madan, A., Rodrigues, S., Sanchez, A., Whiting, M., Young, A.C., Shevchenko, Y., Bouffard, G.G., Blakesley, R.W., Touchman, J.W., Green, E.D., Dickson, M.C., Rodriguez, A.C., Grimwood, J., Schmutz, J., Myers, R.M., Butterfield, Y.S., Krzywinski, M.I., Skalska, U., Smailus, D.E., Schnerch, A., Schein, J.E., Jones, S.J., and Marra, M.A. (2002). Generation and initial analysis of more than 15,000 full-length human and mouse cDNA sequences. *Proc. Natl Acad. Sci. USA* 99(26):16899–16903.

Sugawara, I., Kataoka, I., Morishita, Y., Hamada, H., Tsuruo, T., Itoyama, S., and Mori, S. (1988). Tissue distribution of P-glycoprotein encoded by a multidrug-resistant gene as revealed by a monoclonal antibody, MRK 16. *Cancer Res.* 48(7):1926–1929.

Sugawara, M., Huang, W., Fei, Y.J., Leibach, F.H., Ganapathy, V., and Ganapathy, M.E. (2000). Transport of valganciclovir, a ganciclovir prodrug, via peptide transporters PEPT1 and PEPT2. *J. Pharm. Sci.* 89(6):781–789.

Sun, D., Landowski, C.P., Chu, X., Wallsten, R., Fleisher, D., and Amidon, G.L. (2001). Drug inhibition of Gly-Sar uptake and hPepT1 localization using hPepT1-GFP fusion protein. *AAPS PharmSci* 3:Article 2.

Sun, D., Yu, L.X., Hussain, M.A., Wall, D.A., Smith, R.L., and Amidon, G.L. (2004). In vitro testing of drug absorption for drug 'developability' assessment: forming an interface between in vitro preclinical data and clinical outcome. *Curr. Opin. Drug Discov. Devel.* 7(1):75–85.

Sundaram, M., Yao, S.Y., Ng, A.M., Cass, C.E., Baldwin, S.A., and Young, J.D. (2001). Equilibrative nucleoside transporters: mapping regions of interaction for the substrate analogue nitrobenzylthioinosine (NBMPR) using rat chimeric proteins. *Biochemistry* 40(27):8146–8151.

Suzuki, H., and Sugiyama, Y. (1998). Excretion of GSSG and glutathione conjugates mediated by MRP1 and cMOAT/MRP2. *Semin. Liver Dis.* 18(4):359–376.

Suzuki, H., and Sugiyama, Y. (1999). Transporters for bile acids and organic anions. *Pharm. Biotechnol.* 12:387–439.

Swaan, P.W., and Tukker, J.J. (1997). Molecular determinants of recognition for the intestinal peptide carrier. *J. Pharm. Sci.* 86:596–602.

Taipalensuu, J., Tornblom, H., Lindberg, G., Einarsson, C., Sjoqvist, F., Melhus, H., Garberg, P., Sjostrom, B., Lundgren, B., and Artursson, P. (2001). Correlation of gene expression of ten drug efflux proteins of the ATP-binding cassette transporter family in normal human jejunum and in human intestinal epithelial Caco-2 cell monolayers. *J. Pharmacol. Exp. Ther.* 299:164–170.

Takahashi, K., Masuda, S., Nakamura, N., Saito, H., Futami, T., Doi, T., and Inui, K.-I. (2001). Upregulation of the H+-peptide cotransporter PepT2 in rat remnant kidney. *Am. J. Physiol. Renal Physiol.* 281:F1109–F1116.

Tamai, I., Yabuuchi, H., Nezu, J., Sai, Y., Oku, A., Shimane, M., and Tsuji, A. (1997). Cloning and characterization of a novel human pH-dependent organic cation transporter, OCTN1. *FEBS Lett.* 419:107–111.

Tamai, I., Ohashi, R., Nezu, J.I., Yabuuchi, H., Oku, A., Shimane, M., Sai, Y., and Tsuji, A. (1998). Molecular and functional identification of sodium ion-dependent, high affinity human carnitine transporter OCTN2. *J. Biol. Chem.* 273:20378–20382.

Tamai, I., Nezu, J., Uchino, H., Sai, Y., Oku, A., Shimane, M., and Tsuji, A. (2000a). Molecular identification and characterization of novel members of the human organic anion transporter (OATP) family. *Biochem. Biophys. Res. Commun.* 273:251–260.

Tamai, I., Ohashi, R., Nezu, J.I., Sai, Y., Kobayashi, D., Oku, A., Shimane, M., and Tsuji, A. (2000b). Molecular and functional characterization of organic cation/carnitine transporter family in mice. *J. Biol. Chem.* 275:40064–40072.

Tamai, I., China, K., Sai, Y., Kobayashi, D., Nezu, J.I., Kawahara, E., and Tsuji, A. (2001). Na(+)-coupled transport of L-carnitine via high-affinity carnitine transporter OCTN2 and its subcellular localization in kidney. *Biochim. Biophys. Acta* 1512:273–284.

Tanaka, H., Miyamoto, K.-I., Morita, K., Haga, H., Segawa, H., Shiraga, T., Fujioka, A., Kouda, T., Taketani, Y., Hisano, S., Fukiji, Y., Kitagawa, K., and Takeda, E. (1998). Regulation of the PepT1 peptide transporter in the rat small intestine in response to 5-fluorouracil-induced injury. *Gastroenterology* 114:714–723.

Tanaka, T., Uchiumi, T., Hinoshita, E., Inokuchi, A., Toh, S., Wada, M., Takano, H., Kohno, K., and Kuwano, M. (1999). The human multidrug resistance protein 2 gene: functional characterization of the 5-flanking region and expression in hepatic cells. *Hepatology* 30:1507–1712.

Tein, I. (2003). Carnitine transport: pathophysiology and metabolism of known molecular defects. *J. Inherit. Metab. Dis.* 26:147–169.

Tein, I., De Vivo, D.C., Bierman, F., Pulver, P., De Meirleir, L.J., Cvitanovic-Sojat, L., Pagon, R.A., Bertini, E., Dionisi-Vici, C., Servidei, S., and DiMauro, S. (1990). Impaired skin fibroblast carnitine uptake in primary systemic carnitine deficiency manifested by childhood carnitine-responsive cardiomyopathy. *Pediatr. Res.* 28:247–255.

Temple, C.S., Bronk, J.K., Bailey, P.D., and Boyd, C.A. (1995). Substrate charge dependence of stoichiometry shows membrane potential is driving force for proton-peptide cotransport in rat renal cortex. *Pflugers Arch. Eur. J. Physiol.* 430:825–829.

Temple, C.S., Bailey, P.D., Bronk, J.R., and Boyd, C.A.R. (1996). A model for the kinetics of neutral and anionic dipeptide-proton cotransport by the apical membrane of rat kidney cortex. *J. Physiol.* 494(Pt 3):795–808.

Terada, T., Sawada, K., Saito, H., Hashimoto, Y., and Inui, K.-I. (1999). Functional characteristics of basolateral peptide transporter in the human intestinal cell line Caco-2. *Am. J. Physiol.* 276:G1435–G1441.

Terada, T., Saito, H., Sawada, K., Hashimoto, Y., and Inui, K.-I. (2000a). N-terminal halves of rat H+/peptide transporters are responsible for their substrate recognition. *Pharm. Res.* 17:15–20.

Terada, T., Sawada, K., Tatsuya, I., Saito, H., Hashimoto, Y., and Inui, K.-I. (2000b). Functional expression of novel peptide transporter in renal basolateral membranes. *Am. J. Physiol. Renal. Physiol.* 279:F851–F857.

Terao, T., Hisanaga, E., Sai, Y., Tamai, I., and Tsuji, A. (1996). Active secretion of drugs from the small intestinal epithelium in rats by P-glycoprotein functioning as an absorption barrier. *J. Pharm. Pharmacol.* 48(10):1083–1089.

Thamotharan, M., Bawani, S.Z., Zhou, X., and Adibi, S.A. (1998). Mechanism of dipeptide stimulation of its own transport in a human intestinal cell line. *Proc. Assoc. Am. Physicians* 110:361–368.

Thamotharan, M., Bawani, S.Z., Zhou, X., and Adibi, S.A. (1999a). Functional and molecular expression of intestinal oligopeptide transporter (PepT-1) after a brief fast. *Metabolism* 48:681–684.

Thamotharan, M., Bawani, S.Z., Zhou, X., and Adibi, S.A. (1999b). Hormonal regulation of oligopeptide transporter pept-1 in a human intestinal cell line. *Am J. Physiol.* 276:C821–C826.

Theis, S., Doring, F., and Daniel, H. (2001). Expression of the myc/His-tagged human peptide transporter hPepT1 in yeast for protein purification and functional analysis. *Protein Exp. Purif.* 22:436–442.

Thiebaut, F., Tsuruo, T., Hamada, H., Gottesman, M.M., Pastan, I., and Willingham, M.C. (1987). Cellular localization of the multidrug resistance gene product P-glycoprotein in normal human tissues. *Proc. Natl Acad. Sci. USA* 84:7735–7738.

Thiebaut, F., Tsuruo, T., Hamada, H., Gottesman, M.M., Pastan, I., and Willingham, M.C. (1989). Immunohistochemical localization in normal tissues of different epitopes in the multidrug transport protein P170: evidence for localization in brain capillaries and cross-reactivity of one antibody with a muscle protein. *J. Histochem. Cytochem.* 37(2):159–164.

Tian, Q., Zhang, J., Tan, T.M., Chan, E., Duan, W., Chan, S.Y., Boelsterli, U.A., Ho, P.C., Yang, H., Bian, J.S., Huang, M., Zhu, Y.Z., Xiong, W., Li, X., and Zhou, S. (2005). Human multidrug resistance associated protein 4 confers resistance to camptothecins. *Pharm. Res.* 22(11):1837–1853;

Tirona, R.G., and Kim, R.B. (2002). Pharmacogenomics of organic anion-transporting polypeptides (OATP). *Adv. Drug Deliv. Rev.* 54:1343–1352.

Tirona, R.G., Leake, B.F., Merino, G., and Kim, R.B. (2001). Polymorphisms in OATP-C: identification of multiple allelic variants associated with altered transport activity among European- and African-Americans. *J. Biol. Chem.* 276:35669–35675.

Torok, M., Gutmann, H., Fricker, G., and Drewe, J. (1999). Sister of P-gp expression in different tissues. *Biochem. Pharmacol.* 57(7):833–835.

Torok, H.P., Glas, J., Tonenchi, L., Lohse, P., Muller-Myhsok, B., Limbersky, O., Neugebauer, C., Schnitzler, F., Seiderer, J., Tillack, C., Brand, S., Brunnler, G., Jagiello, P., Epplen, J.T., Griga, T., Klein, W., Schiemann, U., Folwaczny, M., Ochsenkuhn, T., and Folwaczny, C. (2005). Polymorphisms in the DLG5 and OCTN cation transporter genes in Crohn's disease. *Gut* 54:1421–1427.

Tortora, G.J., and Grabowski., S.R. (1993). *Principles of Anatomy and Physiology*, Harper Collins College Publishers, New York, p. 768.

Triscari, J., O'Donnell, D., Zinny, M., and Pan, H.Y. (1995). Gastrointestinal absorption of pravastatin in healthy subjects. *J. Clin. Pharmacol.* 35:142–144.

Trompier, D., Baubichon-Cortay, H., Chang, X.B., Maitrejean, M., Barron, D., Riordon, J.R., and Di Pietro, A. (2003). Multiple flavonoid-binding sites within multidrug resistance protein MRP1. *Cell Mol. Life Sci.* 60:2164–2177.

Urakami, Y., Nakamura, N., Takahashi, K., Okuda, M., Saito, H., Hashimoto, Y., and Inui, K. (1999). Gender differences in expression of organic cation transporter OCT2 in rat kidney. *FEBS Lett.* 461(3):339–342.

Urakami, Y., Okuda, M., Saito, H., and Inui, K. (2000). Hormonal regulation of organic cation transporter OCT2 expression in rat kidney. *FEBS Lett.* 473(2):173–176.

Utoguchi, N., and Audus, L.L. (2000). Carrier-mediated transport of valproic acid in BeWo cells, a human trophoblast cell line. *Int. J. Pharm.* 195:115–124.

Vaidyanathan, J.B., and Walle, T. (2001). Transport and metabolism of the tea flavonoid (−)-epicatechin by the human intestinal cell line Caco-2. *Pharm. Res.* 18(10): 1420–1425.

Valdes, R., Ortega, M.A., Casado, F.J., Felipe, A., Gil, A., Sanchez-Pozo, A., and Pastor-Anglada, M. (2000). Nutritional regulation of nucleoside transporter expression in rat small intestine. *Gastroenterology* 119:1623–1630.

Valdes, R., Casado, F.J., and Pastor-Anglada, M. (2002). Cell-cycle-dependent regulation of CNT1, a concentrative nucleoside transporter involved in the uptake of cell-cycle-dependent nucleoside-derived anticancer drugs. *Biochem. Biophys. Res. Commun.* 296:575–579.

Van, L., Pan, Y.X., Bloomquist, J.R., Webb, K.E., Jr., and Wong, E.A. (2005). Developmental regulation of a turkey intestinal peptide transporter (PepT1). *Poult. Sci.* 84(1):75–82.

Van Aubel, R.A., Hartog, A., Bindels, R.J., Van Os, C.H., and Russel, F.G. (2000). Expression and immunolocalization of multidrug resistance protein 2 in rabbit small intestine. *Eur. J. Pharmacol.* 400(2–3):195–198.

Van Aubel, R.A., Smeets, P.H., Peters, J.G., Bindels, R.J., and Russel, F.G. (2002). The MRP4/ABCC4 gene encodes a novel apical organic anion transporter in human kidney proximal tubules: putative efflux pump for urinary cAMP and cGMP. *J. Am. Soc. Nephrol.* 13:595–603.

Velpandian, T., Jasuja, R., Bhardwaj, R.K., Jaiswal, J., and Gupta, S.K. (2001). Piperine in food: interference in the pharmacokinetics of phenytoin. *Eur. J. Drug Metab. Pharmacokinit.* 26:241–247.

Vickers, M.F., Mani, R.S., Sundaram, M., Hogue, D.L., Young, J.D., Baldwin, S.A., and Cass, C.E. (1999). Functional production and reconstitution of the human equilibrative nucleoside transporter (hENT1) in Saccharomyces cerevisiae. Interaction of inhibitors of nucleoside transport with recombinant hENT1 and a glycosylation-defective derivative (hENT1/N48Q). *Biochem. J.* 339(1):21–32.

Vickers, M.F., Kumar, R., Visser, F., Zhang, J., Charania, J., Raborn, R.T., Baldwin, S.A., Young, J.D., and Cass, C.E. (2002). Comparison of the interaction of uridine, cytidine, and other pyrimidine nucleoside analogs with recombinant human equilibrative nucleoside transporter 2 (hENT2) produced in Saccharomyces cerevisiae. *Biochem. Cell. Biol.* 80(5):639–644.

Vig, B.S., Lorenzi, P.J., Mittal, S., Landowski, C.P., Shin, H.C., Mosberg, H.I., Hilfinger, J.M., and Amidon, G.L. (2003). Amino acid ester prodrugs of floxuridine: synthesis and effects of structure, stereochemistry, and site of esterification on the rate of hydrolysis. *Pharm. Res.* 20(9):1381–1388.

Vijayalakshmi, D., and Belt, J.A. (1988). Sodium-dependent nucleoside transport in mouse intestinal epithelial cells. Two transport systems with differing substrate specificities. *J. Biol. Chem.* 263:19419–19423.

Visser, F., Vickers, M.F., Ng, A.M., Baldwin, S.A., Young, J.D., and Cass, C.E. (2002). Mutation of residue 33 of human equilibrative nucleoside transporters 1 and 2 alters sensitivity to inhibition of transport by dilazep and dipyridamole. *J. Biol. Chem.* 277:395–401.

Wagner, C.A., Lükewille, U., Kaltenbach, S., Moschen, I., Broer, A., Risler, T., Broer, S., and Lang, F. (2000). Functional and pharmacological characterization of the human Na+/carnitine cotransporter hOCTN2. *Am. J. Physiol. Renal. Physiol.* 279:F584–F591.

Walker, D., Thwaites, D.T., Simmons, N.L., Gilbert, H.J., and Hirst, B.H. (1998). Substrate upregulation of the human small intestinal peptide transporter, hPepT1. *J. Physiol.* 507:697–706.

Walters, H.C., Craddock, A.L., Fusegawa, H., Willingham, M.C., and Dawson, P.A. (2000). Expression, transport properties, and chromosomal location of organic anion transporter subtype 3. *Am. J. Physiol. Gastrointest. Liver Physiol.* 279:G1188–G1200.

Walter-Sack, I., and Klotz, U. (1996). Influence of diet and nutritional status on drug metabolism. *Clin. Pharmacokinet.* 31(1):47–64.

Wang, J., and Giacomini, K.M. (1999). Serine 318 is essential for the pyrimidine selectivity of the N2 Na+-nucleoside transporter. *J. Biol. Chem.* 274:2298–2302.

Wang, J., Su, S.F., Dresser, M.J., Schaner, M.E., Wahington, C.B., and Giacomini, K.M. (1997). Na+-dependent purine nucleoside transporter from human kidney: cloning and functional characterization. *Am. J. Physiol.* 273:F1058–F1065.

Wang, D.S., Jonker, J.W., Kato, Y., Kusuhara, H., Schinkel, A.H., and Sugiyama, Y. (2002). Involvement of organic cation transporter 1 in hepatic and intestinal distribution of metformin. *J. Pharmacol. Exp. Ther.* 302:510–515.

Wang, D.S., Kusuhara, H., Kato, Y., Jonker, J.W., Schinkel, A.H., and Sugiyama, Y. (2003). Involvement of organic cation transporter 1 in the lactic acidosis caused by metformin. *Mol. Pharmacol.* 63:844–888.

Wang, Q., Bhardwaj, R.K., Herrera-Ruiz, D., Hanna, N.N., Gudmundsson, O.S., Buranachokpaisan, T., Hidalgo, I.J., and Knipp, G.T. (2004). Expression of multiple drug resistance conferring proteins in normal Chinese and Caucasian small and large intestinal tissue samples. *Mol. Pharm.* 1(6):447–454.

Ward, J.L., Leung, G.P.H., Toan, S.V., and Tse, C.M. (2003). Functional analysis of site-directed glycosylation mutants of the human equilibrative nucleoside transporter-2. *Arch. Biochem. Biophys.* 411(1):19–26.

Watanabe, K., Sawano, T., Jinriki, T., and Sato, J. (2004). Studies on intestinal absorption of sulpiride (3): intestinal absorption of sulpiride in rats. *Biol. Pharm. Bull.* 27:77–81.

van de Waterbeemd, H. (2002). High-throughput and in silico techniques in drug metabolism and pharmacokinetics. *Curr. Opin. Drug Discov. Devel.* 5(1):33–43.

Watkins, P.B. (1997). The barrier function of CYP3A4 and P-glycoprotein in the small bowel. *Adv. Drug Deliv. Rev.* 27:161–170.

Weller, S., Blum, M., Doucette, M., Burnette, T., Cederberg, D.M., de Miranda, P., and Smiley, M.L. (1993). Pharmacokinetics of the aciclovir pro-drug valaciclovir after escalating single- and multiple-dose administration to normal volunteers. *Clin. Pharmacol. Ther.* 54:595–605.

Wenzel, U., Gebert, I., Weintraut, H., Weber, W.M., Clauss, W., and Daniel, H. (1996). Transport characteristics of differently charged cephalosporin antibiotics in oocytes expressing the cloned intestinal peptide transporter PepT1 and in human intestinal Caco-2 cells. *J. Pharmacol. Exp. Ther.* 277(2):831–839.

Wenzel, U., Diehl, D., Herget, M., Kuntz, S., and Daniel, H. (1999). Regulation of the high-affinity H+/peptide cotransporter in renal LLC-PK1 cells. *J. Cell. Physiol.* 178:341–348.

Wenzel, U., Kuntz, S., Diestel, S., and Daniel, H. (2002). PepT1-mediated cefixime uptake into human intestinal epithelial cells is increased by Ca2+ channels blockers. *Antimicrob. Agents. Chemother.* 46:1375–1380.

Wielinga, P.R., Reid, G., Challa, E.E., van der Heijden, I., van Deemter, L., de Haas, M., Mol, C., Kuil, A.J., Groeneveld, E., Schuetz, J.D., Brouwer, C., De Abreu, R.A., Wijnholds, J., Beijnen, J.H., and Borst, P. (2002). Thiopurine metabolism and identification of the thiopurine metabolites transported by MRP4 and MRP5 overexpressed in human embryonic kidney cells. *Mol. Pharmacol.* 62:1321–1331.

Wijnholds, J., Mol, C.A., van Deemter, L., de Haas, M., Scheffer, G.L., Baas, F., Beijnen, J.H., Scheper, R.J., Hatse, S., De Clercq, E., Balzarini, J., and Borst, P. (2000). Multidrug-resistance protein 5 is a multispecific organic anion transporter able to transport nucleotide analogs. *Proc. Natl Acad. Sci. USA* 97:7476–7481.

Williams, J.B., Rexer, B., Sirripurapu, S., John, S., Goldstein, R., Phillips, J.A., Haley, L.L., Sait, N., Shows, T.B., Smith, C.M., and Gerhard, D.S. (1997). The human HNP36 gene is localized to chromosome 11q13 and produces alternative transcripts that are not mutated in multiple endocrine neoplasia, type 1 (MEN I) syndrome. *Genomics* 42(2):325–330.

Wu, X., Yuan, G., Brett, C.M., Hui, A.C., and Giacomini, K.M. (1992). Sodium-dependent nucleoside transport in choroids plexus from rabbit. Evidence for a single transporter for purine and pyrimidine nucleosides. *J. Biol. Chem.* 267:8813–8818.

Wu, X., Prasad, P.D., Leibach, F.H., and Ganapathy, V. (1998). cDNA sequence, transport function, and genomic organization of human OCTN2, a new member of the organic cation transporter family. *Biochem. Biophys. Res. Commun.* 246:589–595.

Wu, X., Huang, W., Prasad, P.D., Seth, P., Rajan, D.P., Leibach, F.H., Chen, J., Conway, S.J., and Ganapathy, V. (1999). Functional characteristics and tissue distribution pattern of organic cation transporter 2 (OCTN2), an organic cation/carnitine transporter. *J. Pharmacol. Exp. Ther.* 290:1482–1492.

Wu, X., George, R.L., Huang, W., Wang, H., Conway, S.J., Leibach, F.H., and Ganapathy, V. (2000a). Structural and functional characteristics and tissue distribution pattern of rat OCTN1, an organic cation transporter, cloned from placenta. *Biochim. Biophys. Acta* 1466:315–327.

Wu, X., Huang, W., Ganapathy, M.E., Wang, H., Kekuda, R., Conway, S.J., Leibach, F.H., and Ganapathy, V. (2000b). Structure, function, and regional distribution of the organic cation transporter OCT3 in the kidney. *Am. J. Physiol. Renal. Physiol* 279:F449–F458.

Wu, X.C., Whitfield, L.R., and Stewart, B.H. (2000c). Atorvastatin transport in the Caco-2 cell model: contributions of P-glycoprotein and the proton-monocarboxylic acid co-transporter. *Pharm. Res.* 17:209–215.

Xia, C.Q., Liu, N., Yang, D., Miwa, G., and Gan, L.S. (2005). Expression, localization, and functional characteristics of breast cancer resistance protein in Caco-2 cells. *Drug Metab. Dispos.* 33(5):637–643.

Xie, W., Radominska-Pandya, A., Shi, Y., Simon, C.M., Nelson, M.C., Ong, E.S., Waxman, D.J., and Evans, R.M. (2001). An essential role for nuclear receptors SXR/PXR in detoxification of cholestatic bile acids. *Proc. Natl Acad. Sci. USA* 98:3375–3380.

Xu, J., Liu, Y., Yang, Y., Bates, S., and Zhang, J.T. (2004). Characterization of oligomeric human half-ABC transporter ATP-binding cassette G2. *J. Biol. Chem.* 279:19781–19789.

Yabuuchi, H., Shimizu, H., Takayanagi, S., and Ishikawa, T. (2001). Multiple splicing variants of two new human ATP-binding cassette transporters, ABCC11 and ABCC12. *Biochem. Biophys. Res. Commun.* 288(4):933–999.

Yamashita, T., Shimada, S., Guo, W., Sato, K., Kohmura, E., Hayakawa, T., Takagi, T., and Tohyama, M. (1997). Cloning and functional expression of a brain peptide/histidine transporter. *J. Biol. Chem.* 272:10205–10211.

Yang, C.Y. (1998). Studies on the human intestinal di-/tripeptide transporter HPT-1 as a potential carrier for orally administered drugs. Thesis, Purdue University.

Yang, C.Y., Dantzing, A.H., and Pidgeon, C. (1999). Intestinal peptide transport of systems and oral drug availability. *Pharm. Res.* 16:1331–1343.

Yao, S.Y., Ng, A.M., Muzyka, W.R., Griffiths, M., Cass, C.E., Baldwin, S.A., and Young, J.D. (1997). Molecular cloning and functional characterization of nitrobenzylthioinosine (NBMPR)-sensitive (es) and NBMPR-insensitive (ei) equilibrative nucleoside transporter proteins (rENT1 and rENT2) from rat tissues. *J. Biol. Chem.* 272(45):28423–28430.

Yao, S.Y., Ng, A.M., Sundaram, M., Cass, C.E., Baldwin, S.A., and Young, J.D. (2001a). Transport of antiviral 3′-deoxy-nucloside drugs by recombinant human and rat equilibrative, nitriobenzylthioinosine (NBMPR)-insensitive (ENT2) nucleoside transporter proteins produced in Xenopus oocytes. *Mol. Membr. Biol.* 18:161–167.

Yao, S.Y., Sundaram, M., Chomey, E.G., Cass, E.E., Baldwin, S.A., and Young, J.D. (2001b). Identification of Cys140 in helix 4 as an exofacial cysteine residue within the substrate-translocation channel of rat equilibrative nitrobenzylthioinosine (NBMPR)-insensitive nucleoside transporter rENT2. *Biochem. J.* 353(Pt 2):387–393.

Yao, W.Y., Ng, A.M., Vickers, M.F., Sundaram, M., Cass, C.E., Baldwin, S.A., and Young, J.D. (2002). Functional and molecular characterization of nucleobase transport by recombinant human and rat equilibrative nucleoside transporters 1 and 2. Chimeric constructs reveal a role for the ENT2 helix 5–6 region in nucleobase translocation. *J. Biol. Chem.* 277:24938–24948.

Yeung, A.K., Basu, S.K., Wu, S., Chu, C., Okamoto, C.T., Hamm-Alvarez, S., von Grafenstein, H., Shen, W.C., Kim, K.J., Bolger, M.B., Haworth, I.S., Ann, D., and Lee, V.H. (1998). Molecular identification of a role for tyrosine 167 in the function of the human intestinal proton-coupled dipeptide transporter (hPepT1). *Biochem. Biophys. Res. Commun.* 250:103–107.

Yoon, H., Fanelli, A., Grollman, E.F., and Philp, N.J. (1997). Identification of a unique monocaroxylate transporter (MCT3) in retinal pigment epithelium. *Biochem. Biophys. Res. Commun.* 234:90–94.

You, G. (2004). The role of organic ion transporters in drug disposition: an update. *Curr. Drug Metab.* 5:55–62.

Zelcer, N., Saeki, T., Reid, G., Beijnen, J.H., and Borst, P. (2001). Characterization of drug transport by the human multidrug resistance protein 3 (ABCC3). *J. Biol. Chem.* 276:46400–46407.

Zelcer, N., Reid, G., Wielinga, P., Kuil, A., van der Heijden, I., Schuetz, J.D., and Borst, P. (2003). Steroid and bile acid conjugates are substrates of human multidrug-resistance protein (MRP) 4 (ATP-binding cassette C4). *Biochem. J.* 371(Pt 2):361–367.

Zeng, H., Liu, G., Rea, P.A., and Kruh, G.D. (2000). Transport of amphipathic anions by human multidrug resistance protein 3. *Cancer Res.* 60(17):4779–4784.

Zhang, L., Dresser, M.J., Chun, J.K., Babbitt, P.C., and Giacomini, K.M. (1997a). Cloning and functional characterization of a rat renal organic cation transporter isoform (rOCT1A). *J. Biol. Chem.* 272:16548–16554.

Zhang, L., Dresser, M.J., Gray, A.T., Yost, S.C., Terashita, S., and Giacomini, K.M. (1997b). Cloning and functional expression of a human liver organic cation transporter. *Mol. Pharmacol.* 51:913–921.

Zhang, L., Schaner, M.E., and Giacomini, K.M. (1998). Functional characterization of an organic cation transporter (hOCT1) in a transiently transfected human cell line (HeLa). *J. Pharmacol. Exp. Ther.* 286:354–361.

Zhang, Y., Han, H., Elmquist, W.F., and Miller, D.W. (2000). Expression of various multidrug resistance-associated protein (MRP) homologues in brain microvessel endothelial cells. *Brain. Res.* 876:148–153.

Zhang, E.Y., Knipp, G.T., Ekins, S., and Swaan, P.W. (2002a). Structural biology and function of solute transporters: implications for identifying and designing substrates. *Drug Metab. Rev.* 34(4):709–750.

Zhang, E.Y., Mitch, A.P., Cheng, C., Ekis, S., and Swaan, P.W. (2002b). Modeling of active transport systems. *Adv. Drug Del. Rev.* 54:329–354.

Zhang, E.Y., Fu, D.J., Pak, Y.A., Stewart, T., Mukhopadhyay, N., Wrighton, S.A., and Hillgren, K.M. (2004). Genetic polymorphisms in human proton-dependent dipeptide transporter PEPT1: implications for the functional role of Pro586. *J. Pharmacol. Exp. Ther.* 310(2):437–445.

Zhao, C., Wilson, M.C., Schuit, F., Halestrap, A.P., and Butter, G.A. (2001). Expression and distribution of lactate/monocarboxylate transporter isoforms in pancreatic islets and the exocrine pancreas. *Diabetes* 50:361–366.

Zhou, X., Thamotharan, M., Gangopadhyay, A., Serdikoff, C., and Adibi, S.A. (2000). Characterization of an oligopeptide transporter in renal lysosomes. *Biochim. Biophys. Acta* 1466:372–378.

Ziegler, T.R., Fernandez-Estivariz, C., Gu, L.H., Bazargan, N., Umeakunne, K., Wallace, T.M., Diaz, E.E., Rosado, K.E., Pascal, R.R., Galloway, J.R., Wilcox, J.N., and Leader, L.M. (2002). Distribution of the H+/peptide transporter PepT1 in human intestine: up-regulated expression in the colonic mucosa of patients with short-bowel syndrome. *Am. J. Clin. Nutr.* 75:922–930.

Zimmermann, C., Gutmann, H., Hruz, P., Gutzwiller, J.P., Beglinger, C., and Drewe, J. (2005). Mapping of multidrug resistance gene 1 and multidrug resistance-associated protein isoform 1 to 5 mRNA expression along the human intestinal tract. *Drug Metab. Dispos.* 33(2):219–224.

8
Bioavailability and Bioequivalence

Sam H. Haidar, Hyojong (Hue) Kwon, Robert Lionberger,
and Lawrence X. Yu

8.1 Introduction

For systemically acting drugs, absorption is a prerequisite for therapeutic activity when drugs are administered extravascularly. Factors affecting drug absorption have been discussed in previous chapters. This chapter will cover general methods to evaluate bioavailability and bioequivalence. Scientific principles as well as regulatory perspectives related to these two topics will be discussed. Historically, the development of sensitive and precise bioanalytical methods in the 1960s and 1970s allowed for the first time the measurement of very low levels of drug concentrations in biological fluids. As a result, pharmacokinetic profiles of drugs, describing absorption, distribution, and clearance, could be determined. Regulations related to bioavailability and bioequivalence were put into place, considering the latest advances in the science. Currently, bioavailability and bioequivalence play a significant role in the discovery, development, and regulation of new drug products. Additionally, bioequivalence studies are a crucial component of abbreviated new drug applications (ANDAs), leading to market access of safe, effective, and low cost generic drugs.

8.2 Bioavailability and Bioequivalence

8.2.1 Bioavailability and its Utility in Drug Development and Regulation

The therapeutic action of a drug is usually correlated with the delivery of the active substance to the site or more accurately, sites, of pharmacological response. US Federal regulations (21 Code of Federal Regulations (CFR), 2006) define bioavailability (BA) as

"the rate and extent to which the active ingredient or active moiety is absorbed from a drug product and becomes available at the site of action".

Most drugs are systemically acting, meaning that they reach their sites of action through the systemic circulation. Thus, it is common for pharmaceutical scientists to evaluate the bioavailability of a drug product as the fraction of the dose reaching the systemic circulation. BA can depend on the physicochemical properties of the drug substance and the route of administration, in addition to drug product excipients and manufacturing process.

The BA of a drug is an important attribute that is investigated early in drug development and used throughout development. In many cases, it is the deciding factor for whether or not a drug candidate is selected for further development (Sun *et al.*, 2004). As stated in the FDA bioavailability and bioequivalence guidance (FDA, 2003a) BA studies help elucidate the process by which a drug is released from its dosage form and reaches the sites of action including the impact of presystemic metabolism and/or transporters. BA studies also provide information about the drug's pharmacokinetic properties such as dose proportionality, linearity, and effect of food on absorption. When multiple formulations are used in the clinical development program, relative bioavailability studies can link observations of safety and efficacy to drug exposure and provide a basis for labeling or formulation optimization. BA studies can also be used in establishing an exposure–response relationship (FDA Exposure–Response guidance, 2003b).

8.2.2 Bioequivalence and its Utility in Drug Development and Regulation

Federal regulations (21 CFR, 2006) define bioequivalence as:

> The absence of a significant difference in the rate and extent to which the active ingredient or active moiety in pharmaceutical equivalents or pharmaceutical alternatives becomes available at the site of drug action when administered at the same molar dose under similar conditions in an appropriately designed study.

Bioequivalence (BE) studies are a major component of ANDAs. They verify that the active ingredient in a generic drug product will be absorbed into the body to the same extent and at the same rate as its corresponding reference listed drug (RLD) product. The significance of BE studies is that when two pharmaceutically equivalent products are shown to be bioequivalent, the two products are judged to be therapeutically equivalent. Therapeutically equivalent products are expected to have the same safety and efficacy profiles, when administered under the conditions listed in the product labeling. This is the basis for the approval and use of generic drug products.

BE studies are not only performed as part of the ANDA process, but also conducted by new drug manufacturers to confirm equivalence between formulations when it is necessary to make manufacturing and/or formulation changes. For example, often the marketed drug product is different in formulation or method of manufacture from the product used in the safety and efficacy clinical trials. These differences may be the result of formulation changes necessary to scale up the product from a small (laboratory or pilot) scale size to a large scale (commercial) size. After approval, the New Drug Application or NDA sponsor may significantly modify the scale of product runs, equipment, manufacturing process, formulation and dosage forms, ingredient specifications, source of supplies, and method of synthesis of the active ingredient. In these cases, the marketed or reformulated product must demonstrate bioequivalence to the original formulation to link the safety and efficacy data of the original product to the new product.

8.2.3 Bioavailability and Bioequivalence Studies: General Approaches

There are several acceptable approaches for the determination of BA and BE. Title 21 of the Code of Federal Regulations (21 CFR, 2006) §320.24 lists the *in vivo* and *in vitro* methods of determining BA or BE for a drug product.

They are:

1. (a) An *in vivo* test in humans in which the concentration of the active ingredient or active moiety, and, when appropriate, its active metabolite(s), in whole blood, plasma, serum, or other appropriate biological fluid is measured as a function of time; (b) an *in vitro* test that has been correlated with and is predictive of human *in vivo* bioavailability data.
2. An *in vivo* test in humans in which the urinary excretion of the active ingredient or active moiety, and, when appropriate, its active metabolite(s), is measured as a function of time.
3. An *in vivo* test in humans in which an appropriate acute pharmacological effect of the active moiety, and, when appropriate, its active metabolite(s), is measured as a function of time if such effect can be measured with sufficient accuracy, sensitivity, and reproducibility.
4. Well-controlled clinical trials that establish the safety and effectiveness of the drug product, for purposes of measuring bioavailability, or appropriately designed comparative clinical trials, for purposes of demonstrating BE. This approach is the least accurate, sensitive, and reproducible of the general approaches for determining BA or BE.
5. A currently available *in vitro* test acceptable to the FDA (usually a dissolution rate test) that ensures human *in vivo* BA.
6. Any other approach deemed adequate by FDA to measure BA or establish BE.

For most systemically acting drugs, the active moiety can be detected and accurately measured in the plasma over time. Therefore, the first (pharmacokinetic) method (1) listed above is preferred. This pharmacokinetic method is generally considered as the most sensitive, accurate, and reproducible method for the assessment of BA and BE. Section 8.3 describes the conduct of these types of studies in detail.

For some drug products, there is sufficient understanding of the physicochemical properties and biological factors that affect BA that there is no need for *in vivo* BE studies. Section 8.4 describes the situations in which the *in vivo* studies can be waived.

Drugs that do not reach their sites of action through the systemic circulation are defined as locally acting drugs. For locally acting drugs, the bioequivalence method of choice is usually dependent on the attributes of the drug and drug product including physicochemical properties, BA, route of administration, site of action, and ability to detect/measure the active moiety. FDA recommends the most sensitive, accurate, and reproducible method for a particular product be used. Section 8.5 provides examples of the selection of BE methods for locally acting drugs

8.3 Pharmacokinetic Bioavailability and Bioequivalence Studies

8.3.1 Bioavailability Studies: General Guidelines and Recommendations

BA for systemically acting, orally administered drug products is usually determined by measuring the concentration of the active ingredient and, when appropriate, its active metabolites over time in samples collected from the systemic circulation. Figure 8.1 shows a typical concentration–time profile. The profile determines the following important parameters:

1. $C_{\max}$ is the maximum observed plasma concentration.
2. AUC_{0-t} is the area under the concentration–time curve. It is calculated using the trapezoid rule on the actual data points.
3. $T_{\max}$ is the time at which $C_{\max}$ is observed.
4. k_{e} is the terminal elimination rate constant determined from fitting the tail of the profile to a linear elimination model: $\mathrm{d}C/\mathrm{d}t = -k_{\mathrm{e}}\, C$.
5. $t_{1/2}$ is the terminal half-life. $t_{1/2} = 0.693/k_{\mathrm{e}}$. It is the time it takes for the concentration to be reduced by half due to drug elimination.
6. AUC_{∞} is the area under the concentration–time curve extrapolated to infinity. $\mathrm{AUC}_{\infty} = \mathrm{AUC}_{0-t} + C_{\mathrm{last}}/k_{\mathrm{e}}$. C_{last} is the last observed concentration.

$C_{\max}$ is usually correlated with the rate of absorption and AUC reflects the extent of absorption, and thus are considered the parameters most relevant to safety and efficacy.

In most cases, BA studies are conducted as a single dose comparison between a test product and a reference product. These studies are generally conducted using a cross-over design in which each subject receives both treatments in a random order. Treatments should be separated by a washout period exceeding five half-lives of the active moieties measured. Following administration of the test or reference product, blood samples are collected to obtain a profile of the time-course

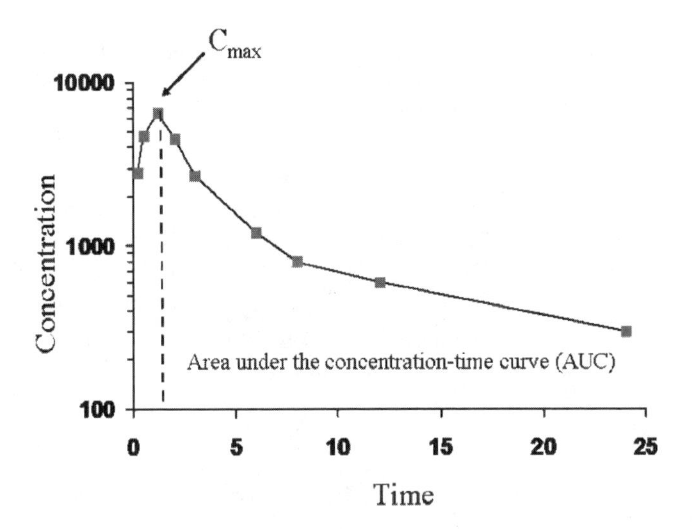

FIGURE 8.1. Example of a plasma-concentration time profile

of the drug for each subject. Drug or active metabolite concentrations in the urine may be used when they are not detectable in the blood or plasma. The sampling schedule, which may differ for each drug, should be of sufficient frequency to ensure precise estimation of the drug's pharmacokinetic parameters. It is generally recommended that samples should be collected over at least three times the elimination half-life of the drug.

As stated previously, a drug's BA can be impacted by its route of administration. By definition, a drug which is administered intravenously (i.v.) has 100% BA. However, BA generally decreases when the drug is administered by extravascular routes (e.g., oral, transdermal, etc.). This decrease is usually a function of incomplete absorption, and/or presystemic metabolism or degradation.

BA may be classified as absolute or relative. Absolute BA is the fraction of the administered dose that reaches the systemic circulation relative to an intravenous dose, while relative bioavailability is the fraction of the dose of a test product that reaches the systemic circulation relative to a non-i.v. reference product. For example, a tablet may have an absolute BA of 60% (or some other value); however, the same tablet would have a *relative* BA of 100%, if the drug in the tablet is absorbed to the same extent as an oral solution of the same drug. The relative BA of the tablet in this case is specific to the oral solution, and may differ relative to other dosage forms.

Absolute BA is calculated by taking the ratio of the dose-corrected AUC of the test product (oral) divided by AUC of the i.v. reference product. Mathematically, the relationship is expressed as

$$\frac{AUC_{PO} \times Dose_{IV}}{AUC_{IV} \times Dose_{PO}} \tag{8.1}$$

where AUC_{PO} and AUC_{IV} are the area under the concentration–time curve after oral and intravenous administration, and $Dose_{po}$ and $Dose_{iv}$ are the amount of dose given orally and i.v., respectively.

Similarly, relative BA can be measured using a test product and a reference (non-i.v.) drug, and calculated by

$$F = \frac{AUC_{test} \times Dose_{ref}}{AUC_{ref} \times Dose_{test}}. \tag{8.2}$$

8.3.2 Bioequivalence Studies: General Guidelines and Recommendations

It is generally recommended that BE studies be conducted using a single dose, cross-over design. Parallel and replicate designs are also acceptable, and may be more appropriate under certain circumstances. Treatments are usually administered to healthy subjects, representative of the general population. Samples of an accessible biologic fluid, usually blood or urine, are analyzed for drug concentrations. Pharmacokinetic parameters, such as AUC and C_{max}, are determined from the resulting concentration–time profiles.

8.3.2.1 Study Design

In the standard cross-over design for *in vivo* BE studies, subjects receive a single dose of test and reference products on separate occasions with random assignment to the two possible sequences of product administration. Treatments are separated by a washout period exceeding five half-lives of the active moieties measured. Parallel designs in which separate groups of subjects receive the test and reference products require larger numbers of subjects and are recommended only in special cases when the half life of the drug is so long that the cross-over design is not feasible. The use of replicate designs for highly variable drugs is discussed in Sect. 8.3.3.3. Single dose studies are recommended over multiple dose studies because single dose studies are generally more sensitive "in assessing release of the drug substance from the drug product into the systemic circulation. . ." (FDA BA/BE guidance, 2003a).

8.3.2.2 Dose

For a product with multiple strengths, the highest strength is usually recommended for use in a BE study. The pharmacokinetics of most drugs is well described by linear absorption, distribution, and clearance processes. The rate of linear processes increases proportionally to the amount of drug or the dose. Thus a bioequivalence conclusion for one of these drugs will be same at any dose.

For drugs with nonlinear pharmacokinetics, the dose used in the BE study should be the most sensitive to differences in formulation. The most common source of nonlinear pharmacokinetics is saturable metabolism, where the rate of metabolism reaches a maximum that is independent of drug concentration.

Other potential causes for nonlinear pharmacokinetics are solubility limited absorption or saturable uptake mediated by transporters. If there are safety concerns with administration of a single dose of the highest strength in healthy subjects, FDA will recommend use of a lower dose.

8.3.2.3 Subjects

Because determination of BE is dependent on statistical methods, the number of subjects in the study should be sufficient to ensure adequate power. The typical number of subjects is 24–36 with the minimum number of subjects in the study being 12. Healthy subjects are recommended for BE studies for two main reasons: patients are more variable and patients require continuous treatment that does not allow for a washout period. The greater variability observed in patients has a direct impact on the sensitivity of BE testing. Patients are generally used only when drug is not safe to administer in healthy subjects.

The physical processes of drug absorption for solid oral dosage forms are usually the same in patients as they are in healthy subjects. Given that BE studies compare the relative performance of two formulations, any conclusion drawn in healthy subjects will also apply to patients. This is true even for products where there is a known difference in BA between patients and healthy subjects.

8.3.2.4 Statistical Analysis of Bioequivalence

Pharmacokinetic parameters AUC and C_{max} are analyzed statistically because of the variability inherent in human subjects. This variability may be observed when the same subject receives the *same* drug product on two different occasions, i.e., the resulting plasma concentrations will not be exactly the same. Because of this inherent variability, an individual who takes two *different* products on separate occasions may show a measurable difference in the pharmacokinetic parameters. In this situation, it is not clear whether this difference is the result of a difference between the products, or the result of normal within subject variability. Thus, FDA recommends that ANDA or NDA applicants use statistical methods to estimate more accurately those differences in pharmacokinetics that result from the two product formulations. When considering the results from BE studies, it is important to understand what statistical tests are used and how FDA uses the results of these statistical tests to determine whether two products are bioequivalent. The following is a qualitative description, drawing on FDA's responses to citizen petitions related to BE (FDA, 2004). Details of the statistical calculations can be found in the FDA guidance statistical approaches to establishing bioequivalence (FDA, 2001b).

The mean is the average of all the differences in pharmacokinetic values observed in the small group of study subjects. For example, in a study the mean AUC of the test product might be 99% of the AUC of the reference product. The mean difference in this case would be 1% and the mean ratio would be 99%. However, if the same BE study is repeated in another small group of subjects, the second study's mean may be different from the first study's mean. Therefore, FDA uses a statistical confidence interval to provide an estimated range that is likely to contain the mean if the drug were given to the entire population. In our example study with a mean ratio of 99%, the confidence interval might be 89–109%. This confidence interval shows that for the entire population, the ratio of the mean AUC between test and reference products is likely (with a 90% probability) to be between 90 and 118%. If the small study used a greater number of subjects to more accurately reflect the general population's results, then the 90% confidence interval would be smaller (i.e., a smaller range of the possible pharmacokinetic values in the general population, such as 93–105%).

FDA determines whether a study shows that two products are bioequivalent based on the confidence interval and not on the mean value of the study. The results of a study are expressed as a confidence interval for the ratio of test to reference products. To decide whether two products are bioequivalent, the calculated confidence interval is compared to an acceptance interval. The acceptance interval (also referred to as acceptance limits) is expressed as two numbers that provide upper and lower limits on the confidence interval. If the confidence interval is contained within this acceptance interval, then FDA concludes that the study demonstrates BE; if not, then the study does not demonstrate BE. The acceptance interval is a fixed standard, while the confidence interval is determined from the data in a particular study.

FDA considers two drug products equivalent when the 90% confidence intervals of the geometric mean ratio for C_{max} and AUC are entirely within 80–125% (see Fig. 8.2). The choice of the 80–125% acceptance limits is based on medical opinion and FDA experience which determined that a difference of 20% or less in drug exposure was not clinically significant for most drugs. Thus, the limits of 80–125% were set around a difference of less than 20% in geometric mean ratio (test/reference) for C_{max} and AUC, although in practice this difference rarely

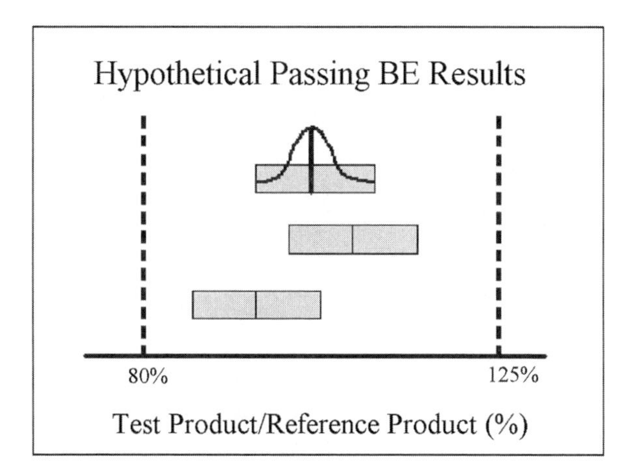

FIGURE 8.2. Hypothetical results from bioequivalence (BE) tests for approved generic drugs

exceeds 10%. FDA has not found any clinical problems resulting from the thousands of drug products approved with the current BE criteria.

As mentioned above, the 80–125% boundaries are acceptance limits for the confidence interval and not a judgment about the acceptable mean differences between test and reference products. The sample mean ratio of the pharmacokinetic values for the test and reference products lie at the center of the confidence interval. Because this confidence interval must fall within the 80–125% boundaries, these statistical criteria limit the acceptable range in which the mean values can stray from the 100% ratio. The actual mean differences FDA found for drugs tested and analyzed under this statistical procedure were much smaller than the 80–125% boundaries. In the 1980s, FDA reviewed 224 BE studies that passed the 80–125% criterion (Nightingale and Morrison, 1987). In these studies, the observed mean difference in AUC between the brand name and the generic product was approximately 3.5%. This analysis was repeated for the 127 BE studies conducted for generic drugs approved in 1997 (Henney, 1997). The average observed difference in AUC in these studies was approximately 3.3%. Recently, FDA surveyed the BE data for the 10-year period of 1996–2005. Results from more than 1,500 BE studies were analyzed. Once again, the mean difference in AUCs between generic products and their brand name counterparts averaged less than 4%.

Figure 8.2 graphically illustrates the relationship between the mean value obtained from a BE study, the 90% confidence interval for that BE study, and

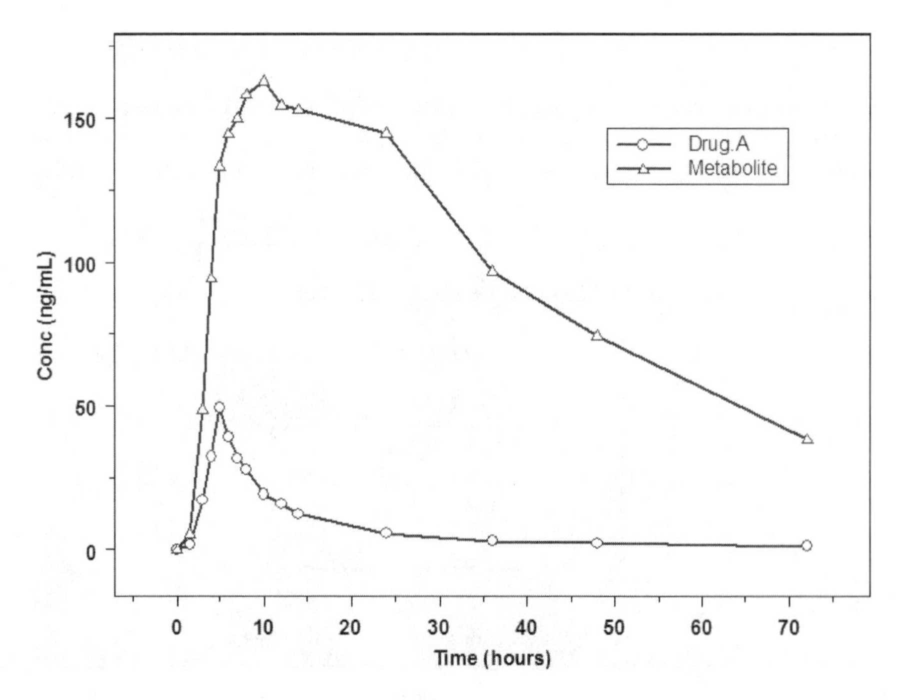

FIGURE 8.3. Plasma concentrations over time for parent drug and major metabolite, for a drug (Drug A) with extensive presystemic metabolism

FDA's acceptance limits of 80–125%. The center of each box is the mean value from a BE study, while the entire box represents the confidence interval from the same BE study. Because the 80–125% acceptance limits are bounds on the confidence intervals, the mean values from passing BE studies must be closer to 100%. As can be seen in Fig. 8.2, the actual mean differences between test and reference listed products will be much smaller than FDA's BE acceptance limits of 80–125%.

8.4 Bioequivalence: Challenging Topics

Some situations where the evaluation of BE presents a challenge include drugs with active metabolites, enantiomers, endogenous substances, and highly variable drugs. Each of these conditions will be discussed in some detail below.

8.4.1 Drugs with Active Metabolites

Following administration, drugs generally undergo biotransformation or metabolism to facilitate their elimination from the body. Metabolism can be systemic or presystemic. In systemic metabolism, drug in circulating blood is exposed to metabolic enzymes as it passes through the liver and other tissue. Presystemic metabolism occurs when the drug is exposed to metabolic enzymes found in the gut wall, skin, or other absorption sites. Additionally, presystemic metabolism occurs when the drug is metabolized by the liver immediately after oral absorption, prior to reaching the site of action (hepatic first pass effect). The relevance of this distinction to BE is that systemic metabolism is determined by the concentration of drug in the systemic circulation, while the presystemic metabolism can be affected by the rate and extent of the release of the absorption of the drug and its rate of release from the drug product.

The biotransformation of drugs can lead to the formation of compounds (metabolites) which may be active or inactive pharmacologically. The metabolites' activity may impact the efficacy of the drug or its side effects. The guidelines for using metabolites in BA or BE studies differ depending on the circumstances, i.e., whether the studies are part of a drug development program supporting an NDA, or they are BE studies supporting an ANDA for a generic product.

During drug development (investigational new drug (IND) or NDA), the objective is to learn as much as possible about a new compound that may become a marketed drug product. Thus, the recommendations for BA studies include measurement of the parent drug and all active metabolites. Determination of the activity of each metabolite relative to the parent is also desired.

On the other hand, by the time a patent expires on a drug and it becomes a candidate for generic competition, much is known about the drug's attributes and clinical performance. As a consequence, the objectives of a BE study supporting an ANDA are different from those for an NDA. For example, the goal of a BE study in the ANDA is to evaluate formulation performance of a generic candidate relative to that of the RLD. Given that the concentration–time profile of the parent drug is generally more sensitive to differences in formulation performance, it is recommended that only concentrations of parent drug released from the dosage form be measured. This is true even if the drug has active metabolites.

The BA/BE Guidance (FDA, 2003a), however, does provide situations where metabolites should be measured in a BE study. These are quoted below:

1. Measurement of a metabolite may be preferred when parent drug levels are too low to allow reliable analytical measurement in blood, plasma, or serum for an adequate length of time. We recommend that the metabolite data obtained from these studies be subject to a confidence interval approach for BE demonstration. If there is a clinical concern related to efficacy or safety for the parent drug, we also recommend that sponsors and/or applicants contact the appropriate review division to determine whether the parent drug should be measured and analyzed statistically.
2. A metabolite may be formed as a result of gut wall or other presystemic metabolism. If the metabolite contributes meaningfully to safety and/or efficacy, we also recommend that the metabolite and the parent drug be measured. When the relative activity of the metabolite is low and does not contribute meaningfully to safety and/or efficacy, it does not have to be measured. We recommend that the parent drug measured in these BE studies be analyzed using a confidence interval approach. The metabolite data can be used to provide supportive evidence of comparable therapeutic outcome.

In the first case, a metabolite is measured in a BE study when levels of the parent drug are too low for accurate measurement, as is the case with some prodrugs. The statistical criteria for BE determination is applied to the metabolite(s) in this case. For example, the 90% confidence interval of the geometric mean ratio of test/reference AUC and C_{max} of the metabolite must fall within 80–125%. This may be the only situation where the confidence interval approach is used with the metabolite for the demonstration of bioequivalence.

In the second case, measurement of the active metabolite is recommended when there is evidence of hepatic first pass metabolism and/or gut presystemic formation of the metabolite. The active metabolite must also contribute significantly to the efficacy and/or safety profile of the drug. Unlike BE studies involving prodrugs, metabolite concentrations in this case are not subject to the BE statistical criteria (confidence interval approach), but summary statistics of the PK parameters serve as supporting evidence of bioequivalence. The parent drug, however, is evaluated statistically using confidence intervals as it would be in studies that do not include measurement of the metabolite(s).

To apply the second case to a particular drug product, there must be presystemic metabolism. One indicator suggesting presystemic metabolism is the early appearance of metabolite levels in the plasma, usually preceding parent drug levels. Plasma levels of the metabolite may also be significantly higher, relative to the parent. Figure 8.3 illustrates the early appearance, as well as higher levels, of a major metabolite relative to that of the parent, for a sample drug (Drug "A") that has significant presystemic metabolism.

8.4.2 Enantiomers vs. Racemates

Enantiomers are stereoisomers, i.e., molecules that are nonsuperimposable mirror images of each other, with identical chemical and physical properties (Wade, 2003). A mixture of equal parts of an optically active isomer and its enantiomer is called a "racemate."

Enantiomers may differ in pharmacological activity and pharmacokinetic properties. This may be related to their three-dimensional fit within cell receptors or enzymes, leading to possible differences in pharmacological responses and potencies, as well as differences in absorption and clearance.

It is generally recommended that chiral assays (which can distinguish individual enantiomers) be used in BA studies during drug development. As stated above, enantiomers could potentially have different pharmacokinetics and pharmacological activities. This recommendation, however, does not extend to BE studies supporting ANDAs. The reason is, racemate levels of the parent drug, or in some cases the metabolite, may be adequate at detecting differences in formulation performance. There are exceptions, however, to this general rule.

The BA/BE Guidance lists four conditions, all of which have to be met, before measurement of the individual enantiomers are needed in BE studies. They are (1) the enantiomers exhibit different pharmacodynamic properties, (2) the enantiomers exhibit different pharmacokinetic properties, (3) the primary safety and efficacy resides in the minor enantiomer (one with the lowest concentration), and (4) there is evidence of nonlinear absorption for at least one of the enantiomers. In rare cases where a drug product meets all of the above four conditions, then a chiral assay is recommended to measure the concentration of individual enantiomers in a BE study supporting an ANDA.

8.4.3 Endogenous Substances

Drug products whose active ingredient is an endogenous substance (one that naturally occurs in the body) present a challenge to evaluating bioequivalence because a measurement of plasma concentration would include both the endogenous concentration plus the amount added by administration of the drug product (exogenous source). This may act to bias the results of a BE study.

To illustrate the potential error we use a model drug where B is the baseline level, R is the change in drug level above the baseline due to the reference product, T is the change in drug level above the baseline due to the test product. The apparent test to reference ratio is

$$\frac{B+T}{B+R}, \tag{8.3}$$

which should be compared to the true test to reference ratio of T/R. To estimate the error in this case let $B = 200$, $R = 200$, $T = 150$, and $T/R = 0.75$. The apparent ratio (unadjusted for baseline) is 0.875 compared to a true ratio (adjusted for baseline) of 0.75. This is a significant reduction in the ability to identify product differences.

One approach that has been successful for some drug products is a baseline correction method. In this approach, the measured predose concentration of the endogenous substance is subtracted from the measured concentration profiles after administration of the drug product. The remaining concentration should better reflect the amount delivered by the drug product.

When using this approach it should be noted that production of many endogenous substances is under feedback control and may be altered by the administered dose and thus there may be a nonlinear dependence on the external dose. Potassium chloride is an example drug for which the feedback control of the endogenous drug concentration is effective to the extent that there is no significant change in potassium concentration after administration of normal doses. For this drug, FDA recommends a urinary recovery study (FDA, 2002).

8.4.4 Highly Variable Drugs

The BE of highly variable drugs and drug products has been discussed in many conferences and meetings, nationally and internationally (Blume and Midha, 1993; Shah *et al.*, 1996). Highly variable drugs are generally defined as drugs or drug products which exhibit within subject variability of 30% or greater.

Drugs with high within subject variability can present challenges in BE studies because of impact on sample size. For example, when comparing the BA of a highly variable reference product with itself, the sample size needed to demonstrate BE can exceed 100 subjects, although there are no true differences in BA between test and reference in this case. Table 8.1 (adapted from Patterson *et al.*, 2001) provides more precise information about the association between within subject variability and sample size needed to achieve adequate statistical power.

Evaluating the BE of highly variable drugs using the standard criteria may present ethical concerns, i.e., unnecessary human testing, in addition to the practical difficulties of large BE studies. For this reason, the FDA has been evaluating different approaches for determining BE that would decrease sample size, without increasing patient risk. Several papers have been published on this topic. Some of the methods studied are summarized below.

8.4.4.1 Static Expansion of the BE Limits

Sample size in BE studies is generally determined by the BA parameter with the highest within subject variability. In most cases, this parameter is C_{max}. The greater variability observed with C_{max} may result from the fact that this parameter is a single point measurement, which is highly dependent on the sampling time/frequency and the elimination rate of the drug. Arbitrary widening of the BE

TABLE 8.1. Sample sizes providing 90% power in two-way, cross over bioequivalence studies

CV_W%	% difference in true BA ratio	Number of subjects needed
30	0	40
	5	54
	10	112
45	0	84
	5	112
	10	230
60	0	140
	5	184
	10	384
75	0	200
	5	264
	10	554

CV_W%, within subject coefficient of variation; BA, bioavailability

limits for C_{max} has been proposed as one approach to reduce sample size when evaluating the BE of highly variable drugs. This entails increasing the 90% confidence interval limits for C_{max} from 80–125% (current FDA criterion) to 75–133%, or even 70–143%.

8.4.4.2 Expansion of Bioequivalence Limits Based on Fixed Sample Size

The basis of this approach is the belief that only a reasonable number of subjects should be used in BE studies (Boddy *et al.*, 1995). To conduct a study with the above method, a fixed number of subjects, e.g., 24, is used in a standard two-period, cross-over design comparing the reference product with itself. For a highly variable drug, this study is likely to fail the 80–125% criteria, because of low power with only 24 subjects. However, the 90% confidence interval obtained from the reference product would become the new "goalpost," or criteria, for subsequent studies comparing a test product (proposed generic) with the reference product (RLD), using the same number of subjects (24).

According to Boddy *et al.* (1995), a drawback of this method is that the wider acceptance limits are based on controlling the sample size, instead of a meaningful measure of formulation differences (Boddy *et al.*, 1995). Additionally, when the test product differs from the RLD by a small measure, there is no guarantee that the confidence interval for the test vs. reference product will fall within the "goalposts" set by the first study, where the reference is compared with itself.

8.4.4.3 Scaled Average Bioequivalence

Finally, a third proposed approach that is currently favored by the FDA is scaled average BE. This method entails widening of the BE limits as a function of the within subject variability of the reference product. In this case, instead of using fixed limits, i.e., 80–125%, to determine if a test product is BE to the reference product, the limits expand as within subject variability of the reference increases.

The variability of the reference is determined in a replicate or partial replicate design, where the reference product is administered twice to the same subject at different periods. Mathematically, this approach may be described by

$$\text{BE limits, upper, lower} = \exp\left(\pm\frac{0.223}{\sigma_{\mathrm{w0}}}\sigma_{\mathrm{wr}}\right), \tag{8.4}$$

where σ_{wr} is the within subject standard deviation of the reference, and σ_{w0} is a constant set by the regulatory agency.

One concern about the use of average scaling for BE purposes is the lack of sensitivity of this method to differences in the point estimate, or the test/reference ratio of the geometric means. At least in theory, the lack of sensitivity may lead two products with unacceptably large differences in formulation performance to be declared BE. For this reason, the FDA has proposed the use of point estimate constraints in conjunction with scaled average BE. For example, if a BE study passes the confidence interval criteria (scaling), but the mean ratio between test and reference exceeds a predefined limit (e.g., ±20), then the two products may not be judged BE for regulatory purposes.

In a simulation-based study conducted at the FDA, the impact of scaled average BE on power was evaluated, and compared to the power of average BE (traditional criteria). Using a sample size of 36 subjects, one million studies were simulated for each variable tested. As can be seen in Fig. 8.4, the scaled average BE can have a significant impact on power even when the point estimate constraint is applied. For example, at within subject variability of 60%, the power of the study using average BE is about 24%, when the test and reference have zero differences in BA. Applying scaled average BE with point estimate constraint under the same test conditions, the power increases to >90%. Thus, scaled average BE appears to have a great practical advantage over traditional BE methods, when it comes to highly variable drugs.

8.5 Biowaivers

For some drug products, there is sufficient understanding of the physicochemical properties and biological factors that affect BA that there is no need for *in vivo* BE studies. Sponsors may request waivers of BE studies (biowaivers) for solutions, products with a range of strengths, and biopharmaceutical classification system (BCS) Class 1 drugs.

8.5.1 Solutions

In vivo BA/BE is self-evident for certain drug products, such as topical solutions, solution nasal spray, oral solutions, elixirs, syrups, tinctures, or other solubilized forms of the drug. For these products, *in vivo* BA/BE can be waived, according to 21 CFR 320.22(b).

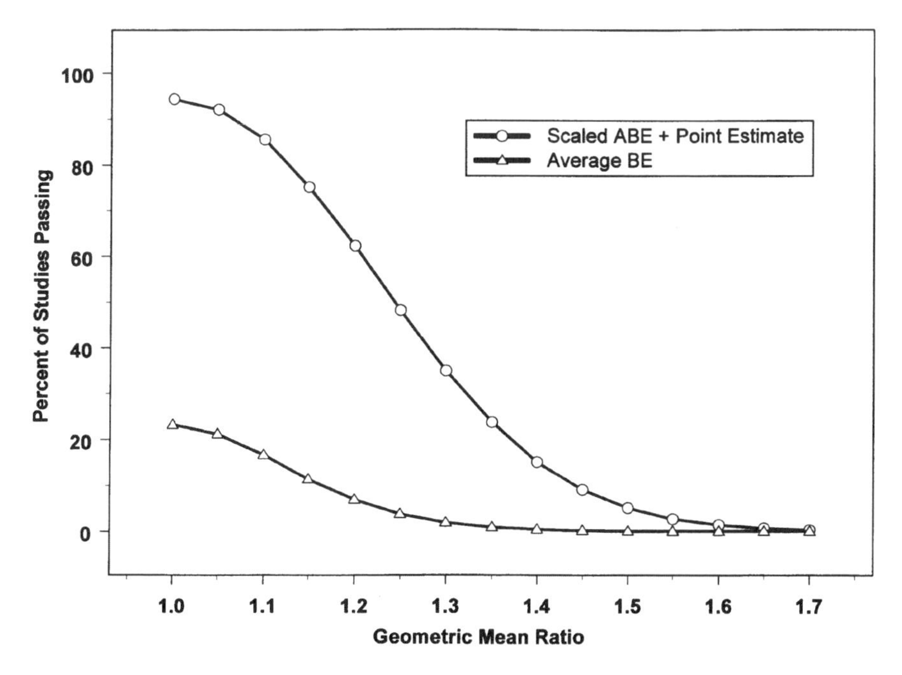

FIGURE 8.4. The difference in power, or percent of studies passing BE, between scaled average BE with a point estimate constraint and average BE (traditional criteria). A million BE studies with a sample size of 36 subjects were simulated. The geometric mean ratio reflects differences between the test and reference products

This waiver assumes that release of the drug substance from the drug product is self-evident and that the solutions do not contain any excipient that significantly affects drug absorption (21 CFR 320.22(b)(3)(iii)). The FDA can deny a biowaiver request if differences in excipients have the potential to change BE. For example, xylitol, sorbitol, and mannitol are commonly used formulation excipients for drug products (Fassihi *et al.*, 1991; Fukahori *et al.*, 1998). They are also used as artificial sweeteners in the food industry. These excipients are not well absorbed in the gastrointestinal (GI) tract. Additionally, they increase the osmotic pressure in the intestine, which changes the flux of water in the GI tract. This osmotic stress can change the gastric emptying time and the intestinal transit times through both the upper and lower parts of the intestine. Transit times in the GI can impact drug absorption. The total amount of drug absorbed depends on the rate of absorption from the intestine and the total time that the drug is present in the intestine.

When transit or emptying times are decreased, there is less time available for drug molecules in solution to be absorbed and thus, the total absorption may be decreased. Scintigraphic evidence suggests that osmotic agents can have minor effects on the residence time in the upper intestinal tract, but significantly reduce the residence time in the lower intestinal tract (Adkin *et al.*, 1995; Kruger *et al.*, 1992). The osmotic pressure changes may also affect the rate of transport across the intestinal wall in addition to changing transit times, which could lead to changes in absorption of low permeability drugs (Polli *et al.*, 2004). As an example of this effect, Chen *et al.* (2007) measured the pharmacokinetics of ranitidine in a four way cross-over study of ranitidine oral solution dosed with various amounts of sorbitol (Chen *et al.*, 2007). Doses of sorbitol greater than 1.25 g significantly reduced the BA of ranitidine from an oral solution.

Another example of a pharmaceutical excipient with a demonstrated effect on drug absorption is polyethylene glycol 400 (PEG 400). Several studies investigated the effect of PEG 400 on the absorption characteristics of ranitidine from the gastrointestinal tract (Basit *et al.*, 2002; Schulze *et al.*, 2003). These studies show that there is no significant effect of PEG 400 on gastric emptying; however, the presence of PEG 400 reduced the mean small intestinal transit times of the ranitidine solutions containing PEG 400. This resulted in changes in drug absorption that depended upon the amount of PEG 400. Low concentrations of PEG 400 increased the absorption of ranitidine, presumably due to changes in intestinal permeability of ranitidine, whereas high concentrations of PEG 400 reduced ranitidine absorption possibly due to shorter small intestinal transit time.

8.5.2 Lower Strength

Waiver of *in vivo* studies for different strengths of a drug product can be granted under §320.22(d)(2) when (1) the drug product is in the same dosage form, but in a different strength; (2) this different strength is *proportionally similar* in its active and inactive ingredients to the strength of the product for which the same manufacturer has conducted an appropriate *in vivo* study; and (3) the new strength meets an appropriate *in vitro* dissolution test. The FDA guidance (FDA, 2003a) defines *proportionally similar* in the following ways:

1. All active and inactive ingredients are in exactly the same proportion between different strengths (e.g., a tablet of 50-mg strength has all the inactive ingredients, exactly half that of a tablet of 100-mg strength, and twice that of a tablet of 25-mg strength).
2. Active and inactive ingredients are not in exactly the same proportion between different strengths as stated above, but the ratios of inactive ingredients to total weight of the dosage form are within the limits defined by the SUPAC-IR (FDA, 1995) and SUPAC-MR (FDA, 1997a) guidances up to and including Level II changes.
3. For high potency drug substances, where the amount of the active drug substance in the dosage form is relatively low, the total weight of the dosage form remains nearly the same for all strengths (within ±10% of the total weight of the strength on which a biostudy was performed), the same inactive ingredients are used for all strengths, and the change in any strength is obtained by altering the amount of the active ingredients and one or more of the inactive ingredients. The changes in the inactive ingredients are within the limits defined by the SUPAC-IR (FDA, 1995) and SUPAC-MR (FDA, 1997a) guidances up to and including Level II changes.

8.5.3 Biopharmaceutical Classification System

The BCS is a scientific framework for classifying a drug substance based on its aqueous solubility and intestinal permeability (Amidon *et al.*, 1995). The solubility classification of a drug in the BCS is based on the highest dose strength in an IR product. A drug substance is considered highly soluble when the highest strength is soluble in 250 ml or less of aqueous media over the pH range of 1.0–7.5. Otherwise, the drug substance is considered poorly soluble. The volume estimate of 250 ml is derived from typical BE study protocols that prescribe administration of a drug product to fasting human volunteers with a glass (about 8 oz) of water.

The permeability classification is based on the extent of intestinal absorption of a drug substance in humans. The BCS guidance indicates for permeability that "In the absence of evidence suggesting instability in the gastrointestinal tract, a drug substance is considered to be highly permeable when gastrointestinal absorption in humans is determined to be 90% or more of an administered dose based on a mass balance determination or in comparison to an intravenous reference dose." The BCS guidance also provides for high permeability to be determined in *in-vitro* epithelial cell culture studies using suitable high and low permeability reference standards. Otherwise, the drug substance is considered to be poorly permeable. Solubility and permeability classifications result in four classes of drug substance:

BCS Class 1: highly soluble and highly permeable
BCS Class 2: poorly soluble and highly permeable
BCS Class 3: highly soluble and poorly permeable
BCS Class 4: poorly soluble and poorly permeable

An IR drug product is considered rapidly dissolving when not less than 85% of the labeled amount of the drug substance dissolves within 30 min using USP Apparatus I at 100 rpm or USP Apparatus II at 50 rpm in a volume of 900 ml or less of each of the following media: (a) acidic media, such as 0.1N HCl or USP simulated gastric fluid without enzymes (SGF); (b) a pH 4.5 buffer; and (c) a pH 6.8 buffer or USP simulated intestinal fluid without enzymes (SIF). Otherwise, the drug product is considered to be a slow dissolution product. When combined with the *in vitro* dissolution characteristics of the drug product, the BCS takes into account three major factors: solubility, intestinal permeability, and dissolution rate. These factors govern the rate and extent of oral drug absorption from IR solid oral dosage forms (FDA, 2001a).

The FDA BCS guidance indicates that sponsors of NDAs and ANDAs may request biowaivers for highly soluble and highly permeable drug substances (Class 1) formulated in IR solid oral dosage forms that exhibit rapid and similar *in vitro* dissolution. Rapid and similar dissolution is demonstrated by

1. Both drug products do not have less than 85% dissolution in 30 min in 900 ml at pHs of 1.2, 4.5, and 6.8

2. Similarity is demonstrated by an f2 comparison in all three pH conditions (the f2 test is not necessary if both products have 85% dissolution in 15 min or less).

Other conditions that should be met to qualify for a biowaiver are (a) the drug must be stable in the gastrointestinal tract, (b) excipients used in the IR solid oral dosage forms have no significant effect on the rate and extent of oral drug absorption, (c) the drug must not have a narrow therapeutic index, and (d) the product is designed not to be absorbed in the oral cavity.

The BCS guidance is generally considered to be conservative with respect to the class boundaries of solubility and permeability, and the dissolution criteria. Thus, the possibility of modifying these boundaries and criteria to allow waivers of *in vivo* BE studies, i.e., biowaivers, for additional drug products has received increasing attention. There are possible opportunities to expand the BCS-based biowaivers to drugs that are not BCS Class 1.

8.5.3.1 Biowaivers for BCS Class 2 Drugs with pH Dependent Solubility

Some drugs (weak bases) are classified as BCS Class 2 because they are highly soluble at low pH, but fail to meet the BCS solubility limit at higher pH. For these BCS Class 2 drugs, their absorption is complete before they reach a pH where their solubility is decreased significantly. For other BCS Class 2 drugs (weak acids), limited solubility at low pH (acid) may not be physiologically relevant and the solubility at higher pH (e.g., pH $>$ 5) is more appropriate. This may be true because most drugs are absorbed in the intestinal region (Yazdanian *et al.*, 2004; Rinaki *et al.*, 2004). It was questioned by Polli *et al.* (2004) that a solubility of the highest strength in 250 ml over the range pH 1–7.5 is conservative and solubility may need to be conducted only between pH 4.5 and 6.8 considering the pH range of the small intestine. Yu *et al.* (2002) suggest a potential intermediate solubility class for drugs that are soluble either in the stomach or in the intestine, because BCS Class 2 drugs would be absorbed in the intestine due to high permeability as long as the drugs are dissolved before or at the time when they reach the absorbing region of the intestine.

8.5.3.2 Biowaivers for BCS Class 3 Drugs

For rapidly dissolving dosage forms of BCS Class 3 drugs (high solubility, low permeability), intestinal permeability is considered to be the major rate-controlling step in oral drug absorption. Thus, these rapidly dissolving BCS Class 3 drug products are expected to behave like an oral solution. Drug dissolution and other formulation differences are unlikely to have effect on the rate and extent of drug absorption, as long as excipients do not alter intestinal permeability or intestinal residence time of the drug. To ensure rapid dissolution *in vivo* using *in vitro* dissolution, Yu *et al.* (2002) and Polli *et al.* (2004) suggest a more rapid *in vitro* dissolution criterion (not less than 85% within 15 min), because the sink condition common in *in-vitro* dissolution may not exist *in vivo*.

8.6 Locally Acting Drugs

Locally acting drug products require exploration of alternative bioequivalence methods because plasma concentration profiles of these products are not always appropriate surrogates of pharmacological activity. Examples of locally acting products include topical dermatological products, inhalation and nasal products. For many of these products, FDA recommends a BE study with clinical endpoints.

A BE study with clinical endpoints will use a product-specific clinical indication recommended by FDA. Patients in the study would be given the test product, the reference product, and/or a placebo. The placebo arm ensures that the study and its conduct are sufficiently sensitive to differences between treatments. If the reference product is labeled for multiple indications, then the indication that is most sensitive to difference in local delivery of drug is usually preferred.

Most clinical endpoint BE studies have a dichotomous endpoint; the treatment either succeeds or fails. To decide if the test product is bioequivalent to the reference, the success proportion for each treatment is calculated, and if the 90% confidence interval for the difference in success is within −20% to +20%, then the test product passes. For dichotomous endpoints, there is no meaning to between-subject variability and all studies must enroll approximately 200–600 subjects to ensure sufficient power.

Some clinical endpoints are continuous variables or can be treated as such. For example, a reduction in a symptom score is a categorical endpoint, but for equivalence purposes may be treated as continuous data if certain assumptions are made. For these studies, the 90% confidence interval of test/reference ratio must be within 80–125%. The number of subjects required will depend on the between subject variability of the particular clinical endpoint.

Because clinical endpoint BE studies can be larger than studies conducted in support of the initial NDA and the insensitivity of some clinical endpoints to formulation differences, there is much interest in developing new BE methods that are more efficient and more sensitive at detecting product differences. A recent addition to the Federal Food Drug and Cosmetic Act at Section 505(j)(8)(A)(ii) indicates that “For a drug that is not intended to be absorbed into the bloodstream, the Secretary may assess bioavailability by scientifically valid measurements intended to reflect the rate and extent to which the active ingredient or therapeutic ingredient becomes available at the site of drug action.” In the sections below, we will identify some alternative approaches to BE that have been employed, and point out some scientific challenges in developing new BE methods for locally acting products.

8.6.1 Topical Dermatological Products

Topical dermatological products are intended to treat conditions of the skin by direct application of the drug product to the skin. The skin consists of multiple layers beginning with the approximately 10 μm thick stratum corneum, the ≈100 μm thick living epidermis, and the ≈1,000 μm thick dermis. Depending on the drug, topical dermatological products can act at any of these layers, but because the drug is applied to the surface of the skin it must diffuse across each layer sequentially. Thus drug can reach at sites of action without ever being systemically absorbed. This definition excludes products such as transdermal delivery systems that are intended to deliver drug systemically.

There are a variety of BE approaches that are or can be used for topical dermatological products:

1. Biowaivers (topical solutions)

This is discussed in the section on biowaivers. For topical products, a biowaiver usually requires that the test and reference product contain equivalent amounts of the same inactive ingredients. This requirement is necessary because many excipients in topical products are penetration enhancers that can alter the permeability of the skin. Differences in excipients can also change how a topical product spreads or adheres to the skin, which can alter its efficacy.

2. *In vitro* tests (no current topical dermatological products qualify)

The SUPAC-SS (FDA, 1997b) guidance discusses the role of *in vitro* release testing for semisolid dosage forms. The guidance states that "An *in vitro* release rate can reflect the combined effect of several physical and chemical parameters, including solubility and particle size of the active ingredient and rheological properties of the dosage form. In most cases, *in vitro* release rate is a useful test to assess product sameness between pre-change and post-change products" (FDA, 1997b). The use of *in vitro* release tests is currently limited to evaluating changes in manufacturing process, or scale-up by the same manufacturer and is not used for BE. However, when the test and the reference product have identical compositions, the only differences are in manufacturing and process scale.

3. *In vivo* pharmacodynamic studies (topical corticosteroids)

For topical corticosteroids, there is an FDA guidance (1995b) that describes a pharmacodynamic BE study. As in all pharmacodynamic BE studies, it is necessary to establish sufficient sensitivity in the dose–response curve to detect differences between products. For the skin blanching study, the dose is varied by changing the amount of time the topical product is applied to the skin. A pilot study using a range of application times of the reference product is used to identify patients in whom skin blanching is sensitive to differences in application time. Then, these patients are used to compare the test and reference products.

4. *In vivo* pharmacokinetic studies (some topical anesthetics)

For some topical dermatological products such as lidocaine/prilocaine cream,[1] a pharmacokinetic BE study has been the only recommended study. For other topical products, a pharmacokinetic study is requested in addition to other studies

[1] Publicly available at http://www.fda.gov/cder/foi/nda/2003/076453.pdf

TABLE 8.2. Examples of study results[a] using a clinical endpoint to demonstrate BE

N	% cure test	% cure ref	90% CI
728	50	48	[−12, +16]
453	46	40	[−8, +20]
447	29	27	[−9, +13]

[a] All of these studies had three arms and used the difference in cure rate as the endpoint with an acceptable 90% confidence interval of −20 to +20 percentage points

to evaluate whether the test and reference products provide equivalent systemic exposure. This additional BE study is usually requested because there are known safety issues related to systemic exposure.

5. Clinical endpoint BE studies (most other products)

As mentioned before, most other topical products establish BE through a clinical endpoint study. Table 8.2 provides some example results of clinical studies used in support of ANDAs (Lionberger, 2004). All of these studies had three arms and used the difference in cure rate as the endpoint with an acceptable 90% confidence interval of −20 to +20 percentage points. The results show that even with relatively large numbers of patients in each study, the confidence intervals were close to the limits defined by FDA. This suggests that these studies were at a high risk of failure.

8.6.2 Locally Acting Nasal and Oral Inhalation Drug Products

Locally acting nasal and inhalation drug products present significant BE challenges that have limited generic competition in these product categories. FDA has published a draft BE guidance for nasal spray products (FDA, 2003c) and approved one suspension nasal spray product following the publication of this guidance. Currently (2006), there is no BE guidance for inhalation products such as metered dose inhalers (MDI) and dry powder inhalers (DPI).

Given that the performance of these products is determined by the properties of both the formulation and the delivery device, the general approach to BE includes demonstration of:

1. Qualitatively the same, and quantitatively essentially the same, formulations
2. Container and closure systems that are as close as possible in critical attributes
3. Equivalent drug product performance (through *in vitro* tests)
4. Equivalent local delivery (through clinical bioequivalence or pharmacodynamic studies)
5. Equivalent systemic exposure (through a pharmacokinetic study)

8.6.2.1 Nasal Spray Products

For nasal spray solutions that are qualitatively and quantitatively the same as the reference product, and for which the container and closure system are as close as possible in critical attributes including metering chamber volume, the local delivery, and systemic exposure tests can be waived and BE can be demonstrated through equivalent *in vitro* drug product performance. Nasal spray suspension must demonstrate equivalence in three categories listed above (local delivery, systemic exposure, and device performance).

There are six measurable properties identified for use in comparing the drug product-device performance of nasal spray products:

1. Single actuation content (SAC) through container (product) life
2. Droplet size distribution by laser diffraction
3. Drug in small particles/droplets, or particle/droplet size distribution by cascade impactor
4. Spray pattern
5. Plume geometry
6. Priming and repriming

The tests require the use of 10 units from each of three lots of test and three lots of reference products. The FDA guidance (FDA, 2003c) did not specify the criteria for establishing equivalence. Prior to August 2005, FDA had evaluated *in vitro* studies on nasal sprays based on the ratios of geometric means of the test and reference products falling between acceptance limits of 90–111% and evidence for comparable variability of the test and reference products. Since 2005, FDA has been using population bioequivalence (PBE) to compare test and reference products. Information regarding the PBE methodology has been posted on the Agency Web site since April 11, 2003. Inherent in the PBE method is the principle that the BE acceptance limits depend upon the relative variability of the test and reference products observed in the study. This ensures that the acceptance limits are appropriate for the specific products being compared and are based on the characteristics of the approved RLD. In the case of low variability data for the reference product, the acceptance limits narrow toward the 90–111% criteria used in the previous geometric mean method, enabling only test products with comparable variability to meet the acceptance criteria. Conversely, in the case of a high variability reference product, the acceptance limits might be wider.

Usually local delivery BE studies of inhalation products require either pharmacodynamic or clinical studies with a demonstrated dose–response. However, nasal sprays for treatment of seasonal allergic rhinitis (SAR), perennial allergic rhinitis (PAR), and perennial nonallergic rhinitis (PNAR) indications have very limited dose–response. For these indications, equivalent local delivery is assumed to occur when products meet the formulation and device recommendations listed above, and when drug-device performance demonstrates equivalence by showing similar effectiveness in a comparative clinical trial of test and reference products. This clinical trial involves a large number of subjects, which may range from 500 to

1,000 patients. The lowest dose is usually recommended to increase the sensitivity of the study to potential differences between test and reference products, assuming an E_{max} model, as described later in this section. The test product demonstrates equivalent systemic exposure in a pharmacokinetic study in healthy subjects.

Because locally acting nasal spray products are not designed to deliver drug systemically, the conduct of the pharmacokinetic study can be challenging due to very low plasma concentrations. Significant analytical development effort may be required to develop a sensitive method to quantify the low plasma concentrations.

8.6.2.2 Oral Inhalation Products

BE of MDI and DPI products follow the same general approach recommended for nasal spray products (equivalent *in vitro* drug product performance, local delivery, and systemic exposure). As opposed to solution formulation nasal sprays for which *in vivo* studies are not deemed necessary to establish BE, *in vivo* studies are recommended for both solution and suspension MDIs, as well as DPIs.

For the drug product performance tests, a complete list of required tests is not available. However, the critical product quality attributes of MDI and DPI products have been discussed in the scientific literature and in product labeling.

Almost all DPIs currently approved in the United States are breath-actuated so patients' aspiratory effort provides energy for delivery of the powder formulation. There is population variation in the ability of asthma patients to induce flow through the device (Frijlink and De Boer, 2004). A generic sponsor should try to match the performance of the reference product. Localization of delivery in the lung is clearly related to the distribution of particle sizes emitted from the device. *In vitro* tests utilizing a cascade impactor can measure the aerodynamic particle size distribution in a way that is related to deposition in the lung, although it may not be predictive of *in vivo* deposition (Mitchell and Nagel, 2003).

All of these aspects of product performance are affected by the properties of the particles in the formulation. Part of the challenge in designing a DPI is that performance also depends on the device used and the interactions of the particles in the formulation with the device. The aerosol produced by a given patient will depend on the design of the device (its resistance to airflow). The patient effort results in a velocity gradient applied to the dry powder formulation. The effect of a specific velocity gradient on a powder formulation (how much aerosol will be created) will depend on the particle properties. Thus, manufacturing of the powders for DPI may involve critical process parameters to control particle size and shape and surface properties (Telko and Hickey, 2005).

Local delivery studies in asthma patients are challenging because many asthma drugs have a very shallow dose–response curve (the difference between one puff and two puffs may not be clinically detectable) and high between-subject variability, which requires a very large number of subjects to be used in a PD equivalence test. This is particularly true of dose–response for inhaled corticosteroids, as described below. For the nasal spray products, the establishment of dose–response in the clinical endpoint was not required, but because of the more complex nature of particle delivery to the lung, FDA considers establishing the dose sensitivity of the local delivery study of inhalation products essential to establishment of equivalence.

Equivalence of local delivery for a test orally inhaled product relative to the reference orally inhaled product may be evaluated using a "dose-scale" method. The relative BA is determined in terms of "delivered" dose of the test formulation required to produce a PD response of the same magnitude as exhibited by the reference formulation, and its calculation takes into consideration the within-study dose–response. In the dose-scale method, the PD response (E_{R}) to varying doses of the reference product is fit to an $E_{\max}$ model to determine the function, ϕ_{R},

$$\phi_{\mathrm{R}} = E_{0\mathrm{R}} + \frac{E_{\max\mathrm{R}}\mathrm{Dose}_{\mathrm{R}}}{ED_{50\mathrm{R}} + \mathrm{Dose}_{\mathrm{R}}}. \tag{8.5}$$

The relative BA "F" of a dose of the test product relative to that of the reference product can be calculated by applying the inverse of ϕ_{R} to the mean of the response data of the PD response (E_{T}) of test product:

$$F = \phi_{\mathrm{R}}^{-1}(E_{\mathrm{T}})/\mathrm{Dose}_{\mathrm{T}}. \tag{8.6}$$

The Office of Generic Drugs (OGD) has previously used this approach for statistical evaluation of PD BE studies in ANDAs for albuterol MDIs. The applications contained studies based on bronchoprovocation model (PD measure: Histamine PC_{20}) or the bronchodilatation model (PD Measures: AUEC-FEV1 and $\mathrm{FEV1}_{\max}$).

The number of subjects needed to establish BE of a product using the dose-scale method is a function of the slope of the dose–response curve and the variability in the pharmacodynamic response. As the ratio of the variance relative to the slope decreases, fewer subjects are needed to determine the relative BA. BE using a histamine challenge study has been demonstrated for an albuterol inhalation aerosol with 24 subjects (Stewart *et al.*, 2000). The number of subjects required estimating relative BA of an inhaled corticosteroid (ICS) is large; however, estimates suggest that hundreds of subjects would be needed to establish BE for a parallel study design of an ICS. Traditionally, a cross-over study design could not be used for BE studies of ICS due to the long washout period required between treatments. In the literature, an asthma stability model has been suggested as a method for comparing local delivery of ICS using an FEV1 endpoint in a cross-over design. This is estimated to require many fewer subjects to meet the BE requirements (Ahrens *et al.*, 2001).

Another challenge to the design of local delivery studies is that several inhalation products contain two active ingredients, a short acting ICS and long acting beta-agonist. To demonstrate BE, local delivery of both components must be equivalent. However, the FEV1 endpoint used in the asthma stability model is affected by both components. Exhaled nitric oxide (eNO) has been suggested as a pharmacodynamic endpoint for ICS (Silkoff *et al.*, 2001), as eNO levels appear to be unaffected by concomitant administration of beta-agonists. This endpoint, combined with an FEV1 endpoint for the beta-agonist, may enable equivalence of both components to be established.

8.7 Conclusions

Evaluation of BE for systemically acting drugs using pharmacokinetics is well established. Unusual cases such as endogenous substances and highly variable drugs sometimes require new study designs and new statistical analysis procedures. The knowledge available about formulation development and formulation performance for oral dosage forms has allowed the FDA to determine that *in vitro* testing in some cases can provide adequate evidence of BE. Drug companies can now request waivers of *in vivo* BE studies (biowaivers) for some of their products, greatly reducing the cost of such studies. Opportunities for future expansion of biowaivers have been identified and discussed above.

Locally acting drugs are more complex in terms of BA/BE. An appropriate BE method often needs to be established based on a scientific analysis of each drug product. As illustrated in the case studies, all of the following types of studies have been used by FDA to evaluate bioequivalence of locally acting drugs:

1. Clinical endpoint BE study
2. Pharmacodynamic endpoint BE study
3. Pharmacokinetic BE study
4. *In vitro* BE study

The development of generic versions of topical dermatological, nasal, and inhalation products can be significantly impacted by the required BE testing. Development of new and more efficient methods for evaluating BE of locally acting drugs could lead to faster development of generic drugs and facilitate formulation and manufacturing changes.

Acknowledgment. The authors would like to thank Dr. Wallace Adams for his valuable suggestions.

References

Adkin, D. A., Davis, S. S., Sparrow, R. A., Huckle, P. D., Phillips, A. J., and Wilding, I. R. (1995). The effect of different concentrations of mannitol in solution on small intestinal transit: implications for drug absorption. *Pharm. Res.* 12(3):393–396.

Ahrens, R. C., Teresi, M. E., Han, S. H., Donnell, D., Vanden Burgt, J. A., and Lux, C. R. (2001). Asthma stability after oral prednisone: a clinical model for comparing inhaled steroid potency. *Am. J. Respir. Crit. Care Med.* 164:1138–1145.

Amidon, G. L., Lennernas, H., Shah, V. P., and Crison, J. R. (1995). A theoretical basis for a biopharmaceutic drug classification: the correlation of in vitro drug product dissolution and in vivo bioavailability. *Pharm. Res.* 12:413–420.

Basit, A. W., Podczeck, F., Newton, J. M., Waddington, W. A., Ell, P. J., and Lacey L. F. (2002). Influence of polyethylene glycol 400 on the gastrointestinal absorption of ranitidine. *Pharm. Res.* 19(9):1368–1374.

Blume, H., and Midha, K. K. (1993). Bio-international 92, conference on bioavailability, bioequivalence and pharmacokinetic studies. *J. Pharm. Sci.* 82(11):1186–1189.

Boddy, A. W., Snikeris, F. C., Kringle, R. O., Wei, G. C., Oppermann, J. A., and Midha, K. K. (1995). An approach for widening the bioequivalence acceptance limits in the case of highly variable drugs. *Pharm. Res.* 12:1865–1868.

Chen, M. L., Straughn, A. B., Sadrieh, N., Meyer, M., Faustino, P. J., Ciavarella, A. B., Meibohm, B., Yates, C. R., and Hussain, A. S. (2007). A modern view of excipient effects on bioequivalence: case study of sorbitol. *Pharm. Res.* 24:73–80.

Fassihi, A. R., Dowse, R., and Robertson, S. S. D. (1991). Influence of sorbitol solution on the bioavailability of theophylline. *Int. J. Pharm.* 72:175–178.

Food and Drug Administration. (1995a). SUPAC-IR: immediate-release solid oral dosage forms: scale-up and post-approval changes: chemistry, manufacturing and controls, in vitro dissolution testing, and in vivo bioequivalence documentation.

Food and Drug Administration. (1995b). Topical dermatological corticosteriods: in vivo bioequivalence.

Food and Drug Administration. (1997a). SUPAC-MR: modified release solid oral dosage forms scale-up and postapproval changes: chemistry, manufacturing, and controls; in vitro dissolution testing and in vivo bioequivalence documentation.

Food and Drug Administration (1997b). Guidance for industry: SUPAC-SS: nonsterile semisolid dosage forms scale-up and postapproval changes: chemistry, manufacturing, and controls; in vitro release testing and in vivo bioequivalence documentation.

Food and Drug Administration (2001a). Guidance for industry: waiver of in-vivo bioavailability and bioequivalence studies for immediate release solid oral dosage forms based on a biopharmaceutical classification system.

Food and Drug Administration (2001b). Guidance for industry: statistical approaches to establishing bioequivalence.

Food and Drug Administration (2002). Draft guidance for industry: potassium chloride modified-release tablets and capsules: in vivo bioequivalence and in vitro dissolution testing.

Food and Drug Administration. (2003a). Guidance for industry: bioavailability and bioequivalence studies for orally administered drug products – general considerations.

Food and Drug Administration. (2003b). Guidance for industry exposure–response relationships – study design, data analysis, and regulatory applications.

Food and Drug Administration. (2003c). Draft guidance for industry bioavailability and bioequivalence studies for nasal aerosols and nasal sprays for local action.

Food and Drug Administration. (2004). Response to levothyroxine citizen petition. Docket number 2003P-0387.

Frijlink H. W., and De Boer A. H. (2004). Dry powder inhalers for pulmonary drug delivery. *Expert. Opin. Drug Deliv.* 1:67–86.

Fukahori, M., Sakurai, H., Akatsu, S., Negishi, M., Sato, H., Goda, T., and Takase, S. (1998). Enhanced absorption of calcium after oral administration of maltitol in the rat intestine. *J. Pharm. Pharmacol.* 50:1227–1232.

Henney, J. E. (1999). Review of generic bioequivalence studies. *JAMA* 282(21):1995.

Kruger, D., Grossklaus, R., Herold, M., Lorenz, S., and Klingebiel, L. (1992). Gastrointestinal transit and digestibility of maltitol, sucrose and sorbitol in rats: a multicompartmental model and recovery study. *Experientia* 15:733–740.

Lionberger, R. A. (2004). Presentation to FDA advisory committee for pharmaceutical science, April 2004.

Mitchell, J. P., and Nagel, M. W. (2003). Cascade impactors for the size characterization of aerosols from medical inhalers: their uses and limitations. *J. Aerosol Med.* 16:341–377.

Nightingale, S., and Morrison, J. C. (1987). Generic drugs and the prescribing physician. *JAMA* 258(9):1200–1204.

Patterson, S. D., Zariffa, N. M.-D, Montague, T. H., and Howland, K. (2001). Non-traditional study designs to demonstrate average bioequivalence for highly variable drug products. *Eur. J. Clin. Pharmacol.* 57:663–670.

Polli, J. E., Yu, L. X., Cook, J. A., Amidon, G. L., Borchardt, R. T., Burnside, B. A., Burton, P. S., Chen, M. L., Conner, D. P., Faustino, P. J., Hawi, A. A., Hussain, A. S., Joshi, H. N., Kwei, G., Lee, V. H., Lesko, L. J., Lipper, R. A., Loper, A. E., Nerurkar, S. G., Polli, J. W., Sanvordeker, D. R., Taneja, R., Uppoor, R. S., Vattikonda, C. S., Wilding, I., and Zhang, G. (2004). Summary workshop report: biopharmaceutics classification system – implementation challenges and extension opportunities. *J. Pharm. Sci.* 93(6):1375–1381.

Rinaki, E., Dokoumetzidis, A., Valsami, G., and Macheras, P. (2004). Identification of biowaivers among Class II drugs: theoretical justification and practical examples. *Pharm. Res.* 21:1567–1572.

Schulze, J. D., Waddington, W. A., Eli, P. J., Parsons, G. E., Coffin, M. D., and Basit, A. W. (2003). Concentration-dependent effects of polyethylene glycol 400 on gastrointestinal transit and drug absorption. *Pharm. Res.* 20(12):1984–1988.

Shah, V. P., Yacobi, A., Barr, W. H., Benet, L. Z., Breimer, D., Dobrinska, M. R., Endrenyi, L., Fairweather, W., Gillespie, W., Gonzalez, M. A., Hooper, J., Jackson, A., Lesko, L. J., Midha, K. K., Noonan, P. K., Patnaik, R., and Williams, R. L. (1996). Evaluation of orally administered highly variable drugs and drug formulations. *Pharm. Res.* 13(11):1590–1594.

Silkoff, P. E., McClean, P., Spino, M., Erlich, L., Slutsky, A. S., and Zamel, N. (2001). Dose–response relationship and reproducibility of the fall in exhaled nitric oxide after inhaled beclomethasone dipropionate therapy in asthma patients. *Chest* 119:1322–1328.

Stewart, B. A., Ahrens, R. C., Carrier, S., Frosolono, M., Lux, C., Han, S. H., and Milavetz, G. (2000). Demonstration of in vivo bioequivalence of a generic albuterol metered-dose inhaler to Ventolin. *Chest* 117(3):714–721.

Sun, D., Yu, L. X., Hussain, M. A., Wall, D. A., Smith, R. L., and Amidon, G. L. (2004). In vitro testing of drug absorption for drug "developability" assessment: forming an interface between in vitro preclinical data and clinical outcome. *Curr. Opin. Drug Discov. Devel.* 7(1):75–85.

Telko, M. J., and Hickey, A. J. (2005). Dry powder inhaler formulation. *Respir. Care* 50:1209–1227.

Wade, L. G. (2003). Organic Chemistry, 5th Edition. Pearson Education, Inc., Upper Saddle, NJ.

Yazdanian, M., Briggs, K., Jankovsky, C., and Hawi, A. (2004). The "high solubility" definition of the current FDA guidance on biopharmaceutical classification system may be too strict for acidic drugs. *Pharm. Res.* 21: 293–299.

Yu, L. X. (2006). Evolving bioavailability/bioequivalence regulatory standards, strategies in oral drug delivery 2005, Garmisch-Partenkirchen, Germany.

Yu, L. X., Ellison, D. C., Conner, D. P., Lesko, L. J., and Hussain, A. S. (2001). Influence of drug release properties of conventional solid dosage forms on the systemic exposure of highly soluble drugs. *AAPS Pharm. Sci.* 3:E24.

Yu, L. X., Amidon, G. L., and Polli, J. E. (2002). Biopharmaceutics classification system: the scientific basis for biowaiver extensions. *Pharm. Res.* 19:921–925.

9 A Biopharmaceutical Classification System Approach to Dissolution: Mechanisms and Strategies

William E. Bowen, Qingxi Wang, W. Peter Wuelfing, Denise L. Thomas, Eric D. Nelson, Yun Mao, Brian Hill, Mark Thompson, Kimberly Gallagher, and Robert A. Reed

9.1 Introduction

Dissolution testing is a common characterization method employed by the pharmaceutical industry to design formulations and assess product quality. It is a required performance test by many regulatory authorities for solid oral dosage forms, transdermal patches, stents, and oral suspensions. Dissolution testing is unique in that it is the only finished product test method in routine use that measures the effect of the formulation and physical properties of the active pharmaceutical ingredient (API) on the *in vitro* rate of drug solubilization. As a result, dissolution testing is the only test that monitors the impact of environmental storage conditions and manufacturing process upon the rate of drug release from the dosage form. These sensitivities have led to the use of the dissolution test as a measure of formulation bioperformance.

9.2 Biopharmaceutical Classification System Approach to Dissolution

The BCS was introduced by Amidon *et al.* (1995) as a means to predict oral drug absorption based on the contribution of solubility, permeability, and dissolution to oral drug absorption (FDA, 2000; European Agency for the Evaluation of Medicinal Products, 2001). A high solubility drug is defined by the BCS as the highest dose that is soluble in 250 mL or less in aqueous media from pH 1 to 7.5. A compound is considered highly permeable by the BCS if its extent of absorption in humans is 90% or more of an administered dose. Lastly, a drug is considered rapidly dissolving if no less than 85% of label claim is released within 30 min across the aforementioned pH range in either USP dissolution apparatus

I (baskets) at 100 rpm or USP dissolution apparatus II (paddles) at 50 rpm. Drugs are then defined based on these solubility, permeability, and dissolution characteristics to fall within one of four BCS classifications (Fig. 9.1).

According to the BCS approach, the rate of drug absorption is determined by the dissolution rate (k_d) and the permeability rate (k_p) (Fig. 9.2). The rate of dissolution is a function of drug solubility and formulation characteristics, while the permeability rate is largely a function of a drug compound's chemical structure (polarity, functional groups, salt form, etc.). Three distinct scenarios are defined by this BCS absorption model. In the simplest case both k_d and k_p are large. This results in the drug being rapidly solubilized and well absorbed. When k_d is much greater than k_p, the drug is rapidly released and solubilized from

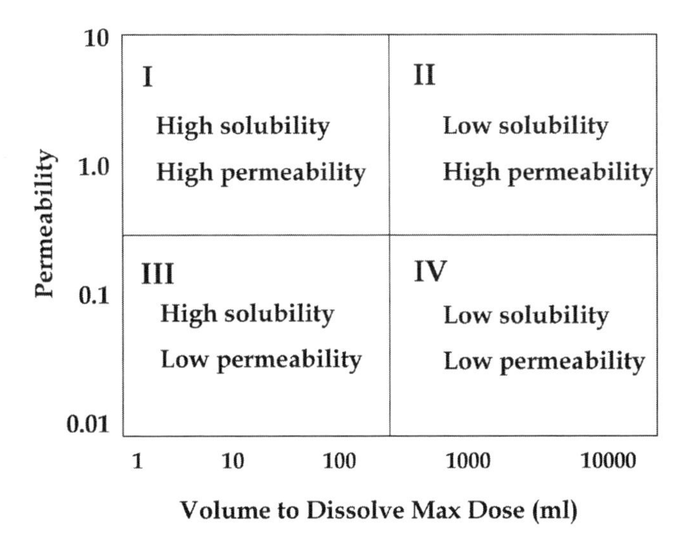

FIGURE 9.1. Biopharmaceutical Classification System (BCS) drug classification matrix

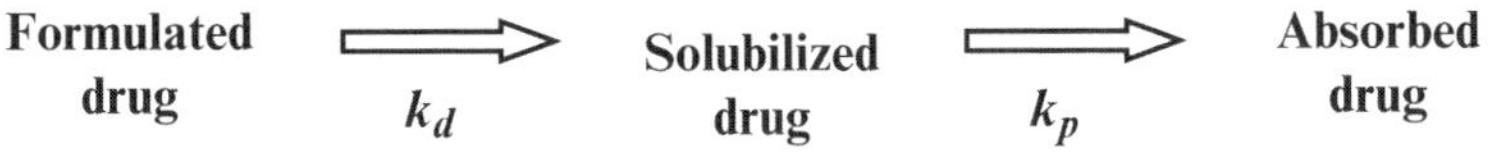

Formulated drug ⟹ k_d **Solubilized drug** ⟹ k_p **Absorbed drug**

Where:

Scenario 1: both k_d & k_p are fast - well absorbed

Scenario 2: $k_d >> k_p$ - permeation control

Scenario 3: $k_d << k_p$ - dissolution controlled

FIGURE 9.2. A typical BCS-based drug absorption process

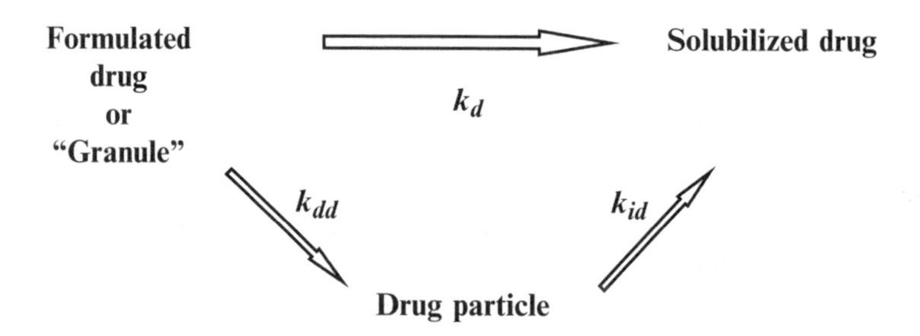

FIGURE 9.3. Kinetics of the dissolution process for a typical drug product

the formulation and the permeation of the drug controls drug absorption. Lastly, if k_p is much greater than k_d, drug dissolution is the rate limiting step to drug absorption.

Conventional oral drug formulations are typically not thought of as affecting the permeability of the drug substance. As a result, the overall absorption is generally only influenced by changes in the concentration of solubilized drug available, as governed by the overall dissolution rate k_d, and thus serves as the primary focus of a formulation scientist. Furthermore, the generation of solubilized drug can be thought of as involving a two-step process, dosage form disintegration followed by intrinsic solubilization of the drug particles (Fig. 9.3). The liberation of drug substance from the formulation matrix is dependent upon the cohesive properties of the formulation and can be described with a characteristic rate k_{dd}. The timescale of k_{dd} ranges from several seconds in the case of rapidly disintegrating immediate release (IR) dosage forms to many hours for controlled release (CR) dosage forms. The second step of the dissolution process, the solubilization of drug particles, is largely governed by the physical properties (e.g., solubility, particle size, surface area) of the API and takes place with a characteristic rate k_{id}. The timescale of k_{id} also ranges from minutes to hours and is largely dependent on the API solubility at physiological pH range and hydrodynamic forces that the drug product experiences, both *in vitro* and *in vivo*.

The overall dissolution rate k_d can either be limited by the rate of disintegration ($k_{dd} \ll k_{id}$), the intrinsic rate of drug solubilization ($k_{dd} \gg k_{id}$), or a combination of both disintegration and drug solubilization rates ($k_{dd} \sim k_{id}$). Overall, the assessment of the relative significance of disintegration and solubilization (k_{dd} versus k_{id}) provides a mechanistic framework to assess where the dissolution test adds value or whether alternative measures provide increased specificity for the critical quality attribute(s) governing the dissolution process.

9.3 *In Vitro–In Vivo* Dissolution Correlation

The objective of dissolution testing varies during the life cycle of a dosage form (Brown *et al.*, 2004). The primary objective in early development phases is to establish working knowledge of the dissolution mechanism and begin to understand the potential *in vivo* predictability of the method (e.g. *in vitro in vivo* relationship (IVIVR)). The objective shifts during later phase development to build an understanding of the impact of critical formulation and process parameters on dissolution and to identify the method's potential to provide *in vitro in vivo* correlation (IVIVC), IVIVR, or other biorelevant information. The ultimate goal is to identify a quality control (QC) dissolution testing method by the time of product registration to verify process and product consistency and, if applicable, utilize *in vitro* dissolution to address the impact of subsequent formulation or process change to *in vivo* performance. An important aspect of dissolution during the life cycle of a dosage form is its utility as a potential surrogate of *in vivo* performance *via* the establishment of IVIVC or IVIVR.

The use of dissolution testing as an indicator of *in vivo* absorption can speed the initial stages of formulation development particularly in the case of CR products. Regulatory guidance categorizes IVIVC into Level A, Level B, and Level C correlations (FDA, 1997b). Level A is the most informative, as it represents a generally linear, point-to-point relationship between *in vitro* dissolution and *in vivo* absorption profiles. In a level B correlation, the dissolution time is compared with the mean residence time or *in vivo* dissolution time. A level C correlation establishes a single point relationship of a dissolution parameter (drug released at one specific timepoint) and one PK parameter. A variety of convolution/deconvolution-based method have been used to determine the fraction of dose absorbed and its relationship with fraction of the dose dissolved *in vitro*. A typical one-compartment model is described below by the Wagner–Nelson equation (Wagner 1975):

$$F_a = \frac{1}{f_a}\left(1 - \frac{\alpha}{\alpha - 1}(1 - F_d) + \frac{\alpha}{\alpha - 1}(1 - F_d)^{\alpha}\right),$$

where F_a is the fraction of total amount of drug absorbed at time t; f_a, the fraction of the dose absorbed at time infinity; α, the permeation rate constant/dissolution rate constant ($\alpha = K_p/K_d$); and F_d is the fraction of the dose dissolved *in vitro* at time t.

This model quantitatively demonstrates the dependence of IVIVC on the relative rates of dissolution, intestinal permeation, and the fraction of dose absorbed. The fraction of total amount of drug absorbed at time $t\,(F_a)$ is determined from a plasma concentration–time profile and assumes the absence of nonlinear postabsorption and presystemic metabolism. It can also be calculated using mass balance dependent and independent mathematical deconvolution techniques (Polli *et al.*, 1996; Wagner, 1975; Dunne *et al.*, 1977; Buchwald, 2003). The fraction of the dose absorbed at time infinity (f_a), again assuming no postabsorption or presystemic metabolism, is bioavailability. Fraction of the dose dissolved at time $t\,(f_d)$ is determined from the *in vitro* dissolution profile and assumes first-order dissolution kinetics. Clearly for orally administered drugs, dissolution and intestinal permeation are recognized as two possible absorption rate limiting phenomena (Fig. 9.4). Dissolution rate-limited absorption ($k_d \ll k_p$) is emphasized in the literature and

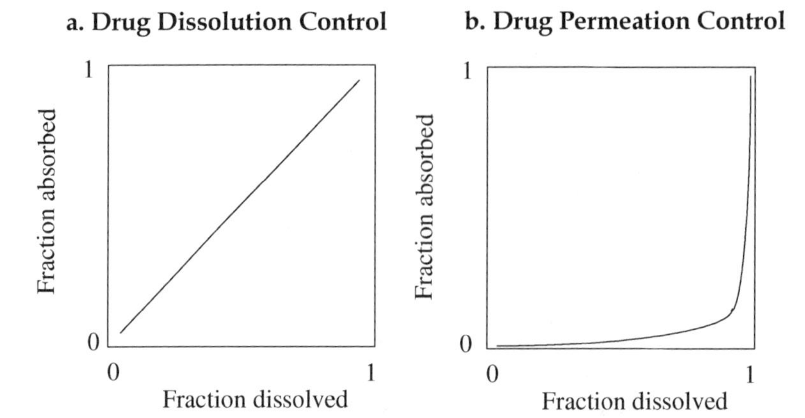

FIGURE 9.4. Theoretical fraction dose absorbed (f_a) versus fraction of dose dissolved (f_d) plots under (a) drug dissolution control and (b) drug permeation control

provides the greatest opportunity for IVIVC (Mojaverian *et al.*, 1992; Hussein and Friedman, 1990; Devi *et al.*, 1989; Aiche *et al.*, 1989; Llabres and Farina, 1989; Dhopeshwarker *et al.*, 1994; Caramella *et al.*, 1993; Civiale *et al.*, 1991; Harrison *et al.*, 1993; Fassihi and Ritschel, 1993; Humbert *et al.*, 1994; Dietrich *et al.*, 1988; Johnson and Swindell, 1996). Drug absorption in this scenario is directly related to dosage form properties (disintegration and/or solubility) and therefore, the dissolution test adds value to support formulation development, biowaiver justification, and setting of acceptance criteria. In many circumstances, the dissolution rate can be tuned through formulation design to establish IVIVC. Alternatively, the intestinal permeability of an API can be so low (dictated by API molecular structure) that the *in vivo* performance of the dosage form is controlled by the intrinsic biopharmaceutical properties of the drug ($k_d \gg k_p$). In this case, the opportunity for IVIVC is very low and the dissolution test of the dosage form does not add value other than as a quality control test for process consistency.

9.4 Recent Climate: Pharmaceutical Quality Assessment

A science-based approach to pharmaceutical development is encouraged in international conference on harmonization (ICH) draft consensus guidance Q8 and has been published for comment in the Code of Federal Register (International Conference on Harmonization, 2004; *Fed. Regist*, 2005). Companies that apply science-based principles to pharmaceutical development and share the resultant information with regulatory agencies may receive enhanced science and risk-based regulatory quality assessment that could affect setting of acceptance criteria, reduce the volume of data submitted at filing, and result in a flexible postapproval process to support continuous improvement and innovation. To this end, the intended use of the product should dictate the product's design, its specifications, and ultimately its manufacturing process and controls. The product design specifications should ensure the product's performance to reliably and consistently meet the therapeutic objective. The manufacturing process should be capable of reliably and consistently meeting the target product design specifications, thus ensuring the therapeutic objective is met. The knowledge gained through a pharmaceutical quality development process provides the basis for "science-based submissions and their regulatory evaluation." (International Conference on Harmonization, 2004)

The alignment of drug BCS class and drug dissolution testing requires a thorough exploration of the dissolution design space (Fig. 9.5). Conventional dissolution testing remains the most general method for assessment of drug release and solubilization (FDA, 1997a; International Conference on Harmonization, 1999). However, it often lacks the specificity required to measure the key formulation or API properties critical to drug release. Consideration of the key formulation or API properties as the guiding principle of drug dissolution design leads to improved product quality by providing increased specificity of the test and in-process controls. For example, characterization of the drug release as disintegration or erosion controlled allows for more scientifically aligned process design, development, and testing approach, such as tablet disintegration. Similarly, drug release limited by the intrinsic rate of API solubilization lends itself to API particle size monitoring as opposed to traditional nonspecific dissolution approaches. The quality attribute can be further defined to more specific measures such as tablet porosity or tablet hardness in the case of disintegration control or API surface area in the case of API solubilization control should the data allow. Instances where drug release is a combination of both disintegration and intrinsic rate of API solubility may be

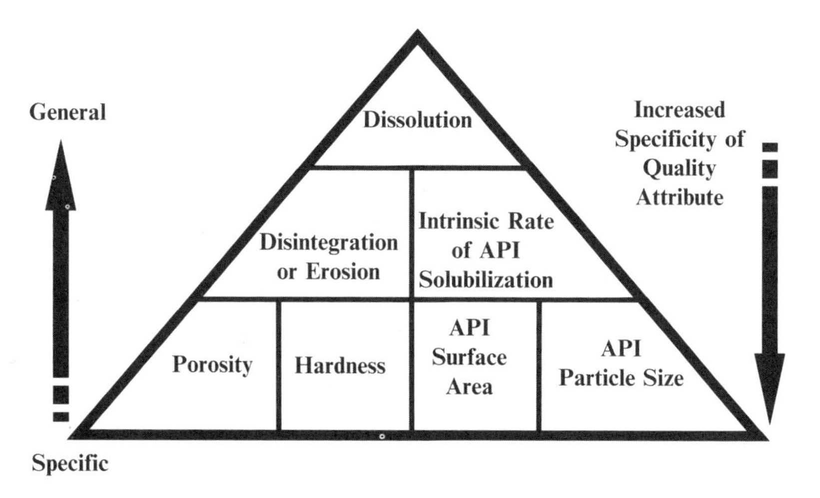

FIGURE 9.5. Drug dissolution design space

best handled by the dissolution test if no mechanistically limiting properties can be clearly defined. In this chapter, case studies are used to explore the application of increased specificity to assess drug release attributes.

9.5 Discussion

Solubility and Caco-2 P_{app} data have been generated in these laboratories for a number of drug substances across the BCS matrix (Fig. 9.6). Note that the classifications resulting from this analysis assume that good correlation between the Caco-2 permeabilities and the human fraction dose absorbed exists. The breadth in API solubility and permeability values found in modern drug development are readily apparent from this plot and serve as a matrix to consider the identification of quality attributes shown in Fig. 9.5. Case studies from this experience base are selected to illustrate both where the dissolution test adds value and when a specific alternate test for a key formulation or API property provides a better indication of product performance.

9.5.1 BCS Class I and III Case Studies

The differentiation of high solubility compounds into BCS classes I or III is due to their permeability characteristics alone, thus it is logical to treat both classes in the same manner when considering *in vitro* drug release analysis. Furthermore,

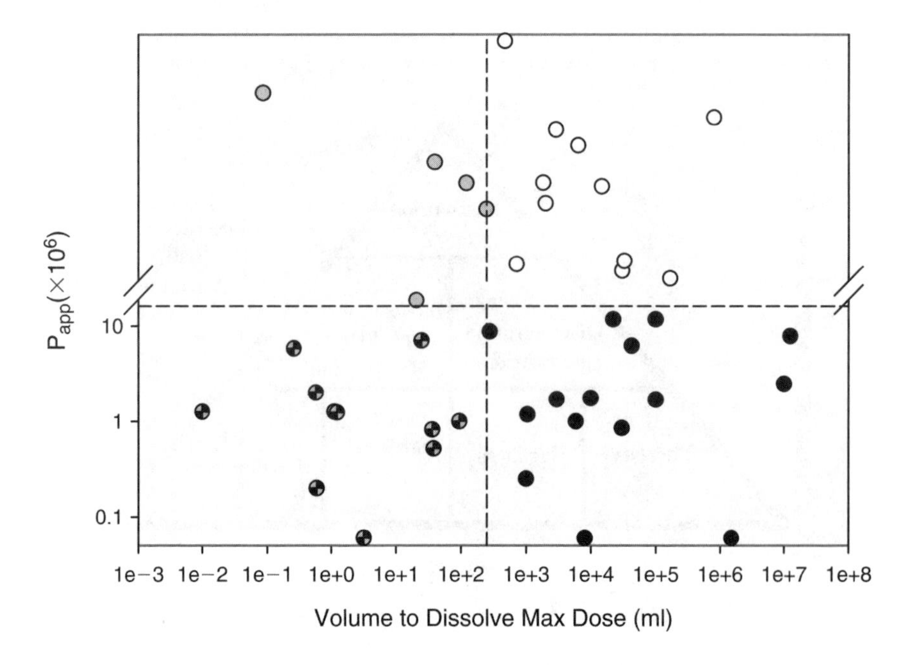

FIGURE 9.6. Solubility and P_{app} values generated in these laboratories

immediate release formulations of BCS class I/III drugs are designed to ideally lead to the formation of an oral solution in the stomach prior to gastric emptying. Accordingly, the target dissolution release for all BCS I/III IR formulations is defined as no less than 85% of label claim released in 15 min (FDA, 1997a). Immediate release formulations exhibiting dissolution of no less than 85% of label claim in 30 min allow for potential biowaiver opportunities based on comparative dissolution criteria (FDA, 2000). Nevertheless, dissolution of BCS class I/III IR formulations is typically where the intrinsic dissolution rate of the drug is fast relative to the disintegration of the formulation ($k_{dd} \ll k_{id}$) and the scientific value added by traditional dissolution testing is minimal. In this context, disintegration presents a more specific measure of dosage form performance since it limits drug solubilization in gastric media as shown in Fig. 9.3. This general approach to dissolution evaluation for BCS class I and III IR dosage forms is captured in the quality attribute decision tree shown in Fig. 9.7. A thorough characterization of the dissolution mechanism should be initiated when drug release does not meet the target dissolution criterion to understand what formulation property(ies) controls drug release. The outcome of this mechanistic evaluation then provides guidance for the definition of an appropriate quality attribute and testing strategy. The following case studies of BCS class I/III formulations further illustrate the decision tree pathways highlighting real world situations where critical quality attribute driven drug release testing are applied.

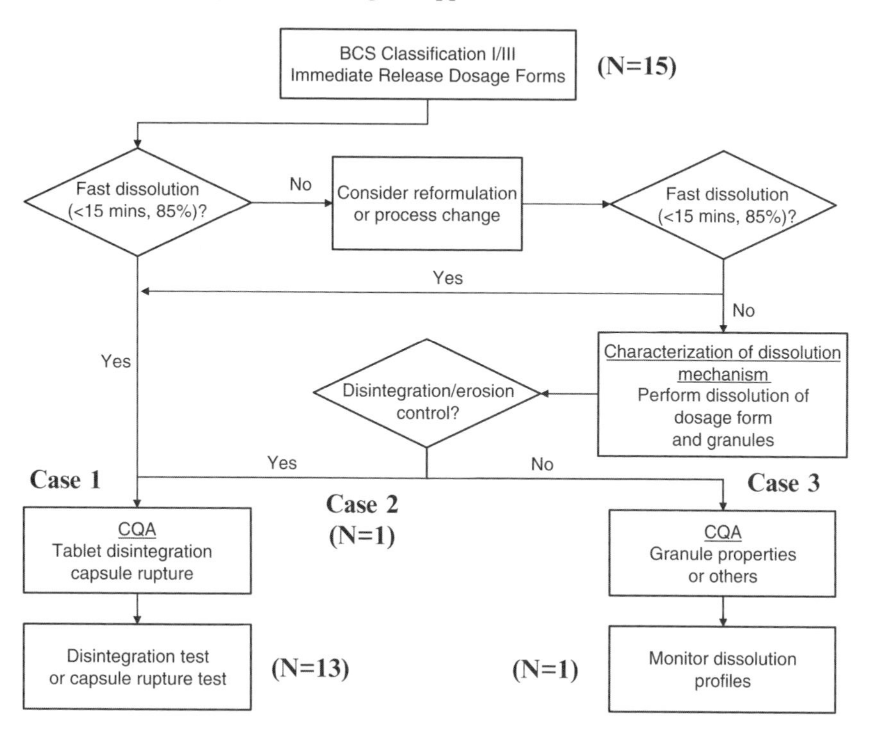

FIGURE 9.7. Decision tree for BCS Class I and III formulations

9.5.1.1 Case Study 1: Fast Release (>85% Release in 15 min) with Disintegration Controlled Dissolution

This case study presents a compound exhibiting >40 mg/mL solubility across pH 1–7.5 solutions and good permeability *via in vitro* Caco-2 analysis. The API is formulated as an IR tablet formulation utilizing a direct compression process and contains unit strengths up to 200 mg. Typical excipients including microcrystalline cellulose, dicalcium phosphate, and croscarmellose sodium are used in the formulation. The dissolution mechanism of the dosage unit is clearly tablet disintegration followed by rapid solubilization of drug particles as determined by analysis of API and formulation dissolution properties. The dissolution profiles in USP I (baskets) at 100 rpm in several media are shown in Fig. 9.8.

All profiles in physiologically relevant media yield greater than 90% release in 10 min. Developmental work indicated that dissolution cannot detect differences in tablet hardness (as set by varying compression force) even at the 10 min timepoint. However, a clear correlation to hardness is observed with disintegration testing thus making it a more sensitive technique for physical attributes (Fig. 9.9).

While this dissolution method can be used as a test to ensure product drug release, it is clear that the technique does not directly measure the primary quality attribute leading to drug dissolution. In this instance, disintegration is the more sensitive method and serves as a better indicator of product quality to ensure that the product continues to be like dosing an oral solution as intended in the formulation design.

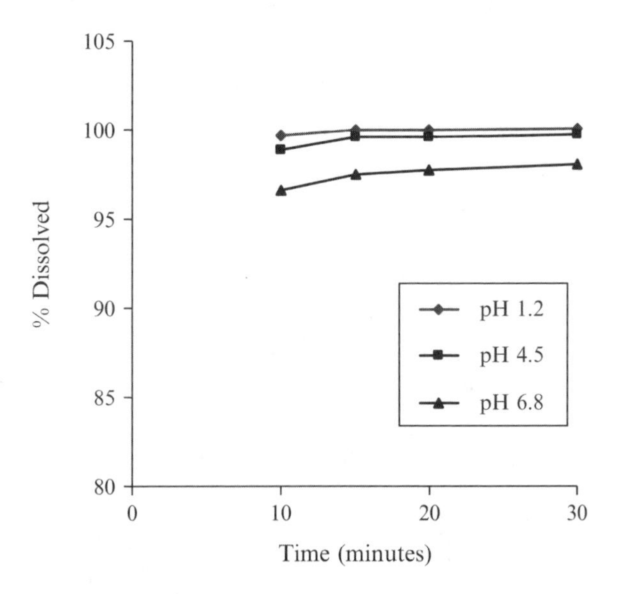

FIGURE 9.8. Dissolution results for Case Study 1 generated using USP Apparatus I (baskets) at 100 rpm in 900 mL of physiological relevant dissolution media

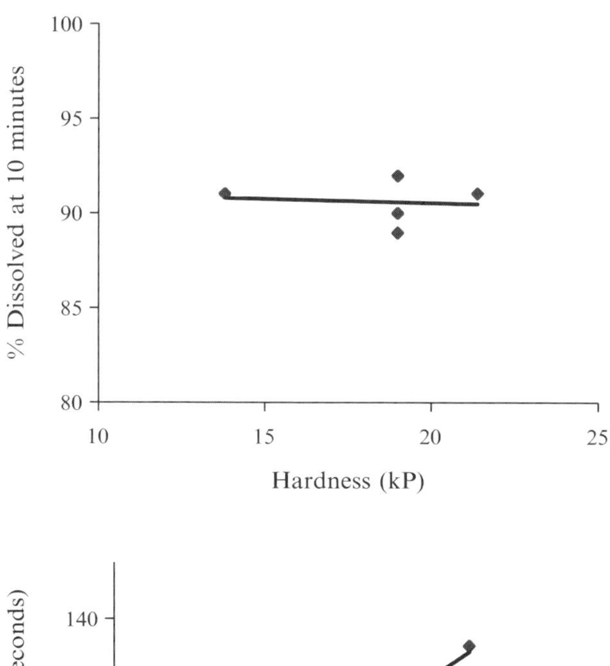

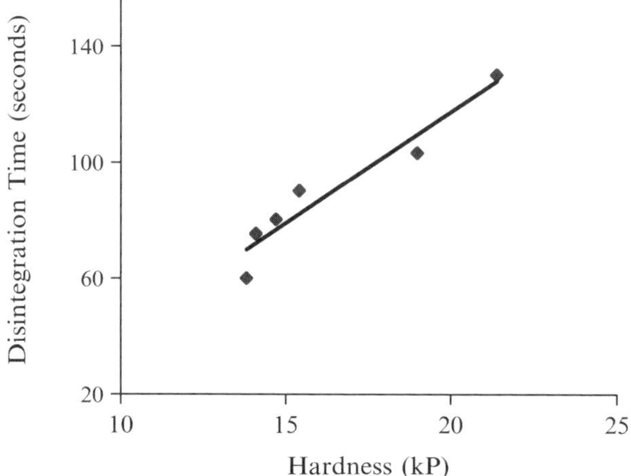

FIGURE 9.9. Dissolution at the 10 min time point (top) and disintegration (bottom) results for Case Study 1 as a function of tablet hardness generated using USP Apparatus I (baskets) at 100 rpm in 900 mL of water or using a USP disintegration apparatus with water, respectively

9.5.1.2 Case Study 2: <85% Release in 15 min with Disintegration/Erosion Controlled Dissolution

This case study involves a dual active combination product where both actives have high aqueous solubility (>40 mg/mL) across pH 1–7.5, and as in Case Study 1, possess good permeability based on *in vitro* Caco-2 analysis. A high shear wet granulation process is used to formulate a tablet dosage form containing the excipients microcrystalline cellulose, sodium lauryl sulfate, and polyvinylpyrrolidone. Dissolution results for tablets of differing hardness were generated using media providing the slowest drug release rate within the physiologically relevant range (Fig. 9.10). The higher paddle speed is required to remove "coning" effects resultant of the large tablet size (~1,000 mg). In this case, the dissolution and disintegration are sensitive to changes in tablet hardness. Complete drug release is

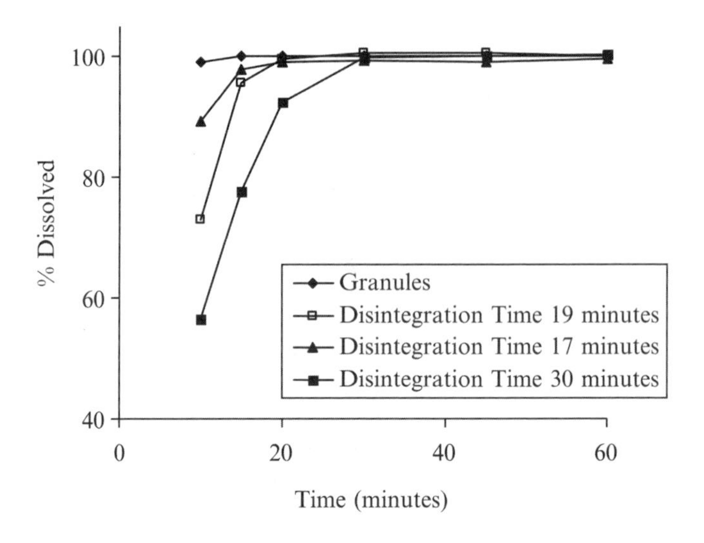

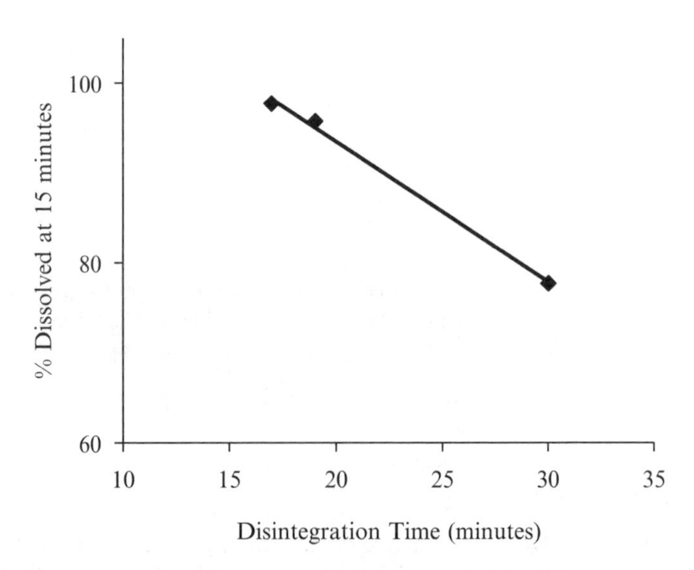

FIGURE 9.10. Dissolution (top) and disintegration (bottom) results for Case Study 2 for tablets of varying hardness generated using USP Apparatus II (paddles) at 75 rpm in 900 mL of water or using a USP disintegration apparatus with water, respectively

achieved immediately after tablet disintegration, hence the dissolution mechanism is attributed to tablet erosion (readily confirmed by visual observation) followed by rapid solubilization of drug particles and make this case amenable to disintegration analysis.

As in the previous example, tablet disintegration is sensitive to physical differences in the formulation. In fact, the results generated with the rapid disintegration method show good correlation to the dissolution data. In this case, either method is capable of sufficiently monitoring the drug release quality attribute; however, no additional value is provided by labor intensive dissolution analysis, making disintegration the logical choice for monitoring product performance.

9.5.1.3 Case Study 3: Dissolution Mechanism not Dependent on Disintegration/Erosion

This case study involves a high solubility drug (0.1 mg/mL) across pH 1–7.5 and high permeability based on *in vitro* Caco-2 analysis. A high shear wet granulation process was used to formulate a tablet dosage form containing typical excipients including lactose and dicalcium phosphate. The resultant granule particle size has significant influence on the rate of drug solubilization, clearly indicating that the dissolution mechanism is controlled in part by the release of drug from larger granules after initial tablet disintegration (Fig. 9.11). At this point, two options can be considered from a formulation/process design perspective: (1) control the granule distribution to less than 150 μm or (2) allow the granule distribution to exceed 150 μm.

In the first case, disintegration is an acceptable quality attribute. In the second case, disintegration is not appropriate as a quality attribute as tablet breakdown is

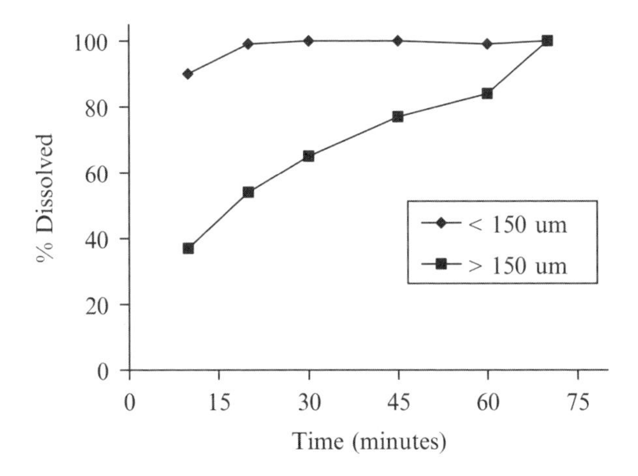

FIGURE 9.11. Dissolution results for Case Study 3 generated using USP Apparatus II (paddles) at 50 rpm in 900 mL of water. Each profile represents release rates from granules that are less than 150 μm or greater than 150 μm in size. The paddle speed was increased to 150 rpm for the final 15 min to measure the full drug content of the granules

not the rate limiting step to drug release. Dissolution testing is a better monitor for quality control purposes. Clearly, an understanding of the mechanistic aspects of the product performance should guide the decision to define the best means of quality control.

The previous case studies are three examples from the authors' experience with 15 BCS class I/III compounds. The dissolution results of 13 of these compounds showed greater than 85% drug release in 15 min with tablet disintegration as the rate limiting step to dissolution, thus making them amenable to disintegration testing. One drug product did not meet the definition of rapid release, but drug release was still under disintegration control and thus still a candidate for disintegration testing. Only one drug product out of the 15 was found to exhibit a drug dissolution mechanism appropriate for monitoring *via* traditional dissolution analysis (Case Study 3). It is important to note that even in this instance, drug release could be attenuated through control of the manufacturing process to yield another instance of tablet disintegration controlled dissolution.

It should be the goal of formulation development to keep BCS class I/III IR formulations rapidly dissolving if the desire is to simulate dosing of an oral solution. High solubility drug formulated as drug filled/drug blend in capsules also fit into this paradigm as ultimately the rupture of a gelatin capsule will be the rate limiting step to drug release. Application of disintegration during development should be supplemented with dissolution analysis when concerns over form changes, agglomeration (drug in capsule), or adherence to particular excipients exist. In addition, it is worthwhile to consider that the application of the decision tree in Fig. 9.7 to BCS class II/IV drug compounds that are highly soluble at gastric pHs provided a clear understanding of their dissolution mechanism and *in vivo* absorption are developed (*vide infra* Case Study 10).

9.5.2 BCS Class II and IV Case Studies

Twenty five BCS class II and IV compounds are present in the BCS matrix shown in Fig. 9.6. The overall dissolution rate (k_d) of these formulations is generally governed by the low solubility of the drug substance and corresponding slow intrinsic drug solubilization (k_{id}). The disintegration of such a dosage form will initially produce a suspension of the API, whose dissolution then depends mainly on the intrinsic solubilization of the drug, its particle size distribution, and the driving force for solubilization created by the absorption of drug in the small intestine. In most cases, the bioavailability of a well-formulated immediate release dosage form of this type will be similar to that of a drug suspension. Liquid filled capsule (LFC) formulations, when the API is dissolved in a liquid vehicle to maximize bioavailability, are an exception to this suspension model as the dissolution of the dosage form is simply the capsule rupture. In addition, some formulations result in an overall dissolution rate that is a combination of both formulation disintegration and intrinsic drug solubilization rates. Thus, the value added by traditional dissolution testing of BCS class II/IV compounds varies depending upon the difference between k_{id} and k_{dd}. This variability is captured in the BCS class II/IV

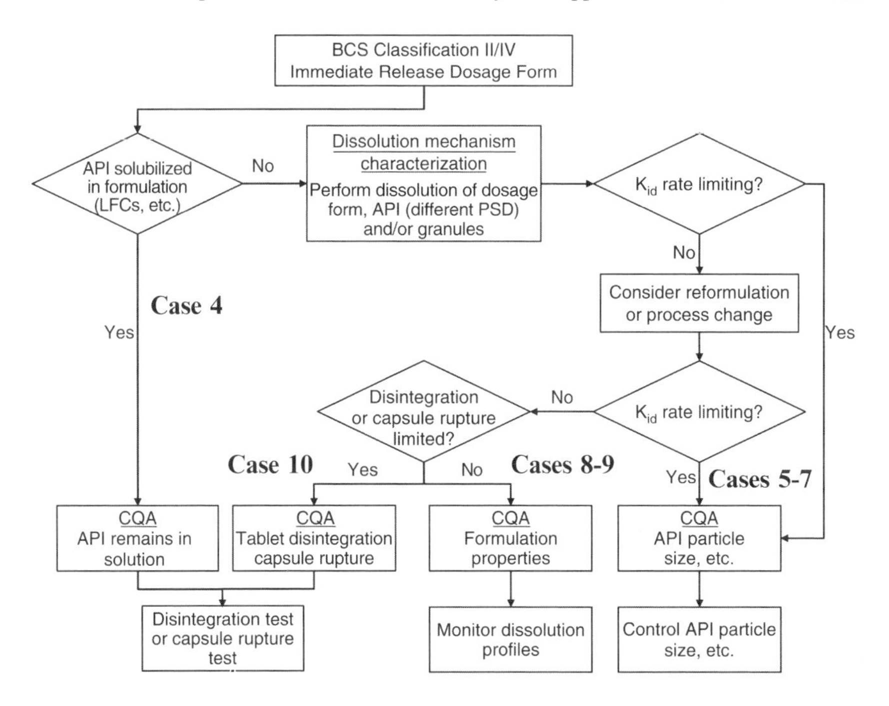

FIGURE 9.12. Decision tree for BCS class II and IV formulations

dissolution quality attribute decision tree shown in Fig. 9.12. The following case studies of BCS class II/IV drugs compare formulation data highlighting real situations where critical quality attribute driven drug release testing can be applied.

9.5.2.1 Case Study 4: Liquid Filled (True Solution) Capsules

LFCs, while typically formulated for BCS class II/IV drugs, can fall into two general classes: (1) true solution filled and (2) API suspension filled. Drug release in true solution filled products is dominated by capsule rupture as the drug is already solubilized in the formulation. Suspension filled capsules involve capsule rupture and likely a subsequent intrinsic drug solubilization step. In the case of the true solution filled capsules (Fig. 9.12), the formulation design and development is to ensure that the drug remains solubilized such that one is dosing a true solution. The fact that the drug is already in solution heightens the point that rupture is the critical quality attribute for these formulations.

9.5.2.2 Case Studies 5, 6, and 7: Intrinsic Rate of Drug Solubilization Controlled Dissolution

A common scenario with low solubility drugs (BCS class II/IV) is an *in vitro* release profile that is dominated by the intrinsic rate of solubilization. The following three case studies are examples of intrinsic drug solubilization controlled dissolution. Case Study 5 is a BCS class II compound with pH independent low aqueous solubility (<5 μg/mL across pH 1–7.5) formulated in a lyophilized tablet formulation that fully disintegrates in 1–2 s in the mouth. Such dosage forms achieve release of the drug particles from the formulation prior to entering the stomach, thus they closely mimic drug suspension formulations, such as suspension filled capsules. Predictably, the dissolution performance of this dosage form is far more dependent on API properties than on batch-to-batch formulation or process variations (Fig. 9.13). In fact, conventional IR tablets and orally disintegrating tablets manufactured from the same API lot give comparable dissolution profiles at times >20 min. The slowed dissolution at the earlier time points for the conventional tablet can be attributed to the longer disintegration time observed for this formulation (Fig. 9.14). Overall, the rapid permeation rate for this drug leads to comparable *in vivo* extent of absorption for the two formulations with an observed decrease in C_{max} and increase in T_{max} for the conventional tablet, as suggested by the *in vitro* results. Hence for the lyophilized tablet formulation, the compound API particle size is the appropriate quality attribute.

The intrinsic properties of the drug substance can also control dissolution rate in submicron sized drug containing formulations. Such dosage forms are produced from surfactant and polymer-stabilized colloidal drug dispersions, and are explicitly designed to release submicron sized API upon formulation disintegration to enhance dissolution kinetics and improve bioavailability. Case Study 6 involves a BCS class IV compound with extremely low aqueous solubility across pH 1–7.5. The formulation consists of high intensity wet milled API and stabilizing excipients coated onto microcrystalline cellulose beads and filled into

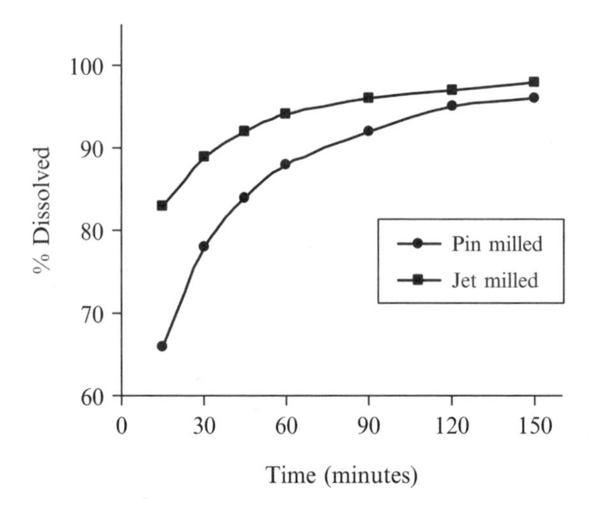

FIGURE 9.13. Dissolution results for Case Study 5 formulated in an orally disintegrating tablet. The jet milled API and the pin milled API had mean particle sizes of 4.8 and 12.0 μm, respectively. Dissolution profiles were generated using USP Apparatus II (paddles) at 100 rpm in 900 mL of 0.6% SDS

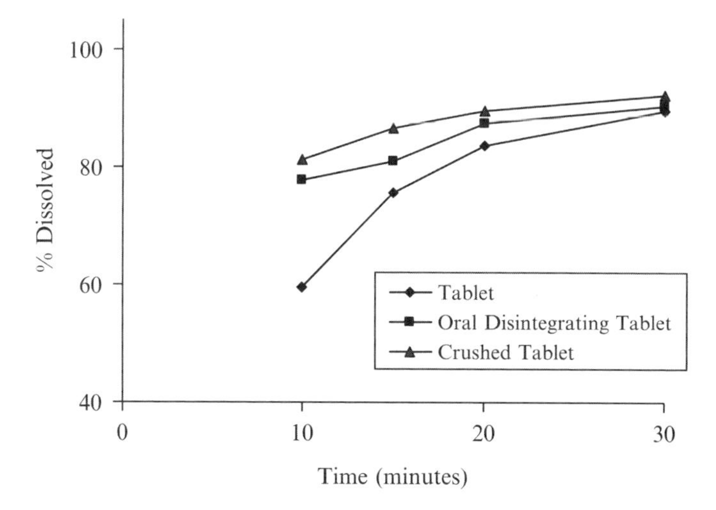

FIGURE 9.14. Dissolution results for Case Study 5 conventional, crushed conventional, and orally disintegrating tablets manufactured from the same API lot. Dissolution profiles were generated using USP Apparatus II (paddles) at 50 rpm in 900 mL of 2.0% SDS

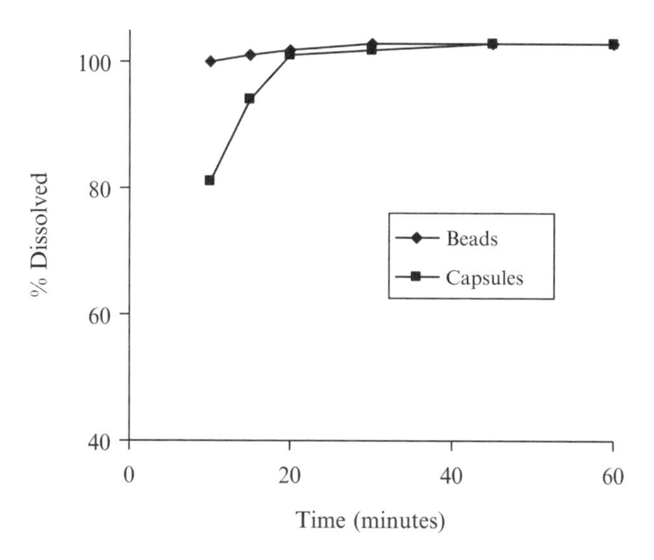

FIGURE 9.15. Dissolution of Case Study 6 capsules and capsule contents, which are coated beads with <200 nm particle size API. The capsule product clearly reflects the capsule rupture as defining the dissolution curve. Dissolution profiles were generated using USP Apparatus II (paddles) at 100 rpm in 900 mL of 2.2% SDS media

a hard gelatin capsule shell. Rupture of the gelatin capsule shell initially delays the release and dispersion of the coated beads (Fig. 9.15). However, once the coated beads are exposed to the dissolution media, the dissolution profile is very rapid and essentially superimposable with that of the coated beads themselves.

Furthermore, one anticipates that the *in vivo* release of the formulation is not strongly dependent upon rupture of the hard gelatin capsule shell but rather the intrinsic solubility of the API which is particle size dependent. The particle size distribution generated in the product manufacture is narrow and reproducible, does not change with subsequent processing or final product storage through shelf expiry. It is for this reason that in-process end point determination of API particle size for the milling unit operation is an appropriate strategy for ensuring product quality.

Case Study 7 presents a BCS class IV compound exhibiting low aqueous solubility and *in vitro* permeability. The drug is formulated *via* dry blending with microcrystalline cellulose and magnesium stearate followed by encapsulation. Capsule dissolution profiles display a significant dependence on API particle size (Fig. 9.16). In fact, the API D90 results show good correlation to the %drug released at 60 min (Fig. 9.17). Clearly, API particle size is the quality attribute governing the dissolution of this product. Therefore, control of the API particle size *via* the manufacturing process is appropriate from the pharmaceutical quality assessment perspective, to ensure product quality.

9.5.2.3 Case Studies 8 and 9: Mixed Contribution of Formulation Colligative Properties and Intrinsic Rate of Drug Solubilization

Immediate release dosage forms of BCS class IV compounds are not always limited by the intrinsic rate of drug solubilization. Case Study 8 presents an example where formulation properties and composition impact the disintegration, and

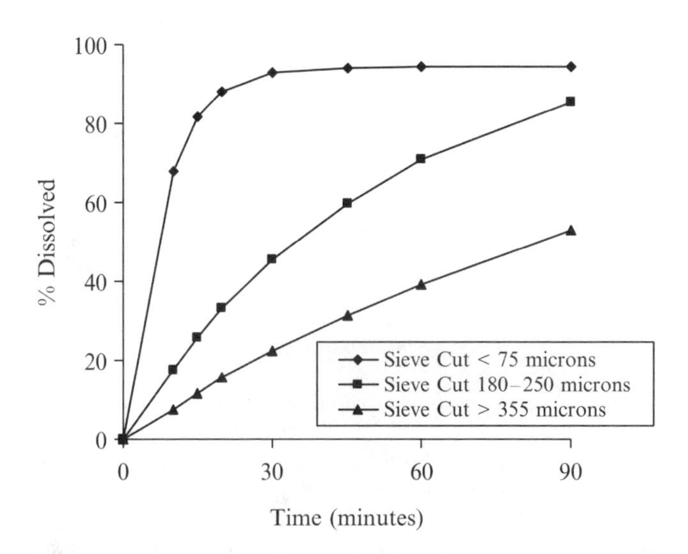

FIGURE 9.16. Dissolution results for Case Study 7 capsules as a function of API particle size. Dissolution profiles were generated using USP Apparatus II (paddles) at 100 rpm in 900 mL of 2% Tween 80 media

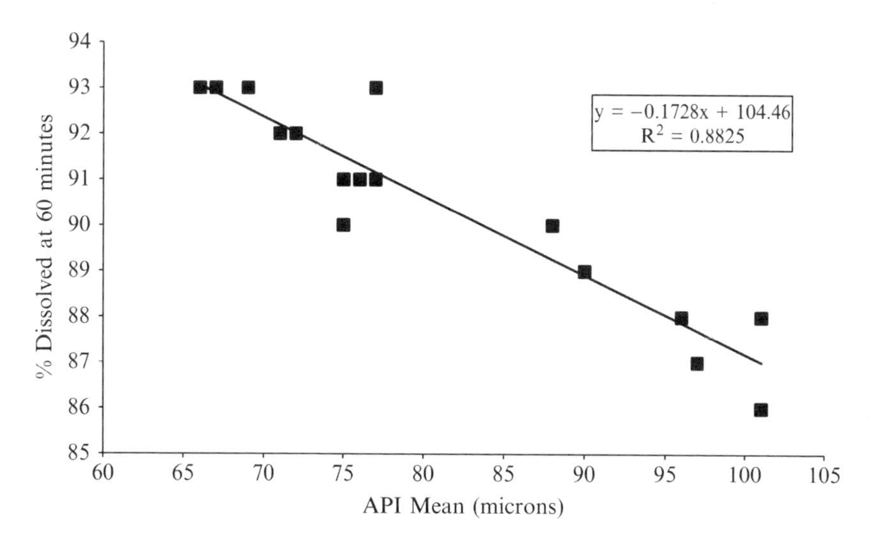

FIGURE 9.17. Percentage dissolved in 60 min for Case Study 7 capsules as a function of API D90 value. Dissolution profiles were generated using USP Apparatus II (paddles) at 100 rpm in 900 mL of 2% Tween 80 media

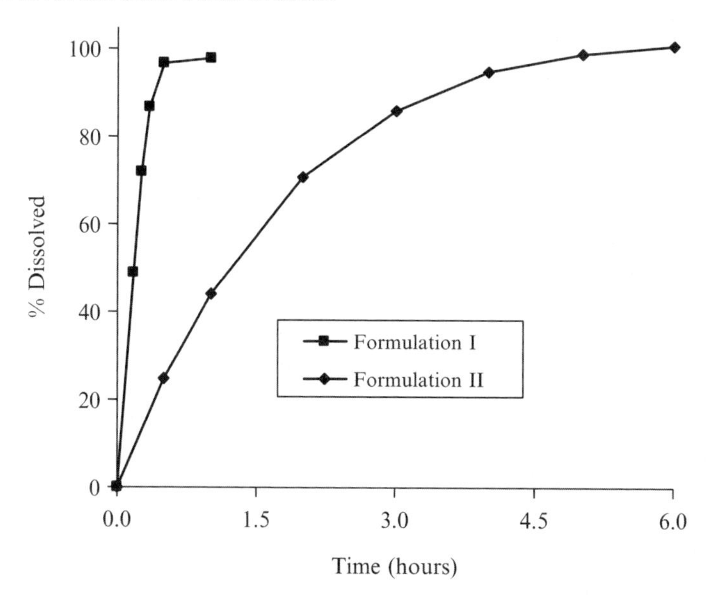

FIGURE 9.18. Dissolution results for Case Study 8 tablet formulations. Dissolution profiles were generated using USP Apparatus II (paddles) at 100 rpm in 900 mL of pH 6.8 buffer

ultimately the overall dissolution rate. The dissolution profiles for two formulation designs of a BCS class IV drug are given in Fig. 9.18. Formulation I is intended to rapidly disintegrate while Formulation II is designed to increase the impact of the formulation on the overall release rate. These same tablet formulations

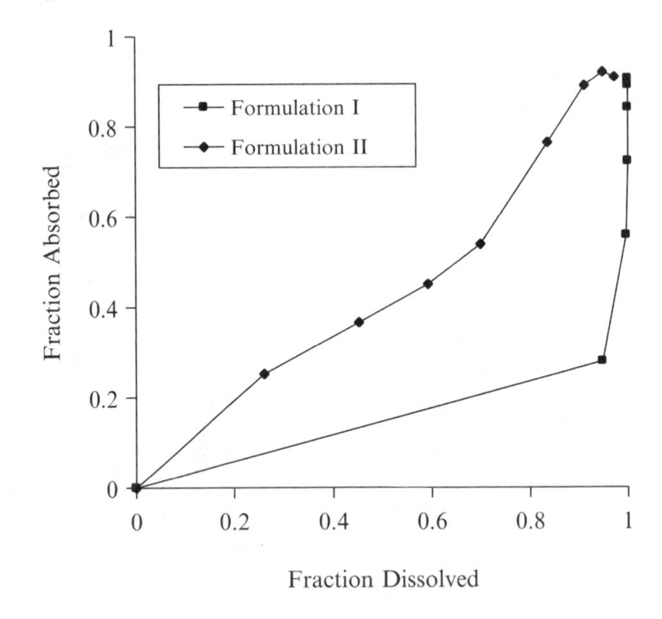

FIGURE 9.19. Wagner–Nelson deconvolution of human *in vivo* and *in vitro* dissolution profiles shown in Fig. 9.18 for Case Study 8 Formulation I and II. Formulation I exhibits more permeation control character and Formulation II exhibits primarily dissolution control character

were also dosed in humans and pharmacokinetic (PK) data collected. As introduced earlier, Wagner–Nelson models were then applied to compare the *in vitro* and *in vivo* performance of these two formulations (Fig. 9.19). The results of this analysis suggest that Formulation I is largely exhibiting permeation control. Here dissolution testing provides limited value and a disintegration test would be a more appropriate measure of product performance. However, as might be anticipated, Formulation II exhibits dissolution control behavior *in vivo*, thus dissolution profiles would add value to characterize the quality of this product. Clearly, a Wagner–Nelson type analysis is a useful tool for identifying the value of the dissolution testing.

Case Study 9 is another example of where the combination of the formulation properties and the process parameters can greatly impact drug release profiles. In this case, two drug substances are incorporated in a single capsule formulation. The first API is a BCS class IV compound with low aqueous solubility from pH 3.0–7.5 but high solubility at low pH. The second API is a BCS class II compound with poor solubility across the pH range of 1.0–7.5. The two APIs are mixed with lactose and microcrystalline cellulose and granules of the blend formed through a roller compaction process. The granules are then encapsulated. In exploring the process design space for the roller compaction unit operation, varying the roll pressure had a significant impact on the release profile (note both APIs had similar release rates) (Fig. 9.20). Furthermore, it was demonstrated that the capsule shell

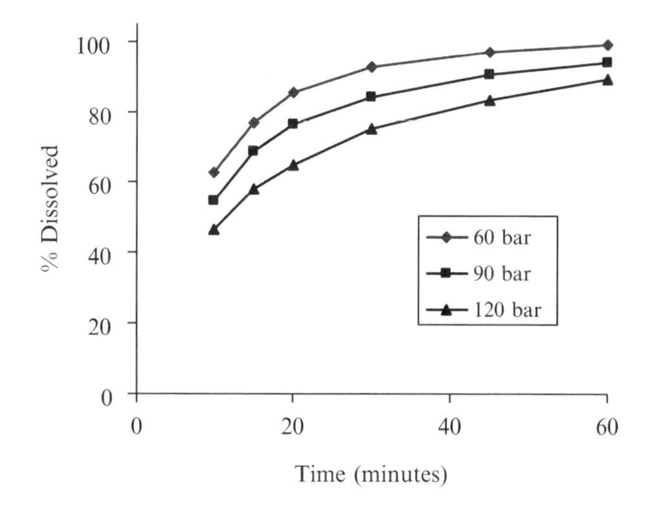

FIGURE 9.20. Dissolution results for Case Study 9 capsules as a function of the roll compaction pressure applied ranging from 60 to 120 bar. Dissolution profiles were generated using USP Apparatus II (paddles) at 50 rpm in 900 mL of 2% SDS media

had no impact on the overall release rate and that the drug release from the roller compacted granules is the rate determining step. Hence, neither the capsule disintegration nor the API particle size distribution was contributing to the *in vitro* dissolution rate. In this case, monitoring the drug release by dissolution testing presents the best measure of product quality.

9.5.2.4 Case Study 10: API with High Solubility at Gastric pHs

BCS class II/IV compounds can also have pH dependent solubility, exhibiting high aqueous solubility in gastric pHs (e.g., amines) or intestinal pHs (e.g., many organic acids). The API used in Case Study 10, while defined as BCS Class IV because of its low solubility across the pH range of pH 3.0–7.5 and low permeability, behaves like a BCS Class III drug as it is highly soluble at gastric pH's (>200 mg/mL). At gastric pHs, the *in vitro* release profile is rapid (>90% release at 20 min) for both granules and the encapsulated product. Therefore, the drug behaves like a rapidly dissolving BCS class III compound as the formulation mimics dosing an oral solution. Dissolution testing results of the encapsulated product and the unencapsulated granulation show that the drug is rapidly released from the granules and the dissolution profile of the encapsulated material is dominated by the capsule shell rupture time (Fig. 9.21). As is typical of many hard gelatin encapsulated products, the key quality concern is gelatin crosslinking that may ultimately slow the drug dissolution rate. As dissolution measures capsule rupture for this product, disintegration is a suitable surrogate test to monitor the product quality.

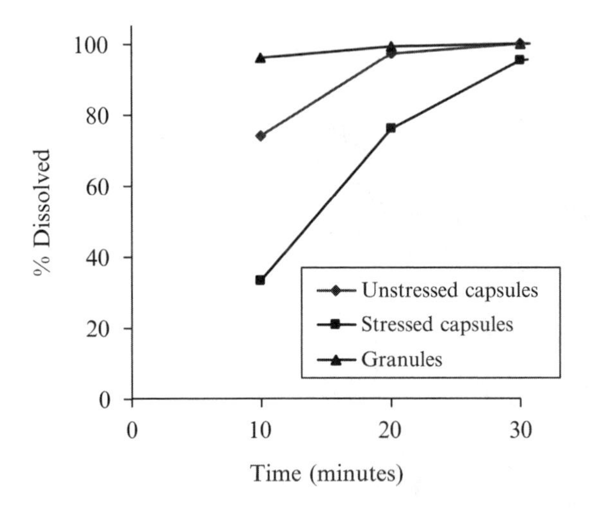

FIGURE 9.21. Dissolution results for Case Study 10 granules, unstressed capsules, and stressed capsules. The slowing observed for stressed capsules is readily attributable to hard gelatin crosslinking. Dissolution profiles were generated using USP Apparatus II (paddles) at 50 rpm in 0.1N HCl

In summary, BCS class II/IV compounds are subject to the same mechanistic dissolution considerations as described for BCS class I/III compounds. It should be the goal of formulation development to keep BCS II/IV IR formulations rapidly disintegrating if the desire is to simulate dosing of an oral suspension. API properties (e.g., particle size) of BCS II/IV compounds are frequently identified as quality attributes and are controlled in lieu of dissolution monitoring, provided a correlation with dissolution is established during development. In the case of LFCs, capsule shell rupture presents a sufficient quality measure as the API is solubilized, and rupture in the gut releases a drug solution. Lastly, for BCS II/IV compounds with high solubility in gastric pH, consideration should be given a simple disintegration test if the product dissolution is demonstrated to be disintegration controlled.

9.5.3 Controlled Release Dosage Form Case Study

Controlled release formulations, often called sustaining release formulations, may include BCS class I–IV drug compounds and are specifically designed such that the formulation "controls" the overall drug release rate. At the extreme of a high solubility and high permeability drug, one can expect that as the drug is released, it is rapidly dissolved and absorbed and an IVIVC can be established. However, at the other extreme, a low solubility and low permeability drug, one can envision a CR product that is still dominated by intrinsic rate of solubilization or drug permeation and the ability to generate an IVIVC is greatly challenged. Controlled release products are often developed to maintain therapeutic activity for an extended time or to reduce high plasma concentrations to minimize adverse effects. Therefore, it is desirable to develop an IVIVC so that an *in vitro* dissolution test can be used to indicate or predict changes which may have an effect on the efficacy or safety of the product.

Case Study 11 involves a BCS class I compound with high permeability and high solubility across pH 1–7.5. A CR gel extrusion module (GEM) tablet was developed to reduce the dosing frequency of the drug product. The GEM tablet consists of a core tablet of drug, a water swellable carbomer polymer, sodium phosphate, lactose, microcrystalline cellulose, and magnesium stearate. The core tablet is manufactured by a roller compaction process and coated to provide a rigid, water impermeable cellulose polymer membrane. Holes are then drilled through the coating. The release mechanism involves water ingress through the drilled holes, then polymer hydration followed by extrusion of drug particles present in the core tablets through holes in the coating, and rapid intrinsic solubilization of the drug particles. Subsequently, the Case Study 11 drug is rapidly solubilized and readily absorbed.

Formulations with high, medium, and low release rates were generated by varying the number of holes drilled through the outer coating to give corresponding *in vitro* dissolution profiles (Fig. 9.22). These formulations were also evaluated in a single dose pharmacokinetic (PK) study in humans and their PK behavior compared against that of an immediate release tablet formulation (Fig. 9.23). The mean plasma profiles of the drug were deconvoluted according to the Wagner–Nelson method to generate *in vivo* fraction absorbed versus time profiles. An IVIVC model was then established by plotting the mean fraction

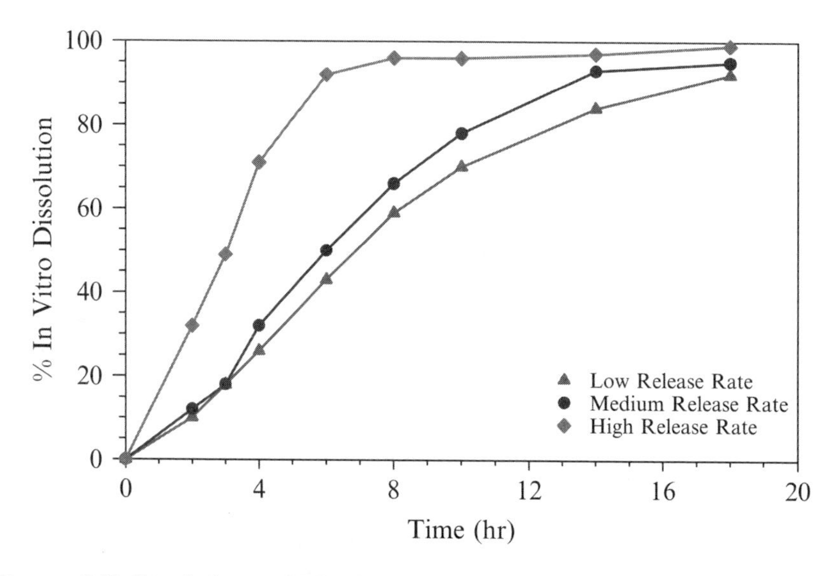

FIGURE 9.22. Dissolution results for Case Study 11 in designed controlled release tablets with low, medium, and high release rates. Dissolution profiles were generated using USP Apparatus II (paddles) at 100 rpm in 900 mL of water

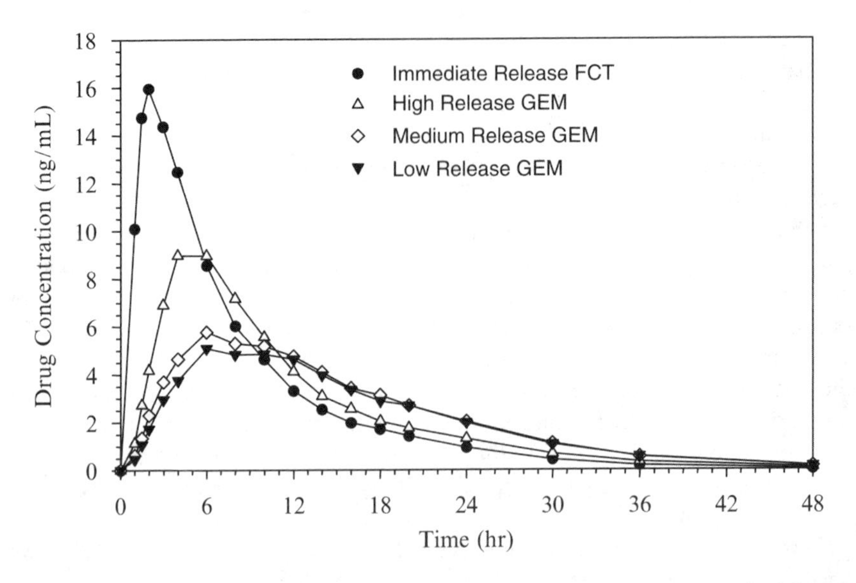

FIGURE 9.23. Mean plasma concentration of Case Study 11 drug versus time for an immediate release (IR) formulation and three designed controlled release tablets with low, medium, and high *in vitro* release rates

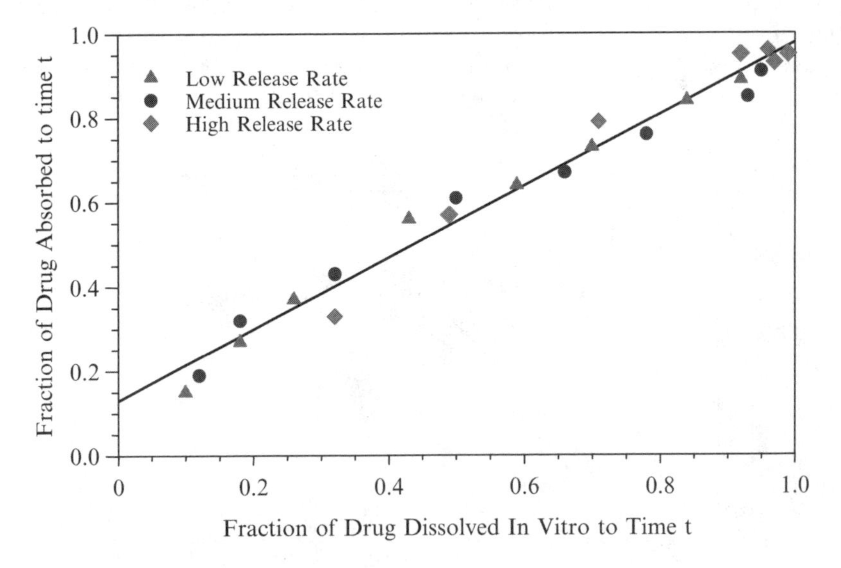

FIGURE 9.24. Fraction absorbed versus fraction dissolved for Case Study 11 three designed controlled release tablets with low, medium, and high *in vitro* release rates

absorbed *in vivo* versus the mean fraction dissolved *in vitro* (Fig. 9.24). A good linear relationship was obtained with correlation coefficient of 0.98. Evaluation of the predicted error in *in vivo* parameters (data not shown here) meets FDA criteria for valid Level A IVIVC. In this instance, the number and size of the holes in the impermeable membrane coating are an appropriate quality attribute and can be used as a surrogate for dissolution.

Clearly, dissolution profiles need to be monitored during the development of controlled release dosage forms. However, once the dissolution mechanism is well understood, the impact of formulation properties on dissolution profiles well characterized and an IVIVC established, the identification of quality attribute alternatives to dissolution may be explored. These quality attributes can then be monitored to ensure product quality in lieu of dissolution. In this instance, the number and size of the holes in the impermeable membrane coating are an appropriate quality attribute and can be used as a surrogate for dissolution.

9.5.4 Pharmaceutical Quality Assessment Implications of Dissolution

The aforementioned case studies illustrate the value of understanding the relationship between a drug's solubility characteristics and its mechanism of dissolution once formulated. This understanding provides a quality attribute framework to assess biorelevant features of formulated drugs. Further, this knowledge can then be employed to guide formulation design and selection and process scale-up and optimization.

The overall mechanistic guidance from these laboratories is provided in Table 9.1. As previously noted, formulations where the dissolution rate limiting step is governed by API intrinsic solubilization ($k_{dd} \gg k_{id}$) are better controlled by monitoring critical quality attributes such as API particle size or surface area as opposed to less specific dissolution testing. Fast disintegrating tablets, capsules, and submicron drug containing formulations are where these characteristics would most frequently be observed. Formulations where the rate of

TABLE 9.1. Critical quality attributes controlling dissolution performance as a function of dissolution mechanism

Scenarios	Dissolution rate-limiting step	Critical quality attributes and test	Formulation types
$k_{dd} \gg k_{id}$	API intrinsic solubilization	API particle size, surface area solubility, etc.	Fast disintegrating tablets, capsules, nanoformulation
$k_{dd} \ll k_{id}$	Formulation disintegration	Disintegration, porosity, hardness, capsule rupture test	IR tablets, LFCs, API or powder blend in capsules
$k_{dd} \sim k_{id}$ both fast	Formulation disintegration	Disintegration, porosity, hardness, capsule rupture test	IR tablets, LFCs, API or powder blend in capsules
$k_{dd} \sim k_{id}$ both slow	API intrinsic solubilization and formulation disintegration	Dissolution profiles, disintegration, API particle size, surface area solubility, etc.	Tablets

dissolution is limited by the rate of disintegration ($k_{dd} \ll k_{id}$) are best monitored by critical quality attributes such as disintegration, porosity, hardness, or capsule rupture. IR tablets, LFCs, and API or powder blend filled capsules are all formulations subject to disintegration control. In instances where drug intrinsic solubility and disintegration are competitive and both fast, the disintegration of the formulation is the most specific measure of dissolution as drug is solubilized as quickly as it is exposed to the dissolution media. Appropriate critical quality attributes in this case include disintegration, porosity, hardness, or capsule rupture. When drug intrinsic solubility and disintegration are competitive and both slow, the intrinsic rate of drug solubilization and disintegration affect the overall dissolution rate. Appropriate critical quality attributes include dissolution profiles, disintegration, API particle size, and surface area depending upon the specifics of the case in question.

Application of the principles discussed herein has the potential to improve product quality and process consistency by increasing the specificity in quality attributes definition and measurement. It is envisioned that this approach to dissolution is in line with recent science-based pharmaceutical development guidance from the ICH. Further, this approach guides the industry away from often meaningless dissolution testing approaches into BCS aligned strategies that are rational and scientifically defendable.

9.6 Conclusion

The BCS provides a science-based framework for the identification of drug dissolution quality attributes and the development of rational methods for ensuring product quality. Experience generated in this laboratory for BCS class I–IV compounds in a variety of formulation types clearly shows that a thorough mechanistic understanding of drug dissolution combined with BCS class information can be employed to determine rational product quality testing strategies. In this context, the application of traditional dissolution testing may not be an appropriate pharmaceutical quality assessment approach as the test often measures a combination of cohesive properties of the formulation and intrinsic properties of the API. As a result, alternative methods may be suitable for measurement of key formulation quality attributes provided that the appropriate studies have been performed to establish a correlation between the quality attribute and the formulation. Application of the concepts discussed herein to pharmaceutical development should be utilized to help guide formulation and process development in an efficient and focused manner.

References

Aiche, J. M.; Pierre, N.; Beyssac, E.; Prasad, V. K.; Skelly, J. P. (1989). New Results on an In Vitro Model for the Study of the Influence of Fatty Meals on the Bioavailability of Theophylline Controlled-Release Formulations. *J. Pharm. Sci.* **78**: 261–263.

Amidon, G. L.; Lennernas, H.; Shah, V. P.; Crison, J. R. (1995). A Theoretical Basis for a Biopharmaceutic Drug Classification: The Correlation of In Vitro Drug Product Dissolution and In Vivo Bioavailability. *Pharm. Res.* **12**: 413–420.

Brown, C. K.; Chokshi, H. P.; Nickerson, B.; Reed, R. A.; Rohrs, B. R.; Shah, P. A. (2004). Acceptable Analytical Practices for Dissolution Testing of Poorly Soluble Compounds. *Pharm. Technol.* **28**: 56–65.

Buchwald, P. (2003). Direct, Differential-Equation-Based In-Vitro–In-Vivo Correlation (IVIVC) Method. *J. Pharm. Pharmacol.* **55**: 495–504.

Caramella, C.; Ferrari, F.; Boneroni, M. C.; Sangalli, M. E.; Debernardi di Valserra, M.; Feletti, F.; Galmozzi, M. R. (1993). In Vitro/In Vivo Correlation of Prolonged Release Dosage Forms Containing Diltiazem HCl. *Biopharm. Drug Dispos.* **14**: 143–160.

Civiale, C.; Ritschel, W. A.; Shiu, G. K.; Aiache, J. M.; Beyssac, E. (1991). In Vivo–In Vitro Correlation of Salbutamol Release from a Controlled Release Osmotic Pump Delivery System. *Methods Find. Exp. Clin. Pharmacol.* **13**: 491–498.

Devi, K. P.; Rao, K. V. R.; Baveja, S.; Fathi, M.; Roth, M. (1989). Zero-Order Release Formulation of Oxprenolol Hydrochloride with Swelling and Erosion Control. *Pharm. Res.* **6**: 313–317.

Dhopeshwarker, V.; O'Keefe, J. C.; Zatz, J. L.; Deeter, R.; Horton, M. (1994). Development of an Oral Sustained-Release Antibiotic Matrix Tablet Using In-Vitro/In-Vivo Correlations. *Drug Dev. Ind. Pharm.* **20**: 1851–1867.

Dietrich, R.; Brausse, R.; Benedikt, G.; Steinijans, V. W. (1988). Feasibility of In Vitro/In Vivo Correlation in the Case of a New Sustained-Release Theophylline Pellet Formulation. *Arzneimittelforschung/Drug Res.* **38**: 1229–1237.

Dunne, A.; O'Hara, T.; Devane, J. (1977). Level A In-Vivo–In-Vitro Correlation: Nonlinear Models and Statistical Methodology. *J. Pharm. Sci.* **86**(11): 1245–1249.

European Agency for the Evaluation of Medicinal Products. (2001). Note for Guidance on the Investigation of Bioavailability and Bioequivalence. (CPMP/EWP/QWP/1401/98).

Fassihi, R. A.; Ritschel, W. A. (1993). Multiple-Layer, Direct-Compression, Controlled-Release System: In Vitro and In Vivo Evaluation. *J. Pharm. Sci.* **82**: 750–754.

FDA. (1997a). Guidance for Industry: Dissolution Testing of Immediate Release Solid Oral Dosage Forms. U.S. Department of Health and Human Services, Food and Drug Administration, Center for Drug Evaluation and Research (CDER), Rockville, MD.

FDA. (1997b). Guidance for Industry: Extended Release Oral Dosage Forms: Development, Evaluation, and Application of In Vitro/In Vivo Correlation. (1997b). U.S. Department of Health and Human Services, Food and Drug Administration, Center for Drug Evaluation and Research (CDER), Rockville, MD.

FDA. (2000). Guidance for Industry: Waiver of In Vivo Bioavailability and Bioequivalence Studies for Immediate-Release Solid Oral Dosage Forms Based on a Biopharmaceutics Classification System. U.S. Department of Health and Human Services, Food and Drug Administration, Center for Drug Evaluation and Research (CDER), Rockville, MD.

Fed. Regist. (2005). **70**: 6888–6889.

Harrison, L. I.; Mitra, A. K.; Kehe, C. R.; Klinger, N. M.; Wick, K. A.; McCarville, S. E.; Cooper, K. M.; Chang, S. F.; Roddy, P. J.; Berge, S. M.; Kisicki, J. C.; Dockhorn, R. (1993). Kinetics of Absorption of New Once-a-Day Formulation of Theophylline in the Presence and Absence of Food. *J. Pharm. Sci.* **82**: 644–648.

Humbert, H.; Cabiac, M.-D.; Bosshardt, H. (1994). In Vitro-In Vivo Correlation of a Modified-Release Oral Form of Ketotifen: In Vitro Dissolution Rate Specification. *J. Pharm. Sci.* **83**: 131–136.

International Conference on Harmonization. (1999). Specifications: Test Procedures and Acceptance Criteria for New Drug Substances and New Drug Products: Chemical Substances, Q6A.

International Conference on Harmonization. (2004). Pharmaceutical Development, Q8.

Johnson, K. C.; Swindell, A. C. (1996). Guidance in the Setting of Drug Particle Size Specification to Minimize Variability in Absorption. *Pharm. Res.* **13**: 1795–1798.

Hussein, Z.; Friedman, M. (1990). Release and Absorption Characteristics of Novel Theophylline Sustained-Release Formulations: In Vitro–In Vivo Correlation. *Pharm. Res.* **7**: 1167–1171.

Llabres, M.; Farina, J. B. (1989). Gastrointestinal Bioavailability Assessment of Commercially Prepared Sustained-Release Lithium Tablets Using a Deconvolution Technique. *Drug Dev. Ind. Pharm.* **15**: 1827–1841.

Mojaverian, P.; Radwanski, E.; Lin, C.; Cho, P.; Vadino, W. A.; Rosen J. M. (1992). Correlation of In Vitro Release Rate and In Vivo Absorption Characteristics of Four Chlorpheniramine Maleate Extended-Release Formulations. *Pharm. Res.* **9**: 450–456.

Polli, J. E.; Crison, J. R.; Amidon, G. L. (1996). Novel Approach to the Analysis of In-Vitro–In-Vivo Relationships. *J. Pharm. Sci.* **85**: 753–760.

Wagner, J. G. (1975a). Application of the Loo-Reigelaman Absorption Method. *J. Pharmacokinet. Biopharm.* **3**: 51–57.

Wagner, J. G. (1975b). *Fundamentals of Clinical Pharmacokinetics, 1st Edition*. Drug Intelligence Publications: Hamilton, IL.

10 Food Effects on Drug Bioavailability: Implications for New and Generic Drug Development

Barbara Myers Davit and Dale P. Conner

10.1 Introduction

10.1.1 Objectives

The US FDA's mission is to ensure that safe and effective drugs are approved for use by American consumers. Among the many factors that the FDA examines to ensure that drug products are safe and effective is the effect of food on the bioavailability of the active drug from the drug product. Thus, the FDA asks applicants seeking marketing approval of new or generic drug products to conduct human pharmacokinetic studies examining the effects of food on drug absorption and bioavailability. The objectives of this chapter are to (1) describe several mechanisms by which food affects drug bioavailability; (2) present examples from FDA-approved drug product labeling to show how food effects are considered in optimizing therapy; (3) distinguish between food effects on drug substance versus drug product; and (4) explain the role of food-effect studies in new and generic drug development.

10.1.2 Oral Bioavailability Defined

Oral bioavailability of drugs is determined by the administered dose that is absorbed from the gastrointestinal (GI) tract. Bioavailability is defined in the FDA's regulations as "the rate and extent to which the active ingredient or active moiety is absorbed from a drug product and becomes available at the site of action" (21 CFR Sect. 320.21(a)).

10.1.3 How Food Can Affect Drug Bioavailability

Food can affect drug bioavailability either by interacting directly with drug substance or indirectly by altering drug release and subsequent absorption from the drug product. The first part of this chapter will discuss various types of food effects on the drug substance, emphasizing the study of food effects in the context of new drug evaluation. Examples will be used to illustrate each type of food effect. Each example will be followed by a brief discussion of how the food effect should be considered in optimizing the treatment regimen. The FDA-approved labels will be used to illustrate this concept. The latter part of this chapter will discuss food effects on drug product, with an emphasis on the role of food-effect studies in generic drug product approval.

10.2 Food Interactions with Drug Substance

10.2.1 Pharmacokinetic Parameters Used to Characterize Food Effects on Drug Bioavailability

Food interactions with drug substance can be manifested by the following effects on oral bioavailability: prolonged rate of absorption, decreased absorption, increased absorption, and unaffected absorption. Food effects on rate of drug absorption are reflected as changes in peak plasma concentrations (C_{max}) and time to reach peak plasma concentrations (T_{max}). Food effects on extent of drug absorption are reflected as changes in the area under the drug plasma concentration versus time curve (AUC_{0-t} and AUC_{∞}). Therefore, in characterizing food effects for regulatory purposes, applicants generally conduct human pharmacokinetic studies in which the investigator administers the drug to healthy subjects[1] with food and determines any changes in C_{max}, T_{max}, and AUC.

10.2.2 Prolonged Rate of Drug Absorption in the Presence of Food

A prolonged or decreased rate of absorption in the presence of food is generally observed for drugs that show high solubility, high GI permeability, and rapid oral absorption (Fleisher *et al.*, 1999; Martinez and Amidon, 2002; Wu and Benet, 2005). The major mechanism responsible for delayed drug absorption in the fed state is delayed gastric emptying rate, thought to occur by a feedback mechanism from receptors in the proximal small intestine (Singh and Malhotra, 2004). Thus, the presence of food in the stomach delays the onset of the intestinal absorption process, which is manifested by a decrease in C_{max} and prolongation of T_{max}. Generally, there will be no effect on AUC unless the drug is unstable in the GI tract. Two such examples are the antiretroviral nucleoside analogs lamivudine and zidovudine (Moore *et al.*, 1999). For both drugs, food slows the rate of absorption, delaying the T_{max}, and reducing the C_{max}. The extent of absorption of both

[1] In some cases, for reasons of safety, it may be appropriate to evaluate food effects on drug pharmacokinetics in patients rather than healthy subjects.

lamivudine and zidovudine is not altered by administration with food. In these cases, the food effects on T_{max} and C_{max} are not considered clinically important, and the FDA-approved labels for lamivudine and zidovudine drug products briefly describe the food effects and states that these products may be administered with or without food (GlaxoSmithKline, Combivir® label, ©2006a).

10.2.3 Decreased Drug Absorption in the Presence of Food

10.2.3.1 Overview

In decreased drug absorption, the extent of absorption (AUC) is significantly decreased. The decrease in AUC is also reflected in a decrease in C_{max}. Three main mechanisms underlying decreased absorption are instability in gastric acids, physical/chemical binding with food, and increased first-pass metabolism/clearance (Welling, 1996; Fleisher *et al.*, 1999; Singh and Malhotra, 2004).

10.2.3.2 Instability in Gastric Acids

One mechanism underlying decreased absorption is instability in gastric acids. Prolonged gastric residence time in the fed state can accelerate the rate of hydrolysis for acid-labile drugs. The antiretroviral nucleoside analog didanosine is such an example. Due to its acid lability, didanosine is formulated as buffered tablets and enteric coated tablets, although these formulations do not provide complete protection from these food effects. When the buffered tablet was given 2 h after a meal, the didanosine C_{max} and AUC both were decreased by approximately 55% compared to administration in the fasting state (Bristol-Myers Squibb, Videx® label, 2006a). When the enteric coated tablet was given with food, C_{max} and AUC decreased by approximately 46% and 19%, respectively, when compared to administration in the fasting state (Bristol-Myers Squibb, Videx® label, 2006b). Consequently, the FDA-approved labels for these products state that they should be taken on an empty stomach.

10.2.3.3 Physical or Chemical Binding with Food Components

A second mechanism causing decreased drug absorption is physical or chemical binding of drug with food components. One example is ciprofloxacin, a broad spectrum quinolone antimicrobial agent. Multivalent cations, such as aluminum, iron, magnesium, and calcium, chelate with ciprofloxacin in the GI tract, resulting in reduced bioavailability (Lomaestro and Bailie, 1995). Neuhofel *et al.* (2002) showed that both ciprofloxacin AUC and C_{max} were reduced by about 21% and 22%, respectively, by calcium-fortified orange juice compared with nonfortified orange juice. This reduced bioavailability can result in loss of antibacterial efficacy (Fleisher *et al.*, 1999). As a result, the FDA-approved label cautions that ciprofloxacin should not be taken with dairy products (like milk or yogurt) or calcium-fortified juices since ciprofloxacin absorption may be significantly reduced (Bayer, Cipro® label, ©2005a,b). However, the label also states that ciprofloxacin may be taken without regard to meals, and may be taken with a meal that contains dairy products or calcium-fortified juices.[2]

10.2.3.4 Increased First-Pass Metabolism and Clearance

A third mechanism causing decreased drug absorption is increased first-pass metabolism and clearance. For example, repeated intake of specific food constituents, such as cruciferous vegetables and charcoal-broiled meats, both of which induce cytochrome P450 1A2 (CYP1A2) can enhance first-pass metabolism and clearance of certain drugs, such as clozapine, imipramine, and theophylline, by inducing drug-metabolizing enzymes (Martinez and Amidon, 2002; Harris *et al.*, 2003). In general, the effects of these diet–drug interactions appear to be small. Because diets are so diverse and complex, it can be very difficult to assess the frequency and clinical relevance of these interactions (Harris *et al.*, 2003). However, of note is the fact that the FDA-approved label for the narrow therapeutic index drug theophylline (used to treat chronic asthma) states that higher doses of theophylline may be required to achieve the desired effect when it is coadministered with St John's Wort, a dietary supplement which is a potent inducer of drug metabolizing enzymes (Purdue Pharma, Uniphyl® label, ©2004). The theophylline label also cautions against abruptly stopping St John's Wort as this may result in theophylline toxicity.

10.2.4 Increased Drug Absorption in the Presence of Food

10.2.4.1 Inhibition of First-Pass Effect

In increased drug absorption, the AUC is significantly increased and is usually associated with an increased C_{max}. One possible mechanism underlying increased drug absorption is the inhibition of the first-pass effect. One of the best-known such food–drug interactions is with grapefruit juice, which is recognized to enhance bioavailability of a number of drugs by inhibiting *in vivo* both the efflux transporter P-glycoprotein (a multidrug resistance protein) and CYP3A (Martinez and Amidon, 2002; Deferme and Augustijns, 2003; Harris *et al.*, 2003). Such interactions appear to be more marked and prevalent for drugs that undergo intestinal metabolism (Maka and Murphy, 2000; Singh and Malhotra, 2004). Grapefruit juice markedly increases concentrations of some orally administered medications by as much as 300% (Harris *et al.*, 2004). Sponsors developing drugs determined to be CYP3A and/or P-glycoprotein substrates may conduct studies investigating effects of grapefruit juice on pharmacokinetics; the results of these

[2] According to the FDA-approved label for ciprofloxacin extended-release tablets (Bayer, Cipro® XR label, ®2005), ciprofloxacin bioavailability is not affected when coadministered with the FDA standard high-fat meal, which contains dairy products (FDA Guidance for Industry, *Food-Effect Bioavailability and Fed Bioequivalence Studies*, 2003).

studies are usually summarized in the FDA-approved label. The nature of the labeling comments depends on the magnitude of the drug–grapefruit interaction and the potential effects of increased drug systemic exposure on safety or efficacy. The FDA-approved label for the immunosuppressant cyclosporine,[3] for example, states that patients on cyclosporine should avoid grapefruit juice and grapefruit (Novartis, Neoral® label, ©2005). On the other hand, the FDA-approved label for the calcium channel blocker felodipine[4] notes that the bioavailability of felodipine is increased approximately twofold when taken with grapefruit juice, but provides no specific recommendations about grapefruit juice consumption and felodipine dosing other than a statement that caution should be used when CYP3A4 inhibitors are administered with felodipine (AstraZeneca, Plendil® label, ©2003).[5]

Intake of a high-protein meal has been shown to increase the bioavailability of highly extracted drugs such as propranolol when compared to a high-carbohydrate meal or administration in the fasted state. Various mechanisms have been proposed for this effect, including increase in splanchic (hepatic) blood flow (McLean *et al.*, 1978); metabolic inhibition by amino acids (Semple and Fangming, 1995); and saturation of hepatic first-pass metabolism due to increased GI absorption (Tam, 1993). The FDA-approved propranolol label notes a food effect on propranolol bioavailability and advises patients to take the drug product (at bedtime) consistently on either an empty stomach or with food (Reliant, InnoPran® XL label, ©2005).

10.2.4.2 Physicochemical and Physiological Effects

For many lipophilic drugs, co administration with a high-fat meal will significantly enhance the drug's GI absorption and oral bioavailability (Welling, 1996; Fleischer *et al.*, 1999; Singh and Malhotra, 2004). A classic example is isotretinoin, a retinoid used to treat severe recalcitrant nodular acne (Colburn *et al.*, 1983). Food increases peak and total isotretinoin plasma exposure by over 200%, and because food improves drug bioavailability, the isotretinoin label stipulates that the drug product be given with food (Roche, Accutane® label, 2005). Such food effects can be physicochemical or physiological. Physicochemical effects would include increased drug dissolution and solubility in the components of the meal (Fleischer *et al.*, 1999; Singh and Malhotra, 2004). Higher lipid solubility in the presence of a fatty meal may therefore lead to improved partitioning of the drug into the intestinal membrane. This is thought to be one of the mechanisms contributing to the increased absorption of the lipophilic antiparasitic drug atovaquone (Rolan

[3] Cyclosporine is a substrate of both CYP3A and P-glycoprotein (Soldner *et al.*, 1999).

[4] Felodipine is a substrate of CYP3A only (Soldner *et al.*, 1999).

[5] The Plendil® label, in the PRECAUTIONS section, states that increases in felodipine concentrations resulting from coadministration with CYP3A4 inhibitors [such as grapefruit juice] may lead to increased effects such as lower blood pressure and increased heart rate (AstraZeneca, Plendil® label, ©2003).

et al., 1992). Dietary fat taken with atovaquone increases the rate and extent of absorption, increasing AUC 2–3 times and C_{max} five times over fasting. The FDA-approved label for Malarone®, an atovaquone-containing combination drug product, states that the drug product must be taken with food or a milky drink (GlaxoSmithKline, Malarone® label, ©2006b).

10.2.4.3 Effects of Bile Release

Physiological mechanisms in response to a fatty meal mainly include increased bile output, enhanced lymphatic effect, and increased gastric residence time due to the fact that a fatty meal delays gastric emptying (Fleischer *et al.*, 1999; Singh and Malhotra, 2004). Increased bile output can lead to an increase in bile salts in the GI tract, which in turn can enhance the dissolution rate of low-solubility drugs by micellar solubilization. Increased transport of drugs *via* the lymphatic system could thus minimize first-pass metabolism through the GI tract and/or liver. It is thought that bile release is at least partially responsible for increased absorption of the cefalosporin antibiotic cefuroxime in the presence of food (Williams and Harding, 1984). The FDA-approved label for cefuroxime axetil notes that the absolute bioavailability of cefuroxime increases from 37 to 52% when the tablet formulation is taken with a meal (GlaxoSmithKline, Ceftin® label, ©2003). As this increase is relatively modest, the label states that cefuroxime axetil can be administered without regard to meals.

10.2.4.4 Effects of Longer Gastric Residence Time

The longer gastric residence time of drug particles may substantially increase the extent of drug dissolution of poorly soluble drugs, which exhibit higher solubility in the lower gastric pH (Fleischer *et al.*, 1999; Singh and Malhotra, 2004). Fluids ingested along with the meal as well as food-induced biliary and pancreatic secretion could contribute to a postprandial increase in bioavailability of such poorly water-soluble drugs. The bioavailability of the antibacterial drug nitrofurantoin is increased in the presence of food because of delayed gastric emptying resulting in increased dissolution, and therefore, absorption (Maka and Murphy, 2000). The FDA-approved nitrofurantoin label states that the drug product should be given with food to improve drug absorption (Procter & Gamble, Macrobid® label, 2003).

10.2.5 Drug Absorption Unaffected by Food

Finally, for many drugs, there is no significant change in the rate or extent of absorption. Drugs that are not affected by food are (a) those that are relatively insensitive to food-induced changes in the GI tract; (b) those that are rapidly and completely absorbed from the GI tract, e.g., finasteride (an inhibitor of Type II 5α-reductase indicated for the treatment of male pattern baldness); and (c) those that are well-absorbed from both the large and small intestine, e.g., bicalutamide (a nonsteroidal antiandrogen used in the treatment of prostate cancer) (Singh and Malhotra, 2004). The FDA recommends that, when such absence of a food effect is documented, the product label should state that a food effect is not present (Marroum and Gobburu, 2002). Accordingly, the FDA-approved labels for both products provide this information. The finasteride tablet label states that finasteride bioavailability is not affected by food, and that the drug may be administered with or without meals (Merck, Propecia® label, ©1997). The bicalutamide tablet label states that coadministration with food has no clinically significant effect on rate or extent of absorption, and that the drug may be taken with or without food (AstraZeneca, Casodex® label, 2005a).

10.2.6 FDA Guidance for Industry on Characterizing Food Effects in Drug Development

10.2.6.1 Objectives

In practice, it is difficult to determine the exact mechanism by which food changes the bioavailability of a drug product without performing specific mechanistic studies. For this reason, the food-effect bioavailability studies conducted during new drug development are generally exploratory and descriptive, and where appropriate, are used to make labeling claims for the drug product. FDA's Guidance for Industry entitled *Food-Effect Bioavailability and Fed Bioequivalence Studies* (2003) gives information to sponsors and/or applicants about food-effect bioavailability studies for orally administered drug products. The guidance provides recommendations for when food-effect studies are appropriate, as well as recommendations on study design, data analysis, and product labeling.

10.2.6.2 Recommended Designs for Food-Effect Bioavailability Studies

In the *Food-Effect Bioavailability and Fed Bioequivalence Studies* Guidance (2003), the FDA asks sponsors to conduct food-effect bioavailability studies for all new chemical entities as early in drug development as possible, i.e., during the investigational new drug (IND) phase of development. The FDA recommends a randomized, balanced, single-dose, two treatment (fed versus fasting), two period, two sequence crossover design for studying food effects on drug bioavailability. The formulation should be tested on an empty stomach (fasting conditions) in one period and following a test meal (fed conditions) in the other period. The FDA also recommends use of high calorie and high-fat meals in these studies.

10.2.6.3 Recommendations for Drug Product Labeling

Regulatory Guidance

The FDA asks investigators to compare the following pharmacokinetic parameters in these studies: the area under the plasma concentration versus time curve (AUC_{∞}, AUC_{0-t}); peak exposure (C_{max}); time to peak exposure (T_{max}); lag-time

(T_{lag}) for modified-release products, if present; termination elimination half-life ($T_{1/2}$); and any other relevant parameters. The effect of food on the absorption and bioavailability of the drug should be described in the Clinical Pharmacology section of the drug product labeling. The Dosage and Administration section of the labeling should provide instructions for drug administration in relation to food based on clinical relevance. For example, the label should state whether changes in systemic exposure caused by coadministration with food results in safety or efficacy concerns, or when there is no important change in systemic exposure but there is a possibility that the drug substance causes GI irritation when taken without food.

Labeling Examples

How FDA-approved labels deal with effects of food on drug efficacy has been illustrated with several of the examples presented above. How the FDA-approved label deals with food effects on drug safety is well-illustrated by the label for the antiretroviral nonnucleoside reverse transcriptase inhibitor efavirenz. The efavirenz label states that a high-fat/high-calorie meal or a reduced-fat, normal caloric meal given with the efavirenz capsule was associated with a mean increase of 22 and 17% in the mean AUC_{∞} and a mean increase of 39 and 51% in efavirenz C_{max}, respectively (Bristol-Myers Squibb, Sustiva® label, 2005). Administration of the efavirenz tablet with a high-fat/high-caloric meal was associated with a 28% in the mean AUC and a 79% increase in the mean C_{max}. Consequently, the label recommends that efavirenz be taken on an empty stomach.

The respective labels for the antiprotozoal agents tinidazole and mefloquine illustrate how FDA-approved labels deal with food effects and GI irritation. The tinidazole label states that food has a minimal effect on C_{max} and T_{max} and no effect on AUC and $T_{1/2}$ (Mission, Tindamax® label, 2006). However, the label recommends that tinidazole taken with food minimizes the incidence of epigastric discomfort and other GI side effects. Similarly, the label for mefloquine cautions that the drug should not be taken on an empty stomach (Roche, Lariam® label, 2004).

10.3 Food Interactions with Drug Product

10.3.1 Introduction

Food effects on absorption of the drug substance should be distinguished from food effects on drug absorption arising *via* interactions with the formulation. For example, many excipients used to get extended-release behavior depend on the environmental conditions of the GI tract to provide the extended delivery of the drug (Gai *et al.*, 1997). Thus, by producing physiological changes in the GI tract, such as changes in gastric pH, gastric emptying rate, intestinal motility, and intestinal secretions, food can interact with components of the formulation and therefore play an important role in drug bioavailability.

10.3.2 Issues with Modified-Release Drug Products: Potential for Dose-Dumping

One way in which food can interact with formulation components is by causing an extended-release formulation, meant to release drug slowly over a prolonged period, to "dose-dump." Dose-dumping is a term that describes the rapid release of the active ingredient from the dosage form into the GI tract (FDA, 2005). Food effects on dose-dumping were first observed for extended-release formulations of theophylline. One of the most dramatic interactions with food occurred with some extended-release theophylline formulations. Shortly after the approval of the theophylline extended-release product Theo-24® by the FDA,[6] Hendeles *et al.* (1985) showed that, when the product was given to human subjects together with a high-fat meal, the theophylline C_{max} was over twice as high as when given to fasted subjects. Karim *et al.* (1985) reported similar findings when the extended-release theophylline tablet product Uniphyl® was given to healthy male subjects with a high-fat breakfast.

Gai *et al.* (1997) compared the effects of fasting versus "normal," high-fat, and high-fat/high-protein meals on theophylline rate (C_{max}) and extent (AUC) of absorption from two different extended-release tablet formulations, one based on a Carbopol 974P hydrophilic matrix, the other based on Cutina HR hydrogenated castor oil lipid matrix. Compared with fasting, any class of meal given with the hydrophilic matrix tablet produced a higher theophylline C_{max} but not AUC. By contrast, when given with the lipid matrix tablet, the high-fat and high-fat/high/protein meals increased both the theophylline AUC and C_{max}, whereas AUC and C_{max} following a normal meal were comparable to values in fasting subjects. The authors suggested that, for the hydrophilic matrix tablet, food increased the rate but not extent of theophylline absorption due to the delay in gastric emptying, whereas, for the lipid matrix tablet, the surface-active effect of bile salts together with erosion promoted by lipase action were responsible for the increases in both the rate and extent of theophylline absorption.

Su *et al.* (2003) found that coadministration of a standard Chinese breakfast (rich in carbohydrate) with the theophylline porous matrix extended-release tablet product Euphyllin Retard® increased the rate but not extent of theophylline absorption. Su also found that within-subject variability in C_{max} was greater in fed subjects than in fasted subjects. The authors suggested that the delay in gastric emptying was responsible for the increase in C_{max}.

---

[6] Theo-24® was approved by the FDA in 1983 (Orange Book, 2006). At that time, the FDA asked applicants developing theophylline sustained-release drug products to characterize pharmacokinetics in fasting subjects (Weinberger, 2004).

Because theophylline is a narrow therapeutic index drug,[7] the rapid absorption of relatively large amounts of the drug can cause serious toxicity. The findings from the above studies underscore how important it is for dosage forms to release drug predictably regardless of whether taken with food or on an empty stomach.

10.3.3 Issues with Modified-Release Drug Products: Formulation-Dependant Food Effects

10.3.3.1 *In Vitro* Drug Release Predictive of Food Effects

Schug *et al.* (2002a) observed formulation-dependent food effects on relative bioavailability from two extended-release nifedipine (a calcium channel blocker) products approved for marketing in the European Union. The two products were Slofedipine® XL, a monolithic modified-release tablet with an erosive polymer matrix, and Adalat® OROS formulated with the osmotically driven Gastrointestinal Therapeutic System (GITS) (Grundy and Foster, 1996). Nifedipine bioavailability from the two formulations was comparable in fasted subjects. However, food caused a pronounced delay in nifedipine absorption from Slofedipine XL but not from Adalat. As a result, the relative bioavailability of nifedipine from Slofedipine XL was only 28% that of Adalat OROS over the intended dosing interval of 24 h. The two formulations had different dissolution properties *in vitro*. Drug release from Slofedipine XL was minimal at low pH and more rapid at pH 6.8, suggesting that retention of the relatively acid resistant Slofedipine XL formulation in the acidic medium of the stomach in the presence of food led to reduced *in vivo* bioavailability over the 24-h dosing interval. By contrast, drug release from Adalat OROS was more rapid in acidic media, suggesting that dissolved nifedipine could still leave the stomach even if coadministration with food prolonged capsule retention by the stomach.

10.3.3.2 *In Vitro* Drug Release Profiles Not Predictive of Food Effects

An entirely different food effect on another nifedipine bioavailability was observed with a third extended-release formulation marketed in Europe, Nifedicron®. Nifedicron consists of a capsule containing several mini tablets. Food had a pronounced effect on nifedipine bioavailability from Nifedicron, compared with fasted conditions, resulting in a pronounced increase in C_{max} values. The resulting point estimate for C_{max} (fed/fasted) was 340%. Food also increased the mean AUC_{0-24} by 57% compared with the mean values in fasted subjects. In the same study, the investigators compared nifedipine bioavailability from Adalat OROS in fasted and fed subjects and found no effect on AUC_{0-24h} and a 21% increase in

[7] In its Guidance for Industry, *Bioavailability and Bioequivalence Studies of Orally Administered Drug Products – General Considerations* (2003), the FDA defines a narrow therapeutic index or narrow therapeutic range drugs as "drug products containing certain drug substances subject to therapeutic drug concentration or pharmacodynamic monitoring, and/or where product labeling indicates a narrow therapeutic range designation."

C_{max}. Both products were robust toward differing pH values of the media during *in vitro* drug release testing. Thus, the observed potential for dose-dumping in the presence of food was not anticipated.

10.3.4 Implications for Development of Generic Modified-Release Drug Products

10.3.4.1 Introduction

The studies summarized above illustrate two problems related to formulation-dependent food interactions with modified-release[8] dosage forms; first, dose-dumping interactions between food and extended-release formulations can sometimes result in serious adverse events; and second, formulation-dependent food interactions are not always predictable from the *in vitro* characteristics of the dosage form. Because the result of the interaction between the food and the system used to control the liberation of the drug is difficult to predict, it becomes essential to test possible interactions with each new formulation.

These potential formulation-dependent food interactions have implications for the generic drug industry. Under the FDA's regulations, inactive ingredients in a generic solid oral dosage form drug product can differ from the inactive ingredients used in the corresponding innovator drug product (or reference drug product) (21 CFR Sect. 320.1(c)). In addition, generic modified-release products may be formulated with a different release mechanism than their corresponding reference products (Pfizer v. Shalala, 1998). Thus, because (1) each generic modified-release drug product can have a different formulation and release mechanism than its corresponding reference drug product; and (2) the relative direction and magnitude of food effects on modified-release formulations may be difficult to predict, the FDA asks generic drug applicants to conduct studies comparing the relative bioavailability of the generic and corresponding reference drug products under fed conditions. Such studies are called fed bioequivalence studies.

10.3.4.2 Role of *In Vivo* Fed Bioequivalence Studies

In its regulations (21 CFR Sect. 320.1(e)), the FDA defines bioequivalence as

> The absence of a significant difference in the rate and extent to which the active ingredient or active moiety in pharmaceutical equivalents or pharmaceutical alternatives becomes available at the site of drug action when administered at the same molar dose under similar conditions in an appropriately designed study.

Each new generic drug product must demonstrate bioequivalence to its corresponding reference drug product as a condition of approval.[9] Where the focus

---

[8] Modified-release products include delayed-release products and extended- (controlled-) release products (FDA Guidance for Industry, *Bioavailability and Bioequivalence Studies of Orally Administered Drug Products – General Considerations*, 2003).

[9] The FDA can waive *in vivo* bioequivalence study requirements for those generic drug products that meet the criteria of 21 CFR §320.22.

is on release of the drug substance from the drug product into the systemic circulation, the FDA asks applicants submitting Abbreviated New Drug Applications (ANDAs) to determine bioequivalence in an *in vivo* human study with pharmacokinetic endpoints in which drug rate and extent of absorption is determined. The study should be conducted between the generic product and corresponding reference drug product using the strength listed in the FDA *Approved Drug Products with Therapeutic Equivalence Evaluations* (2006) or "Orange Book." In its Guidance for Industry, *Bioavailability and Bioequivalence Studies for Orally Administered Drug Products: General Considerations* (2003), the FDA recommends that the parameters AUC_{0-t}, AUC_{∞}, and C_{max} be used as measures for bioequivalence. The FDA concludes that the generic product and corresponding reference product are bioequivalent when the 90% confidence interval for the ratio of population geometric means – based on ln-transformed data – is contained within the bioequivalence limits for AUC and C_{max}.

Bioequivalence studies with pharmacokinetic endpoints are always conducted in healthy normal human subjects under fasting conditions unless precluded for reasons of safety. For the reasons described above, for new generic modified-release drug products, the FDA also asks applicants to conduct fed bioequivalence studies. Thus, the FDA asks that each ANDA for a systemically available solid oral modified-release dosage form be supported by an acceptable fasting and fed bioequivalence study.[10] By approving a new generic drug product, the FDA is determining that the generic is therapeutically equivalent to the corresponding reference drug product. Thus, if the new generic modified-release drug product and corresponding reference product are bioequivalent under both fasted and fed conditions, the two will produce the same therapeutic response whether or not taken with food, and the two can be switched under similar conditions of use.

10.3.5 Implications for Development of Generic Immediate-Release Drug Products

10.3.5.1 BCS Class I Drugs

The FDA considers a different set of criteria in deciding whether to ask for fed bioequivalence studies (in addition to fasting bioequivalence studies) of new generic immediate release drug products.[11] The FDA Guidance for Industry *Food-Effect Bioavailability and Fed Bioequivalence Studies* (2003) notes that important

[10] These studies are generally conducted on the highest strength of modified-release tablet or capsule of the product line. The FDA may request additional fasting studies on lower strengths of a modified-release tablet product line if it makes a determination that one or more of the lower strengths are not proportionally similar to the strength for which bioequivalence was determined.

[11] The FDA requests applicants submitting ANDAs to conduct bioequivalence studies in fasted subjects for all immediate-release drug products, unless safety precludes administration on an empty stomach. For example, for drugs such as mefloquine, which should be taken with food for safety reasons, the FDA asks generic drug applicants to conduct only fed bioequivalence studies.

food effects on bioavailability are least likely to occur with many rapidly dissolving, immediate-release drug products containing highly soluble and highly permeable drug substances (Biopharmaceutics Classification System Class I, or BCS Class I drug substances). Because BCS Class I drugs and drug products are rapidly dissolving across a wide pH range, formulation effects on absorption are minimized (FDA Advisory Committee for Pharmaceutical Sciences, 2002). However, for some BCS Class I drugs, food can influence bioavailability when there is a high presystemic metabolism effect, extensive absorption, complexation, or instability of the drug substance in the GI tract. In some cases, excipients or interactions between excipients and food-induced changes in gut physiology can contribute to these food effects and influence drug bioavailability. However, results of fed bioequivalence studies of representative BCS Class I drug products have showed that food had no effect on the drug bioavailability from different formulations of the same BCS Class I drug (FDA Advisory Committee for Pharmaceutical Science, 2002). Thus, if the FDA determines that a particular drug substance is in BCS Class I based on satisfaction of the criteria detailed in the FDA Guidance for Industry *Waiver of In Vivo Bioavailability and Bioequivalence Studies for Immediate-Release Solid Oral Dosage Forms Based on a Biopharmaceutics Classification System* (2000), the applicant submitting an ANDA can request a waiver of both fasting and fed bioequivalence studies.

10.3.5.2 Label-Driven Criteria for Requesting Fed Bioequivalence Studies

Regulatory Guidance

For other immediate-release drug products (BCS Class II, III, and IV),[12] food effects can result from a complex combination of factors that influence the *in vitro* dissolution of the drug product and/or the absorption of the drug substance. As with modified-release drug products, the FDA focuses on whether a new generic immediate-release drug product will be bioequivalent to the corresponding reference drug product under conditions of use. As a result, the question of whether a fed bioequivalence study should be conducted or not is driven by the drug product label. Thus, the FDA asks applicants submitting ANDAs for immediate-release drug products to conduct a fed bioequivalence study in addition to the fasted study, unless one of three exceptions is met. As stated in the FDA Guidance for Industry, *Food-Effect Bioavailability and Fed Bioequivalence Studies* (2003), these exceptions are as follows: (1) the drug is in BCS Class I; (2) the Dosage and Administration section of the FDA-approved reference drug product label states that the product should be taken only on an empty stomach; or (3) the RLD label does not make any statements about the effect of food on drug absorption or administration.

---

[12] BCS Class II drug substances are low solubility, high permeability, and exhibit dissolution rate-limited absorption; BCS Class III drug substances are high solubility, low permeability, and exhibit permeability rate-limited absorption; BCS Class IV drug substances are low solubility, low permeability, and exhibit very poor oral bioavailability (Martinez and Amidon, 2002).

Examples

How the FDA decides whether or not to conduct a fed bioequivalence study for potential generic immediate-release drug products can be illustrated by several examples. The most commonly occurring situation is one in which the FDA-approved label summarizes food effects on bioavailability, and provide instructions for whether the drug product should be given with food. This is typified by the label for the zidovudine capsules, described above, which states that the product may be administered with or without food and that the extent of zidovudine absorption was similar when compared in human subjects (GlaxoSmithKline, Retrovir® label, ©2006c). Since none of the above three exceptions apply in this case, the FDA asks generic companies developing zidovudine capsules to conduct both fasting and fed bioequivalence studies against the corresponding reference.

The label-driven criteria for whether or not a fed bioequivalence study should be conducted means that the FDA asks for a fed bioequivalence study even if the label documents no food effect on drug absorption. This is illustrated by finasteride tablets. As described above, the finasteride label states that food does not affect rate or extent of absorption and that the product may be given with or without food. Since the label makes statements about (lack of) food effects on both absorption and administration, the FDA asks applicants submitting ANDAs for generic finasteride tablets to conduct both fasting and fed bioequivalence studies.

A case in which the FDA would request only a fasting bioequivalence study for a generic drug product is illustrated by the drug didanosine. As described above, the FDA-approved label for didanosine delayed-release pellets states that, in the presence of food, the didanosine C_{max} and AUC were reduced by approximately 46% and 19%, respectively, compared to the fasting state. The Dosage and Administration section of the didanosine delayed-release pellets label states that the product should be taken on an empty stomach (Bristol-Myers Squibb, Videx® EC, 2006a). Since this product meets exception (2), above, the FDA asks applicants developing generic versions of didanosine delayed-release pellets to conduct only fasted bioequivalence studies.

10.3.6 Recommendations for Designing Fed Bioequivalence Studies

With respect to fed bioequivalence study design, the FDA recommends a similar design to that recommended for food-effect bioavailability studies, except that the treatments should consist of both test and reference product formulations following a test meal. The FDA asks that both food-effect bioavailability and fed bioequivalence studies be conducted using meal conditions that are expected to provide the greatest effects on GI physiology. It is thought that this will provide the greatest effect on systemic drug availability. A high-fat (approximately 50% of the total caloric content of the meal) and high-calorie (approximately 800–1,000 kcal) meal is recommended as a test meal. The test meal should derive approximately 150, 250, and 500–600 kcal from protein, fat, and carbohydrates, respectively. The FDA asks applicants to provide the caloric breakdown of the meal in the study report submitted to the ANDA.

10.3.7 Food Effects and Generic Drug Product Labeling

Thus, characterizing food effects on the drug product formulation is an important part of generic drug development.[13] The conclusion of bioequivalence under fed conditions indicates that, with respect to food, the language in the FDA-approved label of the generic drug product can be the same as for the reference drug product.

10.3.8 Sprinkle Studies in New and Generic Drug Product Development

10.3.8.1 Sprinkle Studies in Development of New Modified-Release Capsules

In some cases it may be desirable to offer patients the option of sprinkling the contents of a modified-release drug capsule drug product onto a soft food such as applesauce, and swallowing without chewing. Because, direct contact with food prior to administration can potentially interact with release-controlling excipients, in these cases the FDA asks new drug applicants to conduct additional *in vivo* "sprinkle" relative bioavailability studies. These sprinkle bioavailability studies compare the drug product opened up and sprinkled on the soft food with the drug product given on an empty stomach. The label should indicate whether the drug capsule product can be given by sprinkling the contents on a soft food, and summarize the results of the sprinkle bioavailability study.

10.3.8.2 Sprinkle Studies in Development of Generic Modified-Release Capsules

If a generic modified-release capsule drug product is developed to a corresponding reference drug product labeled for sprinkling on a soft food, then the FDA asks the ANDA applicant to conduct a sprinkle bioequivalence study. The sprinkle bioequivalence studies compare the generic product opened up and sprinkled on the soft food with the reference product opened up and sprinkled on the soft food. As with fasted and fed bioequivalence studies, the FDA concludes that the generic is bioequivalent to the reference under sprinkled conditions if the geometric mean test/reference ratios for AUC and C_{max} fall within the bioequivalence limits of 80–125%. In this case, a conclusion of bioequivalence indicates that the language in the FDA-approved label of the generic drug product can be the same as for the reference drug product with respect to sprinkling on soft food for administration.

---

[13] Studies investigating food effects on drug products also serve an important role in new drug development. For example, when a sponsor develops a modified-release formulation of an already-marketed immediate-release product, it is particularly important to compare food effects on the immediate-release versus the proposed new modified-release formulation.

Sprinkle bioequivalence studies are not necessary for immediate-release drug capsule products, as the capsule contents do not contain release-controlling excipients.

10.3.8.3 Example

The role of sprinkle studies in generic drug product development is illustrated by the case of omeprazole delayed-release capsules. The Dosage and Administration section of the FDA-approved omeprazole label states that the capsule contents can be sprinkled on one tablespoon of applesauce and swallowed with a glass of cool water to ensure complete swallowing of the pellets (AstraZeneca, Prilosec® label, ©2005b). Because this is a modified-release drug product, the FDA asks generic drug applicants to conduct fasting and fed bioequivalence studies to document bioequivalence. Because of the labeling statements about administration with applesauce, the FDA also asks that generic drug applicants conduct a bioequivalence study in which subjects, following an overnight fast, receive the generic or reference product mixed in applesauce and swallowed with 240 mL of water.

10.4 Summary and Conclusions

The various examples presented above show that food can alter the bioavailability of a drug substance in many ways. These changes in bioavailability can sometimes necessitate dosage adjustments or specific dosing instructions in relation to administration with meals (FDA, 2001). The physiological changes occurring in response to food intake can also influence the demonstration of bioequivalence between a new generic drug product and its corresponding reference drug product. The FDA expects developers of both new and generic drug products to characterize these food effects during drug development, and has posted a guidance for industry providing recommendations about when food-effect bioavailability and fed-bioequivalence studies should be conducted, how these studies should be designed/analyzed, and how study results should be reported in drug product labeling. The ultimate objective is to provide an informative drug product label that will convey to clinical practitioners and patients information essential for the safe and effective use of either an innovator or generic drug product.

References

AstraZeneca Pharmaceuticals LP, Wilmington DE 19850 (©2003). Plendil® Extended-Release Tablets package insert.

AstraZeneca Pharmaceuticals LP, Wilmington DE 19850 (2005a). Casodex® Tablets package insert.

AstraZeneca Pharmaceuticals LP, Wilmington DE 19850 (®2005b). Prilosec® Delayed-Release Capsules package insert.

Bayer Pharmaceuticals Corporation, 400 Morgan Lane, West Haven CT 06516 (©2005a). Cipro® Tablets and Cipro® Oral Suspension package insert.

Bayer Pharmaceuticals Corporation, 400 Morgan Lane, West Haven CT 06516 (©2005b). Cipro® XL Tablets package insert.

Bristol-Myers Squibb Company, Princeton, NJ 08543 (2005). Sustiva® Capsules and Tablets package insert.

Bristol-Myers Squibb Virology, Bristol-Myers Company, Princeton, NJ 08543 (2006a). Videx® Buffered Tablets package insert.

Bristol-Myers Squibb Virology, Bristol-Myers Company, Princeton, NJ 08543 (2006b). Videx® EC package insert.

Colburn, W.A., Gibsen, D.M., Wiens, R.E., and Hannigan, J.J. (1983). Food increases the bioavailability of isotretinoin. *J. Clin. Pharmacol.* 23:534–539.

Deferme, S. and Augustijns, P. (2003). The effect of food components on the absorption of P-gp substrates: a review. *J. Pharm. Pharmacol.* 55:153–162.

FDA News (2005). FDA asks Purdue Pharma to withdraw Palladone® for safety reasons (July 12, 2005), http://www.fda.gov/bbs/topics/news/2005/NEW01205.html.

Fleisher, D., Li, C., Zhou, Y., Pao, L., and Karim, A. (1999). Drug, meal and formulation interactions influencing drug absorption after oral administration: clinical implications. *Clin. Pharmacokinet.* 36:233–254.

Gai, M.N., Isla, A., Andonaeugui, M.T., Thielemann, A.M., and Seitz, C. (1997). Evaluation of the effect of three different diets on the bioavailability of two sustained release theophylline matrix tablets. *Int. J. Clin. Pharm. Ther.* 35:565–571.

GlaxoSmithKline, Research Triangle Park, NC 27709 (©2003). Ceftin® Tablets and Ceftin® Oral Suspension package insert.

GlaxoSmithKline, Research Triangle Park, NC 27709 (©2006a). Combivir® Tablets package insert.

GlaxoSmithKline, Research Triangle Park, NC 27709 (©2006b). Malarone® Tablets package insert.

GlaxoSmithKline, Research Triangle Park, NC 27709 (©2006c). Retrovir® Tablets, Capsules, and Syrup package insert.

Grundy, J.S. and Foster, R.T. (1996). The nifedipine gastrointestinal therapeutic system (GITS). Evaluation of pharmaceutical, pharmacokinetic and pharmacologic properties. *Clin. Pharmacokinet.* 30:28–51.

Harris, R.Z., Jang, G.R., and Tsunoda, S. (2003). Dietary effects on drug metabolism and transport. *Clin. Pharmacokinet.* 42:1071–1088.

Hendeles, L., Weingerger, M., Milavetz, G., Hill, M., and Vaughan, L. (1985). Food-induced "dose-dumping" from a once-a-day theophylline product as a cause of theophylline toxicity. *Chest* 87:758–765.

Karim, A., Burns, T., Wearley, L., Streicher, J., and Palmer, M. (1985). Food-induced changes in theophylline absorption from controlled-release formulations. I. Substantial increased and decreased absorption with Uniphyl tablets and Theo-Dur Sprinkle. *Clin. Pharmacol. Ther.* 38:77–83.

Lomaestro, B.M. and Bailie, G.R. (1995). Absorption interactions with fluoroquinolones. *Drug Saf.* 12:314–333.

Maka, D.A. and Murphy, L.K. (2000). Drug-nutrient interactions: a review. AACN Clin. Issues 11:580–589.

Marroum, P.J. and Gobburu, J. (2002). The product label: how pharmacokinetics and pharmacodynamics reach the prescriber. *Clin. Pharmacokinet.* 41:161–169.

Martinez, M.N. and Amidon, G.L. (2002). A mechanistic approach to understanding the factors affecting drug absorption: a review of fundamentals. *J. Clin. Pharmacol.* 42:620–643.

McLean, A.J., McNamara, P.J., du Souich, P., Gibaldi, M., and Laika, D. (1978). Food, splanchnic blood flow, and bioavailability of drugs subject to first-pass metabolism. *Clin. Pharmacol. Ther.* 24:5–10.

Merck & Co., Inc., Whitehouse Station, NJ 08889 (©1997). Propecia® Tablets package insert.

Mission Pharmacal Company, San Antonio, TX 78230-1355 (2006). Tindamax® Tablets package insert.

Moore, K.H.P., Shaw, S., Laurent, Z.L., Lloyd, P., Duncan, B., Morris, D.M., O'Mara, M.J., and Pakes, G.E. (1999). Lamivudine/zidovudine as a combined formulation tablet: bioequivalence compared with lamivudine and zidovudine administered concurrently and the effect of food on absorption. *J. Clin. Pharmacol.* 39:593–605.

Neuhofel, A.L., Wilton, J.H., Victory, J.M., Hejmanowski, L.G., and Amsden, G.W. (2002). Lack of bioequivalence of ciprofloxacin when administered with calcium-fortified orange juice: a new twist on an old interaction. *J. Clin. Pharmacol.* 42:461–466.

Novartis Pharmaceuticals Corporation, East Hanover, NJ 07936 (©2005). Neoral® Soft Gelatin Capsules USP and Neoral® Oral Solution, Modified, USP, package insert.

Pfizer, Inc. v. Shalala, 1 F.Supp.2d 38 (D.D.C. 1998).

Procter & Gamble Pharmaceuticals, Inc., Cincinnati, OH 45202 (2003). Macrobid® Capsules package insert.

Purdue Pharmaceutical Products L.P., Stamford, CT 09601-3431 (©2004). Uniphyl® Tablets package insert.

Reliant Pharmaceuticals, Inc., Liberty Corner, NJ 07938 (©2005). InnoPran® XL package insert.

Roche Laboratories, Inc., Nutley, NJ 07110-1199 (2004). Lariam® Tablets package insert.

Roche Laboratories, Inc., Nutley, NJ 07110-1199 (2005). Accutane® Capsules package insert.

Rolan, P.E., Mercer, A.J., Weatherley, B.C., Holdich, T., Meire, H., Peck, R.W., Ridout, G., and Posner, J. (1992). Investigation of the factors responsible for a food-induced increase in absorption of a novel protozoal drug 566C80. *Br. J. Clin. Pharmacol.* 33:226P–227P.

Schug, B.S., Brendel, E., Chantraine, E., Wolf, D., Martin, W., Schall, R., and Blume, H.H. (2002a). The effect of food on the pharmacokinetics of nifedipine in two slow release formulations: pronounced lag-time after a high fat breakfast. *J. Clin. Pharmacol.* 53:582–588.

Schug, B.S., Brendel, E., Wonnemann, M., Wolf, D., Wargenau, M., Dikngler, A., and Blume, H.H. (2002b). Dosage form-related food interaction observed in a marketed once-daily nifedipine formulation after a high-fat American breakfast. *Eur. J. Clin. Pharmacol.* 58:119–125.

Semple, H.A. and Fangming, X. (1995). Interaction between propranolol and amino acids in the single-pass isolated, perfused rat liver. *Drug Metab. Dispos.* 23:794–798.

Singh, B.N. and Malhotra, B.K. (2004). Effects of food on the clinical pharmacokinetics of anticancer agents: underlying mechanisms and implications for oral chemotherapy. *Clin. Pharmacokinet.* 43:1127–1156.

Soldner, A., Christians, U., Susanto, M., Wacher, V.J., Silverman, J.A., and Benet, L.Z. (1999). Grapefruit juice activates P-glycoprotein-mediated drug transport. *Pharm. Res.* 16:478–485.

Su, Y.-M., Cheng, T.-P., and Wen, C.-Y. (2003). Study of the effect of food on the absorption of theophylline. *J. Chin. Med. Assoc.* 66: 715–721.

Tam, Y.K. (1993). Individual variation in first-pass metabolism. *Clin. Pharmacokinet.* 25:300–328.

U.S. Code of Federal Regulations (2006a). Title 21, Part 320 – Bioavailability and Bioequivalence Requirements, Subpart A – General Provisions, Section 320.1 *Definitions*. U.S. Government Printing Office, Washington, D.C., pp. 184–185.

U.S. Code of Federal Regulations (2006b). Title 21, Part 320 – Bioavailability and Bioequivalence Requirements, Subpart B – Procedures for Determining the Bioavailability or Bioequivalence of a Drug Product, Section 320.22. *Criteria for waiver of evidence of in vivo bioavailability or bioequivalence*. U.S. Government Printing Office, Washington, D.C., pp. 186–188.

U.S. Department of Health and Human Services, Food and Drug Administration, Center for Drug Evaluation and Research (2000). *Waiver of In Vivo Bioavailability and Bioequivalence Studies for Immediate-Release Solid Oral Dosage Forms Based on a Biopharmaceutics Classification System* (August 31, 2000); http://www.fda.gov/cder/guidance/3618fnl.pdf.

U.S. Department of Health and Human Services Food and Drug Administration, Center for Drug Evaluation and Research (2002). Advisory Committee for Pharmaceutical Sciences meeting transcript (May 7, 2002), http://www.fda.gov/ohrms/dockets/ac/02/transcripts/3860T1.pdf.

U.S. Department of Health and Human Services, Food and Drug Administration, Center for Drug Evaluation and Research (2003a). *Guidance for Industry: Bioavailability and Bioequivalence Studies for Orally Administered Drug Products – General Considerations* (March 19, 2003); http://www.fda.gov/cder/guidance/5356fnl.pdf.

U.S. Department of Health and Human Services, Food and Drug Administration, Center for Drug Evaluation and Research (2003b). *Guidance for Industry: Food-Effect Bioavailability and Fed Bioequivalence Studies* (January 31, 2003); http://www.fda.gov/cder/guidance/5194fnl.pdf.

U.S. Department of Health and Human Services, Food and Drug Administration, Center for Drug Evaluation and Research, Office of Pharmaceutical Sciences, Office of Generic Drugs (2006). *Approved Drug Products with Therapeutic Equivalence Evaluations* (Orange Book), 26th Ed. (April 11, 2006); http://www.fda.gov/cder/orange/obannual.pdf.

U.S. Food and Drug Administration (2001). Draft guidance for industry on food-effect bioavailability and fed bioequivalence studies: study design, data analysis, and labeling; availability. *Fed. Regist.* 66:59433.

U.S. Food and Drug Administration (2003). Guidance for industry on food-effect bioavailability and fed bioequivalence studies; availability. *Fed. Regist.* 68:5026–5027.

Weinberger, M.M. (1984). Theophylline QID, BID and now QD? A report on 24-hour dosing with slow-release theophylline formulations with emphasis on analysis of data used to obtain Food and Drug approval for Theo-24. *Pharmacotherapy* 4:181–198.

Welling, P.G. (1996). Effects of food on drug absorption. *Annu. Rev. Nutr.* 6:383–415.

Williams, P.E. and Harding, S.M. (1984). The absolute bioavailabaility of oral cefuroxime axetil in male and female volunteers after fasting and after food. *J. Antimicrob. Chemother.* 13:191–196.

Wu, C-Y. and Benet, L.Z. (2005). Predicting drug disposition via application of BCS: transport/absorption/elimination interplay and development of a Biopharmaceutics Drug Disposition Classification System. *Pharm. Res.* 22:11–23.

11
In Vitro–In Vivo Correlation on Parenteral Dosage Forms

Banu S. Zolnik and Diane J. Burgess

11.1 IVIVC Definition

In vitro and *in vivo* correlation (IVIVC) for drug products, especially for solid oral dosage forms, has been developed to predict product bioavailability from *in vitro* dissolution. Biological properties such as C_{max}, or AUC have been used to correlate with *in vitro* dissolution behavior such as percent drug release in order to establish IVIVC. IVIVC can be used to set product dissolution specifications; and as a surrogate for *in vivo* bioequivalence in the case of any changes with respect to formulation, process, or manufacturing site.

11.2 Modified Release Parenteral Products

Modified release (MR) parenteral products achieve sustained blood levels of therapeutics consequently decreasing dosing frequency and increasing patient compliance. These systems offer advantages over traditional dosage forms due to their sustained release capabilities and therefore more consistent blood levels that can result in a lowering of the systemic toxicity of drugs. The efficacy of chemotherapeutic agents has been reported to improve when steady relatively low blood levels were achieved compared to high dose i.v. bolus injections (Herben *et al.*, 1998; Hochster *et al.*, 1994). This can be accomplished by encapsulation of chemotherapeutics within liposomal and polymeric delivery systems. In addition, modified release parenteral products are used for targeted and localized drug delivery, which also reduces unwanted side effects.

Potential drug candidates for MR parenterals are chemotherapeutics or other drugs with a high incidence of adverse side effects; proteins or other macromolecules due to their instability in the gastrointestinal (GI) tract; drugs with short half-lives; drugs with low solubility; and drugs that are susceptible to high first-pass effect.

Modified release parenterals include: microspheres; liposomes; emulsions; suspensions; implants; drug eluting stents; and dendrimers. Recent developments in synthetic chemistry have been utilized to make dendrimers, liposomes, and other parenteral delivery systems multifunctional through the addition of targeting moieties, and imaging agents. The reader is referred to the detailed reviews on the incorporation of monoclonal antibodies and other ligands to such delivery systems (Torchilin, 2005; Torchilin and Levchenko, 2003; Torchilin and Lukyanov, 2003). In addition, delivery system particle size and drug loading can be manipulated to alter tissue distribution as well as release rates. Surface modification with polyethylene glycol (PEG) or other polymers has been utilized to increase the blood circulation half-life.

11.3 Factors to Consider for Meaningful IVIVC

Strategies to develop meaningful IVIVC for MR products are summarized below. It is important first to obtain *in vivo* data, and then identify the *in vivo* drug release mechanism. The *in vitro* release method can then be designed with consideration to the *in vivo* release profile and mechanism.

11.3.1 Product Related Factors

There are several factors related to the formulation of MR parenterals that may affect the *in vivo* performance of these products when administered *via* parenteral routes (i.m., i.v., s.c., intra-CSF). These factors include formulation dispersibility, stability, injection volume, viscosity, and biocompatibility. To ensure dispersibility and also ease of injection, microspheres, and other dispersed system parenterals can be suspended in a vehicle containing an isotonic solution of carboxymethylcellulose, surfactant prior to administration (http://www.gene.com/gene/products/information/opportunistic/nutropin-depot/insert.jsp). The injection of a homogenous suspension of microspheres should be assured otherwise erroneous dosing may occur that would affect the *in vivo* data and hence the development of an IVIVC. On the other hand, the presence of surfactant could affect the release properties *in vivo* by enhancing drug solubility and diffusion or affecting viscosity. It has been reported that variation in the injection depth for i.m. administration resulted in large variations in plasma drug concentrations (Zuidema *et al.*, 1994). Formulation stability should be monitored prior to injection of dispersed systems since any particle size change may result in adverse effects and alteration of drug release characteristics. Another important factor with respect to microspheres is the reconstitution time since premature drug release may occur in the delivery vehicle due to dissolution of surface associated drug from the microspheres. This may result in an underestimation of the initial dose released (burst release) upon administration.

Nonionic surfactants such as Cremophor®EL (CrEL; polyoxyethyleneglycerol triricinolate 35) and polysorbate 80 (Tween 80) have been used to solubilize a variety of drugs prior to i.v. administration. A detailed review by Tije *et al.* reports adverse effects such as acute hypersensitivity and peripheral neurotoxicity as well as altered pharmacokinetics of chemotherapeutics when administered with these surfactants (ten Tije *et al.*, 2003). In addition, there may be toxicity issues with certain excipients, especially when used at high concentration. For example, administration of propylene glycol at concentrations above 40% has been reported to cause muscle damage. Consequently, *in vivo* markers, such as cytosolic enzymes, creatine kinase and lactate dehydrogenase, should be monitored as these are indicators of tissue damage which may result from either the drug, or the excipients. The encapsulated drug formulation may result in a reduction in toxicity, for example microspheres or liposomes can be used to isolate high concentrations of irritant drugs which are then released slowly at levels that either do not show toxicity or show limited toxicity *in vivo*. For example, it has been shown that encapsulation of tissue irritant drugs into liposome formulations reduced muscle damage considerably (Kadir *et al.*, 1999). Toxicity and irritancy at the *in vivo* site can affect drug release due to resulting edema as well as the presence of increased numbers of neutrophils and macrophages.

The different manufacturing techniques used to prepare polymeric delivery systems as well as liposomes mostly involve the use of organic solvents. The processes of removal of organic solvent and of determining the amount of residual solvent in the product are crucial due to the *in vivo* relevance (toxicity, tolerance, systemic side effects).

11.3.2 Factors Affecting In Vitro Release

In vitro release methods are an integral part of the product development process to establish quality, performance, and batch to batch consistency as well as *in vivo* and *in vitro* relationships. Current uses of *in vitro* release testing are summarized in Table 11.1.

Unfortunately, there is a lack of standards or guidance documents for *in vitro* release testing methods for modified release parenterals. The United States Pharmacopeia (USP) apparatus for dissolution testing methods were developed for solid oral dosage forms and transdermal products. Briefly, USP Apparatus 1 (basket) and 2 (paddle) are suitable for solid dosage forms. Apparatus 3 (reciprocating cylinder) and Apparatus 4 (flow-through cell) were developed for drugs with limited solubility and are useful for MR products. Apparatus 5 (paddle over disc), Apparatus 6 (cylinder), and 7 (reciprocating disk) were developed for transdermal delivery systems. In some cases current USP methods have been modified to overcome limitations of the existing methods for application to MR parenterals

TABLE 11.1. Current uses of *in vitro* release testing method

- Formulation development
- Quality assurance and process control
- Evaluation of the changes in the manufacturing process
- Substantiation of label claims
- Compendial testing

products. For example, USP Apparatus 4 has been adapted for microsphere testing through the inclusion of glass beads in the flow-through cells (Zolnik *et al.*, 2005). The glass beads are interspersed between the microspheres to prevent aggregation during the release study and to more closely simulate the *in vivo* conditions where the microspheres are interspersed among the cells, e.g, at the s.c. site (Zolnik *et al.*, 2005). Moreover, the addition of the glass beads in the flow-through cells allows laminar flow of release media and prevents the formation of channels in the solid bed where the media flow through while other areas in the bed would remain unwetted. Figure 11.1 displays the schematic diagram of flow-through cell containing microspheres and glass beads in the closed mode.

In vitro release testing methods currently used in research and development as well as quality control include: dialysis sac, sample-and-separate, ultrafiltration, continuous flow methods, and microdialysis. The dialysis sac method involves suspending microspheres or other dispersed systems in a dialysis sac with a semi-permeable membrane that allows diffusion of the drug, and then drug concentration is monitored in the receiver chamber. Disadvantages of this method include (a) potential for dispersed system aggregation due to the lack of agitation and (b) violation of sink conditions may result when drug release from the microspheres is faster than drug diffusion through the membrane (Chidambaram and Burgess, 1999). A reversed dialysis method has been developed by Chidambaram and Burgess, where the dispersed phase is placed in the large chamber with the media and the sacs contain only media. The sacs are then sampled at the different time points. This method overcomes the problem of violation of sink conditions (Chidambaram and Burgess, 1999). The sample and separate technique utilizes USP Apparatus 2 (paddle method) where microspheres are dispersed in the media

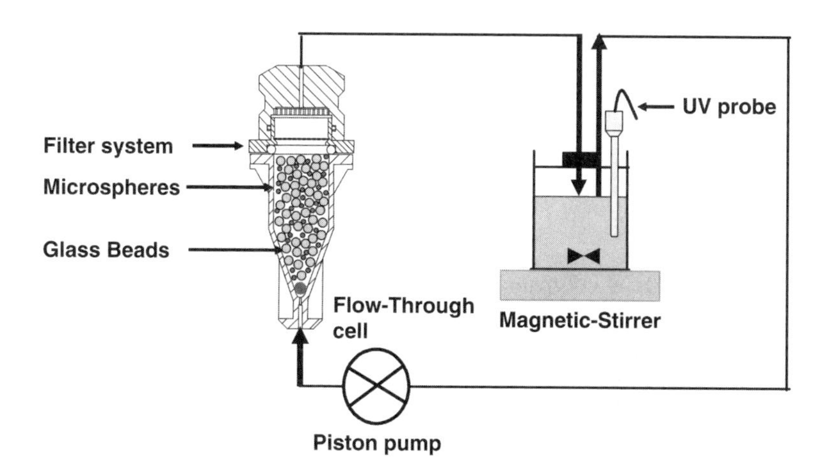

FIGURE 11.1. Schematic diagram of 12 mm flow-through cell containing microspheres and glass beads in the flow-through method (closed system). Placement of the fiber optic probe in the reservoir vessel is also shown

and at different time points samples are withdrawn, separated *via* ultracentrifugation or filtration and the filtrate is analyzed for drug content with an appropriate analytical method. The disadvantages of this method are the difficulty in separation of the delivery system from the media, for example, ultrafiltration requires 1 or 2 h at high centrifugational force (150, 000 × *g*) and this often is an undesirable method due to disruption of the delivery system and consequent alteration in the release pattern (Chidambaram and Burgess, 1999). As an alternative, low pressure ultrafiltration has been used to prevent disruption. The disadvantage of this method is the lack of available membranes with appropriate cut off points since some delivery systems are in the submicron and micron size range (Magenheim *et al.*, 1993). The continuous flow method (USP Apparatus 4) consists of a reservoir, a pump and flow-through cells where the microspheres or other dispersed systems are contained. The continuous flow method avoids problems associated with separation of the dispersed system from the media since the dispersed system is isolated in the flow-through cells and the media can be sampled from the reservoir. Another advantage of the flow-through method for dispersed systems is that since the dispersed system is isolated from the media reservoir this allows *in situ* monitoring. UV fiber optic probes can be placed in the media reservoir vessel thus avoiding the potential problem of interference from dispersed system particles sticking to the probe. *In situ* monitoring has the advantage that multiple time points can be analyzed to allow for complete characterization of the release profile. For example, this method has been used to characterize the burst release phase from microspheres. Schematic showing the placement of the *in situ* probes is shown in Fig. 11.1 (Zolnik *et al.*, 2005). In addition, violation of sink conditions for drugs with limited solubility is not an issue with the continuous flow USP 4 method due to the ease of media replacement.

Microdialysis has been used to study pharmacokinetics of drugs in peripheral tissues (Boschi and Scherrmann, 2000; de la Pena *et al.*, 2000). Recently microdialysis has been used to monitor drug release *in vitro* (Dash *et al.*, 1999). The basic principle of this technique is to measure drug release continuously from an implant site by mimicking a capillary blood vessel with a thin dialysis tube. An advantage of this technique is that the flow rate of the media can be adjusted to as low as 0.5 μl/min. Other advantages of this technique are (a) small volume (b) continuous monitoring of drug release, and (c) online analysis (Dash *et al.*, 1999). Dash *et al.* had compared a microdialysis method with the USP Apparatus 3 method to monitor ciprofloxacin release from PLGA implants and reported that both these methods were in close agreement. Researchers have also evaluated miniaturized methods where small volumes of media are employed due to the *in vivo* relevance (volume at the s.c. site is low). However, the disadvantages of this method are violation of sink conditions and the potential for dispersed system aggregation due to the limited volume and lack of agitation.

In order to develop meaningful IVIVC, study design for *in vitro* release should be performed after *in vivo* data are available, so that media conditions can be manipulated to mimic the *in vivo* behavior. To this end, researchers have investigated different media conditions to aid in the development of a relationship between *in vivo* and *in vitro* release data. The use of cosolvent, addition of surfactants and enzymes, variation in pH, ionic strength, agitation and temperature have been investigated (Agrawal *et al.*, 1997; Aso *et al.*, 1994; Blanco-Prieto *et al.*, 1999; Hakkarainen *et al.*, 1996; Jiang *et al.*, 2002; Li *et al.*, 2000; Makino *et al.*, 1986). For example, acidic media have been used to mimic drug release from PLGA microspheres *in vivo* (Blanco-Prieto *et al.*, 1999; Heya *et al.*, 1994a).

There is no single *in vitro* release testing method suitable for all parenterals delivery systems due to their complexities. However, USP Apparatus 4 is recommended for modified release oral formulations and is appropriate for modified release parenterals. USP Apparatus 4 has been recommended for MR microsphere products (Burgess *et al.*, 2004). The physicochemical properties of drugs and delivery systems should be taken into account when choosing an appropriate release method. In addition, the *in vitro* method should be able to discriminate between formulations with different *in vivo* release characteristics.

11.3.2.1 Accelerated *In Vitro* Release Testing

Since MR parenterals may be intended to release drug for days, weeks, and even months, accelerated *in vitro* release testing methods are required for routine testing of these products. Therefore, if the accelerated method is to be used as a surrogate for *in vivo* studies IVIVC must be established using the accelerated method. A problem here is that accelerated methods, by their nature, often change the mechanism of drug release and this can make the establishment of an IVIVC more difficult. For example, elevated temperature accelerated conditions have been shown to alter the mechanism of release from PLGA microspheres from degradation controlled to diffusion controlled (Zolnik *et al.*, 2006). On the other hand, under pH accelerated conditions, release from PLGA microspheres appeared to be degradation controlled eventhough morphological changes occurred during degradation that were distinctly different from those that occur during "real-time" *in vitro* release testing.

11.3.3 Mathematical Models of In Vitro Drug Release

Different models have been developed depending on the governing, rate-limiting step of drug release. For MR systems the mathematical models used can be categorized as: diffusion controlled, swelling controlled, and erosion controlled release systems.

Mathematical models to evaluate drug release have been extensively used, especially for solid dosage forms to understand drug transport through barriers. Fick's second law of diffusion states that the rate of change in concentration is proportional to the rate of change in the concentration gradient at that point where the proportionality constant is equal to the diffusivity "D". The assumption is constant diffusivity.

$$\frac{dC}{dt} = D\left[\frac{d^2C}{dx^2} + \frac{d^2C}{dy^2} + \frac{d^2C}{dz^2}\right] \tag{11.1}$$

Various exact solutions of (11.1) depending on the boundary condition of the system were reviewed in detail by Flynn *et al.* (1974). The commonly used form of (11.1) is below. The assumptions necessary to arrive at (11.2) are (a) sink conditions are maintained; (b) diffusivity is constant; and (c) steady state is reached.

$$\frac{dM}{dt} = \frac{DC_0}{h} \tag{11.2}$$

Higuchi derived the following equation for systems when boundaries change with time, such as drug release from a semisolid ointment. The change in the amount released per unit area, dM, is equal to a change in the thickness of the moved boundary, dh. A is the total amount of drug in the matrix. C_s is the saturation concentration of the drug within the matrix

$$dM = A\,dh - \frac{C_s}{2}dh. \tag{11.3}$$

According to Ficks law, dM is equal to (11.2). The equation which describes the amount released as a linear function of the square root of time can be derived (11.4) after setting (11.2) and (11.3) equal. The assumptions used in this derivation are: initial drug loading is much higher than drug solubility, swelling of the system is negligible, sink conditions are maintained, and edge effects are negligible.

$$M = \sqrt{2C_sDA}. \tag{11.4}$$

There are several mathematical models derived for different systems and different geometries (such as, spheres), as well as for release of drugs suspended in spherical particles, and for systems where the rate of drug release is swelling controlled.

Ritger and Peppas (1987) derived a semiempirical equation known as the power law (11.5) for systems with different geometries (slab, cylinder, and sphere) to describe drug release for diffusion controlled, swelling controlled, and controlled by intermediate anomalous mass transport.

$$\frac{M_t}{M\infty} = kt^n, \tag{11.5}$$

where k is a constant and n is the release exponent indicative of the drug release mechanism. In the case of Fickian diffusion controlled release, n equals to 0.43 for spherical geometry.

In order to identify the drug release mechanism from low molecular weight PLGA microspheres, (11.3) was utilized. Diffusion kinetics were confirmed for different flow rates using modified USP Apparatus 4 (Zolnik *et al.*, 2006). Modeling of drug release from biodegradable polymers such as PLGA is complex since it involves not only diffusion phenomena of drug release but also physicochemical changes in the polymer. Empirical models have been derived based on the assumption that one net mechanism with zero order process can describe all mechanisms involved, such as dissolution, swelling, and polymer degradation. Mechanistic models based on Monte Carlo simulations have been applied to describe polymer degradation and diffusion phenomena (Siepmann and Gopferich, 2001). Drug release from such systems has also been modeled by including the dependence of the diffusion coefficient on the polymer molecular weight change (Faisant *et al.*, 2002). Lemaire *et al.* were able to show the relative dominance between the diffusion and erosion release kinetics when different parameters such as erosion rate, initial pore size, porosity and the diffusion coefficient of the drug were varied (Lemaire *et al.*, 2003).

It has been established that PLGA degradation followed pseudo-first-order degradation kinetics (11.6).

$$Mw(t) = Mw_0 e^{-k_{\deg} t} \tag{11.6}$$

First order degradation kinetics have been observed from PLGA microspheres at elevated temperature. This was used to establish drug release mechanisms under accelerated release conditions where temperature varied between 37 and 70°C (Zolnik *et al.*, 2006). It should be noted that when an initial burst release exists, it is recommended to test the burst phase separately under "real-time" conditions, as under accelerated conditions the burst phase is usually not observed. Likewise it is often necessary to model the release separately from the burst phase. High correlation has been observed for drug release from PLGA microspheres postburst release (Zolnik *et al.*, 2006).

11.3.4 Factors Affecting In Vivo Release

In vivo release from MR parenterals such as microspheres may be affected by the environment at the site of administration for example s.c. or i.m. injected products are generally retained at the administration site depending on the particle size. *In vivo* factors that affect drug release can be classified as delivery system independent and delivery system dependent. Delivery system independent factors include barriers to drug diffusion (e.g., fluid viscosity and connective tissue); drug partitioning at the site (e.g., uptake into fatty tissue); available fluid volume at the site; and in the case of intramuscular injection muscle movement may also be an important factor. For example, factors related to subcutaneous tissue are interstitial fluid volume, blood flow rate, osmotic pressure, and the presence of plasma proteins. It has been reported that the diffusion of macromolecules from the interstitium may be delayed by the fibrous collagen network, and the gel structure of proteoglycans as well as possible electrostatic interaction with components of the interstitium. More information on protein absorption and bioavailability from the subcutaneous tissue can be found in a detailed review article by Porter and Charman (2000). Delivery system dependent factors are those specific to a particular delivery system and include enzymatic degradation of susceptible polymers, protein adsorption, phagocytosis as well as inflammatory reaction. For example, the initial acute phase of inflammation results in an influx of fluid together with phagocytic cells and the increased fluid volume may increase drug release and adsorption. Whereas, the chronic stage of inflammation can lead to fibrosis which in turn results in isolation of the delivery system with consequent reduction in the fluid volume. A major challenge to *in vivo* delivery of drug carriers following IV administration is the rapid removal of these particles from circulation by the reticuloendothelial system (RES) mainly the Kupffer cells of the liver and the macrophages of the spleen and bone marrow. In order to reduce interaction with plasma proteins and consequently prevent RES uptake, and increase blood circulation time, MR parenterals have been surface modified with PEG polymers. A thorough review on this subject can be found in an article by Moghimi *et al.* (2001).

Different drugs have been coencapsulated in microspheres to alter their *in vivo* behavior. For example, dexamethasone was coencapsulated with bupivacaine to increase the concentration of bupivacaine at the local site by decreasing its clearance from the tissues due to the vasoconstrictive nature of dexamethasone. In this case, *in vitro* release of bupivacaine was not altered when dexamethasone was incorporated. Care should be taken to determine the pharmacodynamic effects when drugs are given in combination in such formulations (McDonald *et al.*, 2002).

11.4 *In Vitro–In Vivo* Correlation

IVIVC can be categorized as follows: Level A, point-to-point correlation over the entire release profile and is used to claim biowaivers; Level B, mean *in vitro* dissolution time is compared to either the mean residence time or the mean *in vivo* dissolution time; Level C, single point correlation between a dissolution parameter (for example, the amount dissolved at a particular time or the time required for *in vitro* dissolution of a fixed percentage of the dose) and an *in vivo* parameter (for example, C_{max} or AUC); Multiple Level C correlation, a Level C correlation at several time points in the release profile.

Figure 11.2 summarizes general considerations with respect to *in vivo* release and distribution of protein loaded microspheres for establishing IVIVC. In this scheme Morita *et al.* compartmentalized the events involving *in vivo* pharmacokinetics of protein release from microspheres as: drug release rate constants (K_{rel}) from microspheres, protein degradation constant as K_{deg}, drug absorption to systemic circulation defined as K_a, and distribution to target tissues as K_d while drug elimination from kidney or liver defined as K_{el}. In this scheme, Morita *et al.* have also included the possible immune response effects on *in vivo* pharmacokinetics of proteins due to generation of specific antibodies. The authors indicated that antibodies generated in normal mice may alter the clearance rate of bovine derived superoxide dismutase and this affect was not observed in severe combined immunodeficiency disease mice. In this scheme, there are three output functions which are used to establish IVIVC, X1 *in vitro* release profile correlated to either Y1

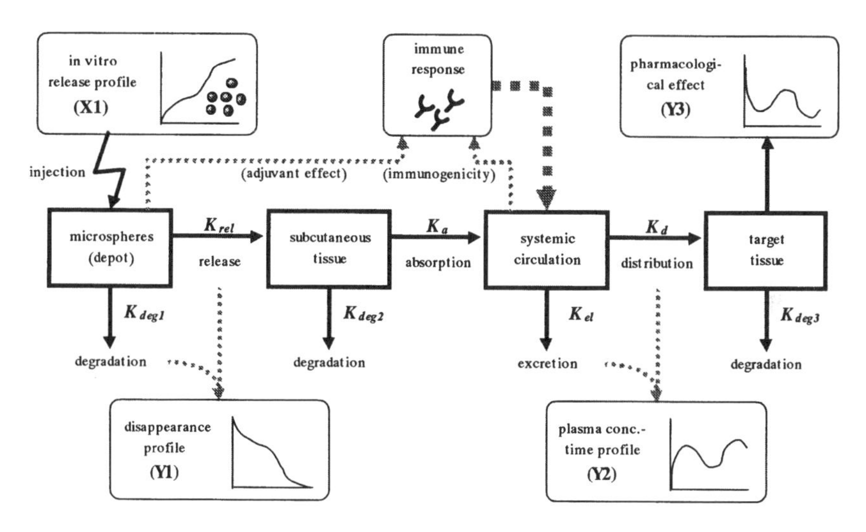

FIGURE 11.2. General considerations for the IVIVC of protein loaded microspheres

defined as disappearance profile from the administration site, or plasma concentration time profile as Y2. The pharmacological effects of drugs at the target tissue are defined as Y3 (Morita *et al.*, 2001). Different levels of correlations can be achieved by comparing X1 (*in vitro* drug release) to Y1 (*in vivo* disappearance) or Y2 (plasma concentration time profile). If Y2 is used, convolution procedure or any other modeling technique can be used to relate plasma concentration time profile to *in vivo* absorption or release rate. If a linear relationship between the *in vitro* and release data does not occur then, IVIVC can be achieved by mathematical modeling (e.g. time variant nonlinear modeling) of the *in vitro* and *in vivo* data (Young *et al.*, 2005).

11.5 Microspheres

Microspheres are polymeric spherical particles in the micron size range. Drug can be entrapped in these particles either in the form of microcapsules with a polymer coating surrounding a drug core or in the form of micromatrices with the drug dispersed throughout the polymer (Burgess and Hickey, 1994). Both natural and synthetic polymers have been used to form microspheres (Cleland, 1997). In this chapter, synthetic polymers such polyesters, poly(lactic acid) (PLA) and poly(lactic-*co*-glycolic acid) polymers (PLGA), will be reviewed. These polymers gained importance in the field of drug delivery due to their biodegradability and relative biocompatibility (Kulkarni *et al.*, 1971). Lupron Depot®, that releases potent analogue of luteinizing hormone–releasing hormone (LH–RH) over periods of 1 and 3 months (Okada, 1997), was the first controlled release microsphere product available on the US market for the treatment of hormone dependent prostate and mammary tumors, and endometriosis. Since 1989, the Food and Drug Administration has approved the following five PLGA microsphere products (Lupron Depot, Sandostatin LAR, Nutropin Depot and Trelstar Depot, Risperdal Consta). Microspheres are designed as modified release drug delivery systems where drug is released in periods of days to months. From a safety and efficacy perspective, it is important to understand drug release kinetics from such formulations. Microsphere systems tend to exhibit complex release kinetics with: an initial burst release, as a result of surface associated drug and this is usually diffusion controlled. Following the burst release phase, the mechanism of release may be diffusion or erosion controlled or a combination of thereof (Gopferich, 1996; Lewis, 1990; Okada, 1997). Drug release from PLGA microspheres typically falls under the combination of diffusion and erosion controlled where an initial burst release is followed by a lag phase and then a secondary, apparent zero order release phase. The lag phase is considered to be a result of the time required for the build up of acid byproducts, and, hence for sufficient bulk erosion to take place, to increase porosity and allow for the subsequent secondary apparent zero order phase (Brunner *et al.*, 1999; Mader *et al.*, 1998; Shenderova *et al.*, 1999).

In the literature different levels of IVIVC have been established for PLGA microspheres. In a study by Zolnik and Burgess, a biorelevant *in vitro* release method (USP 4 method) and Sprague Dawley rat model was utilized to obtain a relationship between *in vitro* and *in vivo* release of dexamethasone from two different PLGA microsphere formulations. A linear IVIVC using the time shifting/scaling method discussed above was established for microsphere formulations prepared with different molecular weights of PLGA. The time scaling/shifting method was applied to *in vitro* data due the observance of faster *in vivo* release of dexamethasone (Zolnik 2005). In addition, the release of dexamethasone was able to control the inflammatory reaction that would otherwise occur to the presence of microspheres and to the tissue damage that occurs due to needle injection and this appeared to result in faster *in vivo* kinetics compared to the *in vitro* kinetics. In a previous publication from our laboratory, it has been reported that release kinetics of vascular endothelial growth factor (VEGF) from PLGA microspheres was slower *in vivo* compared to *in vitro* (Kim and Burgess, 2002). A possible explanation for this is the severe inflammatory reaction that occurred in the presence of these microspheres. This is considered to be a result of both tissue reaction to the PLGA microspheres as well as to the foreign protein (human VEGF was used in a rat model).

It has also been shown for leuprolide that repeated injections of Lupron depot did not alter the bioavailability and urinary excretion of leuprolide and the mean serum levels and AUC of Lupron Depot, correlated linearly with each dose. More detailed information on the formulation, drug release and *in vivo* animal models of leuprorelin depot can be found in a comprehensive review by Okada (1997). The evaluation of different *in vitro* conditions to mimic *in vivo* release of thyrotropin releasing hormone (TRH) from PLGA microspheres were investigated by Heya *et al.* (1994a). Authors concluded that the selection of the media conditions (such as medium pH, buffer concentration, ionic strength) is important to obtain an *in vitro* release profile that mimics *in vivo* release, especially for hydrophilic drugs (Heya *et al.*, 1994a). In the follow-up study, Heya *et al.* examined the pharmacokinetics of TRH from PLGA microspheres and determined that sustained *in vitro* release kinetics were mimicked *in vivo* using 33 mM pH 7 phosphate buffer containing 0.02% Tween 80 (Heya *et al.*, 1994b).

Other examples of Level A correlation were demonstrated by Cheung *et al.* and utilized the continuous flow method in dynamic and static mode to mimic *in vivo* release from locoregionally administered dextran-based microspheres (Cheung *et al.*, 2004). An example of Level B correlation was shown for release of the somatostatin analogue vapreotide from PLA and PLGA microspheres where the mean *in vivo* residence time was correlated with the mean *in vitro* dissolution time (Blanco-Prieto *et al.*, 2004).

A linear IVIVC was demonstrated in a different polymer system by van Dijkhuizen-Radersma (2004) in a study of protein release from poly(ethylene glycol) terephthalate (PEGT)/poly(butylene terephthalate) PBT microspheres. Similar to the properties of polyesters, poly(ether–ester) PEGT/PBT multiblock copolymers exhibit biodegradability and biocompatibility. Three different microsphere formulations with varied PEGT/PBT weight ratio and PEG segment length were investigated. The diffusion coefficient of drugs from PEGT/PBT microspheres was dependent on polymer swelling which in turn was related to its PEG segment length. *In vitro* release from PEGT/PBT microspheres correlated with the volume swelling ratios, faster release was obtained using polymers with higher swelling ratios.

In vivo release kinetics are often not predicted by *in vitro* release methods, possibly due to selection of inappropriate *in vitro* release conditions, and methods (Diaz *et al.*, 1999; Jiang *et al.*, 2003). It should be also noted that inappropriate selection of animal model may result in unsuccessful IVIVC. For example, Perugini *et al.* demonstrated that a rat model was not suitable to induce osteopenia and therefore, IVIVC could not be established (Perugini *et al.*, 2003). The difficulties in determination of drug amounts in the biological matrix also resulted in lack of IVIVC (Yenice *et al.*, 2003). In addition, there are several comprehensive and well-designed research articles in the literature on the evaluation of *in vitro* and *in vivo* release of drugs from microspheres; however, these articles did not attempt to show any mathematical correlation of their *in vivo* and *in vitro* results (Liu *et al.*, 2003).

11.6 Liposomes

Liposomes consist of one or more phospholipid bilayers with enclosed aqueous phase. Depending on the method of preparation of liposomes, different types of liposomes are formed: large multilamellar (MLVs); small unilamellar, (SUVs); or large unilamellar (LUVs). Liposomal drug delivery has advantages over traditional therapy in cancer treatment due to increased tumor uptake *via* enhanced permeation and retention (EPR) effect where tumor tissue has leaky vasculature

TABLE 11.2. MR products in the market

Active drug	Product name	Indications
Microsphere products		
Leuprolide	Lupron	Endometriosis
Octreotide	Sandostatin LAR	Agromegaly
Somatropin	Nutropin depot	Growth therapy
Triptorelin	Triptorelin	Prostate cancer
Abarelix	Plenaxis	Prostate cancer
Liposome products		
Daunorubicin	DaunoXome	Kaposi's sarcoma
Doxurubicin	Mycet	Combinational therapy of recurrent breast cancer
Doxurubicin in PEG-liposomes	Doxil/Caelyx	Refractory Kaposi's sarcoma; ovarian cancer; recurrent breast cancer
Amphotericin B	AmBiosome	Fungal infection
Cytarabine	DepoCyt	Lymphomatous meningitis
Vincristine	Onco TCS	Non-Hodgkin's lymphoma
Emulsion products		
Propofol	Diprivan	Anesthetic
Diazepam	Dizac	Epilepsy

and poor lymphatic drainage (Maeda *et al.*, 2001). Liposomal products with encapsulated daunorubicin, doxorubicin, and vincristine are currently in the market for the treatment of cancer (Table 11.2). Drug release and cell uptake kinetics may depend on the size, charge, surface properties of the liposomes as well as on the types of lipids used. The incorporation of stabilizing lipids with high phase transition temperatures tends to decrease drug release (Anderson and Omri 2004; Bochot *et al.*, 1998; Ruel-Gariepy *et al.*, 2002). One of the major challenges of liposomal drug delivery is the rapid uptake of liposomes by the RES. Sterically stabilized liposomes with PEG chains with increased circulation half-life have been developed to decrease contact with blood components, and consequently avoid recognition by the RES system. In clinical studies, the blood circulation half-life of these "Stealth" liposomes was extended from a few hours to 45 h consequently, altering tissue distribution of drugs compared to free drug controls due to their prolonged circulation. pH sensitive liposomes have been formulated to undergo phase change in the acidic environment resulting in a disruption of the lysosomes, and consequent release of the liposome contents into the cytoplasm (Simoes *et al.*, 2004). Immunoliposomes where immunoglobulins are attached to liposomes *via* covalent binding or by hydrophobic insertion to increase their targeting capabilities have also been formulated. Other types of liposomes have been also where ligands such as folate, transferrin mediated liposomes have been utilized to target tumor tissues. More detailed information on recent advancement on the types of liposomes can be found in a review by Torchilin (2005).

An FDA Draft Guidance document for industry on liposome products states that the characterization of physicochemical properties of liposomes is critical to determine product quality (FDA Draft Guidance, 2002). These tests include determination of morphology, i.e., lamellarity, net charge, volume of entrapment in the vesicles, particle size and size distribution, phase transition temperature, *in vitro* drug release from the liposomes, osmotic properties, and light scattering index. The guidance document states that information on *in vivo* integrity of liposomes should be determined prior to measurement of pharmacokinetic parameters of liposomes. In addition to the information on general pharmacokinetic parameters (i.e., $C_{\max}$, AUC, clearance, volume of distribution, half-life) *in vivo*, comparative mass-balance studies of drug substance and its liposomal formulation were recommended to determine systemic exposure (FDA Draft Guidance, 2002). In order to ensure quality control of the product, chemical stability of liposomes such as phospholipid hydrolysis, nonesterified fatty acid concentration, autooxidation, and drug stability should be identified (Crommelin and Storm, 2003).

Jain *et al.* investigated acyclovir release from multivesicular (MVL) and conventional multilamellar (MLV) vesicles for *in vitro* and *in vivo* studies. They were able to show sustained release in 96 h with MVL liposomes, while MLV's exhibited faster release kinetics in 16 h using dialysis as an *in vitro* release method. Using an *in vivo* rat model, they were able to show sustained plasma levels of drug from MVL up to 32 h, concluding that MVL offered advantages of high drug loading and sustained release with reduced toxicity (Jain *et al.*, 2005).

One of the most commonly used methods to investigate drug release from liposomes is the dialysis method where drug loaded liposomes are placed in dialysis tubes and suspended in a beaker. However, often lack of IVIVC was observed with this method possibly due to violation of sink conditions. Shabbits *et al.* developed an *in vitro* release method using excess amounts of multilamellar vesicles (MLV) as "acceptors" for drug release from "donor" liposomes. They were able to mimic *in vivo* drug release closely using MLV based *in vitro* method which served as a lipid sink (Shabbits *et al.*, 2002). Level A correlation on MVL liposomes was demonstrated by Zhong *et al.* where IVIVC was achieved using plasma as an *in vitro* release medium for drug release (Zhong *et al.*, 2005). However, the use of IVIVC for liposomal products for biowaivers and bioequivalence studies might be difficult since these systems can be very complex. For example, stealth liposomes should remain stable *in vivo*, without any significant release of drug, until uptake into the cells of interest. As expected, such a release profile would be extremely difficult to mimic *in vitro*.

11.7 Emulsions

Emulsions are formed when two or more immiscible liquids with limited mutual solubility are mixed with a high energy input such as *via* ultrasonication, homogenization, or microfluidization. Due to their thermodynamic instability, the use of surfactants is required to improve their stability. Emulsion can be categorized as simple emulsions such as water-in-oil (w/o), or oil-in-water (o/w), or multiple emulsions water-in-oil-in-water (w/o/w) or oil-in-water-in-oil (o/w/o). The most commonly used clinical application of emulsions is for the delivery of parenteral nutrition for patients who can not absorb nutrients *via* the GI route. These nutrients include vitamins, minerals, amino acids, and electrolytes. Emulsions may also be formulated to deliver drugs with low water solubility, for example propofol formulated in o/w emulsion with soybean oil, glycerol and egg lecithin is currently on the market as a sedative-hypnotic agent (http://www.astrazenecaus.com/pi/diprivan.pdf). Other currently available emulsion products can be found in the Table 11.2.

In vitro release of drugs from emulsion systems can be evaluated using the sample and separate method, dialysis and reversed dialysis methods. Since drug release from emulsions is often relatively rapid, the reversed dialysis is recommended so that sink conditions are not violated.

In vivo pharmacokinetic profiles of drugs in emulsion formulations depends on the blood circulation time, the droplet size of the emulsion, the injection volume and the drug lipophilicity (Kurihara *et al.*, 1996; Takino *et al.*, 1994; Ueda *et al.*, 2001). As described above emulsion formulations also suffer rapid uptake by the RES system. In order to improve blood circulation half-life, Reddy *et al.* investigated pegylation of etoposide emulsion. *In vivo* studies using a rat model, showed that pegylated emulsion exhibited a 5.5 times higher AUC compared to the etoposide commercial formulation. The effect of different oxyetylene moieties varied by size on the o/w emulsion blood circulation time was investigated. It was reported that blood circulation half-life was prolonged from approximately 10 min to 100 min when oxyetylene varied from 10 to 20, respectively. Reduction in the liver uptake was observed with emulsions prepared with 20 and higher oxyetylene moieties compared to those with ten oxyetylene moieties (Ueda *et al.*, 2003).

11.8 Hydrogels, Implants

The advantages of hydrogels as depot formulations are their biocompatibility, water permeability, and injectability (*in situ* forming gels) at the site (i.e., tumor site) (Hoffman, 2001; Peppas *et al.*, 2000). One disadvantage of hydrogels is that the drug release rate may not be manipulated, for example: fast release rate of hydrophilic drugs occurs from the hydrogels due to the hydrophilic environment within the hydrogel. Therefore, two phase systems were developed where delivery vehicles (liposomes or microspheres) were entrapped in the hydrogels to control drug release kinetics (Galeska *et al.*, 2005; Moussy *et al.*, 2003; Patil *et al.*, 2004). Patil *et al.* (2004) were able to achieve *in vitro* and *in vivo* controlled release of dexamethasone from microspheres entrapped in a polyvinyl alcohol hydrogel and a linear *in vitro–in vivo* correlation of release rates. Lalloo *et al.* (2006) demonstrated *in vitro* and *in vivo* controlled release of chemotherapeutic topotecan from two phase systems where drug containing liposomes were entrapped in hydrogels. Longer tumor suppression was achieved using with this approach compared to drug alone.

A linear IVIVC was established for methadone release from implants (Negrin *et al.*, 2001). In this study drug release *in vivo* was calculated from the amount of drug remaining inside the implant. However, when *in vivo* methadone release was estimated by deconvolution from serum levels, deviations from linearity occurred at later time points. The authors confirmed the role of possible metabolic induction in the underestimation of *in vivo* release as a consequence of increased methadone clearance with time. Therefore, it should be noted that estimation of *in vivo* release by deconvolution might not be applicable when the drug absorption and disposition function is not linear and not constant with time (Negrin *et al.*, 2004).

11.9 Dendrimers

Dendrimers are synthetic highly branched polymers with a central core with sizes in the nanometer range. The structure and branched topologies of dendrimers resembles a branched tree hence the name is derived from the Greek name *dendra* (meaning tree, tree-like structure). Dendrimers can be categorized based on the number of the branches they possess which are called generations (G-1, G-2, G3, etc.). The molecular weight, chemical composition and size of the dendrimers can be tightly controlled during synthesis of these polymers. Most commonly used dendrimers are based on polyamidoamines (PAMAM), polyamines, and polyesters (Frechet and Tomalia, 2002; Newkome *et al.*, 2001). Dendrimers are ideal candidates as drug/gene delivery carriers, biological imaging agent carriers, and as scaffolds in tissue engineering due to their uniform size, monodispersity, water solubility, modifiable surface characteristics and high drug loading efficiencies (Kobayashi and Brechbiel, 2004; Kukowska-Latallo *et al.*, 1996; Patri *et al.*, 2002). In addition, the surface charge of these polymers can be manipulated to increase biocompatibility and decrease toxicity. For example, it has been shown that the cytotoxicity of cationic PAMAM dendrimers decreased when the surface charge was modified with the addition of lauroyl and PEG chains (Jevprasesphant *et al.*, 2003). It has also been shown that dendrimers can be used as a multifunctional delivery platform loaded with therapeutics, targeting and imaging agents. PAMAM dendrimers loaded with methotrexate (MTX) as a chemotherapeutic, folate as a targeting agent, and fluorescein as an imaging agent accumulated preferentially approximately five times higher than the control in a mouse model with subcutaneous tumors (Kukowska-Latallo *et al.*, 2005). Drugs can be either physically entrapped or conjugated with the dendrimer. However, it has been shown that MTX was readily released in saline when physically entrapped in dendrimers. This was attributed to weak interaction between MTX and the dendrimer when inter molecular forces are neutralized in the PBS solution. However, MTX was retained in the dendrimer and did not exhibit any premature release when conjugated to the dendrimer (Patri *et al.*, 2005). Drug release from the dendrimers can be controlled by change in pH, for example ester terminated half generation PAMAM dendrimers did not release any drugs at pH 7.0. However, drug release occurred at pH 2.0 when internal tertiary amines were protonated (Twyman *et al.*, 1999). In chemotherapy, this pH responsive release mechanism is desired since drug release occurs only in the acidic microenvironment of the tumor tissue not in the systemic circulation. Another example of controlled drug release from dendrimers was the sustained release of indomethacin from dentritic unimolecular micelles (Liu *et al.*, 2000).

Similar to other MR release products mentioned above, different strategies such as modification of the dendrimer surface with polyethyleneoxide, PEG chains, have been successfully applied to decrease their RES uptake (Gillies and Frechet, 2002; Kim *et al.*, 2004; Malik *et al.*, 2000; Wang *et al.*, 2005; Yang and Lopina, 2006). Neutral, generation four (G4) polyester dendrimers did not accumulate in any organ preferentially and they exhibit rapid renal clearance. It was noted that the low molecular weight and compact structure of the neutral G4 dendrimers could pass glomerular filtration (PadillaDeJesus *et al.*, 2002). In a follow-up study, it was shown that highly branched dendrimers (Generation 3) exhibited greater bioavailability and lower renal clearance than that of the compact dendrimers (Generation 2) where the molecular weights of these dendrimers were approximately the same (Gillies *et al.*, 2005). The potential value of dendrimers as a delivery vehicle is promising since biodistribution and pharmacokinetic properties can be manipulated by changing the dendrimer size and conformation (Lee *et al.*, 2005).

References

Agrawal, C. M., Huang, D., Schmitz, J. P., and Athanasiou, K. A. (1997). Elevated temperature degradation of a 50: 50 copolymer of PLA-PGA. *Tissue Engineering* **3**: 345–352.

Anderson M. and Omri, A. (2004). The effect of different lipid components on the in vitro stability and release kinetics of liposome formulations. *Drug Delivery* **11**: 33–39.

Aso, Y., Yoshioka, S., Li Wan Po, A., and Terao, T. (1994). Effect of temperature on mechanisms of drug release and matrix degradation of poly(-lactide) microspheres. *Journal of Controlled Release* **31**: 33–39.

Blanco-Prieto, M. J., Besseghir, K., Orsolini, P., Heimgartner, F., Deuschel, C., Merkle, H. P., Nam-Tran, H., and Gander, B. (1999). Importance of the test medium for the release kinetics of a somatostatin analogue from poly(-lactide-co-glycolide) microspheres. *International Journal of Pharmaceutics* **184**: 243–250.

Blanco-Prieto, M. J., Campanero, M. A., Besseghir, K., Heimgatner, F., and Gander, B. (2004). Importance of single or blended polymer types for controlled in vitro release and plasma levels of a somatostatin analogue entrapped in PLA/PLGA microspheres. *Journal of Controlled Release* **96**: 437–448.

Bochot, A., Fattal, E., Gulik, A., Couarraze, G., and Couvreur, P. (1998). Liposomes dispersed within a thermosensitive gel: a new dosage form for ocular delivery of oligonucleotides. *Pharmaceutical Research* **15**: 1364–1369.

Boschi, G. and Scherrmann, J. (2000). Microdialysis in mice for drug delivery research. *Advanced Drug Delivery Reviews* **45**: 271–281.

Brunner, A., Mader, K., and Gopferich, A. (1999). pH and Osmotic pressure inside biodegradable microspheres during erosion. *Pharmaceutical Research* **16**: 847–853.

Burgess, D. J. and Hickey, A. J. (1994). Microsphere technology and applications. In Swarbrick, J., Boylan, J. C. (eds.), *Encyclopedia of Pharmaceutical Technology*, Vol. 10, Marcel Dekker, New York, pp. 1–29.

Burgess, D. J., Crommelin, D. J. A., Hussain, A. J., and Chen, M.-L. (2004). EUFEPS workshop report, assuring quality and performance of sustained and controlled release parenterals. *European Journal of Pharmaceutical Sciences* **21**: 679–690.

Cheung, R. Y., Kuba, R., Rauth, A. M., and Wu, X. Y. (2004). A new approach to the in vivo and in vitro investigation of drug release from locoregionally delivered microspheres. *Journal of Controlled Release* **100**: 121–133.

Chidambaram, N. and Burgess, D. J. (1999). A novel in vitro release method for submicron-sized dispersed systems. *AAPS pharmSci* **1**: Article 11.

Cleland, J. L. (1997). Protein delivery from biodegradable micropsheres. In Sanders, L. M., Hendren, R. W. (eds.), *Protein Delivery: Physical Systems*, Plenum Press, New York, pp. 1–41.

Crommelin, D. J. and Storm, G. (2003). Liposomes: from the bench to the bed. *Journal of Liposome Research* **13**: 33–36.

Dash, A. K., Haney, P. W., and Garavalia, M. J. (1999). Development of an in vitro dissolution method using microdialysis sampling technique for implantable drug delivery systems. *Journal of Pharmaceutical Sciences* **88**: 1036–1040.

Diaz, R. V., Llabres, M., and Evora, C. (1999). One-month sustained release microspheres of 125I-bovine calcitonin. In vitro-in vivo studies. *Journal of Controlled Release* **59**: 55–62.

Faisant, N., Siepmann, J., and Benoit, J. P. (2002). PLGA-based microparticles: elucidation of mechanisms and a new, simple mathematical model quantifying drug release. *European Journal of Pharmaceutical Sciences* **15**: 355–366.

FDA Draft Guidance: Liposome Drug Products. (August 2002).

Flynn, G. L., Yalkowsky, S. H., and Roseman, T. J. (1974). Mass transport phenomena and models: theoretical concepts. *Journal of Pharmaceutical Sciences* **63**: 479–510.

Frechet, J. M. J. and Tomalia, D. A. (2002). *Dendrimers and other Dendritic Polymers*, Wiley, Chichester, UK.

Galeska, I., Kim, T.-K., Patil, S., Bhardwaj, U., Chatttopadhyay, D., Papadimitrakopoulos, F., and Burgess, D. J. (2005). Controlled release of dexamethasone from plga microspheres embedded within polyacid-containing PVA hydrogels. *AAPS Journal* 7: Article 22.

Gillies, E. R. and Frechet, J. M. (2002). Designing macromolecules for therapeutic applications: polyester dendrimer-poly(ethylene oxide) "bow-tie" hybrids with tunable molecular weight and architecture. *Journal of the American Chemical Society* **124**: 14137–14146.

Gillies, E. R., Dy, E., Frechet, J. M. J., and Szoka, F. C. (2005). Biological evaluation of polyester dendrimer: poly(ethylene oxide) "bow-tie" hybrids with tunable molecular weight and architecture. *Molecular Pharmaceutics* **2**: 129–138.

Gopferich, A. (1996). Polymer degradation and erosion: mechanisms and applications. *European Journal of Pharmaceutics and Biopharmaceutics* **42**: 1–11.

Hakkarainen, M., Albertsson, A.-C., and Karlsson, S. (1996). Weight losses and molecular weight changes correlated with the evolution of hydroxyacids in simulated in vivo degradation of homo- and copolymers of PLA and PGA. *Polymer Degradation and Stability* **52**: 283–291.

Herben, V. M., ten Bokkel Huinink, W. W., Schot, M. E., Hudson, I., and Beijnen, J. H. (1998). Continuous infusion of low-dose topotecan: pharmacokinetics and pharmacodynamics during a phase II study in patients with small cell lung cancer. *Anti-Cancer Drugs* **9**: 411–418.

Heya, T., Okada, H., Ogawa, Y., and Toguchi, H. (1994a). In vitro and in vivo evaluation of thyrotrophin releasing hormone release from copoly(dl-lactic/glycolic acid) microspheres. *Journal of Pharmaceutical Sciences* **83**: 636–640.

Heya, T., Mikura, Y., Nagai, A., Miura, Y., Futo, T., Tomida, Y., Shimizu, H., and Toguchi, H. (1994b). Controlled release of thyrotropin releasing hormone from microspheres: evaluation of release profiles and pharmacokinetics after subcutaneous administration. *Journal of Pharmaceutical Sciences* **83**: 798–801.

Hochster, H., Liebes, L., Speyer, J., Sorich, J., Taubes, B., Oratz, R., Wernz, J., Chachoua, A., Raphael, B., and Vinci, R. Z., et al. (1994). Phase I trial of low-dose continuous topotecan infusion in patients with cancer: an active and well-tolerated regimen. *Journal of Clinical Oncology: Official Journal of the American Society of Clinical Oncology* **12**: 553–559.

Hoffman, A. S. (2001). Hydrogels for biomedical applications. *Annals of the New York Academy of Sciences* **944**: 62–73.

http://www.astrazeneca-us.com/pi/diprivan.pdf

http://www.gene.com/gene/products/information/opportunistic/nutropin-depot/insert.jsp

Jain, S., Jain, R., Chourasia, M., Jain, A., Chalasani, K., Soni, V., and Jain, A. (2005). Design and development of multivesicular liposomal depot delivery system for controlled systemic delivery of acyclovir sodium. *AAPS PharmSciTech* **06**: E35–E41.

Jevprasesphant, R., Penny, J., Jalal, R., Attwood, D., McKeown, N. B., and D'Emanuele, A. (2003). The influence of surface modification on the cytotoxicity of PAMAM dendrimers. *International Journal of Pharmaceutics* **252**: 263–266.

Jiang, G., Woo, B. H., Kang, F., Singh, J., and DeLuca, P. P. (2002). Assessment of protein release kinetics, stability and protein polymer interaction of lysozyme encapsulated poly(d,l-lactide-co-glycolide) microspheres. *Journal of Controlled Release* **79**: 137–145.

Jiang, G., Qiu, W., and DeLuca, P. P. (2003). Preparation and in vitro/in vivo evaluation of insulin-loaded poly(acryloyl-hydroxyethyl starch)-PLGA composite microspheres. *Pharmaceutical Research* **20**: 452–459.

Kadir, F., Oussoren, C., and Crommelin, D. J. (1999). Liposomal formulations to reduce irritation of intramuscularly and subcutaneously administered drugs. In Gupta, P. K., Brazeau, G. A. (eds.), *Injectable Drug Development. Techniques to Reduce Pain and Irritation*, Interpharm Press, Denver, Colorado, pp. 337–354.

Kim, T.-K. and Burgess, D. J. (2002). Pharmacokinetic characterization of ^{14}C-vascular endothelial growth factor controlled release microspheres using a rat model. *Journal of Pharmacy and Pharmacology* **54**: 897–905.

Kim, T. I., Seo, H. J., Choi, J. S., Jang, H. S., Baek, J. U., Kim, K., and Park, J. S. (2004). PAMAM-PEG-PAMAM: novel triblock copolymer as a biocompatible and efficient gene delivery carrier. *Biomacromolecules* **5**: 2487–2492.

Kobayashi, H. and Brechbiel, M. W. (2004). Dendrimer-based nanosized MRI contrast agents. *Current Pharmaceutical Biotechnology* **5**: 539–549.

Kukowska-Latallo, J. F., Bielinska, A. U., Johnson, J., Spindler, R., Tomalia, D. A., and Baker, Jr., J. R. (1996). Efficient transfer of genetic material into mammalian cells using Starburst polyamidoamine dendrimers. *Proceedings of the National Academy of Sciences of the United States of America* **93**: 4897–4902.

Kukowska-Latallo, J. F., Candido, K. A., Cao, Z., Nigavekar, S. S., Majoros, I. J., Thomas, T. P., Balogh, L. P., Khan, M. K., and Baker, Jr., J. R. (2005). Nanoparticle targeting of anticancer drug improves therapeutic response in animal model of human epithelial cancer. *Cancer Research* **65**: 5317–524.

Kulkarni, R. K., Moore, E. G., Hegyeli, A. F., and Leonard, F. (1971). Biodegradable poly (lactic acid) polymers. *Journal of Biomedical Materials Research* **5**: 169–181.
Kurihara, A., Shibayama, Y., Mizota, A., Yasuno, A., Ikeda, M., and Hisaoka, M. (1996). Pharmacokinetics of highly lipophilic antitumor agent palmitoyl rhizoxin incorporated in lipid emulsions in rats. *Biological & Pharmaceutical Bulletin* **19**: 252–258.
Lalloo, A., Chao, P., Hu, P., Stein, S., and Sinko, P. J. (2006). Pharmacokinetic and pharmacodynamic evaluation of a novel in situ forming poly(ethylene glycol)-based hydrogel for the controlled delivery of the camptothecins. *Journal of Controlled Release* **112**: 333–342.
Lee, C. C., MacKay, J. A., Frechet, J. M., and Szoka, F. C. (2005). Designing dendrimers for biological applications. *Nature Biotechnology* **23**: 1517–1526.
Lemaire, V., Belair, J., and Hildgen, P. (2003). Structural modeling of drug release from biodegradable porous matrices based on a combined diffusion/erosion process. *International Journal of Pharmaceutics* **258**: 95–107.
Lewis, D. H. (1990). *Controlled release of bioactive agents from lactide glycolide polymers*, Marcel Dekker, New York.
Li, S., Girard, A., Garreau, H., and Vert, M. (2000). Enzymic degradation of polylactide stereocopolymers with predominant D-lactyl contents. *Polymer Degradation and Stability* **71**: 61–67.
Liu, M., Kono, K., and Frechet, J. M. (2000). Water-soluble dendritic unimolecular micelles: their potential as drug delivery agents. *Journal of Controlled Release* **65**: 121–131.
Liu, F. I., Kuo, J. H., Sung, K. C., and Hu, O. Y. (2003). Biodegradable polymeric microspheres for nalbuphine prodrug controlled delivery: in vitro characterization and in vivo pharmacokinetic studies. *International Journal Pharmaceutics* **257**: 23–31.
Mader, K., Bittner, B., Li, Y., Wohlauf, W., and Kissel, T. (1998). Monitoring microviscosity and microacidity of the albumin microenvironment inside degrading microparticles from poly(lactide-co-glycolide) (PLG) or ABA-triblock polymers containing hydrophobic poly(lactide-co-glycolide) A blocks and hydrophilic poly(ethyleneoxide) B blocks. *Pharmaceutical Research* **15**: 787–793.
Maeda, H., Sawa, T., and Konno, T. (2001). Mechanism of tumor-targeted delivery of macromolecular drugs, including the EPR effect in solid tumor and clinical overview of the prototype polymeric drug SMANCS. *Journal of Controlled Release* **74**: 47–61.
Magenheim, B., Levy, M. Y., and Benita, S. (1993). A new in vitro technique for the evaluation of drug release profile from colloidal carriers – ultrafiltration technique at low pressure. *International Journal of Pharmaceutics* **94**: 115–123.
Makino, K., Ohshima, H., and Kondo, T. (1986). Mechanism of hydrolytic degradation of poly(lactide) microcapsules: effects of pH, ionic strength and buffer concentration. *Journal of Microencapsulation* **3**: 203–212.
Malik, N., Wiwattanapatapee, R., Klopsch, R., Lorenz, K., Frey, H., Weener, J. W., Meijer, E. W., Paulus, W., and Duncan, R. (2000). Dendrimers: relationship between structure and biocompatibility in vitro, and preliminary studies on the biodistribution of 125I-labelled polyamidoamine dendrimers in vivo. *Journal of Controlled Release* **65**: 133–148.
McDonald, S., Faibushevich, A. A., Garnick, S., McLaughlin, K., and Lunte, C. (2002). Determination of local tissue concentrations of bupivacaine released from biodegradable microspheres and the effect of vasoactive compounds on bupivacaine tissue clearance studied by microdialysis sampling. *Pharmaceutical Research* **19**: 1745–1752.

Moghimi, S. M., Hunter, A. C., and Murray, J. C. (2001). Long-circulating and target-specific nanoparticles: theory to practice. *Pharmacological Reviews* **53**: 283–318.

Morita, T., Sakamura, Y., Horikiri, Y., Suzuki, T., and Yoshino, H. (2001). Evaluation of in vivo release characteristics of protein-loaded biodegradable microspheres in rats and severe combined immunodeficiency disease mice. *Journal of Controlled Release* **73**: 213–221.

Moussy, F., Kreutzer, D., Burgess, D., Koberstein, J., Papadimitrakopoulos, F., and Huang, S. (2003). US Patent: apparatus and method for control of tissue/implant interactions.

Negrin, C. M., Delgado, A., Llabres, M., and Evora, C. (2001). In vivo–in vitro study of biodegradable methadone delivery systems. *Biomaterials* **22**: 563–570.

Negrin, C. M., Delgado, A., Llabres, M., and Evora, C. (2004). Methadone implants for methadone maintenance treatment. In vitro and in vivo animal studies. *Journal of Controlled Release* **95**: 413–421.

Newkome, G. R., Moorefield, C. N., and Vogtle, F. (2001). *Dendrimers and Dendrons: Concepts, Syntheses, Applications*, Wiley-VCH, Weinheim, Germany.

Okada, H. (1997). One- and three-month release injectable microspheres of the LH–RH superagonist leuprorelin acetate. *Advanced Drug Delivery Reviews* **28**: 43–70.

PadillaDeJesus, O. L., Ihre, H. R., Gagne, L., Frechet, J. M. J., and Szoka, F. C. (2002). Polyester dendritic systems for drug delivery applications: in vitro and in vivo evaluation. *Bioconjugate Chemistry* **13**: 453–461.

Patil, S. D., Papadimitrakopoulos, F., and Burgess, D. J. (2004). Dexamethasone-loaded poly(lactic-co-glycolic) acid microspheres/poly(vinyl alcohol) hydrogel composite coatings for inflammation control. *Diabetes Technology & Therapeutics* **6**: 887–897.

Patri, A. K., Majoros, I. J., and Baker, J. R. (2002). Dendritic polymer macromolecular carriers for drug delivery. *Current Opinion in Chemical Biology* **6**: 466–471.

Patri, A. K., Kukowska-Latallo, J. F., and Baker, Jr., J. R. (2005). Targeted drug delivery with dendrimers: comparison of the release kinetics of covalently conjugated drug and non-covalent drug inclusion complex. *Advanced Drug Delivery Reviews* **57**: 2203–2214.

de la Pena, A., Liu, P., and Derendorf, H. (2000). Microdialysis in peripheral tissues. *Advanced Drug Delivery Reviews* **45**: 189–216.

Peppas, N. A., Bures, P., Leobandung, W., and Ichikawa, H. (2000). Hydrogels in pharmaceutical formulations. *European Journal of Pharmaceutics and Biopharmaceutics* **50**: 27–46.

Perugini, P., Genta, I., Conti, B., Modena, T., Cocchi, D., Zaffe, D., and Pavanetto, F. (2003). PLGA microspheres for oral osteopenia treatment: preliminary "in vitro"/"in vivo" evaluation. *International Journal of Pharmaceutics* **256**: 153–160.

Porter, C. J. and Charman, S. A. (2000). Lymphatic transport of proteins after subcutaneous administration. *Journal of Pharmaceutical Sciences* **89**: 297–310.

Ritger, P. L. and Peppas, N. A. (1987). A simple equation for description of solute release: I. Fickian and non-Fickian release from non-swellable devices in the form of slabs, spheres, cylinders or discs. *Journal of Controlled Release* **5**: 23–36.

Ruel-Gariepy, E., Leclair, G., Hildgen, P., Gupta, A., and Leroux, J. C. (2002). Thermosensitive chitosan-based hydrogel containing liposomes for the delivery of hydrophilic molecules. *Journal of Controlled Release* **82**: 373–383.

Shabbits, J. A., Chiu, G. N., and Mayer, L. D. (2002). Development of an in vitro drug release assay that accurately predicts in vivo drug retention for liposome-based delivery systems. *Journal of Controlled Release* **84**: 161–170.

Shenderova, A., Burke, T. G., and Schwendeman, S. P. (1999). The acidic microclimate in poly(lactide-co-glycolide) microspheres stabilizes camptothecins. *Pharmaceutical Research* **16**: 241–248.

Siepmann, J. and Gopferich, A. (2001). Mathematical modeling of bioerodible, polymeric drug delivery systems. *Advanced Drug Delivery Reviews* **48**: 229–247.

Simoes, S., Moreira, J. N., Fonseca, C., Duzgunes, N., and de Lima, M. C. (2004). On the formulation of pH-sensitive liposomes with long circulation times. *Advanced Drug Delivery Reviews* **56**: 947–965.

Takino, T., Konishi, K., Takakura, Y., and Hashida, M. (1994). Long circulating emulsion carrier systems for highly lipophilic drugs. *Biological & Pharmaceutical Bulletin* **17**: 121–125.

ten Tije, A. J., Verweij, J., Loos, W. J., and Sparreboom, A. (2003). Pharmacological effects of formulation vehicles: implications for cancer chemotherapy. *Clinical Pharmacokinetics* **42**: 665–685.

Torchilin, V. P. (2005). Recent advances with liposomes as pharmaceutical carriers. *Nature Reviews Drug Discovery* **4**: 145–160.

Torchilin, V. P. and Levchenko, T. S. (2003). TAT-liposomes: a novel intracellular drug carrier. *Current Protein & Peptide Science* **4**: 133–140.

Torchilin, V. P. and Lukyanov, A. N. (2003). Peptide and protein drug delivery to and into tumors: challenges and solutions. *Drug Discovery Today* **8**: 259–266.

Twyman, L. J., Beezer, A. E., Esfand, R., Hardy, M. J., and Mitchell, J. C. (1999). The synthesis of water soluble dendrimers, and their application as possible drug delivery systems. *Tetrahedron Letters* **40**: 1743–1746.

Ueda, K., Ishida, M., Inoue, T., Fujimoto, M., Kawahara, Y., Sakaeda, T., and Iwakawa, S. (2001). Effect of injection volume on the pharmacokinetics of oil particles and incorporated menatetrenone after intravenous injection as O/W lipid emulsions in rats. *Journal of Drug Targeting* **9**: 353–360.

Ueda, K., Yamazaki, Y., Noto, H., Teshima, Y., Yamashita, C., Sakaeda, T., and Iwakawa, S. (2003). Effect of oxyethylene moieties in hydrogenated castor oil on the pharmacokinetics of menatetrenone incorporated in O/W lipid emulsions prepared with hydrogenated castor oil and soybean oil in rats. *Journal of Drug Targeting* **11**: 37–43.

van Dijkhuizen-Radersma, R., Wright, S. J., Taylor, L. M., John, B. A., de Groot, K., and Bezemer, J. M. (2004). In vitro/in vivo correlation for 14C-methylated lysozyme release from poly(ether-ester) microspheres. *Pharmaceutical Research* **21**.

Wang, F., Bronich, T. K., Kabanov, A. V., Rauh, R. D., and Roovers, J. (2005). Synthesis and evaluation of a star amphiphilic block copolymer from poly(epsilon-caprolactone) and poly(ethylene glycol) as a potential drug delivery carrier. *Bioconjugate Chemistry* **16**: 397–405.

Yang, H. and Lopina, S. T. (2006). In vitro enzymatic stability of dendritic peptides. *Journal of Biomedical Materials Research A* **76**: 398–407.

Yenice, I., Calis, S., Atilla, B., Kas, H. S., Ozalp, M., Ekizoglu, M., Bilgili, H., and Hincal, A. A. (2003). In vitro/in vivo evaluation of the efficiency of teicoplanin-loaded biodegradable microparticles formulated for implantation to infected bone defects. *Journal of Microencapsulation* **20**: 705–717.

Young, D., Farrell, C., and Shepard, T. (2005). In vitro/in vivo correlation for modified release injectable drug delivery systems. In Burgess, D. J. (ed.), *Injectable Dispersed Systems: Formulation, Processing and Performance*, Vol. 149, Taylor & Francis, Boca Raton, pp. 159–176.

Zhong, H., Deng, Y., Wang, X., and Yang, B. (2005). Multivesicular liposome formulation for the sustained delivery of breviscapine. *International Journal of Pharmaceutics* **301**: 15–24.

Zolnik, B. S. (2005). In vitro and in vivo release testing of control release parenteral microspheres. *PhD Dissertation.*

Zolnik, B. S., Raton, J. L., and Burgess, D. J. (2005). Application of USP Apparatus 4 and in situ fiber optic analysis to microsphere release testing. *Dissolution Technologies* **12**: 11–14.

Zolnik, B. S., Leary, P. E., and Burgess, D. J. (2006). Elevated temperature accelerated release testing of PLGA microspheres. *Journal of Controlled Release* **112**: 293–300.

Zuidema, J., Kadir, F., Titulaer, H. A. C., and Oussoren, C. (1994). Release and absorption rates of intramuscularly and subcutaneously injected pharmaceuticals (II). *International Journal of Pharmaceutics* **105**: 189–207.

12
In Vitro–In Vivo Correlation in Dosage Form Development: Case Studies

Shoufeng Li, Alan E. Royce, and Abu T. M. Serajuddin

12.1 Introduction

In vitro and *in vivo* correlation (IVIVC) refers to a predictive relationship of the *in vitro* properties of drug substances or dosage forms with their *in vivo* performance. For orally administered drug products, it is usually a correlation between the extent or rate of dissolution of a dosage form and its pharmacokinetic parameters, such as rate, duration, and extent of drug absorption. The physicochemical properties of dosage forms influence their *in vivo* performance in many different ways (Li *et al.*, 2005). Through the establishment of a definitive relationship between certain physicochemical properties of a dosage form with the *in vivo* appearance of its active component, one can establish *in vitro* testing criteria which will predict, its *in vivo* performance. Although IVIVC may be applied to many different types of dosage forms, including topical patches, various injectable forms like microparticulates and depot systems, and different inhalation formulations, the primary purpose of this chapter is to illustrate how the IVIVC concept can be applied to the development of oral dosage forms. IVIVC of oral dosage forms can also help in setting dissolution specifications and in applying *in vitro* data as surrogates for bioequivalence testing in case of certain pre- and postapproval changes (Center for Drug Evaluation and Research, US FDA, 1997). However, the drug product development is a continuous process with increasing physicochemical and pharmacokinetic data being available as it progresses from the early-stage to the late-stage development including life cycle management (LCM). For this reason, the IVIVC should also be a continuous process with more predictability built into it as the product development progresses.

IVIVC has normally been studied for prototype formulations or finished dosage forms, where it involves at least two different formulations and a reference treatment, such as a solution or immediate-release formulation. The formulation properties used must have significantly different *in vitro* or *in vivo* profiles (>10%). The correlation established based on such studies may be categorized into Levels A, B, and C (United States Pharmacopeial Convention, Inc., 1988), and these categories have been discussed in detail in other chapters in this book. Briefly, a Level A IVIVC is generally linear and represents a point-to-point relationship between *in vitro* dissolution rate and *in vivo* input rate (e.g., the *in vivo* dissolution rate of the drug from the dosage form). A Level A relationship may generate predictable plasma profile, including C_{max}, AUC, T_{max}, shape of profile and elimination half-life. It is most useful in product development since it may be used for biowaiver based on *in vitro* dissolution data to qualify changes occurred during manufacture. In a Level B IVIVC, the mean *in vitro* dissolution time is compared either to the mean residence time or the mean dissolution time in the gastrointestinal fluid. However, the entire plasma concentration profile cannot be predicted based on *in vitro* dissolution data, and, therefore, the benefit of a Level B correlation is limited and it is not accepted by Health Agencies for biowaivers.

A Level C IVIVC establishes a single point relationship between a dissolution parameter and a pharmacokinetic parameter, and it does not reflect the complete shape of the plasma concentration versus time curve. A typical example of a Level C IVIVC is the establishment of the correlation between amount of drug released at a certain time point and C_{max}. Level C correlation can be useful to rank order different formulation principles; however, unless a multiple Level C correlation can be established, the usefulness of Level C correlation in predicting full *in vivo* performance may be limited.

Although the IVIVC of drug products based on above guidelines is important to establish specifications for drug products and meet regulatory requirements for product approvals, Li *et al.* (2005) argued that in order to optimize performance of oral dosage forms, the possible relationship between *in vitro* physicochemical attributes and *in vivo* performance must be considered in all phases of drug product development. The product should starting from the identification of new chemical entities for development and continue to the approval of drug products for marketing and even through postapproval changes. Unless the *in vitro–in vivo* relationships are built into the development of drug products, methods such as dissolution testing may not be relevant to the *in vivo* performance of drug products. To address this issue, a four-tier approach, as described below, may be applied for during drug product development process.

12.2 IVIVC in Drug Product Development: A Four-Tier Approach

The application of IVIVC is an evolving process during product development. The scope of IVIVC changes as a NCE progresses from drug discovery phase to early development, full development, and finally to LCM. As shown in Scheme 12.1, the present authors believe a four-tier approach to IVIVC during different phases of drug product development based on the availability of physicochemical and pharmacokinetic data should be utilized. How these tiers may be related to different

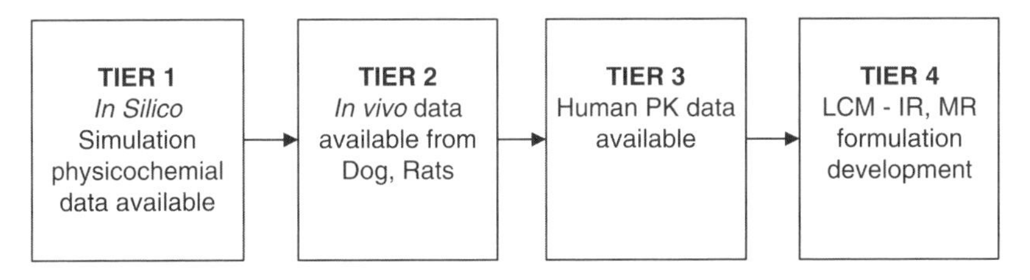

SCHEME 12.1. A four-tier approach to IVIVC. LCM, life cycle management; IR, immediate release; MR, modified release

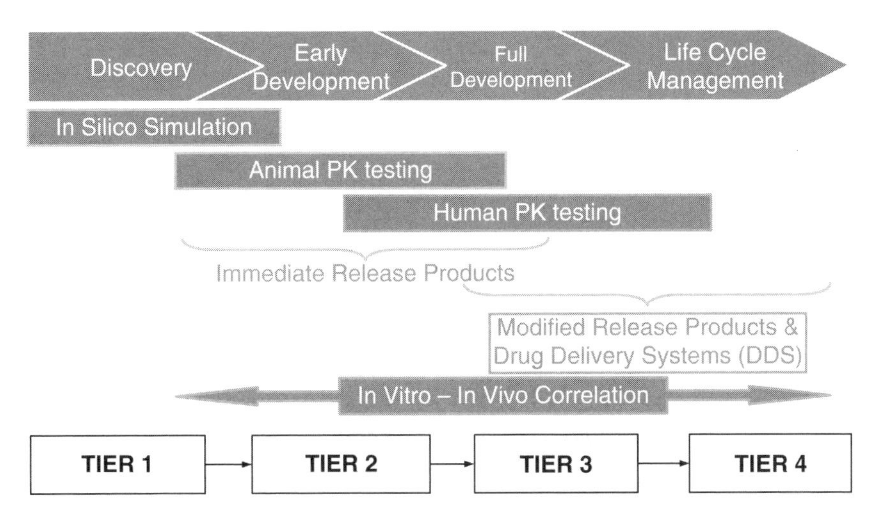

FIGURE 12.1. Discovery and development phases of new chemical entity (NCE) and application of IVIVC in drug development process

stages of drug product development is depicted in Fig. 12.1, and the scopes of IVIVC at different tiers are discussed below.

Tier 1. This tier usually encompasses the drug discovery and the early preclinical phases where the PK data for a compound is not available. Although many *in vivo* studies are carried out in the drug discovery stage, the main focus at this time is to determine pharmacological efficacy and safety of the compound. Since a large number of compounds are studied at this stage to select an NCE for development and only limited physicochemical data are available, the IVIVC at this time is usually conducted through *in silico* simulation of structural properties of a molecule or high-throughput experimental data generated for initial characterization of a compound. The correlation helps in establishing whether a compound is developable or not (Venkatesh and Lipper, 2000; Pudipeddi *et al.*, 2006).

Tier 2. Following the selection of a NCE and during preclinical development leading to the initiation of proof of concept (PoC) or Phase I studies, pharmacokinetic studies in animal models are carried out with emphasis on evaluating biopharmaceutical properties of the NCE and dosage form design strategies. In addition, different physical forms, salts and the particle size of the drug substance can be tested in animal models, which provides the first opportunity to correlate the *in vitro* measurement, i.e., dissolution of the drug substance, with its *in vivo* performance, such as C_{max}, AUC, or *in vivo* absorption profiles. The available PK data also provide the first opportunity to develop a biorelevant dissolution method for the compound. Li *et al.* (2005) proposed a decision tree for selecting dissolution media for compounds of different BCS categories. As shown in Scheme 12.2, the decision tree may be simplified for selecting appropriate dissolution media during a Tier 2 IVIVC in early development. This proposal includes testing of different formulations with different *in vitro* drug release characteristics, such as different salt forms, different particle size distributions, or different formulation principles, in both *in vitro* and *in vivo* settings, and then feed all of the *in vitro* and *in vivo* data into a simulation software, such as GastroPlus™ (Simulations

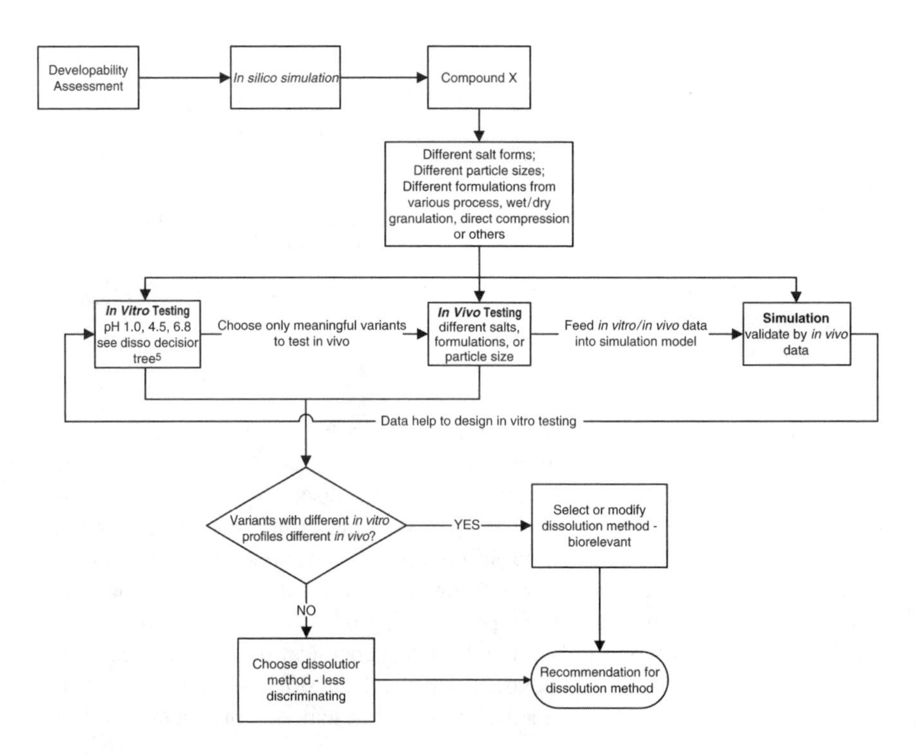

SCHEME 12.2. Decision tree for implementing IVIVC in early product development

Plus, Lancaster, CA). Since formulation development and optimization at later stages of drug product development often rely on the dissolution method established at an earlier stage, it is essential that dissolution method selection in Tier 2 is as meaningful and as biorelevant as possible. If the PK results of different formulations with similar dissolution profiles turn out to be similar, a less discriminating dissolution method may be used. On the other hand, a discriminating and biorelevant dissolution method may be developed if the PK results of different formulations are also different.

Tier 3. The validity of the dissolution method developed in *Tier 2* can be further studied in full development, once human pharmacokinetic data are available. Analysis of all available PK and *in vitro* data must be carried out to determine whether the method needs modification for a better IVIVC. This involves a cross-functional team of scientists from areas of formulation development, dissolution testing, clinical pharmacokinetic testing, and PK modeling.

Tier 4. In the LCM stage of drug development, a large amount of clinical pharmacokinetic data are already available. An IVIVC based on such data has to be considered as part of the development strategy for LCM dosage forms such as modified release oral products or alternative delivery systems including parenteral depot, transdermal patch, etc. The FDA guideline for modified release oral dosage forms clearly defines acceptance criteria for successful IVIVC (Center for Drug Evaluation and Research, 1997).

Case studies at different tiers of *in vitro–in vivo* correlation are presented in the following section.

12.3 Case Studies

12.3.1 Tier 1 – Discovery and Early Preclinical Development: Assessing Developability and Formulation Principles

During the discovery and the early preclinical development stages of a compound intended for oral administration, the major concern is whether it would have acceptable oral absorption or not, since this would determine its developability. The first IVIVC assessment conducted at this time is the classification of the compound according to the Biopharmaceutical Classification System (BCS) proposed by Amidon *et al.* (1995). This identifies potential hurdles in the drug product development and indicates whether any special dosage form design or drug delivery considerations would be necessary. High-throughput methods are in place in most pharmaceutical companies for the generation of experimental data, such as solubility, partition coefficient, membrane permeability, etc., to enable BCS classification. Stability of compounds to certain enzymes present in gut and plasma that could be indicative of their *in vivo* performance is often determined at this time. There are numerous reports in the literature on the application of BCS classification in the early phase of drug development.

Another simplistic but useful approach in predicting potential drug absorption issues is the calculation of maximum absorbable dose (MAD) (Curatolo, 1998) which has been presented elsewhere in this book:

$$\mathrm{MAD} = S \times K_{\mathrm{a}} \times \mathrm{SIWV} \times \mathrm{SITT},$$

where S is solubility (mg/mL) at pH 6.5, K_{a} is transintestinal absorption rate constant (min^{-1}) based on rat intestinal perfusion experiment, SIWV is small intestinal water volume (250 mL), and SITT is the small intestinal transit time (4 h). Chiou *et al.* (Chiou *et al.* 2000) demonstrated that there is a good correlation between absorption rates in humans and rats with a slope near unity. One limitation of the MAD calculation is that only the aqueous solubility in pH 6.5 buffer is taken into consideration, and no considerations are made for possible solubility enhancement by bile salts, surfactants, lypolytic products, etc., present in GI fluids. Although the generation of rat perfusion data in the early development phase is very helpful for the purpose of IVIVC, another limitation could be that the absorption rate constants in rats may not be determined or reliable values may not be available for the calculation of MAD. As shown by Curatolo (1998), two different estimated K_{a} values, one high and one low, may be used under such a circumstance, where MAD will be a dose range instead of a single value.

Both the BCS classification and the MAD calculation rely on the solubility of a drug substance. Although the dissolution rate is dependent on both solubility and surface area, the surface area of drug substance, which is dependent on particle size, is not considered. The effect of particle size on *in vivo* drug absorption has been discussed by Johnson (1996), where the effect of particle size on absorption over a range of important variables, including dose, solubility and absorption rate constant, was simulated. For example, with a fixed absorption rate constant of 0.001 min^{-1}, the relationship between dose and solubility as a function of particle size change could be simulated. In general, the relative effect of particle size on the percent of dose absorbed decreases with an increase in solubility, and particle size becomes practically irrelevant for drugs at a solubility of 1 mg/mL for a dose of 1 mg. Again, when reliable values for absorption rate constants are not available, multiple K_{a} values may be applied for a general estimate of the relative influence of particle size.

The proper *in vivo* assessment of oral absorption requires much time and resource. Therefore, the PK data for IVIVC, whether animal or human, are not usually available during early preclinical development. For this reason, it is desirable to utilize *in silico* modeling approaches for an early assessment of absorbability and to extrapolate results to situations where experimental data are not available. As reported earlier (Li *et al.*, 2005), there are currently two *in silico* approaches for the prediction of oral absorption, statistical models and mechanism-based models. The former is based on a statistical relationship between inputs, typically molecular descriptors derived from molecular structures, and outputs, which could be estimates of oral absorption. Mechanism-based models rely on a good understanding of absorption processes including physiology, GI dissolution, transit, and permeation. The following are a few case studies based on the experience of present authors on mechanism-based modeling; the compounds used are depicted as numbers since detailed chemical structures are not relevant to the objective of this paper and, in some cases, the results are proprietary in nature.

One mechanism-based model that gained popularity in recent years is GastroPlus™ (SimulationsPlus, Lancaster, CA), which simulates and models the gastrointestinal absorption processes based on an "Advanced Compartmental Absorption and Transit" (ACAT) model (Yu *et al.*, 1996a). Inputs to the software include (Li *et al.*, 2005; Agoram *et al.*, 2001):

(a) Oral dose
(b) Physicochemical properties (solubility-pH profile, intestinal permeability, etc.)
(c) Physiological properties (species, GI transit time, GI pH, food status, etc.)
(d) Formulation properties (release profile, particle size, etc.)
(e) Pharmacokinetic parameters (V_d, CL, microscopic kinetic rate constants, etc.) (optional)

All input parameters in the GastroPlus™ simulation are under user control, and the values may change at different stages of the development of a compound. At discovery and preclinical development stages, some of the values could be estimates or based on library data for similar compounds, or from *in silico* calculation.

The output includes:

(a) Fraction of oral dose absorbed (concentration–time profiles in all GI compartments; fractions absorbed from each GI compartment)
(b) Parameter sensitivities (to answer "what-if" questions)
(c) Plasma concentration–time profiles (if the PK parameters are provided) (optional)

In one relatively simple application of GastroPlus™, it was asked whether or not the mean particle size (D_{50}) requirement of Compound I (aqueous solubility: >100 mg/mL) may be relaxed from 35 μm to approximately 100 μm without affecting its oral bioavailability. A simulation suggested that the extent of absorption is not sensitive to changes in particle size in the range of 35–250 μm (Fig. 12.2). This facilitated decision making with respect to dosage form design of Compound I without having to resort to *in vivo* experiments and focused the attention of formulators on the impact of particle size on processibility of dosage forms rather than their bioavailability.

Although simulation is not a replacement for definitive scientific experiments, it provides valuable insight on what one would expect *in vivo* based on the physicochemical properties of a compound. For instance, during the development of a neutral compound (Compound II) with a solubility of 1 μg/mL and high effective permeability (3.0×10^{-4} cm/s), a simulation at different doses and different particle sizes (Fig. 12.3) provided insight into possible strategies for the development of its dosage forms. At a dose of 100 mg, absorption of the compound

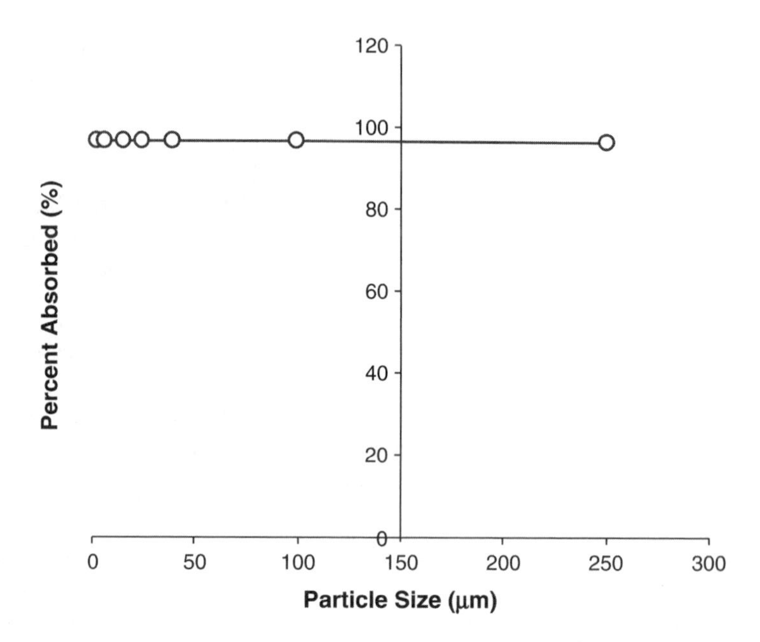

FIGURE 12.2. Simulation on the impact of particle size (35–250 μm) on oral absorption of compound I using GastroPlus^{TM}. Compound I has high aqueous solubility of >100 mg/mL

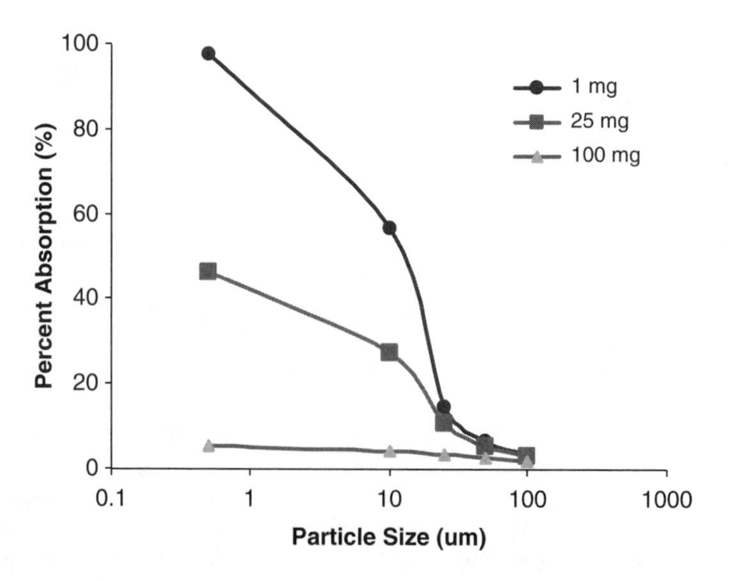

FIGURE 12.3. Simulation for percent absorption of Compound II at doses of 1, 25, and 100 mg over particle size range of 0.5–100 μm. Compound II is a neutral molecule with a pH-independent aqueous solubility of 1 μg/mL

is practically negligible over the particle size range of 0.5–100 μm, indicating that solubility is the limiting factor in absorption of the compound at such a dose and particle size reduction would not be helpful in improving absorption. However, at a dose of 1 mg, a dramatic shift in the dependency of absorption from solubility to particle size is evident from the simulation. Here, the use of micronization or nanoparticle system could provide an advantage. Ultimately, a solid dispersion whereby the drug was liberated in aqueous media as submicron particles was selected for its dosage form development in the expected dosage range from 5 to 80 mg (Dannenfelser *et al.*, 2004).

Although one may not be able to obtain an accurate estimate of dose of a new chemical entity (NCE) until very late in development, formulation scientists could utilize simulation data in a number of ways. At a relatively low dose range, particle size reduction and improvement of wetting properties of drug substances may be quite effective. However, if the molecule is ionizable (for example, a weak base), one may like to choose a salt form that could provide much higher dissolution rate than the free base.

12.3.2 Tier 2 – Preclinical Product Development: Selection of a Meaningful Dissolution Method

One of the questions that are often asked is whether different release properties *in vitro* will result in different *in vivo* absorption rates. It is not uncommon that different formulation approaches such as dry blend/direct compression, dry granulation (roller compaction), or wet granulation may result in formulations with different *in vitro* release characteristics. As shown in Fig. 12.4, Compound III had different *in vitro* release profiles when dry blend and wet granulation formulations were tested *in vitro*. In early time points, the difference was as large as 30%. However, when these formulations were tested *in vivo* in dog, comparable pharmacokinetic profiles were achieved (Fig. 12.5), indicating that the difference in *in vitro* dissolution would not have any significant impact on the *in vivo* performance of the drug product. This would save an enormous amount of time and effort by avoiding later development of a method to match the dissolution profile of the original formulation. In addition, this type of information will be very useful in selecting a biorelevant dissolution method; in this case, a much less discriminating method would be optimal.

As the development phase moves forward, typically to PoC (Proof of Concept) or Phase I stage, the lead candidate is characterized more thoroughly for its physicochemical properties and developability. A preclinical PK study in animal model is typically performed prior to the Phase I clinical trial with problematic compounds. A well-designed preclinical PK study, with input from formulation and clinical experts, may provide an opportunity to set the parameters for drug substance properties, such as particle size, salt forms as well as for formulations. Combined with simulation effort, it would also provide an opportunity to identify the rate-limiting factors for absorption. Further, *in vivo* data obtained at this stage

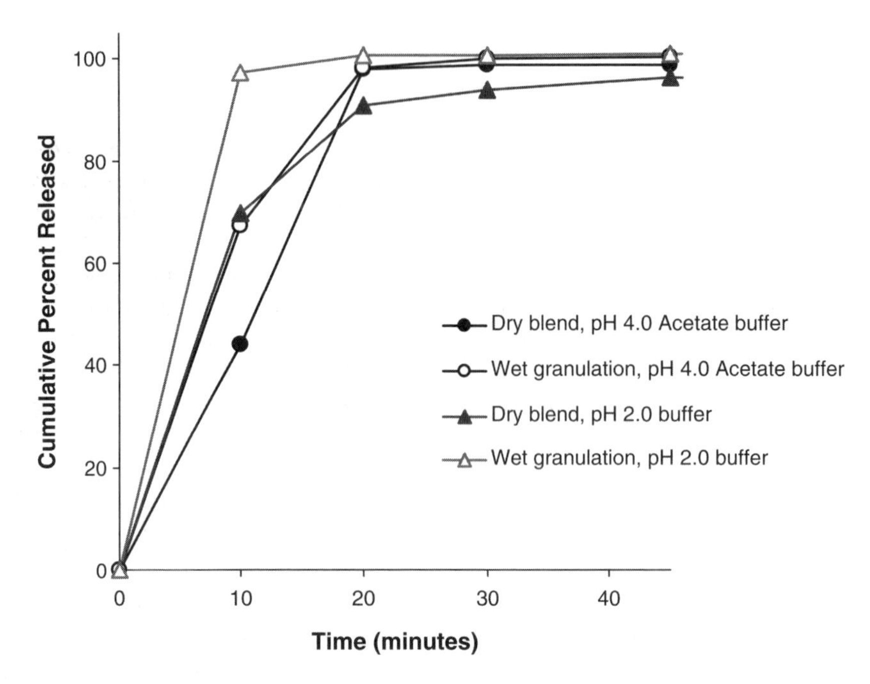

FIGURE 12.4. *In vitro* dissolution profiles of compound III capsule formulations at 0.01N HCl (pH 2.0) and 0.05 M acetate buffer (pH 4.5) using USP I basket method at 100 rpm (37 °C)

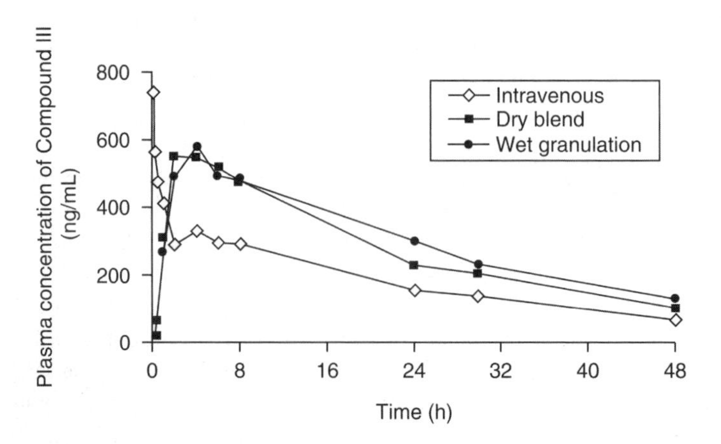

FIGURE 12.5. *In vivo* pharmacokinetic profiles ($C_p - t$) of Compound III in dogs for wet granulation and dry blend formulations at 10 mg/kg. The plasma profile after i.v. administration at 3 mg/kg is also shown

may be used to better justify the development of a biorelevant dissolution method. Among factors that determine the rate and extent of drug absorption following oral administration, dissolution of the solid drug into solution is of primary importance in the drug release/absorption process. Factors affecting drug dissolution has been extensively reviewed by Horter and Dressman (1997) as well as Li *et al.* (2005).

Prediction of the *in vivo* performance of weak bases and their salts could be challenging due to the kinetic nature of the dissolution of salts and the potential for precipitation into their free base forms. In another example, formulations of the free base and different salt forms of a weakly basic compound (Compound IV) were tested in dogs for bioavailability. The formulations tested were an intraveneous formulation containing 3 mg/mL DiHCl salt in 20% HP-β-CD, an oral 0.5% CMC suspension of free base at 2 mg/mL formulation, and drug blend capsule formulations of dihydrochloride (diHCl) and tartrate salts (Table 12.1). The results are summarized in Fig. 12.6. Absolute bioavailability of the diHCl salt had

TABLE 12.1. Summary of formulations administered to dogs for Compound IV

Route	Dose[a] (mg/kg)	Compound I	Volume/no. capsule	Concentration	Formulation
Intravenous	3	di HCl salt	1 mL/kg	3 mg/mL	Solution in 20% hydroxypropyl-beta-cyclodextrin aqueous solution
Oral	10	Free base	5 mL/kg	2 mg/mL	Suspension in 0.5% CMC aqueous solution
Oral	10	diHCl salt	1 cap/dog	Not applicable	Powder-in-capsule
Oral	10	Tartrate salt	1 cap/dog	Not applicable	Powder-in-capsule

[a]All doses are expressed as free base equivalent

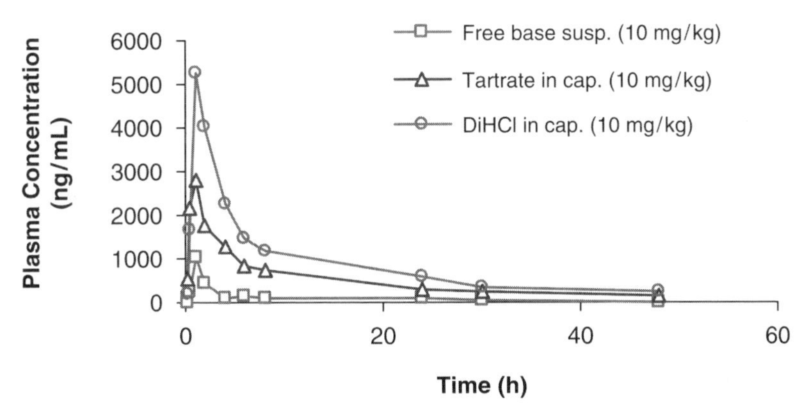

FIGURE 12.6. *In vivo* pharmacokinetic profiles ($C_p - t$) of free base and two different salt forms of Compound IV in dogs: free base suspension, capsule formulation of dihydrochloride (diHCl) and tartrate salt at 10 mg/kg

a mean value of 84%, indicating close to complete absorption. Percent bioavailability of the salt forms and free base suspension was in the order of diHCl (84%) > tartrate (48%) > free base (12%). Both salt forms of Compound IV clearly demonstrated their *in vivo* advantages over that of the free base, while the diHCl salt had higher bioavailability than the tartrate salt. When *in vitro* dissolution profiles of the three oral formulations of Compound IV, i.e., free base suspension, diHCl salt and tartrate salt capsules were determined at pH 2 (Fig. 12.7a),

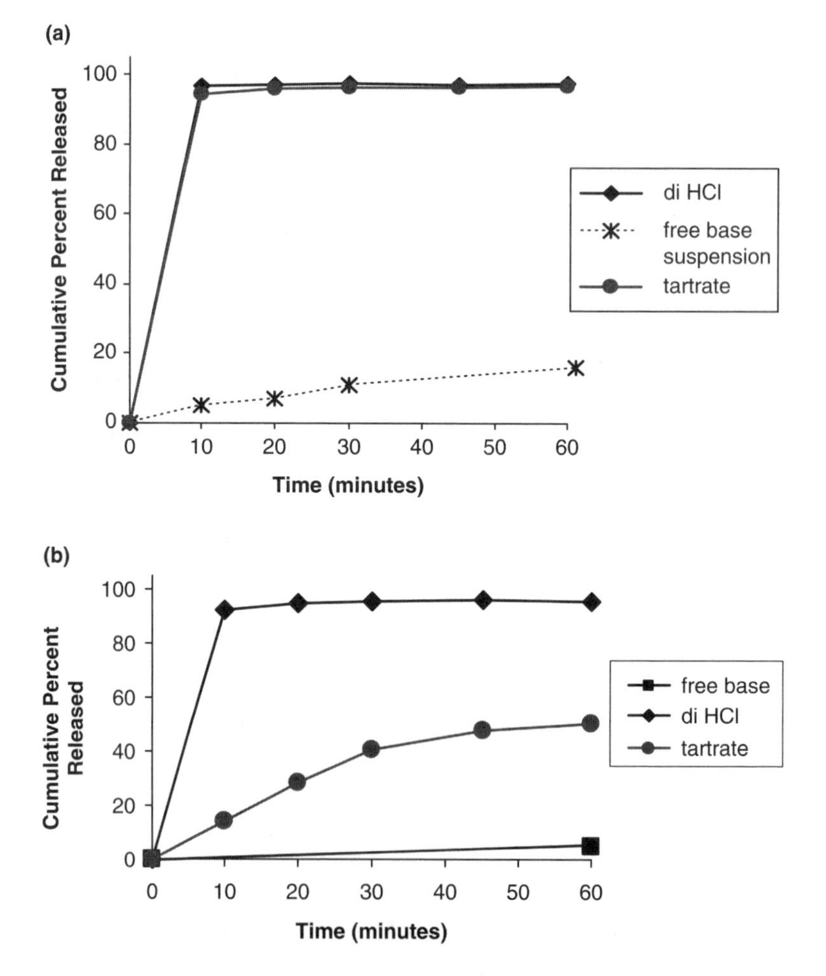

FIGURE 12.7. *In vitro* dissolution profiles of Compound IV suspension (free base) and capsule formulations (diHCl and tartrate salts) at (a) 0.01N HCl (pH 2.0) and (b) 0.05 M Acetate buffer (pH 4.0) using USP I basket method at 100 rpm (37 °C)

comparable profiles are observed for diHCl and tartrate salts, whereas the release profile of the free base suspension was low (<20%). Given the *in vivo* difference observed among the diHCl, tartrate and free base suspension formulation, it warranted an *in vitro* method that could reflect the potential *in vivo* performance, i.e., a biorelevant method. When dissolution rate of these formulations was tested at different pHs, in particular, pH 4.0, distinctly different profiles were evident among the free base, diHCl and tartrate salt formulations (Fig. 12.7b). When amounts dissolved at pH 2 and pH 4 were compared with C_{max} and AUC of these formulations, pH 4 provided a close to linear correlation between the *in vitro* percent released at 60 min and *in vivo* AUC or C_{max} (Fig. 12.8). Therefore, pH 4 was considered to be a biorelevant method since it adequately predicts key parameters for *in vivo* performance. Level C correlation could be established for Compound IV using dissolution method at pH 4. The diHCl salt was selected for development, and no further optimization was performed since Compound IV was only at preclinical stage. Later into development, when there was a potential for lowering the dose for Compound IV, the question was raised whether a salt form was still needed. The above biorelevant dissolution method then provided useful information on the performance of the salt and free base at a much lower dose. This exemplifies the needs

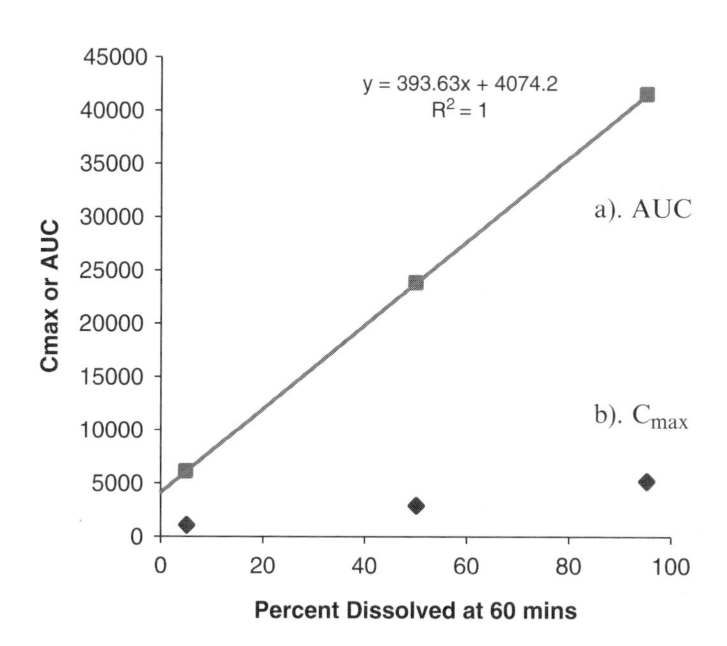

FIGURE 12.8. Level C correlation between *in vivo* plasma C_{max}, AUC, and *in vitro* dissolution of percent released at 60 min for Compound IV: (a) AUC; (b) C_{max}

for developing an IVIVC at early stage of development and its application further into development. The dissolution method may be further optimized based on deconvoluted profiles from the human PK study when such data become available.

12.3.3 Tier 3 – Full Development: Deconvolution of Human Pharmacokinetic Data and Comparison with In Vitro Dissolution Data

Development of IVIVC at the full development stage may be achieved by the following procedure:

1. Mathematically deconvolute *in vivo* plasma concentration profile using model-dependent or model-independent approach into *in vivo* absorption/dissolution profile
2. Compare *in vivo* dissolution profile with *in vitro* dissolution profile
3. Mathematically scale *in vivo* dissolution profile to match *in vitro* dissolution profile or modify *in vitro* dissolution condition if necessary
4. Establish IVIVC using modified *in vivo* or *in vitro* dissolution data.

Before providing a case history, some of the pharmacokinetic concepts inherent in the above procedure are reviewed below.

12.4 Deconvolution and Convolution

Deconvolution in IVIVC is a process where the output (plasma concentration versus time profile) is converted to the input (*in vivo* dissolution of the dosage form). The classical methods of deconvolution of plasma profiles include model-dependent methods, such as Wagner–Nelson (Turner *et al.*, 2004) and Loo–Riegelman (Langenbucher, 2002) and model independent method, such as the numerical deconvolution method (Frick *et al.*, 1998; Mahayni *et al.*, 2000).

The Wagner–Nelson method is a model-dependent method based on one-compartment model, which utilizes the elimination constant and has the advantage of not requiring additional *in vivo* data except oral plasma profile. As shown in equation below:

$$F_{\mathrm{abs}}(t) = \frac{C(t) + k_{\mathrm{e}}\mathrm{AUC}_{0-t}}{k_{\mathrm{e}}\mathrm{AUC}_{0-\infty}}, \tag{12.1}$$

where the fraction absorbed at different time points is estimated by a mass balance approach. Upper portion of (12.1) represents amount of drug in central compartment at time t and amount of drug eliminated in time t, whereas the bottom portion represents total amount of drug that is absorbed in central compartment. In contrast, the Loo–Riegelman method is based on two-compartment model, which requires intravenous dosing data.

Model-independent numerical deconvolution also requires *in vivo* plasma data from an oral solution or intravenous dose as unit impulse function for the application. As represented by (12.2) below:

$$c(t) = \int_0^t c_\delta(t-u) r_{\text{abs}}(u) \mathrm{d}u, \tag{12.2}$$

where the function C_δ in (12.2) represents the concentration–time course that would result from the instantaneous absorption of a unit amount of drug, and it is typically estimated from intravenous injection bolus data or reference oral solution data. In addition, $c(t)$ is the plasma concentration vs. time level of the tested oral formulation, r_{abs} is drug input rate of the oral solid dosage form, and u is variable of integration. In simple terms, the relationship between these terms can be represented as:

$$Y(t) = G(t)X(t), \tag{12.3}$$

where $Y(t)$ is the function describing the plasma concentration–time curve following extravascular administration, $G(t)$ is the function describing the concentration–time curve following bolus intravenous (or impulse) administration, and $X(t)$ is the function describing input, i.e., dissolution from the dosage form.

All three methods have their limitations, but the requirement of data in addition to oral plasma data from a tablet or capsule significantly limit the application of the later two methods.

Convolution in IVIVC is a process where the *in vitro* dissolution profile is converted to a plasma concentration profile (input to output). This can be done in a model-dependent or model-independent manner. In addition, physiology-based model and simulation software can be applied; it uses multiple differential equations which describe the well-characterized physical processes that occur during the controlled release, dissolution, transport, and absorption of drug materials in the gastrointestinal tract (Yu *et al.*, 1996b).

Scaling of data. Since significant difference exists between the *in vivo* and *in vitro* dissolution conditions, it is not uncommon to see time scale difference when comparing *in vivo* dissolution with *in vitro* dissolution profiles. Various approaches can be used to characterize and scale the data. The readers are referred to a book chapter by Li *et al.* (2007) for detailed discussion on characterization of the dissolution profiles.

The introduction of time scale factor is acceptable as long as the same time scale factor is being used for all formulations and for all further applications of the IVIVC model. The time scale factor can be determined by comparing *in vivo* and *in vitro* dissolution profiles or by plotting the time needed for *in vivo* dissolution versus the time needed for *in vitro* dissolution of a particular amount of drug from the dosage form.

In addition to time scale factor, other approaches such as lag time and cut-off factor can be used to account for the possible physiological events, like gastric emptying (lag time) or change in permeability along GI tract (cut-off factor for lack of colon absorption).

Correlation of in vitro and in vivo profiles. Correlation of profiles by means of linear regression is the classical IVIVC method. Altering *in vitro* test conditions systematically by statistical experimental design is an effective tool to match *in vivo* dissolution characteristics of formulations (Huang *et al.*, 2004; Corrigan *et al.*, 2003; Takka *et al.*, 2003). This approach enabled Qiu *et al.* (2003) to achieve a good linear correlation between percent absorbed and percent dissolved of three controlled release formulations.

An alternative method is described by Polli *et al.* (1996), representing an extension of the linear correlation method. For immediate-release formulations having partially permeability-limited or region-dependent absorption, a nonlinear correlation may provide certain advantages (Polli *et al.*, 1997). Further correlation of *in vivo* dissolved dosages with *in vitro* dissolved dosages is described by Dunne *et al.* (1997) by the use of odds, hazard or reversed hazards functions.

Compound V is an acidic compound with limited intrinsic solubility, most ionization of the compound takes place above its p*K*a (~4.3), solubility of Compound V is approximate 1.0 mg/mL at pH 6.8. When Compound V formulation is tested in different pHs, i.e., pH 6.8, 7.4, an 8.0, different release profiles are obtained (Fig. 12.9).

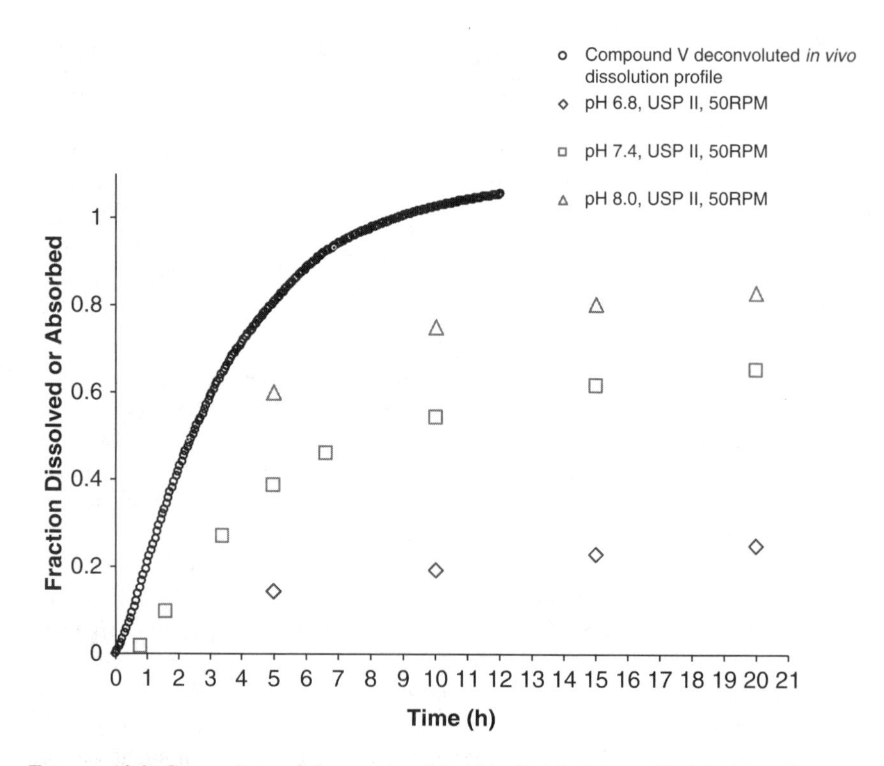

FIGURE 12.9. Comparison of deconvoluted *in vivo* dissolution profile (circle) and *in vitro* dissolution profiles of Compound V (400 mg) at different pH conditions: 0.05 M Phosphate buffer (pH 6.8, pH 7.4) and 0.05 M Borate buffer (pH 8.0) USP II paddle at 50 rpm (37 °C), the *in vitro* dissolution profiles is scaled to match the *in vivo* absorption time

These profiles can be fitted to Weibull function in the following form:

$$F(t) = F^{\infty}\left(1 - \exp\left(\frac{t + t_0}{\alpha}\right)^{\beta}\right), \tag{12.4}$$

where α represents the time at which 63.2% of the drug is dissolved, β is a shape factor that, at values below 1, yields a curve with an initially steep slope followed by a flat course; whereas at a value equal to 1, it describes an exponential curve; and at values greater than 1, yields a curve with a sigmoidal shape. Various shape factors can also be interpreted as different release mechanisms. F^{∞} is the dissolved fraction of the dose after an infinite time. t_0 is lag time that considers the delayed start of dissolution process. A perfect correlation can be achieved if all parameters of the Weibull function of *in vivo* and *in vitro* profiles are identical. The shape factor and lag time obtained from the mathematical fitting of data for Compound V are summarized in Table 12.2. *In vivo* pharmacokinetic profile is obtained and deconvoluted using numeric deconvolution of i.v. data as unit impulse function. When the shape factor of the *in vitro* dissolution profiles and *in vivo* deconvoluted profile is compared (Table 12.2), it is obvious that at pH 7.4 the best similarity between *in vivo* deconvoluted profile (1.1 vs. 1) and *in vitro* dissolution profile was reached and therefore, should be selected as the pH condition for testing the given dosage form. Physiologically, the selection of pH 7.4 is also supported since for the given acidic Compound V, most of the dissolution would occur around 7.4, which is predominant in the lower part of the small intestine, and, most likely, the absorption would also occur in the same region.

For *in vitro* dissolution of Compound V formulations, different rotation speeds were tested where it was noted that at lower rotation speed of 50 or 60 rpm, some coning effect at the bottom of dissolution vessel was observed. This was not preferred since it added an artifact to the results of the dissolution of the dosage form. Instead, rotation speed of 75 rpm was selected. Various surfactant levels, polysorbate 80 at 0.002, 0.01, 0.05, and 0.1%, were also tested. When the *in vitro* profiles were compared with the deconvoluted *in vivo* profile, it was observed that surfactant concentration of 0.05 or 0.1% provided closely matched profiles between *in vitro* dissolution and *in vivo* dissolution (Fig. 12.10). Based on this analysis, a rotation speed of 75 rpm and a pH 7.4 phosphate buffer with 0.1% polysorbate 80 was selected for the *in vitro* dissolution method for Compound V.

TABLE 12.2. Comparison of shape factors of deconvoluted absorption profile and *in vitro* dissolution profiles for Compound IV (fitted to Weibull function with and without log time)

		Dissolution profile		
	Absorption profile	pH 6.8	pH 7.4	pH 8.0
Shape factor	1.1	0.53	1	0.49
Shape factor, no lag time	1.26	0.65	1.34	0.79

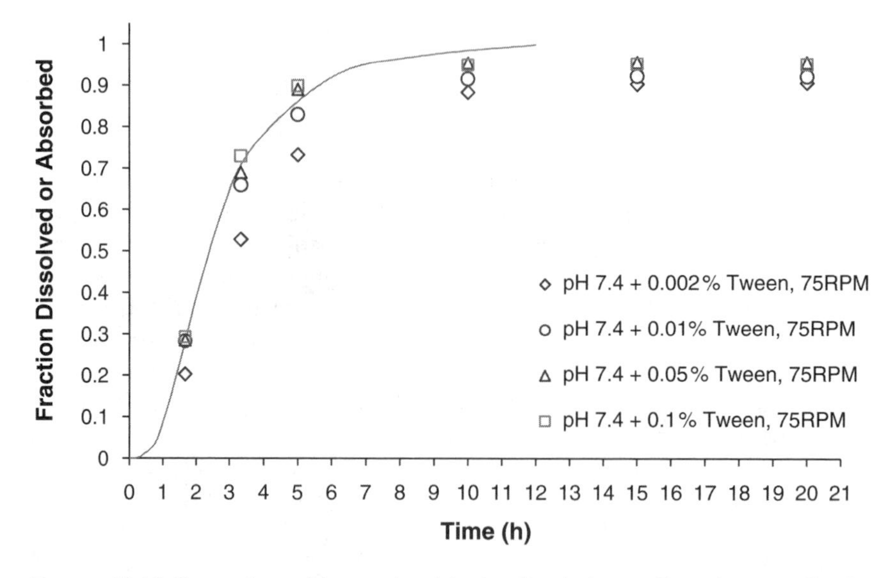

FIGURE 12.10. Comparison of deconvoluted *in vivo* dissolution profile and *in vitro* dissolution profiles of Compound V (400 mg) in 0.05 M Phosphate buffer (pH 7.4) with different concentrations of Tween-80 (0.002, 0.01, 0.05, and 0.1% Tween)

12.4.1 Tier 4: Application of IVIVC in LCM

During LCM, human PK data for IR formulations, often at multiple dose levels, exist. Frequently, development of modified release (MR) formulations is requested at this stage. For this purpose, MR formulations with two or more *in vitro* release rates are tested in a cross-over human PK study. This provides an opportunity to establish IVIVC for MR formulations based on PK data of MR formulations and the already existing PK data on immediate-release oral or intravenuous dosing. The following is a case history where the authors review how a Level A IVIVC was established for MR formulations at two different release rates.

Compound VI is a neutral compound with high solubility (>4 mg/mL) and high permeability (absolute bioavailability >90%, Caco-2 permeability P_{app} > propranol) (BCS Class I), as determined by methods defined by FDA biowaiver guidance (Center for Drug Evaluation and Research, US FDA, 2000). Two MR forms of Compound VI were tested, together with an IR solution, in a cross-over Clinical Pharmacology study. The same batches were used for the *in vitro* dissolution testing. The dosage forms used to establish the *in vitro–in vivo* correlation and for the internal validation are listed in Table 12.3.

The plasma concentration profiles of two disintegrating type of tablets (SR1, SR2) with a target of 80% release after 2–3 (MR1) and 4–6 (MR2) h, respectively, and an orally administered solution were used as a basis for developing a Level A IVIVC. The *in vivo* plasma concentration profiles of the two MR formulations and the solution are presented in Fig. 12.11.

TABLE 12.3. List of formulations tested in clinical pharmacology studies

Strength/name	Study dose(s) (mg)	Characteristics
500 mg/MR tablet (MR1)	500	Target: disintegrating MR variant 80% release after 2–3 h
500 mg/MR tablet (MR2)	500	Target: disintegrating MR variant 80% release after 4–6 h
Solution	500	Powder in tap water

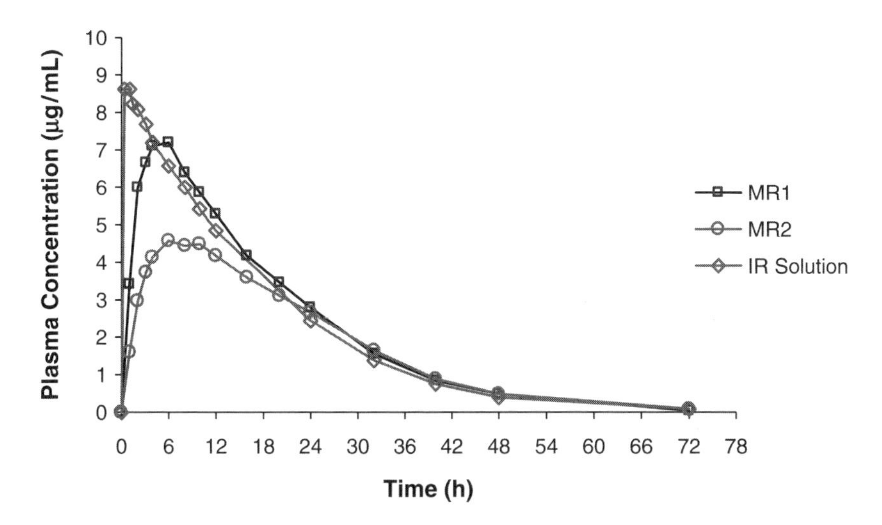

FIGURE 12.11. Compound VI mean plasma concentrations after 500 mg oral administration of a solution and two modified release formulations, MR1 and MR2 tablets, in healthy volunteers. MR1 is a disintegrating variant with 80% release after 2–3 h, and MR2 is a disintegrating variant with 80% release after 4–6 h

Dissolution rates of MR1 and MR2 (500 mg tablets) of Compound VI were tested using USP ⟨711⟩ Apparatus 2 (Paddle Method) at 50 rpm at 37 °C. The medium used for the test is 0.05 M phosphate buffer pH 6.8 (1,000 mL). All tests were performed with $N = 12$. The average data of the *in vitro* dissolution testing were used to develop the IVIVC and internal prediction.

A common approach for deconvolution is to use an orally administered solution as a unit impulse function to obtain the input rate for a SR form. The resulting cumulative input rate represents the kinetics of the *in vivo* dissolution and can be compared with *in vitro* dissolution profiles to obtain an *in vitro–in vivo* correlation. A Level A correlation can be achieved when the curves are superimposable or can be made to be superimposable by the use of a constant time scaling factor for all formulations. The use of time scale factor is acceptable by the regulatory agency according to the Guidance issued by FDA (Center for Drug Evaluation and Research (CDER), 1997).

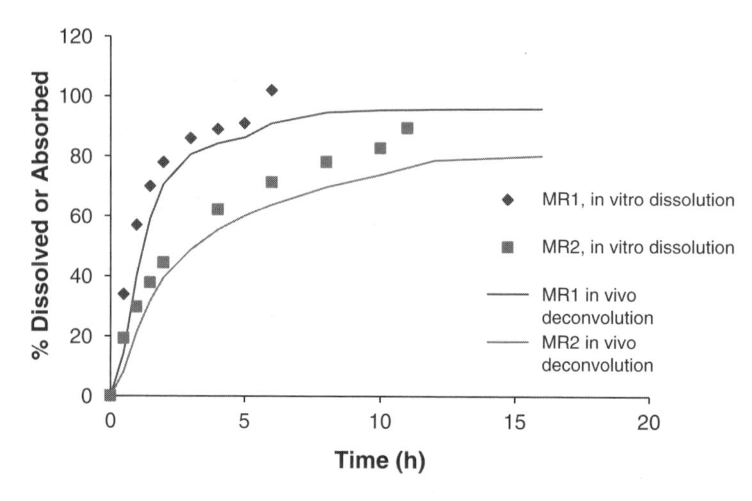

FIGURE 12.12. Mean *in vitro* dissolution profiles of Compound VI MR1 and MR2 tablets ($n = 12$) with USP I, paddle method at 50 rpm (37°C) and comparison of the *in vitro* and *in vivo* dissolution profiles for MR tablets

For Compound VI, a Wagner–Nelson method may be used to obtain the *in vivo* dissolution profiles since, as shown in Fig. 12.11, it follows simple one-compartment pharmacokinetic model. The *in vivo* dissolution data thus generated are then plotted against *in vitro* dissolution profiles in Fig. 12.12.

In Fig. 12.12, a rank order correlation exists for the two MR formulations. However, not surprisingly, there appears to be some time difference between the *in vivo* and *in vitro* dissolution profiles. This type of time scale difference has been reported by other authors as well (Li *et al.*, 2007; Corrigan *et al.*, 2003; Qiu *et al.*, 2003) and is described in the FDA guidance for IVIVC (Center for Drug Evaluation and Research, 1997). It is an acceptable practice to mathematically scale *in vitro* dissolution profile to match *in vitro* dissolution profile or to modify *in vitro* dissolution condition if necessary. In this case, the *in vitro* dissolution profiles were time-scaled according equation below:

$$T_{in\,vivo} = 1.18T_{in\,vitro}.$$

As a result, the correlation between *in vitro* and *in vivo* dissolution improved, which is shown in Fig. 12.13. In order to establish an *in vitro–in vivo* correlation, the fraction dissolved *in vivo* was plotted against the fraction dissolved *in vitro* and a linear regression was applied to find a quantitative relationship. As shown in Fig. 12.14, a linear relationship of close to unity can be established for Compound VI MR formulations with a correlation coefficient of 0.99.

According to FDA guidance (Center for Drug Evaluation and Research, 1997), the predictability of the developed IVIVC model is an integral part of the IVIVC development. During the internal validation stage, the predicted plasma profiles of the formulations that were used to develop the IVIVC model were evaluated

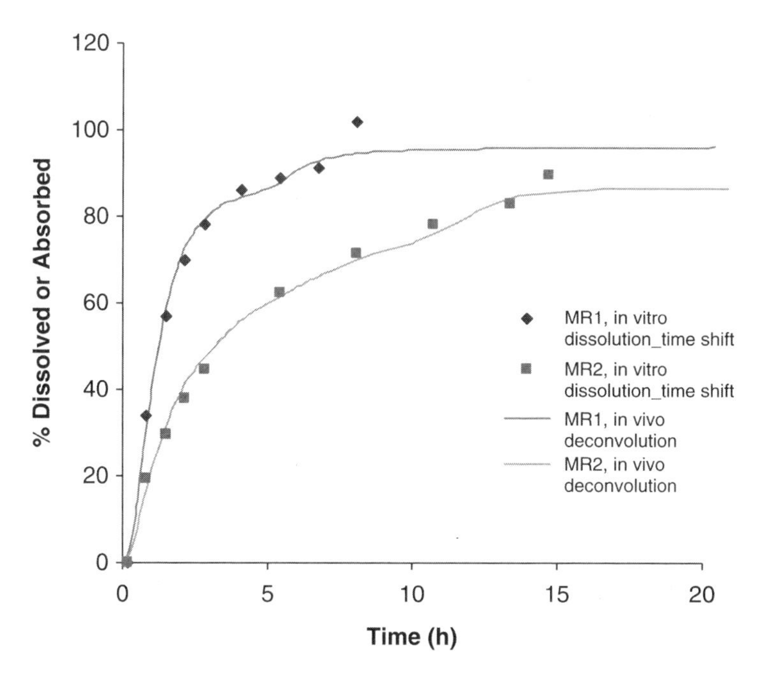

FIGURE 12.13. Comparison of the time-scaled *in vitro* dissolution profile and *in vivo* dissolution profiles of compound VI MR tablets

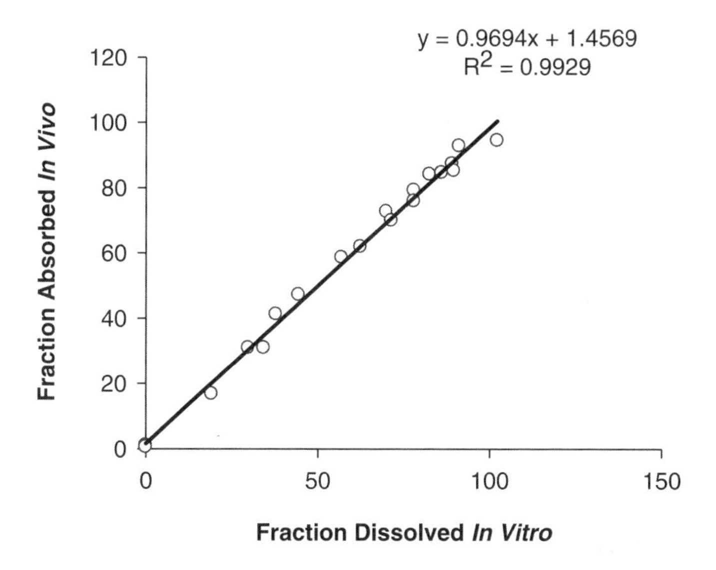

FIGURE 12.14. Relationship between fraction of *in vitro* and *in vivo* dissolution data of compound VI formulations

TABLE 12.4. Comparison of experimental and predicted C_{max} and AUC values for modified released formulations of compound VI

Compound VI	AUC (mg h/mL)			C_{max} (mg/mL)			Requirement (%)
	Experimental	Predicted	% PE	Experimental	Predicted	% PE	
MR 1	151.60	143.20	5.54	7.44	7.20	3.23	<15
MR 2	121.41	123.60	1.80	4.70	4.58	2.55	<15

against the observed values. The *in vitro* dissolution profiles of the formulations were converted to *in vivo* dissolution profiles by applying the IVIVC relationship. The resultant *in vivo* dissolution profiles are then used as input function to simulate/convolute *in vivo* plasma concentration for a given formulation. The resultant convoluted C_{max} and AUC values were then compared with the experimental values. In the case of Compound VI, the human plasma profiles of MR1 and MR2 formulations were predicted based on the established IVIVC model. The internal predictability of AUC_{inf} and C_{max} is shown in Table 12.4 and demonstrates that the predictions are well within the regulatory requirement as defined by the FDA Guidance.

12.5 Conclusions

It has been shown in this chapter that IVIVC may be applied in all stages of drug product development. In a broader sense, IVIVC evolves throughout the full development process. Preliminary IVIVC can be initiated by *in silico* simulation at the drug discovery stage when only physicochemical data and no useful PK data are available. As the drug development progresses, the IVIVC becomes more sophisticated and experiment-based. In the LCM stage, the IVIVC is an integral part of drug product development.

Acknowledgment. The authors would like to thank Dr. Hequn Yin, Dr. Martin Mueller-Zsigmondy, Dr. Colleen Ruegger, and Dr. Yatindra Joshi for assistance and helpful discussion in preparation of this chapter.

References

Agoram, B., Woltosz, W.S., and Bolger, M.B. (2001). Predicting the impact of physiological and biochemical processes on oral drug bioavailability. *Adv. Drug Deliv. Rev.* 50:S41–S67.

Amidon, G.L., Lennernas, H., Shah, V.P., and Crison, J.R. (1995). Theoretical basis for a biopharmaceutical drug classification: correlation of in vitro drug product dissolution and in vivo bioavailability. *Pharm. Res.* 12(3):413–420.

Center for Drug Evaluation and Research (CDER) (1997). Guidance for Industry: Extended release oral dosage forms: development, evaluation, and application of in-vitro/in-vivo correlation. Rockville, MD. US Department of Health and Human Services, Food and Drug Administration.

Center for Drug Evaluation and Research (CDER) (2000). Waiver of in vivo bioavailability and bioequivalence studies for immediate-release solid oral dosage forms based on a biopharmaceutics classification system. Food and Drug Administration.

Chiou, W.L., Ma, C., Chung, S.M., Wu, T.C., Jeong, H.Y. (2000). Similarity in the linear non-linear oral absorption of drugs between human and rat. *Int. J. Clin. Pharmacol. Ther.* 38:532–539.

Corrigan, O.I., Devlin, Y., and Butler, J. (2003). Influence of dissolution medium buffer composition on ketoprofen release from ER products and in vitro–in vivo correlation. *Int. J. Pharm.* 254(2):147–154.

Curatolo, W. (1998). Physical chemical properties of oral drug candidates in the discovery and exploratory setting. *Pharm. Sci. Tech. Today* 1:387.

Dannenfelser, R.-M., He, H., Joshi, Y., Bateman, S., and Serajuddin, A.T.M. (2004). Development of clinical dosage forms for a poorly water soluble drug I: application of polyethylene glycol-polysorbate 80 solid dispersion carrier system. *J. Pharm. Sci.* 93:1165–1175.

Dunne, A., O'Hara, T., and DeVane, J. (1997). Level A in vivo-in vitro correlation: nonlinear models and statistical methodology. *J. Pharm. Sci.* 86(11):1245–1249.

Frick, A., Möller, H., and Wirbitzky, E. (1998). Biopharmaceutical characterization of oral controlled/modified-release drug products. In vitro/in vivo correlation of roxatidine. *Eur. J. Pharm. Biopharm.* 46:313–319.

Horter, D. and Dressman, J.B. (1997). Influence of physicochemical properties on dissolution of drugs in the gastrointestinal tract. *Adv. Drug Deliv. Rev.* 25:3–14.

Huang, Y.-B., Tsai, Y.-H., Yang, W.-C., Chang, J.-S., Wu, P.-C., and Takayama, K. (2004). Once-daily propranolol extended-release tablet dosage form: formulation design and in vitro/in vivo investigation. *Eur. J. Pharm. Biopharm.* 58:607–614.

Johnson, K.C. and Swindell, A.C. (1996). Guidance in the setting of drug particle size specifications to minimize variability in absorption. *Pharm. Res.* 13:1795–1798.

Langenbucher, F. (2002). Handling of computational in vitro/in vivo correlation problems by Microsoft Excel II. Principles and some general Algorithms. *Eur. J. Pharm. Biopharm.* 53:1–7.

Li, S., He, H., Parthiban, L.J., Yin, H., and Serajuddin, A.T.M. (2005). IV-IVC considerations in the development of immediate-release oral dosage form. *J. Pharm. Sci.* 94(7):1397–1417.

Li, S., Mueller-Zsigmondy, M., and Yin, H. (2007). The role of IVIVC in product development and life cycle management in pharmaceutical product development: In vitro–in vivo correlation. In: Chilukuri, D., Sunkara, G., Young, D. (eds.), *Pharmaceutical Product Development (Drugs and the Pharmaceutical Sciences)*, Marcel Dekker Inc., New York.

Mahayni, H., Rekhi, G.S., and Uppoor, R.S. (2000). Evaluation of external predictability of an in vitro–in vivo correlation for an extended-release formulation containing Metoprolol Tartrate. *J. Pharm. Sci.* 89(10):1354–1361.

Polli, J.E., Crison, J.R., and Amidon, G.L. (1996). Novel approach to the analysis of in vitro–in vivo relationships. *J. Pharm. Sci.*, 85:753–760.

Polli, J.E., Rekhi, G.S., Augsburger, L.L., and Shah, V.P. (1997). Methods to compare dissolution profiles and a rationale for wide dissolution specifications for Metoprolol Tartrate tablets. *J. Pharm. Sci.* 86(6):690–700.

Pudipeddi, M., Serajuddin, A.T.M., and Mufson, D. (2006). Integrated drug product development – from lead candidate selection to life-cycle management. In: Smith, C.G. and O'Donnell, J.T. (eds.), *The Process of New Drug Discovery and Development*, 2nd ed. Informa Healthcare USA, Inc., New York, pp. 15–51.

Qiu, Y., Garren, J., Samara, E., Cao, G., Abraham, C., Cheskin, H.S., and Engh, K.R. (2003). Once-a-day controlled-release dosage form of Divalproex Sodium II: development of a predictive in vitro drug release method. *J. Pharm. Sci.* 92(11):2317–2325.

Takka, S., Sakr, A., and Goldberg, A. (2003). Development and validation of an in vitro–in vivo correlation for buspirone hydrochloride extended release tablets. *J. Control. Release* 88:147–157.

Turner, S., Federici, C., Hite, M., and Fassihi, R. (2004). Formulation development and human in vitro–in vivo correlation for a novel, monolithic controlled-release matrix system of high load and highly water-soluble drug niacin. *Drug Dev. Ind. Pharm.* 30(8):797–807.

United States Pharmacopeial Convention, Inc. (1988). In vitro–in vivo correlation for extended release oral dosage forms. *Pharmacopeial Forum* Stimuli Article 4160–4161.

Venkatesh, S. and Lipper, R.A. (2000). Role of the development scientist in compound lead selection and optimization. *J. Pharm. Sci.* 89(2):145–154.

Yu, L.X., Crison, J.R., and Amidon, G.L. (1996a). Compartmental transit and dispersion model analysis of small intestinal transit flow in humans. *Int. J. Pharm.* 140:111–118.

Yu, L.X., Lipka, E., Crison, J.R., and Amidon, G.L. (1996b). Transport approaches to the biopharmaceutical design of oral drug delivery systems: prediction of intestinal absorption. *Adv. Drug Deliv. Rev.* 19:359–376.

Index

A

Abbreviated new drug applications (ANDAs), 263, 271, 328, 330. *See also* Bioequivalence
ABCC. *See* Multidrug Resistance-Associated Protein Family
ABCG2. *See* Breast Cancer Resistance Protein
ABC transporters, 179–182
  BCRP, 193–195
  MRP family, 188–193
  P-gp (*see* P-glycoprotein (P-gp))
Absorption, distribution, metabolism, and elimination, 1, 175
  assessment, *in vitro*/*ex vivo* techniques, 15
Absorption enhancers, action mechanisms of, 142
  membrane components, action on, 143–148
  mucus layer, action on, 143
Absorption potential, in drug permeability, 105, 106
Absorption rate constant, drug absorption and, 82
Accelerated *in vitro* release testing, 341
Accutane® label, 321
2-Acetylaminofluorene, 184
Actinomycin D, 185
Active pharmaceutical ingredient, 3, 290
  with high solubility at gastric pHs, 309
  particle size, 306
  physical properties of, 292, 295, 304
Active transport, 78, 79. *See also* Carrier-mediated intestinal membrane transport
  mechanisms, impact, 10
Acyclovir, 226, 227
Adalat® OROS, 326
ADME. *See* Absorption, distribution, metabolism, and elimination
  and toxicity, 101
Adsorption theory, of mucoadhesion, 148
Advanced Compartmental Absorption and Transit (ACAT) model, 365
Aflatoxin B1, 184
p-Aminohippurate, 192
Amorphous drugs, 13
Antiglucocorticoid/antiprogestin RU486, 190
Antiporter carrier-mediated transport, 78
API. *See* Active pharmaceutical ingredient
Aqueous solubility, factors affecting, 30, 31
Aryl hydrocarbon receptor (AhR), 190
Aspergillus fumigatus, 194
Asthma patients, 285
Atenolol, hydrophilic drug, 143
Atovaquone, 321
ATP-binding cassette, 5
ATR-FTIR spectroscopy, 149

B

Baby hamster kidney (BHK) cells, 197
Basket method, in drug dissolution, 58, 59, 66, 69
BCRP. *See* Breast cancer resistance protein
BCS. *See* Biopharmaceutics classification system
BCS Class I drugs, 328, 329

BCS Class I/III drugs
biowaivers for, 279, 280
case studies of, 296
disintegration/erosion controlled dissolution, 299, 301
fast release with disintegration controlled dissolution, 298
dissolution testing of, 297
immediate release formulations of, 297
BCS Class II/IV drugs
biowaivers for, 279, 280
case studies of, 302
intrinsic drug solubilization controlled dissolution, 303
liquid filled capsules, 303
dissolution testing of, 302
formulation designs of, 307
BE. *See* Bioequivalence
Bicalutamide tablet label, 323
Bile release effects, 322
Bile-Salt Exporting Protein, 183
Bilirubin glucuronides, 190
Bioavailability
absolute, 266
definition of, 262
in drug development and regulation, 262, 263
food effects on, 317
relative, 266, 267
in vivo and *in vitro* methods of, 264
Bioavailability studies
chiral assays in, 273
in clinical development, 17, 18
FDA recommendations of, 265
general approaches of, 264
guidelines and recommendations of, 265, 271
metabolites in, 271, 272
pharmacokinetic method, 265
plasma-concentration time profile, 266
Bioequivalence
challenges of, 271
definition, 263
in drug development and regulation, 263, 264
of highly variable drugs, 274
limits, static expansion of, 274–276
of locally acting drugs, 281 (*see also* Nasal spray products; Oral inhalation products; Topical dermatological products)
tests, hypothetical results from, 269
in vivo and *in vitro* methods of, 264
Bioequivalence studies
biowaivers of
lower strength, 278
solutions, 276
with clinical endpoints, 281, 283
design of, 267
doses used in, 267
FDA recommendations of, 265
general approaches of, 264
guidelines and recommendations of, 267
healthy subjects, 268
metabolites in, 271, 272
of nasal spray products, 284
of oral inhalation products, 285, 286
pharmacokinetic method, 265, 282
sample size in, 274, 275, 277
statistical analysis, 268–271
Biopharmaceutics
classification system, 145, 146, 363
dissolution, 290
drug absorption process, 291
drug classification matrix, 291
drug compounds, 62–64
importance, 61–64
introduction, 6, 7, 30
solubility classification of drug in, 279
definition, 1
drug absorption and transport, 7–10
pharmacokinetic profile developmental strategies, 10–14
studies types, 17, 18
Biopharmaceutics, role in drug development
advanced clinical development, 19, 20
cycles, 14
drug discovery and preclinical development, 15, 16
phase I clinical studies, 16, 17

postapproval process, 20, 21
regulatory process, 21
Biorelevant dissolution media
gastric conditions, 65
intestinal conditions, 67, 68
small intestinal conditions, 69
Biorelevant testing, in drug dissolution, 60, 61
Biowaivers
for BCS Class I, II and III drugs, 276, 280
for topical dermatological products, 282
Biowaivers, 294, 297, 349, 360
Blood–brain barrier (BBB), 206
Breast cancer resistance protein, 193, 194
Bromosulfophthalein (BSP), 205

C

Caco-2, 144–148, 154, 156, 159, 187, 190
analysis, *in vitro,* 298, 301
cell model, 178
in drug absorption, 115, 116
drug permeability, 90–94, 109, 110, 115, 116, 126, 127
hPEPT1 promoter region in, 200
jejunum and, 189, 193
layer transport phenomena, 157
transport, 145
Calcium-fortified orange juice, 319
Canalicular multispecific organic anion transporter (cMOAT), 188
Capsule dissolution profiles, as function of
API D90 value, 307
API particle, 306
roll compaction pressure, 309
Capsule shell rupture, 310
Carbomers. *See* Polyacrylic acid (PAA)
Carbomer–sodium salt (NaC934P), 154
Carbopol 934P/974P, 152, 154, 158
effects, on protease activities, 155
Carcinogens, 192
Carrier-mediated drug absorption, 102
Carrier-mediated intestinal membrane transport
active transport, 79
facilitated diffusion, 78, 79
Carrier proteins, in facilitated diffusion, 78, 79
Casodex® label, 323
Castor oil, hydrogenated, 144
Ceftin® label, 322
Cefuroxime axetil, 322
Center for Drug Evaluation and Research (CDER), 377
[^{14}C]glycylsarcosine, 200, 201
[14C]Gly-Sar, 201
Chelating agent, calcium depletion by, 147, 148
Chemotherapeutic agents, 336
Chenodeoxycholic acid, 190
Chinese hamster ovary (CHO) cells, 192, 221
Chitosan, 143
as absorption enhancer of hydrophilic macromolecular drugs, 157, 158
application, mechanism and safety aspects, 156, 157
Caco-2 cells with, 147
thiolated polymers of, 163
Chitosan–cysteine conjugates, 163
Chitosan hydrochloride, 154
Chitosan-4-thio-butyl-amide (chitosan-TBA) conjugates, 163, 165
Chitosan–thioglycolic acid conjugate, 163
α-Chymotrypsin, 154
Ciprofloxacin, 319
Cipro® label, 320
Clinical developments
advanced development, 19, 20
bioequivalence study, 18
early development, 17–19
[^{14}C]mannitol, 156–159
CMC regulatory systems, 21
CNT (SLC28A), 215–217
Colchicine, 185
Combivir® label, 319
Concentrative nucleoside transporters (CNT), 215
Confocal laser scanning microscopy (CLSM), 147, 155, 159
Constitutive androstane receptor (CAR). *See* NR1I3
Continuous flow method, 340

Controlled release (CR) dosage forms, 292
  case study of, 310–312
  dissolution profiles of, 311
  fraction absorbed *versus* fraction dissolved for, 312
  IVIVC of, 311
  plasma profiles of, 311, 312
Conventional tablet, 304, 305
Convolution. *See* Deconvolution and convolution
Cremophor® EL, 145, 146, 187, 337
Cyclic AMP (cAMP), 147
Cyclic oligosaccharides, 144
Cyclodextrins, 144
Cyclosporine, 185, 187, 321
Cytochrome P450 (CYP), 6
  CYP1A2, 320
  CYP3A, 207
  CYP3A4, drugmetabolizing enzyme, 187, 226
Cytoskeletal F-actin, redistribution of, 147
Cytotoxic T lymphocytes, 184

D

Danckwerts model, 50, 51
Daunorubicin, 348
Deconvolution and convolution, 293, 372–375
  Loo-Riegelman method, 372
  numerical methods, 373
  in vivo and in vitro dissolution profile, comparisons, 374–376
  Wagner-Nelson method, 372
Dehydroepiandrosterone (DHEA), 208
Dendrimers, 351, 352
9-Desglycinamide, 8-L-arginine vasopressin (DGAVP), 153, 154
Desmopressin (1-(3-mercaptopropionic acid)-8-D-arginine vasopressin monoacetate (DDAVP), 164
Dexamethasone, 190, 344, 346
Dialysis sac method, 339
1,4-bis[2-(3,5-dichloropyridyloxy)] benzene (TCPOBOP), 190
Didanosine, 319, 330
Diet–drug interactions, 320
Diffusion controlled release systems, 341
Diffusion layer model, 29, 49, 50
Diffusion theory, of mucoadhesion, 149
Digoxin, 185–187, 226
Digoxin efflux ratio, 127, 128
Dihydrochloride (diHCl) and tartrate salts, 369–371
2,4-Dinitrophenyl-*S*-glutathione, 190, 191
Disintegration/erosion controlled dissolution, 298, 299. *See also* Dissolution process
Dissolution correlation, *in vitro-in vivo,* 292–294
Dissolution process, 362, 367
  BCS approach to, 290
  control of, 294
  design space, 295
  factors affecting, 369
  kinetics of, 292
  media in drug dissolution test, 57–59
  of MR1 and MR2 tablets, 377, 378
  pharmaceutical quality assessment implications of, 313, 314
  test, *in vitro,* 9
  in USP, 298
Dissolution profiles, *in vitro,* 293, 368, 370, 371
Dissolution rate-limited absorption, 293
Dissolution testing, 290
  of BCS class I/III drugs, 297
  of BCS class II/IV compounds, 302
  biorelevant methods, 64–70
  quality control (QC), 293, 302
Dissolution time, *in vitro,* 360
DNP-SG. *See* 2,4-Dinitrophenyl-*S*-glutathione
Docusate sodium (dioctyl sodium sulfosuccinate), 145. *See also* Excipients
Dose-dumping, food effects on, 325
Doxorubicin, 348
Doxorubicin-resistant MCF7 breast cancer cell line (MCF-7/ AdrVp), 193
(D-Pen2, D-Pen5)-enkephalin (DPD PE), 205
Drug carrier-mediated transport classification, 78
Drug compound classification, BCS role, 62, 63
Drug dissolution rate
  factors affecting, 51

physicochemical properties, 52, 53
roles, 55, 56
Drug dissolution testing, for quality control
immediate-release dosage forms, 57, 58
limitations, 59, 60
modified-release dosage forms, 58, 59
Drug–drug interactions
in clinical development, 19
P-gp and, 186, 187
Drug–grapefruit interaction, 321
Drug–nutrient P-gp interactions, 187
Drug permeability
in Caco-2 cells, 90–94
in human intestine, 86–88, 95–97
in rats jejunum, 95–97
Drug permeability and absorption models
physicochemical methods, 105, 106
in silico methods, 113, 114
in situ methods, 111, 112
in vitro methods, 106–111
in vivo methods, 112, 113
Drug products
development, food effects in, FDA Guidance for Industry, 323
development, IVIVC in
decision tree for implementing, 362
four-tier approach of, 360–363
new chemical entity (NCE) and application of, 361
development, life cycle management (LCM) stage of, 363
dissolution rate, factors affecting, 53–55
labeling claims for, 323
labeling, recommendations for, 323, 324
Drug release, *in vitro,* 326
factors affecting, 338
mathematical models of, 341–343
methods of, 339–341
uses of, 338, 339
Drug release, *in vivo,* factors affecting, 343, 344
Drug(s)
absorption and transport, physical/chemical properties, 7–10
with active metabolites, 271
administration, 11
bioavailability and physicochemical properties, 41–43
chemical and physical stability, 3
degradation, chemical reactions, 39
development process, biopharmaceutics role, 14
discovery and preclinical development, 15, 16
dissolution, 8–10, 29
elimination, 5, 6
forms and polymorphs, 2, 3
formulation properties, 4
ionization and p*K*a property, 3, 4
manufacturing processes, 54, 55
membrane transport, 4, 5
with nonlinear pharmacokinetics, 267
permeation control, 294
pharmacokinetic properties of, 263, 267
pharmacological effects of, 345
physiological and biological principles, 4–6
solubility, 2, 35, 38, 82, 83
stability, 3
Drugs absorption, 291, 318
biological factors, 84–86
decreased rate, mechanism of
increased first-pass metabolism and clearance, 320
instability in gastric acids, 319
physical/chemical binding with food components, 319
dosage factors, 86
food effects on, 324
in intestinal membrane, 101
membrane permeability and, 81, 82
path, factors in, 80–86
prolonged rate, mechanism of
effects of bile release, 322
inhibition of first-pass effect, 320
longer gastric residence time and, 322
physicochemical and physiological effects, 321
solid dosage forms dissolution and, 83, 84
solubility, 82, 83
unaffected by food, 322, 323
in vivo, particle size effect on, 364

Drug safety, food effects on, 324
Drugs binding, nonspecific, 125
Drugs bioavailability. *See* Bioavailability
Drugs diffusion, 343
Drug solubilization, 292
  dissolution, 303 (*see also* Dissolution process)
  intrinsic rate of, 306
Dry blend formulations, 368
Dry powder inhalers (DPI), 285

E

Efavirenz label, 324
Efflux ratio, in drug absorption, 127–129
  apical pH effects, 131, 132
  substrate concentration effects, 129, 130
Electroneutral proton/cation exchanger, 197
Electronic descriptors, in intestinal drug absorption, 114
Electronic theory, of mucoadhesion, 148
Emulsions
  IVIVC of, 349, 350
  types of, 349
Enantiomers *vs.* racemates, 273
Endocytosis intestinal membrane transport, 79, 80
Endogenous substances, 273, 274
Enhanced permeation and retention (EPR) effect, 347
ENT (SLC29A), 216, 218, 219
Equilibrative nucleoside transporters (ENT), 215
Erosion controlled release systems, 343
Estradiol-17β-D-glucuronide, 192, 193
Estradiol-17-glucuronide, 190
17β-Estradiol-3-sulfate, 194
Estrone-3-sulfate, 194, 207
Ethylene diamine tetraacetic acid (EDTA), 144, 145
Euphyllin Retard®, 325
Everted gut technique, in drug absorption, 107
Excipients, as absorption enhancers, 139
  action mechanisms of, 142–148
  mucoadhesive polymers
    see Mucoadhesive polymers, as absorption enhancers
  paracellular transport, 142
  transcellular transport, 141, 142
Exhaled nitric oxide (eNO), 286
Extended-release drug products, biorelevant methods, 69
Extracellular matrix metalloproteinase inducer (EMM-PRIN), 222

F

Facilitated diffusion, 78, 79. *See also* Carrier-mediated intestinal membrane transport
F-actin, 157
Farnesoid X receptor (FXR). *See* NR1H4
Fasted state simulated intestinal fluid (FaSSIF), 67–69
Fasting bioequivalence study, 330. *See also* Bioequivalence studies
Fatty meal, 322
Fed bioequivalence studies
  for immediate-release drug products, 330
  label-driven criteria for, 329, 330
  with pharmacokinetic endpoints, 328
  recommendations for design of, 330
  role of, 327
Felodipine, 321
Fexofenadine, 187, 228
Fick's laws
  first law of diffusion, 77, 176, 177
  second law of diffusion, 341
Finasteride label, 323, 330
First-pass metabolism and clearance, 320
Flippase model (FM), 185
Flow-through cell system, 339
  in drug dissolution, 58, 66
Fluo-3, 192
Fluorescein isothiocyanate (FITC-dextran), 155
Folinic acid (leucovorin), 192
Food and Drug Administration (FDA)
  approved drug products, 146
  approved labels, 324
  BE acceptance limits, 270, 271
  guidance, correlation levels, 61
Food-drug interactions, 318, 324

Food-effect bioavailability studies, recommended designs for, 323
Fumitremorgin C, 194

G
Gastric acids, instability in, 319
Gastric emptying time, in drug absorption, 84
Gastric pH values, in dissolution rate, 65
Gastric residence time, 322
Gastrointestinal absorption processes, 365
Gastrointestinal fluids components, volume, and properties, 85
Gastrointestinal (GI) fluids, 364
Gastrointestinal (GI) tract, 175, 277
- drug absorption and
  - biological factors in, 84–86
  - blood flow in, 85
  - surface area in, 84
- physiological changes in, 324
- physiological conditions, factors in, 60, 61
- time in oral drug absorption, 84, 85
- transit and drug ionization, 8

Gastrointestinal Therapeutic System (GITS), 326
GastroPlus™ simulation, 362
- input parameters in, 365
- particle size impact on oral absorption, 366

Gelatin capsule shell, 305
Gelatin crosslinking, 309
Gel extrusion module (GEM) tablet, 311
Generic drug products
- approval and use of, 263
- immediate-release products, development of, 328
- labeling, food effects and, 331
- modified-release products, development of, 327, 328

Genetic polymorphism, 201, 226
Geometric descriptors, in intestinal drug absorption, 114
GF120918, 186, 194
GLUT1, 224
Glutathione (GSH), 191, 205
Glycyl sarcosine, 146
Granules, 309, 310
Grapefruit juice, 320
GW4064, 190

H
HEK 293 cells, OATP-B-transfected, 205
α-Helix TMD, 204
Hepatocyte nuclear factor 1α (HNF-1α), 207
HepG2 cells, 192
H1-histamine receptor, 228
High-throughput screening, 101, 120
HIV-1 protease inhibitors, 225
hOCT1/2, 210, 211
HT29-H cultures, 143
HTS. *See* High-throughput screening
Human Genome Organization (HUGO) Nomenclature Committee, 181, 193, 195, 204
Human Intestinal Peptide Transporter 1 (HPT1), 197
Human OATP-C gene, 207
Human peptide/histidine (hPHT) transporters, 196
Hybrid descriptors, in intestinal drug absorption, 114
Hydrogels, 350
Hydrophilic cyclodextrin (2,6-di-*O*-methyl-β-cyclodextrin), 187
Hydrophilicity/lipophilicity of drug, 2
Hydrophobic vacuum cleaner (HVC) model, 185
Hydroxypropyl methyl cellulose (HPMC), 145. *See also* Excipients

I
ICH. *See* International Council on Harmonization
Immediate release (IR) dosage forms, 292
Immobilized Artificial Membrane (IAM), in drug intestinal absorption, 106
Immunohistochemical staining, 193
Immunoliposomes, 348
Inactive Ingredients Database, 146
Inhaled corticosteroid (ICS), 286
InnoPran® XL, 321
Insulin, 144

Interfacial barrier model, 50, 51
International Council on Harmonization, 21, 294
Intestinal conditions, pH values, 67, 68
Intestinal drug absorption, computational methods, 113
Intestinal membrane transport
  carrier-mediated, 78, 79
  paracellular and endocytosis, 79, 80
  passive diffusion, 76–78
Intestinal motility in oral drug absorption, 85
Intestinal Peptide Transporter (PT1), CDH17, 195
Intestinal permeation, 293
Intestinal transporters impact, on bioavailabilty, 225–229
In vitro and *in vivo* correlation (IVIVC), 9, 10, 36, 56, 61
  case studies, 363
  categories of, 344
  of controlled release tablets, 311
  convolution in, 373
  deconvolution in, 372
  definition of, 336
  in drug product development, 360
  of emulions, 349, 350
  in LCM, 376
  level A, B and C correlations, 293, 344, 347
  of liposomes, 347–349
  of modifed release products, 337, 376
  of oral dosage forms, 359
  PK data for, 364
  of protein loaded microspheres, 345–347
Isotretinoin label, 321

K

"Kinetic solubility" in oral drug solubility, 37
Kupffer cells, 344
Kyte–Doolittle hydropathy plots, 222

L

Lactose monohydrate, 145. *See also* Excipients
Lamivudine, 318
Lariam® label, 324
Leukotriene C4 (LTC4), 190, 191, 205
Leuprolide, 346
Lidocaine/prilocaine cream, 282
Lipinski's Rule of Five states, 177
Lipophilic compounds transport, 76, 77
Lipophilicity/hydrophilicity, of drugs, 2
Lipophilicity, in oral drugs, 105, 124
Liposomes, 338
  for cancer treatment, 348
  IVIVC of, 347–349
  pharmacokinetic parameters of, 349
  physicochemical properties of, 348
  types of, 347
Liquid filled capsule (LFC), 303
Liver X receptor (LXR), 207
Loo–Riegelman method, 372
Low molecular weight heparin (LMWH), 155, 162
Lung resistance-associated protein (LRP), 226
Lupron Depot®, 345
Luteinizing hormone–releasing hormone (LH–RH), 345
L-Val-acyclovir (valacyclovir), 226, 227
LY335979, 186
Lyophilized tablet formulation, 304

M

Macrobid® label, 322
Malarone®, 322
Mannitol, 277
Mardin-Darby canine kidney, 102
Maximum absorbable dose (MAD)
  calculation, 364
  concept, 10, 36
  estimation in human intestinal permeability, 88, 89
  in vivo drug absorption estimation, 88
MCF-7 cells, 194
MCT1, 222
MDCK. *See* Mardin-Darby canine kidney
MDCK cell in drug permeability, 111
MDR2/3, 183
mdr1a and *mdr1b* gene, in rodents, 183
MDR1 gene, human, 226
Mean resident time, 84, 85

Mebendazole anthelmintic drug, 59
Mefloquine, 324
Membrane permeability, drug absorption and, 81, 82
Metabolism. *See* Presystemic metabolism; Systemic metabolism
Metered dose inhalers (MDI), 285
Methotrexate (MTX), 185, 191, 351
Methoxyflavones, 187
4-(Methylnitrosamino)-1-(3-pyridyl)-1-butanol, 192
1-Methyl-4-phenylpyridium (MPP^+), 210
Michaelis–Menten equation, 197
Microdialysis, 340
Micronization, 367
Microspheres, 337
  drug delivery systems and, 346
  IVIVC of, 345–347
  testing of, 339
Mitoxantrone resistance protein *(MXR),* 193
MK571, antagonist of MRP2, 192
Modified-release capsules development, sprinkle studies in, 331
Modified-release drug products, issues with
  formulation-dependant food effects, 325
  potential for dose-dumping, 325
Modified release (MR) formulations
  fraction of *in vitro* and *in vivo* dissolution data of, 379
  IVIVC for, 376
  in vivo plasma concentration profiles of, 376, 377
Modified release (MR) parenteral products, 336, 343, 348
  factors related to formulation of, 337
  IVIVC for, 337
  in vitro release from, 338
  in vivo release from, 343
Modified release (MR) tablets
  dissolution rates of, 377, 378
  human plasma profiles of, 380
  time-scaled *in vitro* dissolution profile and *in vivo* dissolution profiles of, 379
Monocarboxylate transporters (MCT), 221
  expression of, 223, 224
  molecular and structural characteristics of, 222
  regulation of, 224, 225
  substrate specificity of, 222, 223
Monocarboxymethylated chitosan (MCC), 161, 162
mRNA expression, of MRPs, 189, 190
MRP7/8/9, 189
MRP2/3 (ABCC2/3), 188, 189, 191, 192
MRP4/5/6 (ABCC4/5/6), 189, 192
MRP1 and *MRP2,* 226
MRP family. *See* Multidrug Resistance-Associated Protein Family
MRP isoform expression regulation, 190
MRP/MRP2 gene, 188, 190
MRPs. *See* Multidrug resistance proteins
MRT. *See* Mean resident time
Mucin, 143, 149
Mucin glycoproteins, 156
Mucoadhesive polymers, as absorption enhancers
  classes of, 152–165
  material properties of, 150–152
  mucoadhesion, theories of, 148–150
Mucus–mucoadhesive chains, 151
Mucus/polymer interface, 149, 150
Multiangle laser light scattering (MALLS), 158
Multidrug Resistance-Associated Protein Family
  expression of, 188, 189
  regulation of, 190
  substrate specificity of, 190–193
Multidrug Resistance 1 *(MDR1)* gene, 183
Multidrug resistance proteins, 5
Multivesicular (MVL) liposomes, 349. *See also* Liposomes

N

N-acetylglucosamine, 156
Na^+/H^+ antiporter, 197
β-Naphthoflavone, 190
Nasal spray products, 283, 284
Natural killer cells, 184
Neoral® label, 321
Nernst–Brunner equation, 50
Net effective permeability (P_{eff}) in pass perfusion method, 112
New chemical entity (NCE), 175, 361, 362

New Drug Application (NDA), 264
Nifedicron®, 326
Nifedipine, 326
Nitrofurantoin, 322
N-methylquinine, 210
N,N,N,-Trimethyl Chitosan Hydrochloride (TMC), 162, 164
  as absorption enhancer of peptide drugs, 159–161
  synthesis and characterization of, 158, 159
Nonsteroidal antiinflammatory drugs (NSAIDs), 204, 205
Noyes–Whitney equation
  in oral drug absortion, 29
  in solid dosage forms dissolution, 83
NR1H4, 190
NR1I2, 190
NR1I3, 190
N-trimethyl chitosan chloride, 159
Nuclear factor-E2-related factor 2 (Nrf2), 190
Nucleoside transporters, 214
  expression of, 219, 220
  molecular and structural characteristics of, 215, 216
  regulation of, 220, 221
  substrate specificities of, 217–219
Nucleotide binding domain(s) (NBDs), 179, 183, 193

O

OATP-B, 228
OATP/Oatp family
  expression of, 206, 207
  isoform mediated transport, 205
  regulation of, 207–209
  substrate specificity of, 204, 205
OAT (SLC22A), 202–204
Occludin, tight junction's membrane protein, 147
Ochratoxin A, fungal toxin, 192
Octanol/water partition coefficient, 30
OCT family, 228
Office of Generic Drugs (OGD), 286
Oligopeptide transporter, PepT1, 179
Omeprazole label, 332
Oral absorption prediction, *in silico* approaches for, 364
Oral bioavailability, 317
Oral dosage forms, IVIVC of, 359
Oral drug absorption
  bioavailability, 27, 28, 75, 76
  chemical stability, 38, 39
  development of, 12, 13
  food, 85
  importance of dissolution processes, 47, 48
  molecules stereochemistry and bioavailability, 29
  pH-solubility profile, 31–34
  solid state properties, 39–41
  steps in, 48
Oral drug absorption assessment, in humans
  drug bioavailability and peameability, 95–97
  in vitro data, 90–94
  in vivo data, 86–90
Oral drug bioavailability, factors affecting, 41
Oral drug chemical stability, 38, 39
Oral drug dissolution
  factors affecting, 51–55
  importance, 47, 48
  theories, 49–51
Oral drug dosage forms, 13, 14
Oral drug solid state properties
  amorphous material, 40, 41
  particle size, 41
  polymorphism, 39, 40
Oral drug solubility
  definition, 30
  determination, 36, 37
  in gastric and intestinal fluid, 35
  pH role, 31–34
  prediction, 38
  temperature effects, 34, 35
Oral inhalation products, 285, 286
Orally disintegrating tablet, 304
Organic Anion Transporters (OAT)
  SLC22A, 202–204
  SLCO, 204–209
Organic anion transporting polypeptides (OATPs), 202–204

Organic Cation Transporters (OCT), 209
  expression of, 211, 212
  regulation of, 212–214
  substrate specificity of, 210, 211
  transport, 211
Osmotic pressure, 277

P

PAA gel, 149
Paddle method, in drug dissolution, 58, 59, 66, 69
PAMPA. *See* Parallel artificial membrane permeability assay
PAMPA model in drug absorption, 114, 115
Paracellular intestinal membrane transport, 76, 79
Paracellular transport, 142
Parallel artificial membrane permeability assay, 102–104, 114, 115
  and Caco-2 cell models comparison, 116–123, 125, 126
Passive diffusion, in intestinal membrane transport, 76–78
Passive drug absorption, 102
Passive transcellular diffusion, 176
PepT1 cDNA sequence, 196
Peptide/Histidine Transporters 1/2 (PHT1/2), 195–199
Peptide Transporters 1 (PepT1), 195–198, 226–228
  transport activity, 200
Permeability coefficient (Pc) in Caco-2 cells, 116
Peroxisome proliferator-activated receptor α (PPARα), 190
P-glycoprotein (P-gp), 146, 226, 320. *See also* Excipients
  Caco-2 cells studies, 127–129
  drug transport and, 185
  expression of, 183, 184
  regulation of, 184
  role in oral drug absorption, 103
  substrate specificity of, 185–188
Pharmaceutical product, bulk properties, 3
Pharmaceutical quality assessment, 294–296
  implications of dissolution, 313, 314
  science-based principles, 294
Pharmacoat 606, 152
Pharmacokinetic profiles, *in vivo,* 368, 369
Pharmacokinetics
  chemical modification in, 11, 12
  drug absorption, 4, 5
  drug distribution and body metabolism, 5, 6
  parameters, 268, 318
  and pharmacodynamics (PD), 175, 179
Pharmacokinetic studies
  in animal models, 362, 367
  in clinical development, 18, 19
Phenobarbital, 190
Phosphatidylcholine, 183
9-(2-Phosphonylmethoxyethyl)adenine, anti-HIV drug, 192
pH-partition hypothesis, 176, 177, 202
pH sensitive liposomes, 348. *See also* Liposomes
Physical pharmacy, physical–chemical properties, 2
Piperine, 187
PK. *See* Pharmacokinetics
Placental ABC protein *(ABCP),* 193
Plasma concentration time curve (AUC), 184, 186, 226
Plasma drug concentrations, 337
Plendil® label, 321
Polyacrylates, 152
  as absorption enhancers, 153, 154
  enzyme inhibitory effects of, 154, 155
  thiolated polymers of, 162, 163
Poly(acrylates) polycarbophil, 154
Poly(acrylic acid)-Ca^{2+} complex, 155
Poly(acrylic acid)/mucin interface, 149
Polyacrylic acid (PAA), 148, 152, 153
Polyamidoamines (PAMAM), 351
Polyamines, 351
Poly[β-(1-4)-2-amino-2-deoxy-D-glucopyranose]. *See* Chitosan
Polycarbophils (PCPs), 152–154, 162, 163
  effects, on protease activities, 155
Polychlorinatedbiphenyl126 (PCB126), 190
Polyesters, 345, 351
Polyethylene glycol (PEG), 337, 344
Polyethylene glycol 400 (PEG 400), 145, 278. *See also* Excipients

Poly(ethylene glycol) terephthalate (PEGT)/poly(butylene terephthalate) PBT microspheres, 347
Poly(2-hydroxyethyl methacrylate), 153
Poly(lactic acid) (PLA), 345
Poly(lactic-*co*-glycolic acid) polymers (PLGA) microspheres, 341, 345, 346. *See also* Microspheres
Polymeric delivery systems, 338
Polysorbate 80, 144
Polyvinylidene fluoride, 115
Poly vinyl pyrrolidone, 41, 54
Population bioequivalence (PBE), 284
Potassium chloride, 274
Potential excipient effect, on bioavailability, 145–147
Powder X-ray diffraction, 37
Preclinical drug development, phase I clinical studies, 16, 17
Preclinical drug product development, 367
Pregnane X receptors (PXR), 190, 207
Pregnenalone-16α-carbonitrile (PCN), 208
Pregnenolone-16α-carbonitrile (PCN), 207
Presystemic metabolism, 271, 272
Prilosec® label, 332
Prodrug in biopharmaceutics, 11, 12
Propecia® label, 323
Propranolol label, 321
Propylene glycol, 145, 338. *See also* Excipients
Protein kinase C (PKC), 146, 207
Proton/oligopeptide transporters (POT), 195
  peptide transporter mediated transport, 197, 198
  PTR2 family, 196
  regulation, 200–202
  substrate specificity, 198, 199
PSAd in drug absorption, 113, 114
PTR2 family, 196
PVDF. *See* Polyvinylidene fluoride
PVP. *See* Poly vinyl pyrrolidone
PXRD. *See* Powder X-ray diffraction

Q

Quantitative structure property relationship (QSPR), 113

R

Ranitidine, 278
Reciprocating cylinder, in drug dissolution, 58, 59, 66
Reticuloendothelial system (RES), 344
Retinoic acid, 184
Retrovir® label, 330
Reversed dialysis method, 339
Rhodamine 123, 146
  in rat pituitary cells, 184
Rifampin, 184, 187, 190
RNAse protection assay, 206
Roller compaction process, 311
RT-PCR, 189
"Rule of 5" in oral drug absortion, 27, 28, 113

S

Scaled average bioequivalence, 275. *See also* Bioequivalence
Shake flask method in solubility measurement, 36, 37
Simulated gastric fluid (SGF), 62
Simulated intestinal fluid (SIF), 62
Single nucleotide polymorphisms (SNPs), 178, 201, 209, 226
Single pass perfusion method, 111, 112
SITT. *See* Small intestinal transit time
SIWV. *See* Small intestinal water volume
Skin treatment, 281
SLC22A. *See* Organic Cation Transporters (OCT)
SLC15A family. *See* Proton/oligopeptide transporters (POT)
Slofedipine® XL, 326
Small intestinal transit time, 36, 364
Small intestinal water volume, 36, 364
SN-38, 191, 192, 194
Sodium butyrate, 184
Sodium carboxymethylcellulose (NaCMC), 153, 163
Sodium dodecyl sulfate (SDS), 144
Sodium lauryl sulfate (SLS), 145. *See also* Excipients
Sodium taurodeoxycholate, 144
Sodium taurodihydrofusidate (STDHF), 144
Solid dosage forms, 341
  dissolution, 83, 84

Solubility
  in drug absorption, 82, 83
  and permeability values, 296
Solute carrier family 21A (SLC21A). *See* OATP/Oatp family
Solute carrier (SLC) transporters
  MCT (SLC16A), 221–225
  OATP (SLCO), 204–209
  OAT (SLC22A), 202–204
  OCT, 209–214
  SLC15A, 195–202
  SLC28A and SLC29A, 214–221
Solution hydrodynamics, in drug dissolution rate, 55
Sorbitol, 277
Stealth liposomes, 348
Stokes–Einstein equation, 55
Stressed and unstressed capsules, 310
Superporous hydrogels composite (SPHC), 160
Superporous hydrogels (SPH), 160
Surfactant, 337
Suspension filled capsules, 303, 304
Sustiva® label, 324
Swelling controlled release systems, 342
Symporter carrier-mediated transport, 78
Systemic metabolism, 271

T

Tablet disintegration, 301
Tablet formulations, dissolution profiles for, 307
Tablet hardness, 298–300
Talinolol, 187
Taurochenodeoxycholate-3-sulfate, 192
Taurochenodeoxycholic acid, 190
Taurocholate/HCO_3 exchange, 205
Taurocholic acid, 190
Taurolithocholate-3-sulfate, 192
Taurolithocholic acid, 190
TC-7 clone, 110
TEER. *See* Transepithelial electrical resistance
2,3,7,8-Tetrachlorodibenzo-*p*-dioxin (TCDD), 190
Tetraethylammonium (TEA), 210, 211
Theo-24®, 325
Theophylline, 320, 325
Thiolated polymers, 164
  of chitosan, 163
  of polyacrylates and cellulose derivatives, 162, 163
Thyrotropin releasing hormone (TRH), 346
Tight junction, 156, 178
  potential excipient effect on, 147, 148
  zonula occludens-1 protein of, 157
Tight junctional complex, 176
Tindamax® label, 324
Tinidazole, 324
Tissue irritant drugs, 338
Topical corticosteroids, 282
Topical dermatological products
  biowaivers for, 282
  clinical endpoint BE studies, 283
  for skin treatment, 281
  in vitro tests, 282
  in vivo studies, 282
Topological descriptors, in intestinal drug absorption, 114
Transcellular transport, 76, 141, 142
Transepithelial electrical resistance, 110, 178
Transmembrane domains (TMD), 188, 193, 209, 210, 222
Transporter-and paracellular-mediated drug absorption, 123–125
Trimethyl Chitosan 60 (TMC60), 147, 160–162
Tween 80, 145, 146, 187. *See also* Excipients

U

Ultrafiltration, 340
Uniphyl®, 320, 325
Uniporter carrier-mediated transport, 78
United States Pharmacopeia (USP) apparatus, 338
  dissolution profiles in, 298–301
Ussing chamber technique in drug absorption and permeability, 108
UV fiber optic probes, 340

V

Valspodar (PSC833), 186
Vascular endothelial growth factor (VEGF), 346

Verapamil, 191
Videx®, 319, 330
Vinblastine-binding site, 185
Vincristine, 348
Vitamin E TPGS, 145, 146
Volume of distribution (Vd), 5

W

Wagner–Nelson method, 293, 308, 311, 372, 378. *See also* Deconvolution and convolution
Wet granulation formulations, 368
Wet granulation of drugs, 54
Wetting theory of mucoadhesion, 148, 149

X

Xenopus oocytes, 198
Xylitol, 277

Z

Zidovudine, 318, 330

Printed in the United States of America.